AF904923

RESEARCH IN BIODIVERSITY – MODELS AND APPLICATIONS

Edited by **Igor Ya. Pavlinov**

Research in Biodiversity – Models and Applications
Edited by Igor Ya. Pavlinov

Published by InTech
Janeza Trdine 9, 51000 Rijeka, Croatia

Copyright © 2011 InTech
All chapters are distributed under the Creative Commons Non Commercial Share Alike Attribution 3.0 license, which permits to copy, distribute, transmit, and adapt the work in any medium, so long as the original work is properly cited. After this work has been published by InTech, authors have the right to republish it, in whole or part, in any publication of which they are the author, and to make other personal use of the work. Any republication, referencing or personal use of the work must explicitly identify the original source.

Statements and opinions expressed in the chapters are these of the individual contributors and not necessarily those of the editors or publisher. No responsibility is accepted for the accuracy of information contained in the published articles. The publisher assumes no responsibility for any damage or injury to persons or property arising out of the use of any materials, instructions, methods or ideas contained in the book.

Publishing Process Manager Dragana Manestar

Technical Editor Teodora Smiljanic

Cover Designer InTech Design Team

Additional hard copies can be obtained from orders@intechopen.com

Research in Biodiversity – Models and Applications, Edited by Igor Ya. Pavlinov
p. cm.
ISBN 978-953-307-794-9

We are IntechOpen, the world's leading publisher of Open Access books

Built by scientists, for scientists

4,000+
Open access books available

116,000+
International authors and editors

120M+
Downloads

151
Countries delivered to

Our authors are among the
Top 1%
most cited scientists

12.2%
Contributors from top 500 universities

WEB OF SCIENCE™

Selection of our books indexed in the Book Citation Index
in Web of Science™ Core Collection (BKCI)

Interested in publishing with us?
Contact book.department@intechopen.com

Numbers displayed above are based on latest data collected.
For more information visit www.intechopen.com

For EU product safety concerns:
IN TECH d.o.o., Prolaz Marije Krucifikse Kozulić 3, 51000 Rijeka, Croatia,
info@intechopen.com or visit our website at intechopen.com.

Meet the editor

Dr. Igor Ya. Pavlinov, born on March 26, 1950 in Moscow, Russia. His DrS dissertation was "Cladistic approach in phylogenetics and taxonomy: theoretical foundations of evolutionary cladistics" (1997). He is currently affiliated with Zoological Museum, Moscow Lomonosov State University (Russia) where he is a leading researcher and curator of mammls. His principal research interests are in taxonomy of mammals (mainly rodents); morphometrics, theoretical taxonomy and phylogenetics. His principal published scientific books are: "Systematics of Mammals of the USSR" (1998), "A Short Guide to Terrestrial Mammals of Russia" (2002). "Systematics of Recent Mammals" (2003), "Introduction to Contemporary Phylogenetics" (2005), "A review of Phylogeny and Classification of Gerbillnae (Mammalia: Rodentia)" (2007), "Biological Systematics: Evoluiton of Ideas" (2011). He also published several popular books on natural history including "The Wondering Animal Life" (2000), "The History of the Life on the Earth" (2001).

Contents

Preface

Biodiversity has became some time ago one of the key notions in various fields of human activity, from biological disciplines dealing directly with the kinds of living beings up to strategy of sustainable development of the humankind. Such a high rating of biodiversity theme has been resulting in publication of many books on these subjects in the past decades.

The present book is one of them. Its content is as wide as the biodiversity itself, beginning from the theoretical models of biodiversity and finishing with the latter's uses for human needs. It contains 17 chapters grouped into 5 sections, each dealing with certain aspect of biodiversity researches and uses.

The first section,, "Models of Biodiversity", contains several important theoretical chapters, some of which consider the ways in which biodiversity could be thought about, while others develop issues of more specific "biological" sound. Of particular interest are chapters (R. Rozzi; B.P. Anthony & oth.) concerning some aspects of interrelation between biological and socio-cultural components of the entire life on Earth at any scale of its consideration. Noteworthy is, as well, the chapter on pretty original "creativity theory of ecosystem" (Liu & oth.). Under consideration in this section are also: theoretical model of relation between ecological niches and diversity maintenance (P. Chesson), numerical model of the factors underlying avian species imperilment in sub-Saharan Africa and Europe (R.E. Crafton & B.P. Anthony), and the diversity turnover rate model for a beetle group undergone periodical catastrophic impacts of forest fire (E.T. Azeria & oth.).

A small section "History of Biodiversity Researches" contains the only chapter of C. Bradshaw & N. Sodhi reviewing in short some key publications on this subject that appeared during the last hundred years. It allows to trace how the very understanding of importance of biodiversity has been emerging.

The section "Evaluating biodiversity" is also small to include just one chapter describing an approach for monitoring the conservation of regional tropical phytodiversity in India (Giriraj A. et oth.). It is to be pointed out however that such a general issue as biodiversity evaluation is considered, to a higher or lesser degree, in most of the chapters in this book, as it is impossible to study and protect biodiversity without its measuring.

A number of chapters gathered under the title "Regional Biodiversity" deal with biodiversity assessment and protection under particular conditions. The latter might be quite different being encompassed either by the national and provincial governmental polices and strategies (S. Schindler et oth. on biodiversity in Austria; J. López-Pujol & oth. on plant diversity in China; B. Kohlmann on biodiversity in Costa Rica; F. Vessella & oth. on reforestation in Europe; L.F.A. Montag on a local aquatic biota in Brazil), by human-made environments (Y. Natori & oth. on rural Japan), or by certain natural sources availability (R.E. Diaz-Caravantes & E. Sánchez-Flores on Mexico, water supply is of concern). This section overlaps in part by its content with the previous ones, as the ideas and models discussed over there are illustrated by the data on regional biodiversity (Europe, sub-Saharan and in particular South Africa, India, Canada)

Another small section entitled "Uses of Biodiversity" contains a chapter on so called bioprospecting (A. Harvey). This relatively new notion accentuates a rather old subject of the use of diverse organisms in biomedical issues, but now it is considered under commercial circumstances.

The concluding section "Morphological Disparity" contains I. Pavlinov's chapter, in which some theoretical concepts and notions concerning that disparity are considered. The principle message of this paper is that morphological disparity should not be delimited by just between-species differences (nowadays predominating concept), but is be considered in its most widened sense to include all its forms and aspects.

Igor Ya. Pavlinov, DrS
Leading Researcher, Zoological Museum of Moscow M.V.
Lomonosov University
Moscow, Russia

Part 1

Models in Biodiversity

1

Towards Bridging Worldviews in Biodiversity Conservation: Exploring the Tsonga Concept of *Ntumbuloko* in South Africa

Brandon P. Anthony[1], Sylvia Abonyi[2], Petra Terblanche[3] and Alan Watt[1]
[1]Department of Environmental Sciences & Policy, Central European University
[2]Department of Community Health & Epidemiology, University of Saskatchewan
[3]Department of Sports, Arts, and Culture, Limpopo Province
[1]Hungary
[2]Canada
[3]South Africa

1. Introduction

Many scholars and managers now question the traditional top-down, and often hegemonic, approaches of excluding local participation and ignoring local interests in management of biodiversity both within and outside formal protected areas (Johnston, 1995; Kiss, 1990). Greater participatory planning is believed to enhance local support for biodiversity conservation goals and decrease conflicts between local people and conservation authorities (Happold, 1995; Heinen, 1996; Manfredo et al., 2004).

Efforts worldwide to integrate biodiversity conservation and rural development objectives have had mixed results, indicating that synergies between the two are not inherent, and they are not a panacea (Alpert, 1996; Barrett et al., 2005; Hughes & Flintan, 2001; Newmark & Hough, 2000). We argue here that they must more fully incorporate local worldviews in their design and implementation if they ever hope to succeed. For institutions responsible for conservation, detailed knowledge of the people whose lives are affected by conservation policies can be as important as information about the biodiversity to be conserved (Anthony & Bellinger, 2007; Brechin et al., 2002; Veech, 2003). Moreover, it has been noted that in addition to playing a key role in human-environment interactions (Nietschmann, 1992; Smith, 2001), cultural elements of nature protection can be a resource providing insight into development of conservation plans (Kuriyan, 2002; Stevens, 1997) while also reinforcing community identity and, promoting community cohesion and adaptability (Goodland, 1991; Kleymeyer, 1992; Robinson & Redford, 1994). Thus, recognition and understanding of different local cultural systems permits a broader, more appropriate overall policy toward natural resource use (Maffi, 2004).

We report here on research that included a focus on conflicts and synergies between local Tsonga people and conservation authorities in and around the boundaries of Kruger National Park (KNP) in Limpopo Province, South Africa. We begin below with a brief description of official conservation in Limpopo Province, followed by an introduction to *ntumboloko*, the Tsonga worldview that shapes the perspectives on the environment and conservation that was the focus of the research we report on here. We subsequently outline our theoretical approach, and our methods. Our results present perspectives on conservation and its context in the *ntumbuloko* worldview of the Tsonga. In our discussion we identify potential conservation synergies and conflicts between the two worldviews: that of 'western science', which predominantly influences official conservation management and practices in Limpopo Province, and that of Tsonga communities bordering the KNP. We conclude with some thoughts about the importance for conservation agencies to philosophically and practically understand and integrate local worldviews into their biodiversity conservation and socio-economic objectives.

1.1 Official conservation in Limpopo Province

1.1.1 Within protected areas: Kruger National Park

The Kruger National Park (KNP), situated in the Republic of South Africa (Figure 1), is approximately 350 km from north to south and covers nearly 2 million ha (Mabunda et al., 2003). Established in 1926, KNP is home to an unparalleled diversity of wildlife and maintained by a very sophisticated management system (Braack, 2000). Internationally, KNP functions as a major tourism destination with up to one million visitors annually, and serves as an important socioeconomic and ecological component of the Great Limpopo Transfrontier Park. Traditionally, KNP management has strictly followed 'western' scientific principles, often dismissing other forms of knowledge systems (see Wolpert, 1993), with the aim of single species management and manipulating ecosystems to meet prescribed goals (Carruthers, 1995). Following the dynamic economic and political transformations within South Africa since 1994, South African National Parks, including KNP, has witnessed a transformation in its policies which seek to integrate conservation and socioeconomic objectives of neighboring communities, including community resource use policies and the establishment of community fora (Mabunda et al., 2003).

1.1.2 Outside protected areas

Outside protected areas in Limpopo Province, environmental management is primarily the responsibility of the Department of Finance and Economic Development – Environmental Affairs (DFED) (Limpopo Provincial Government, 2005). DFED is operationally sub-divided into municipal districts which provide conservation extension services, control damage-causing animals outside KNP, and monitor and regulate the use of natural resources. DFED activities are largely governed by the *Limpopo Environmental Management Act* (LEMA) No. 7 of 2003, which is analogous with national legislation. DFED is also mandated to promote sustainable development by creating partnerships with communities, NGOs, the private sector, and other government departments. The DFED was created by the post-Apartheid regime as a part of new provisional governmental structures intended to introduce greater democracy to South Africa. Until that time, traditional authorities (TAs), based on chieftanship, were the effective representatives of communities. The legal competences of the TAs are ambiguous because, while they are recognized in the constitution as legitimate centers of authority, their actual rights and responsibilities are not clearly spelled out (Ntsebeza & Hendricks, 1998).

1.2 The people of Limpopo Province

1.2.1 Tsonga

Communities in the study area comprise almost exclusively (96.2 - 99.1%) people from the Tsonga people group (Statistics South Africa 2003). Tsonga are a diverse population, and in the mid-1990s numbered about 1.5 million in South Africa, and at least 4.5 million in southern Mozambique and Zimbabwe (1UpInfo, 1996). In the 18th century, ancestors of the Tsonga lived in small, independent chiefdoms. Most Tsonga relied on fishing for subsistence, although goats, chickens, and crop cultivation were also important. Because their coastal lowland habitat was tsetse-fly infested, cattle were rare in their economies (1UpInfo, 1996).

During the *mfecane*[1] and subsequent turmoil of the 19th century the history of the Tsonga was dominated by invasions of Zulu conquerors who left Chaka and enslaved the Ama-Thonga of the coast (Junod, 1912). Many Tsonga emigrated inland to the Transvaal from 1835 to 1840. Some successfully maintained their independence from the Zulu, while others were conquered by Zulu warriors even after they had fled. One Zulu military leader, Soshangane, established his authority over a large Tsonga population in the northern Transvaal in the mid-19th century (1UpInfo, 1996). The descendants of some of the conquered populations are known as the Shangaan, or Tsonga-Shangaan.

Tsonga who migrated inland brought new sources of food into the Transvaal, including cassava, certain kinds of groundnuts, potatoes and sorghum. Particularly important were the maize and fowls introduced in their new settlement areas. Agricultural work was performed almost exclusively by women, except for initial land clearing which was the men's responsibility (Magubane, 1998). Even today, labour division along gender lines still exists: men are traditionally hunters, herdsmen, fishermen, housing constructors, as well as traders; women are agriculturalists, gatherers, and collect water and fuelwood (Ombe, 2003). Crop harvesting was usually cooperative, done on a rotational basis, with area communities gathering to harvest each family's crop in turn.

By the early 1920s, the Tsonga-speakers constituted about 4% of the total South African population (Magubane, 1998). In the north, large chiefdoms, including Xikunda, Mhinga, Xigalo, and Makuleke occupied distinct reserves adjacent to the KNP. The Tsonga-Shangaan homeland, Gazankulu, was carved out of northern Transvaal Province during the 1960s and granted self-governing status in 1973. In the 1980s, the government of Gazankulu established a legislative assembly made up mostly of traditional chiefs. The chiefs opposed homeland independence but favored a federal arrangement with South Africa (1UpInfo, 1996).

Communities were torn apart as families were moved to the Tsonga homeland, and the resulting taxation and overpopulation made people increasingly dependent on migrant labour. This caused men to leave their families for long periods, and today even women in rural areas seek seasonal work on nearby citrus farms (Mathebula, pers. comm.).

---

[1] 'The Crushing' - a series of Zulu and other Nguni wars and forced migrations in the early 19th century that changed the demographic, social, and political configuration of southern and central Africa.

However, traditional Tsonga homesteads (*muti*) still exist: a typical settlement consists of a man, his wife or wives, their children and the families of their married sons (Magubane, 1998). Cylindrical houses with earth walls and conical thatched or reed roofs constitute the generally circular homestead, bordered with a perimeter wall or fence, made from branches and tree stumps. At the homestead center is the cattle kraal (*xivaya* or *tshanga*). A special meeting area (*huvo*), usually enclosed by branches and situated under a tree, exists within the community, as does the *gandzelo* for sacrificial purposes, which may be anywhere in the *muti*. The *vandal*, which may be inside or outside the *muti*, is where the men meet to discuss the administration and the affairs of the *muti*. No woman or child is allowed in this area.

Family authority rests with the father, who is traditionally treated with great respect by the wife and children. Within an extended family, the ranking and status of wives and their children is determined by the order in which they were married (Magubane, 1998). A typical Tsonga-Shangaan Traditional Authority is composed of a chief (*hosi*), under which a hierarchy exists to serve the community at large (Hartman et al., 1993). Junod (1912, pp. 367) states that the role of the chief is tantamount to tribal life as 'the [chief] forms the center of national life. It is through him that the clan becomes conscious of its own unity. Without him, it loses its bearings and it has lost its head'. Chieftainship is hereditary and falls to the most senior member of the oldest lineage in the strongest clan in the group. The new chief must be approved by the council and formally inducted into office.

In the past the *hosi* yielded supreme power. He allocated land and sanctioned the start of initiation rites, harvest ceremonies and rain dances; he mediated between members of the group and ancestral spirits; he made all decisions relating to war and the army; he was also responsible for the administration of the group, and tried serious cases and those on appeal from headmen (*tindhuna*) (Magubane, 1998).

1.2.2 Tsonga and *ntumbuloko*

The concept of *ntumbuloko* dominates the Tsonga worldview and has been defined by Chitlango & Balcomb (2004:183) as 'cultural and social norms, customs, traditions, and institutions that constitute the basis for existence, self-understanding and identity in Tsonga society.' Traditional Tsonga cosmology includes that man has a physical body (*miri*) and a spiritual body with two attributes, *moya* and *ndzuti*. The *moya* (associated with the spirit) enters the body at birth, and on death is released to join the ancestors. According to Magubane (1998) the *ndzuti* is linked to a person's shadow, reflects human characteristics, and on death, leaves the body for the spirit world. The spirit of the dead (*swikwembu*) is imbued with the individual and human characteristics of the person and can hold much power with respect to causing rain to fall or trees to bear fruit (Junod, 1912). Not only is there life after death, but on entering the world of the dead the individual retains links with the living. Thus, for many Tsonga today, 'society'implies a concept including both the living and the dead.

1.3 Theoretical approach: a space for consideration of Tsonga and official views of environment and conservation

Our research primarily draws from Firey's (1960) resource use theory as it provides a comprehensive approach to understanding human dimensions of resource management.

Resource use theory recognizes that ecological, economic, and ethnological/cultural frames of reference all interact with each other in a form of negotiation and trade-offs to optimize each of these frames and, thus, play a role in shaping perceptions of the use and fate of a resource system. This system is socially constructed and viewed differently by different social groups from their own frame of reference, based on personal needs, perceptions, and attitudes regarding a natural resource system (see Gergen, 1994; Hannigan, 1995). According to Firey, *any resource process, to be adopted, must first be accorded some worth by people in terms of their own system of activities* (emphasis ours). Thus, there are some resource complexes[2] which are not valued by a given people and which, consequently, will not be adopted (no matter how superior they may be by other criteria). Resource conservation or sustainability depends on maintenance of a particular social order, because social order provides common expectations and values that make it possible for a group of people to set limits on environmental change by limiting destructive economic opportunism or 'gain-seeking' (Firey, 1978).

2. Data collection methods and analyses

2.1 Face-to-face questionnaire

Household face-to-face questionnaires were administered to 240 randomly selected households of 38 villages (C.I.=6.28; C.L.=95%) within 15 km of KNP in Limpopo Province (Figure 1). Data on socio-demographic variables including age, sex, household income, household size, education level and years family has resided in village were collected by trained local field assistants. Following the Firey framework, a series of questions concerning (i) community needs, (ii) the components and value of *ntumbuloko,* and (iii) costs and benefits of the KNP to local communities were also incorporated. The questionnaires integrated both closed- and open-ended questions, and were manifest (content) coded using a contextual method based on positive/negative or topical classifications (c.f. Weisberg et al., 1996).

Questionnaires were first written in English, translated into Tsonga-Shangaan by a linguist, and then translated back into English by a field assistant. Inconsistencies and/or clarifications in the text were then modified based on pre-testing and discussions with the field assistants (c.f. Sudman, 1983). Whenever possible, household heads were surveyed and were defined as being the individual who assumed responsibility for the household (Budlender, 2003). Before administering the questionnaire, cultural norms were followed, i.e. an introduction of the administrators, the form and rationale of the questionnaire and an explanation of its intended purpose(s).

2.2 Interviews

Semi-structured interviews were conducted with KNP, DFED, TAs and other community representatives. Where necessary, follow-up interviews were carried out to clarify issues and explore further avenues of interest. Fieldwork was conducted from January-November 2004.

---

[2] 'Resource complexes' are man-mind-land structures which show stability and resilience to external influences, and that impose constraints upon humans such that they willingly conform their behavior to the practices which comprise that resource system. These contrast with 'resource congeries', which show little stability and vary widely to external changes.

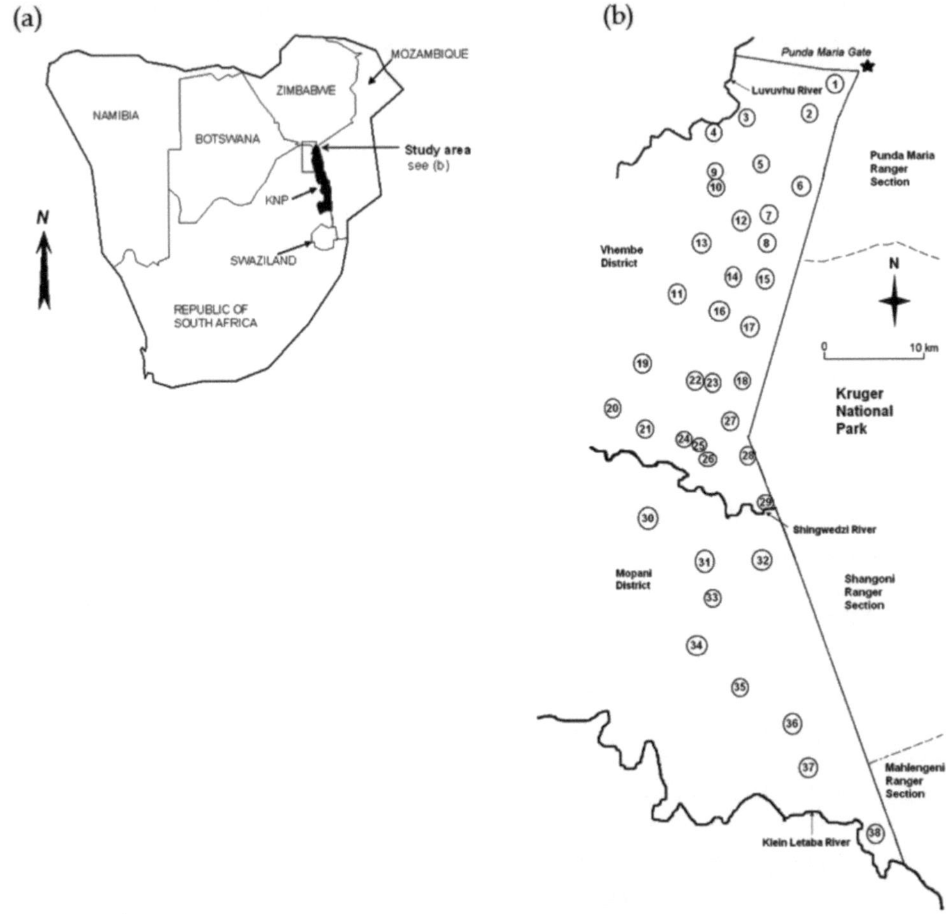

Fig. 1. (a) The location of Kruger National Park in Southern Africa. (b) Expanded view illustrates study area with location of 38 villages (listed below with associated *de jure* Traditional Authorities). **Mhinga TA:** Matiyani (1), Josepha (2), Mhinga (3), Botsoleni (4), Maphophe (5), Maviligwe (6), Makuleke (7), Makahlule (8); **Shikundu TA:** Ximixoni (9), Saselemani (10), Nkovani (11); **Bevhula TA**: Ntlhaveni D (12), Nkavela (13), Makhubele (14), Bevhula (15); **Magona TA**: Nghomunghomu (16), Mashobye (17), Magona (18); **Madonsi TA**: Gijamhandzeni (19), Matsakali (20), Halahala (21), Peninghotsa (22), Govhu (23), Merwe A (24), Shisasi (25), Jilongo (26); **Mtititi TA**: Lombaard (27), Plange (28), Altein (29); **Xiviti TA**: Mininginisi Block 3 (30), Mininginisi Block 2 (31), Muyexe (32), Shitshamayoshe (33), Khakhala (34), Gawula (35), Mahlathi (36), Ndindani (37), Hlomela (38).
Source: Anthony (2007); reproduced with permission from Cambridge University Press.

2.3 Archival research

Limpopo Province DFED records were compiled from both Mopani District, which extends from the Shingwedzi River south through the study area, and Vhembe District, which includes the northern section of the study area. Moreover, relevant records of the KNP and monthly meeting minutes were reviewed of the Hlanganani Forum, which has been the primary liaison between KNP and neighbouring communities in the northern part of the park since 1994. During analyses, we initially utilized a grounded coding process to identify themes in interview and archival data, followed by a more explicit coding process that incorporated Firey's concepts.

3. Results

3.1 Demographic and socio-economic factors

The questionnaire sample consisted of 83 males (34.6%) and 157 females (65.4%), ranging in age from 18 to 102 (mean=39.3±17.63). Household sizes ranged from 1 to 18 persons (mean=5.8±2.65), and families had resided in their village from 1 to 52 years (mean=23.2±12.60).

Respondents were also asked to list the ages and sex of all household members. Men (N=662, mean age=22.1±17.102) represented 47.52% of the sampled households, while women (N=731, mean age=26.5±19.716) constituted 52.48%. The population structure is broad-based with over half of the population <20 yrs of age, and comprises a higher proportion of women compared to men, especially in age classes above 29 yrs.

3.2 Community needs

Survey respondents were asked to rank the five most important community needs from a predefined list, based on interviews with community members and municipal government staff (Table 1). A weighted score was calculated for each need and used as an indicator of its importance. Employment was ranked as the most important community need overall, followed by health, school, electricity and drinking water facilities. Of least importance to respondents were protecting forests and wild animals which, in contrast, are of primary concern for conservation agencies.

Overall Rank	Community need	*n*	mean score
1	Employment	185	3.10
2	Health facilities	164	2.37
3	School facilities	182	2.34
4	Electricity facilities	144	1.95
5	Drinking water facilities	111	1.26
6	Road improvement	81	0.80
7	Training opportunities	86	0.74
8	Protection of crops/livestock	61	0.73
9	Housing	52	0.60
10	Preserving traditional culture	36	0.33
11	Tourism development	27	0.29
12	Protection of forest	29	0.26
13	Protection of wild animals	32	0.26

Table 1. Overall ranking of community needs by community survey respondents (N=238). Mean scores range from 0 (no importance) to 5 (most important).

3.3 Beliefs and attitudes

Respondents were asked what they believed to be components of *ntumbuloko;* responses are summarized in Figure 2. Chi-square and correlation tests were conducted for gender, age, household income and education level but no significant associations were found, suggesting that beliefs in the sampled households regarding the different parts of *ntumbuloko* are independent of these variables.

Based on their concept of *ntumbuloko,* almost all (98.7%) respondents believed that they 'need' *ntumbuloko,* for a variety of reasons which we classified according to McNeely et al. (1990) (Table 2). In addition to more direct utilitarian values, respondents indicated that *ntumbuloko* is highly valued for its socio-cultural, educational, spiritual and historical attributes. When respondents were asked whether they believed they needed to protect *ntumbuloko,* a majority (85.4%) agreed. The need to maintain and enhance utilitarian use values ranked highest for those responding positively to this question, although socio-cultural and spiritual aspects were also noted, including the following: *'it is life'; 'to lose* ntumbuloko *is to lose ourselves';* 'ntumbuloko *dictates that we should continue initiation school* [3]".

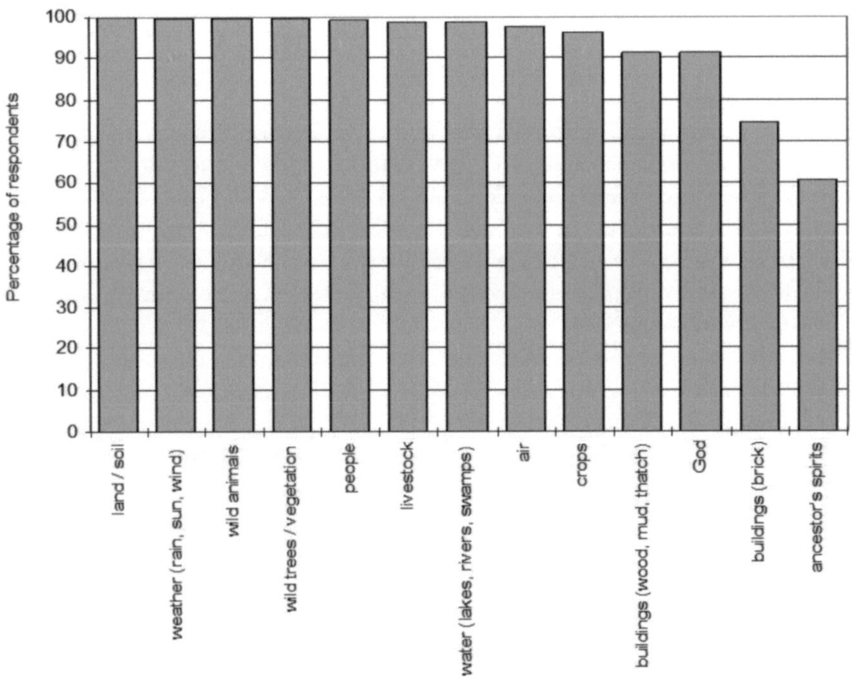

Fig. 2. Frequency of belief about components of *ntumbuloko* (N=240)

Ten percent of the respondents stated that they didn't know whether they should protect *ntumbuloko,* claiming that they didn't know how they could protect it. In contrast, 4.6% indicated that they did not believe they needed to protect *ntumbuloko,* citing that "*it was created long ago*".

[3] In traditional Tsonga culture, puberty marks the end of childhood and the beginning of adolescence. During this time young men and women enter initiation schools. Schools vary, but in principle they perform a similar social function, that of a 'rite of passage' marking the transition from adolescence to adulthood. This is much more than a physical change; it also represents a change in social status.

Direct value	Indirect value
Consumptive/non-market (27.5%): • food • fodder for animals • fuelwood • traditional medicine • construction materials • traditional clothing	*Non-consumptive - ecological functions (19.5%)*: • storm protection • cleaning air • soil protection • sustains environment
Productive/commercial (4.7%): • fodder for animals • traditional medicine • drawing tourists	*Non-consumptive – non-ecological functions (41.6%)*: • part of creation ('I belong to it'; 'makes us aware of God's creation') • education ('we can learn much from it'; 'children learn from it as they grow up') • historical heritage ('it serves as a reminder of the past') • aesthetic ('brings and brightens life for people') • cultural ('it is our culture to love *ntumbuloko*')
	Option (6.7%): • for future generations, 'to build the future'.

Table 2. Categorized responses as to why community members 'need' *ntumbuloko*. Relative percentages of responses are included for each sub-category.

In an informal conversation, one high school teacher in the area stated that he believes the ancestors' spirits can control rain and consequently crop production, therefore those who are still living must continue to honor them through dances, drums, and meetings. However, this cosmology is neither universal nor static amongst the Tsonga. According to one college teacher from the area, the Tsonga primarily define 'beauty' of plants and animals according to their use or utility. He reported that since he began teaching in 1989, his personal perception on nature has changed because 'they [campus management] made it wrong for us to kill any animals on the campus'. He usually would kill a snake on site as is the Tsonga custom, but now he 'tries to chase it away'. He now believes that this 'has helped to keep snake bites down at the college where no one has been bitten in 3 years'.

3.4 Traditional Authorities

Respondents were asked to evaluate both their respective TA and the municipal government, in terms of how well it was doing in its role with respect to land-use, whatever they conceived that to be. More than half of the respondents (51.6%) couldn't comment on the effectiveness of the municipal government, stating that they didn't know of its activities. For those that did evaluate the institution, 23.8% assessed it positively and 24.6% negatively. Negative opinions of the effectiveness of the municipal government were largely governed by housing and water shortages, poor road maintenance, and the belief that it 'does nothing in our area' and 'shows favoritism in its activities'. These data collectively suggest that the performance of municipal government is highly varied in the study area, with specific *de jure* TAs experiencing greater activity than others. In contrast, the roles and responsibilities of Traditional Authorities are much better recognized, with respondents stating that their functions are extensive, ranging from provision of residential and agricultural sites, to protecting forests/wild animals and overseeing people's concerns. Considering that access to land for cultivation was secure for over 70% of respondents, and more than 85% felt their land was 'good', this suggests that *TAs are perceived as largely competent by local communities in securing access to good quality land for agriculture*. Moreover, *TAs have a much higher approval rating compared to local government by respondents,* with less than 12% of respondents reporting negatively overall.

In order to identify what variables might be influencing this evaluation, correlation analysis was used to compare responses with selected demographic and socio-economic variables. Although age (r=0.14, $p<0.05$, N=240) and level of education (r=-0.13, $p<0.05$, N=240) were significantly correlated with responses towards TA effectiveness, linear regression analysis revealed that they are very weak predictors of responses (R^2=0.02), suggesting that the selected variables do not play a decisive role in influencing opinions.

These functional distinctions were also confirmed during interviews with various community members and representatives of TAs. According to one *hosi* (chief), although all communal lands are owned by the state, TAs have authority to grant lands for garden plots and homesteads to their *muganga* (village(s)) members. Mtititi TA representatives stated that they are responsible for access to and control over a number of resources, including allocation of grazing and residential sites, and granting permission to collect fuelwood. They play a judicial role in fining any persons caught illegally collecting any resource that requires a permit, especially those persons who do not reside within the TA area, in which case guilty parties receive a stiffer penalty. They also play an important role in resource monitoring stating, 'In the event that the tribal police see that the amounts of resources are dwindling, they inform the *hosi* who would then inform the community to cease collecting that resource.'

4. Discussion

4.1 Components of *ntumbuloko*

South Africa has undergone dramatic socio-political changes in the last decade, with enhanced opportunities for formal education in the rural areas. However, the extent to which formal education and exposure to alternative views has affected perceptions and attitudes of rural people towards nature and its conservation is still uncertain (see Els, 1994; Mabunda, 2004). *Ntumbuloko* permeates the Tsonga worldview, and our research supports previous work (Chitlango & Balcomb, 2004; Els, 2002; Junod, 1913; Terblanche, 1994) in that the Tsonga perceive *ntumbuloko* as more than just the biophysical environment: there is still strong belief that it also embraces people (*vanhu*), God (*Xikwembu*), ancestors' spirits (*swikembu*), and tradition, and this belief is independent of sex, age, household income and education level. These results are congruent with a study on perceptions regarding causes and treatment of diseases in Northern [now Limpopo] Province (Mabunda, 2001), which found that the notion of supernatural causality associated with many diseases predominated among all groups, but was highest among university students. In our study, supernatural causality still prevails and is manifested in the belief of many respondents, even amongst the young and more highly educated, that rain and associated harvests are strongly linked with appeasing ancestors' spirits, and not solely the product of environmental factors, which western science principles would prescribe.

4.2 Value of *ntumbuloko*

In addition to more direct utilitarian values, *ntumbuloko* is highly valued for its indirect non-consumptive attributes, including non-ecological functions embracing socio-cultural, educational, spiritual and historical qualities (Anthony & Bellinger 2007). The need to maintain and enhance utilitarian use values ranked highest for those responding positively to the question whether they need to protect *ntumbuloko*, although socio-cultural and spiritual aspects were also noted. In addition to holding a broader view of nature, Tsonga also believe in a plethora of practices which they see as being essential for its protection. In addition to reduced consumption of resources, environmental education, and altering practices to protect flora and fauna (which one might expect in western societies), the need to maintain cultural and spiritual traditions which are embedded in the broader definition of nature held by the Tsonga were also noted.

4.3 Community needs

Opinions expressed on nature conservation, i.e. protecting trees and wild animals, lag far below more immediate development needs such as employment, health, education, and improving infrastructure. Employment needs were apparent as we noted male absence in the study area, which is likely attributable to outmigration (or cyclic migration) to larger urban centers or mines where employment opportunities are greater (Bryceson, 1999). Male absence in rural areas can create labor vacuums, especially in cases where domestic responsibilities are sharply divided amongst household members, and may disproportionately increase pressures on households with only women and children. In this research about one in eight households was comprised only of women and children. This constraint is exacerbated by time required for women and children carrying out domestic chores, including almost 20 hours per week for collecting fuelwood and drinking water alone in the study area (Figure 3) (Anthony, 2006). With water scarcity perceived to be widespread in the study area, and fuelwood becoming scarcer in some areas north of the

Fig. 3. Tsonga woman on route to collect drinking water from community tap. Reproduced with permission from Anthony (2006)

Shingwedzi River, the extent of these constraints appears to be worsening (Anthony, 2006). These constraints suggest that opportunities for women and children desiring to secure formal employment, training, and/or education are severely limited. For conservation agencies, recognizing these limitations is an important step in articulating any conservation and/or development programs that seek local relevance. *Time is a precious commodity that should be understood in its local context, and household members are unlikely to engage in activities making extensive demands on their time unless these are directly related to improving livelihoods.*

As we think further about the needs of communities, the question arises, *If local communities are so dependent on local wild resources, why is their protection ranked so low?* The answer may be found in two related concepts of Tsonga beliefs, i.e. values associated with *ntumbuloko*, and the role of humans in the environment.

First, the Tsonga value *ntumbuloko* more for its utilitarian rather than aesthetic qualities, believing that local resources were given by God, and it is their right to use them to maintain human survival (Eckert et al., 2001; Els, 2002). However, 'meaningful and judicious use is not always implied by this inherent right, and this difference in conceptual approach often leads to conflict with nature conservation authorities' (Els, 2002, p.655); thus resource use conflicts are often rooted deeply in culture. Hence, the negotiations of resource users as conceptualized in Firey's theory then become operational: the perceived *aesthetic values* of nature are 'traded off' for more imperative needs of human survival and development. Here, however, distinctions within and between Firey's three frames become blurred, limiting its application in these contexts. Western concepts of the 'ecological frame', developed mainly by ecologists and geographers, are based on the interactions between organisms and their biophysical environments. Conversely, the 'ethnological frame' to resource phenomena has principally been developed by anthropologists and sociologists and focuses on a people's culture. Firey's definition and explanation of these frames treats them as separate entities. However, the Tsonga concept of *ntumbuloko* embodies both ecological and cultural frames; decoupling it into two separate frames, at present, is irrational for most Tsonga. Therefore, *developing nature conservation activities in these contexts have a greater chance of being rejected if they do not incorporate the wider concept of* ntumbuloko *constructed by the Tsonga.* This also has implications for current stakeholders and future researchers in similar contexts: research findings may have lower relevance and/or be more difficult to communicate locally if these distinctions in conceptual definitions are not recognized.

Second, it is inconceivable and irrational for the Tsonga to believe that protection of forests and wild animals is man's responsibility (Els, 2002). On one hand, our research supports Els' view, as most respondents believe that it is God's (*Xikwembu*) responsibility to ultimately ensure the sustainability of resources. On the other hand, although God (*Xikwembu*) and ancestors' spirits (*swikwembu*) are still believed to be components of *ntumbuloko*, such beliefs may not be as widespread as they were in the past. For example, in a study of Tsonga communities in a more densely populated region adjacent to KNP to the south, Hunter et al. (2010) found that environmental concern was strongly related to material needs and livelihoods, and this was gendered and varied substantially by village. This transition may be the result of increasing exposure to Christianity, alternative views of nature in educational institutions (Millar, 2004), economic development opportunities or cultural taboos (Kuriyan, 2002), and/or restrictions on resource use imposed by government and TAs, although such causal relationships were beyond the scope of our research.

Embedded cultural and spiritual beliefs and practices hold value for the Tsonga and should be acknowledged when establishing partnerships in environmental protection. This includes the role that *ntumbuloko* has for the Tsonga in education, spiritual identity and as historical heritage. These beliefs, strongly held by many Tsonga, are thus very resistant to change and are likely to persist. It is these beliefs which have the greatest potential to conflict with western approaches to conservation, as they claim inherent differences with respect to *who* is responsible for protecting flora and fauna, and *how* they are to be used. Practically for conservation initiatives, the two concepts regarding Tsonga beliefs explained above translate into the recognition that *conservation programs are unlikely to be accepted in these contexts if they are based primarily on aesthetic values of nature, or if they do not acknowledge the belief by local communities of the role that God and ancestors' spirits play in nature.*

4.4 Traditional Authorities

The strong role that TAs play in land allocation and resource access and use has a number of far reaching implications. Chiefly authority is ascribed by lineage rather than achieved through elections, and its patriarchal principles ensure that major decisions on land allocation are almost invariably taken by men. However, this research shows that many people, irrespective of gender, still look to their chiefs for land allocation and are satisfied with it. Indeed, only 10.2% of women respondents felt that their TAs are not doing a good job, compared with 14.5% of men. These results concur with Campbell & Shackleton (2001) and Ntsebeza (1999), who showed that TAs still maintain strong positive influence in South Africa's communal areas.

The role of DFED in the study area is uncertain and ambiguous. Although the primary body responsible for implementing and enforcing LEMA 2003 regulations, its activities are limited. Indeed, TAs are *de facto* principally controlling access to natural resources and enforcing LEMA 2003 stipulations, with tribal courts functioning in part to fine transgressors. Perceptions of the DFED by local TAs are generally negative, as this agency is seen only within its role in enforcement. It is also criticized for its weakness in delivering much-needed environmental education and awareness to communities on the role of the provincial government. In addition, there is widespread criticism of the poor control of damage-causing animals by DFED and the withholding of compensation for damages caused by these animals (Anthony, 2007).

Similar to criticisms launched at the ineffectiveness of local government, weaknesses in co-operative governance between DFED and TAs are inhibiting resource conservation, leading to situations in which opportunities are established for 'gain-seekers' to exploit resources at unsustainable rates. DFED managerial staff acknowledge that discussion and co-operation regarding land use, including biodiversity conservation, between provincial and municipal governments and TAs is practically non-existent, and needs to be strengthened (Anthony, 2006). In light of the increasing pressures on natural resources and the aspirations of some communities to engage in conservation agreements with the KNP, it would be wise for these institutions to heed these trends and seek co-operative ways to halt resource over-exploitation before conditions render it practically impossible to effectively pursue any community-based conservation initiatives at all.

4.5 'Gain-seekers' and resource exploitation

The Firey model contends that resource conservation is possible only when people share expectations that others will forego opportunistic practices threatening sustainability.

Firey's predictions may indeed be materializing in South Africa. Political transformation processes have led in many cases to *de facto* open access systems with new forms of opportunism, manifested by perverse incentives for unsustainable resource extraction, especially by 'gain-seeking' outsiders (Figure 4). These are exacerbated by low capacities in the provincial government structures and fueled by the stripping of powers of legitimate TAs (Anthony et al., 2010). According to a KNP internal report, increasing rates and magnitude of *inter alia* deforestation has been observed in areas adjacent to KNP claiming that 'trucks transporting newly cut poles and wood are often observed along the roads in adjacent areas'. In its summary, this report emphasized that 'the rate at which the destruction and degeneration is taking place will render the area useless for future community-based conservation projects.'

Fig. 4. Illegally collected fuelwood (mostly *Colophospermum mopane*) confiscated by Magona Traditional Authority in August 2004. Reproduced with permission from Anthony (2006)

Concerns about increased extraction and use of fuelwood, sand and medicinal plants by 'outsiders' have been observed elsewhere in Limpopo Province (Kirkland et al., 2007; Twine et al., 2003). Similarly, there is widespread belief in our study area that new political freedoms and democracy, coupled with the disintegration of powers of TAs, imply an uncontrolled liberty in which people are allowed to access and use resources as they wish. As early as 1994, DFED staff had noted that with respect to hunting game in rural areas, '…with the current constitutional changes, many people think the old laws are no longer valid and that this is creating problems' (cited in Anthony, 2006). In addition to these misconceptions, one of the key issues in the increased exploitation of resources by external harvesters is the control of access to resources by TAs. Although believed to be imperfect by some government staff, and involving corruption by some current TA personnel, the previous permit and enforcement system under TAs was generally recognised as being effective in limiting the impact of external harvesters. With national political changes, however, TAs no longer have the resources to control land as they previously did and, at best, can only work in co-operation with provincial departments. Juxtaposed with the decreasing power and ability of TAs to control resource use, local and provincial government is, at present, unable to fill this institutional vacuum, especially given other pressing priorities such as provision of water, sanitation and electricity.

The outcome is a situation where, at least in some parts of the study area, external gain-seekers have seized the opportunity to either hire locals or harvest resources themselves at convenient times so as to maximize profit and minimize risks of being caught in illegal activities. This includes sand removal, illegal commercial harvesting of trees and poaching game (Anthony, 2006). Firey posits that, in conditions where the social order begins to disintegrate, incentives to inhibit one's propensity for gainful resource processes may be removed, security will be exchanged for economic efficiency, and resource congeries in the form of calculating opportunism will become the norm. Of further concern is that this new agency, having no determinate structure, can offer little resistance to further change. Therefore, *if left unabated and where sanctions are relatively ineffective, unsustainable resource extraction will continue in these areas and may severely limit future opportunities and environments in which community-based conservation can be implemented or, in a worse scenario, will deplete natural resources from which local communities currently derive much of their livelihoods.* Moreover, this will likely have potential implications for ecological integrity, creating an 'edge effect' along the KNP boundary (Woodroffe & Ginsberg, 1998). The situation calls for returning social stability to the rural areas and the institutions that *de facto* govern resources within them. As Firey (1960, p. 238) reminds us, development that involves cultural stabilization brings about non-gainful-but-likely practices that 'insinuate themselves into people's thinking and, abetted by a stable environment, enter into behavior as elements of a resource complex...and become supports for social order, contributing to its maintenance and resisting its change.' Consequently, the solution we outline below involves working to improve management and helping it to meet the new challenges it faces.

The problem of opportunistic exploitation can be resolved in our context through a number of means. Firstly, increasing capacity of provincial conservation structures to effectively enforce environmental legislation will likely lead to decreased opportunism, but will not adequately address the cultural conundrum. Resource conservation depends on the ability to obscure resource users' perception of private gain, to gratify their incentives for security in personal relationships, and to enlist the willing conformity of all resource users. Plans, including excessive coercion or rule enforcement, which do not win consent on these fronts will usually fail as they are often expensive and considered illegitimate. Indeed, by increasing powers only to municipal and provincial governments and ignoring local customs and traditions in these contexts, a reverse effect may result in which TAs and their devotees may see this as a return to the 'fences and fines' approach to conservation under Apartheid (this time outside the KNP), and further polarize themselves from government objectives (Gibson & Marks, 1995; Michaelidou et al., 2002). A second alternative, which may lead to cultural stabilization, involves devolving natural resource access and use powers to local TAs. The drawbacks here, however, are that not all TAs are considered legitimate, and may not have the required capacity to effectively handle these responsibilities (Anthony, 2006). Moreover, current and potential possibilities of corruption, misrepresentation and elitism are left unabated in devolving powers to this lower level, especially if there are weak mechanisms for accountability (Ribot, 2002).

Instead of these more extreme alternatives, we advocate a more co-operative approach which sees provincial structures striving to work more hand-in-hand with local TAs in both communicating, and enforcing, natural resource legislation. Similarly, *defining what resources should be conserved, and how and for whom they should be managed should be based on interactive dialogue between the DFED and local communities*. This has promise for at least three reasons. First, it would promote citizen involvement, through traditional structures, in government affairs and redistributing power and resources to enable local people to participate in decisions that directly affect their lives (Luckham et al., 2000). Second, by maintaining and utilizing traditional structures, which are largely believed to be 'good' and 'preferable' by local communities, anxiety may be minimized regarding proposed changes in natural resource management (Anthony, 2006). Finally, it would be one tangible avenue through which government could effectively harmonize the institution of traditional leadership within the new system of democratic governance as laid out in the *Traditional Leadership and Governance Framework Act* No. 41 of 2003. Provincial structures in this arrangement would continue to play an overseer role especially in managing external threats (Michaelidou et al., 2002), but would allow TAs (where considered legitimate by local communities) to continue to exercise traditional resource management powers and, where feasible, decentralize enforcement to TAs coupled with corresponding capacity-building. Areas of conflict (e.g. use of specific protected species) would ideally be mutually agreed upon through interactive dialogue, based on research investigating sustainable harvesting of resources, and supported by flexible policies.

5. Conclusion

At an international level it has been recognized that natural resources cannot be managed effectively without the co-operation and participation of resource users to make laws and regulations work (Baland & Platteau, 1996). This makes managing protected areas an even more complex and dynamic undertaking than the traditional 'fences and fines' approach. This is exacerbated in contexts where socio-economic and political forces are also experiencing dramatic transformation. The core of natural resource management in South Africa's communal areas, including the use and value of resources, often lies in deeply rooted and relatively stable concepts which are unlikely to change in the near future, and are often not obvious in their alignment with western conservation principles. For any degree of long-term resource sustainability, compatibility must be sought between western concepts of nature conservation and local worldviews of the intended beneficiaries of any conservation and/or development projects. Moreover, the knowledge system of any culture, including that of western science, is not static, but '[a]ssimilation of "outside" knowledge, and synthesis and hybridisation with existing knowledge, are continuing processes' (Howes & Chambers, 1979, p. 12). For PAs wishing to engage in extending management options to neighboring communities, it is critical to both develop an ongoing understanding of, and recognize, how communities conceptualize humankind's relationship to the environment, rights to resource access and use, and resource management principles.

Another feature indicative of South Africa's emerging democracy is the disintegration of TAs in the rural areas, exacerbated by institutional non-uniformity, and minimal capacity of provincial government in enforcing environmental legislation. This has created *de facto* open access systems exemplified by escalating opportunities for gain-seeking and perverse incentives for illegal exploitation of resources, especially by external forces. If left unabated, these conditions will have increasingly adverse effects on local livelihoods and are likely to jeopardize future conservation initiatives. Where this is occurring, or is imminent, improving social cohesion and circumventing unsustainable resource practices through a more co-operative and adaptive approach to resource management by relevant institutions is needed. This principle applies not only to our study area, but also to conservation agencies elsewhere which face similar challenges, especially in cases characterized by dramatic transformations in institutional responsibilities and increasing financial constraints.

Conservation agencies have a formidable task, both philosophically and practically, in attempting to understand and integrate local worldviews into their biodiversity conservation and socio-economic objectives. Interactions with local people are complex, dynamic, and driven by economic as well as socio-political forces. We offer no single remedy or solution to address conflicts in the study area, but rather a suite of possibilities that should be explored. The question remains as to whether strategies developed by KNP to effectively involve local communities will gain normative weight so that local institutions will be able to meet their biodiversity conservation and socio-economic objectives, or whether these institutions will further lose control to pressures originating from within and from external sources. This research has shed light on these complexities and it is hoped that its findings will contribute to a more stable and sustainable future for both the KNP and its neighbors, and for those in similar contexts elsewhere.

6. Acknowledgement

We thank the CEU Doctoral Research Support Grant for funding assistance; all Traditional Authorities in the study area for support; and the questionnaire respondents for opening their homes to us. We also thank Akua Addo-Boadu and Zebedee-Feka Njisuh for their comments on an earlier draft of this manuscript.

7. References

1UpInfo. (1996). *Country Study & Guide: South Africa* [on-line. Retrieved from http://www.1upinfo.com/country-guide-study/south-africa/

Alpert, P. (1996). Integrated Conservation and Development Projects. *BioScience* Vol.46, No.11, (December 1996), pp. 845-855, ISSN 0006-3568

Anthony, B.P. (2006). *A View from the Other Side of the Fence: Tsonga Communities and the Kruger National Park, South Africa.* Unpublished doctoral dissertation, Central European University, Budapest, Hungary

Anthony, B. (2007). The Dual Nature of Parks: Attitudes of Neighbouring Communities towards Kruger National Park, South Africa. *Environmental Conservation* Vol. 34, No. 3, (September 2007), pp. 236-245, ISSN 0376-8929

Anthony, B.P., & Bellinger, E.G. (2007). Importance Value of Landscapes, Flora and Fauna to Tsonga Communities in the Rural Areas of Limpopo Province, South Africa. *South African Journal of Science* Vol.103, No.3-4, pp. 148-154, ISSN 0038-2353

Baland, J.M., & Platteau, J.P. (1996). *Halting Degradation of Natural Resources: Is There a Role for Rural Communities?* FAO, ISBN 0198289219, Rome, Italy

Barrett, C.B., Lee, D.R., & McPeak, J.G. (2005). Institutional Arrangements for Rural Poverty Reduction and Resource Conservation. *World Development* Vol.33, No.2, (February 2005), pp. 193-197, ISSN 0305-750X

Braack, L.E.O. (2000). *Kruger National Park* (2nd ed.). New Holland, ISBN 1859744710, London

Brechin, S.R., Wilshusen, P.R., Fortwangler, C.L., & West, P.C. (2002). Beyond the Square Wheel: Toward a More Comprehensive Understanding of Biodiversity Conservation as Social and Political Processes. *Society and Natural Resources* Vol.15, No.1, pp. 41-64, ISSN 0894-1920

Bryceson, D.F. (1999). *Sub-Saharan Africa Betwixt and Between: Rural Livelihood Practices and Policies.* ASC Working Paper No. 43. African Studies Centre, Leiden, The Netherlands

Budlender, D. (2003). The Debate about Household Headship. *Social Dynamics: A Journal of African Studies* Vol.29, No.2, pp. 48-72, ISSN 0253-3952

Campbell, B., & Shackleton, S. (2001). The Organizational Structures for Community-Based Natural Resource Management in Southern Africa. *African Studies Quarterly* Vol. 5, No.3., ISSN 215-2448, Retrieved from http://web.africa.ufl.edu/asq/v5/v5i3a6.htm

Carruthers, J. (1995). *The Kruger National Park: A Social and Political History.* University of Natal Press, ISBN 0869809156, Pietermaritzburg, South Africa

Chitlango, A.J., & Balcomb, A. (2004). Ntumbuluko, Theology and Issues of Gender in the Tsonga Worldview: Towards an African Woman's Theology of Liberation in Mozambique. *Scriptura* Vol.86, No. 2, pp. 180-189, ISSN 0254-1807

Eckert, B., de Beer, F.C., & Vorster, L.P. (2001). Worldviews and Decision Making: Natural Resource Management of the Laka of Mapela in an Anthropological Perspective. *South Africa Journal of Ethnology* Vol.24, No. 3, pp. 88-98, ISSN 0379-8860

Els, H. (1994). Werknermers van die Nasionale Krugerwildtuin se waarde-oordele oor aspekte van die natuur en natuurbewaring [Value Judgements of Personnel of the Kruger National Park on Aspects of Nature and Nature Conservation]. *South Africa Journal of Ethnology* Vol.17, No. 4, pp. 109-120, ISSN 0379-8860

Els, H. (2002). The Role of African Rural People within their Environment. In: *Game Ranch Management* (4th ed.), ed. J. du P. Bothma, pp. 645-661, Van Schaik, ISBN 9780627024719, Pretoria, South Africa

Firey, W. (1960). *Man, Mind, and Land: A Theory of Resource Use,* Free Press, ISBN 9780837198347, Glencoe, Illinois

Firey, W. (1978). Environmentalism and the Rediscovery of Limits. In: *Sociocultural Change Since 1950,* eds. T.L. Smith & M. Singh, pp. 273-278. Vikas, ISBN 0706906594, New Delhi

Gergen, K.J. (1994). *Realities and Relationships: Soundings in Social Construction.* Harvard University Press, ISBN 0674749316, Cambridge

Gibson, C.C., & Marks, S. (1995). Transforming Rural Hunters into Conservationists: An Assessment of Community-Based Wildlife Management Programs in Africa. *World Development* Vol.23, No. 6, (June 1995), pp. 941-957, ISSN 0305-750X

Goodland, R. (1991). Prerequisites for Ethnic Identity and Survival. In: *Resident Peoples and National Parks: Social Dilemmas and Strategies in International Conservation*, eds. P.C. West & S.R. Brechin, pp. 301-311, University of Arizona Press, ISBN 0-8165-1128-4, Tucson

Hannigan, J.A. (1995). *Environmental Sociology: A Social Constructionist Perspective*. Routledge, ISBN 0415112559, London

Happold, D.C.D. (1995). The Interaction between Humans and Mammals in Africa in Relation to Conservation: A Review. *Biodiversity and Conservation* Vol.4, No. 4, (June 1995), pp. 395-414, ISSN 0960-3115

Hartman, J.B., Kriel, J.D., Boonzaaier, C.C., Els, H., & Wasserman, I. (1993). *Report on the Development of Tribal Authorities in Gazankulu*. Department of Anthropology and Archaeology, University of Pretoria, Pretoria

Heinen, J.T. (1996). Human Behaviour, Incentives and Protected Area Management. *Conservation Biology* Vol.10, No.2, (April 1996), pp. 681-684, ISSN 1523-1739

Howes, M., & Chambers, R. (1979). *Indigenous Technical Knowledge: Analysis, Implications and Issues*. IDS Bulletin 10.2. Institute of Development Bulletin, University of Sussex, Brighton, UK

Hughes, R., & Flintan, F. (2001). *Integrating Conservation and Development Experience: A Review and Bibliography of the ICDP Literature*. International Institute for Environment and Development, ISBN 1-899825-77-0, London

Hunter, L.M., Strife, S., & Twine, W. (2010). Environmental Perceptions of Rural South African Residents: The Complex Nature of Environmental Concern. *Society & Natural Resources* Vol. 23, No.6, (June 2010), pp. 525-541, ISSN 0894-1920

Johnston, B.R. (1995). Human Rights and the Environment. *Human Ecology* Vol.23, No.2, (June 1995), pp. 111-123, ISSN 0300-7839

Junod, H.A. (1912). *The Life of a South African Tribe* (Volume 1). Imprimerie Attinger Freres, Neuchatel, Switzerland, Retrieved from http://www.archive.org/stream/lifeofsouthafric01junouoft#page/n7/mode/2up

Junod, H.A. (1913). *The Life of a South African Tribe* (Volume 2). Imprimerie Attinger Freres, Neuchatel, Switzerland, Retrieved from http://www.archive.org/stream/lifeofsouthafric02junouoft#page/n5/mode/2up

Kirkland, T., Hunter, L.M., & Twine, W. (2007). 'The Bush is No More': Insights on Institutional Change and Natural Resource Availability in Rural South Africa. *Society and Natural Resources* Vol.20, No.4, pp. 337-350, ISSN 0894-1920

Kiss, A. (ed.). (1990). *Living with Wildlife: Wildlife Resource Management with Local Participation in Africa*. World Bank Technical Paper 130. World Bank, ISBN 0-8213-1697-4, Washington

Kleymeyer, C.D. (1992). Cultural Energy and Grassroots Development. *Grassroots Development* Vol.16, No. 1, pp. 38-39, ISSN 0733-6608

Kuriyan, R. (2002). Linking Local Perceptions of Elephants and Conservation: Samburu Pastoralists in Northern Kenya. *Society and Natural Resources* Vol.15, No. 10, pp. 949-957, ISSN 0894-1920

Limpopo Provincial Government. (2005). Department of Economic Development, Environment and Tourism. Retrieved August 24, 2005, from http://www.limpopo.gov.za/prov_depts/econ_dev_environ_tourism.htm

Luckham, R., Goetz, A.M., & Kaldor, M. (2000). *Democratic Institutions and Politics in Contexts of Inequality, Poverty, and Conflict: A Conceptual Framework.* IDS Working Paper 104. Institute of Development Studies, University of Sussex, Brighton, UK

Mabunda, D. (2004). *An Integrated Tourism Management Framework for the Kruger National Park, South Africa, 2003.* Unpublished doctoral dissertation, University of Pretoria, Pretoria, South Africa.

Mabunda, D., Pienaar, D.J., & Verhoef, J. (2003). The Kruger National Park: A Century of Management and Research. In: *The Kruger Experience: Ecology and Management of Savanna Heterogeneity*, eds. J.T. du Toit, K.H. Rogers, & H.C. Biggs, pp. 3-21. Island Press, ISBN 1-55963-982-2, Washington, D.C.

Mabunda, M.M. (2001). Perceptions of Disease, Illness, and Healing among Selected Black Communities in the Northern Province, South Africa. *South Africa Journal of Ethnology* Vol.24, No. 1, pp. 11-16, ISSN 0379-8860

Maffi, L. (2004). Conservation and the Two 'Cultures' - Bridging the Gap. *Policy Matters* Vol.13, (November 2004), pp. 256-266

Magubane, P. (1998). *Vanishing Cultures of South Africa*, Struik, ISBN 1868259676, London

Manfredo, M., Teel, T., & Bright, A.D. (2004). Application of the Concepts of Values and Attitudes in Human Dimensions of Natural Resources Research. In: *Society and Natural Resources: A Summary of Knowledge,* eds. M. Manfredo, J. Vaske, B. Bruyere, D. Field, & P. Brown, pp. 271-282. Modern Litho, Jefferson, Missouri

Mathebula, M.R. Tribal Secretary, Homu Tribal Authority. Informal interviews. Homo North, Limpopo Province, South Africa.

McNeely, J.A., Miller, K.R., Reid, W.V., Mittermeier, R.A., & Werner, T.B. (1990). *Conserving the World's Biological Diversity.* IUCN, WRI, Conservation International, WWF-US, & World Bank, ISBN 0915825422, Gland, Switzerland and Washington, D.C.

Michaelidou, M., Decker, D.J., & Lassoie, J.P. (2002). The Interdependence of Ecosystem and Community Viability: A Theoretical Framework to Guide Research and Application. *Society and Natural Resources* Vol.15, No.7, pp. 599-616, ISSN 0894-1920

Millar, D. (2004). *Interfacing Two Knowledge Systems: Local knowledge and Science in Africa.* Proceedings of Bridging Scales and Epistemologies: Linking Local Knowledge with Global Science in Multi-Scale Assessments, ISBN 1-59726-037-1, Alexandra, Egypt, 17-20 March 2004

Newmark, W.D., & Hough, J.L. (2000). Conserving Wildlife in Africa: Integrated Conservation and Development Projects and Beyond. *BioScience* Vol. 50, No.7, (July 2000), pp. 585-592, ISSN 0006-3568

Nietschmann, B. (1992). *The Interdependence of Biological and Cultural Diversity.* Occasional Paper No. 21. Centre for World Indigenous Studies, Olympia, Washington

Ntsebeza, L. (1999). *Land Tenure Reform in South Africa: An Example from the Eastern Cape Province.* Paper presented at Workshop on Land Rights and Sustainable Development in sub-Saharan Africa, Berkshire, UK, 16-19 February 1999

Ntsebeza, L., & Hendricks, F.T. (1998). *Democracy and Despotism in Post-Apartheid South Africa: The Role of Chiefs/Traditional Authorities in Rural Local Government.* Paper presented at Conference on Comparing Experiences on Democratization between South Africa and Nigeria, Centre for African Studies, University of Cape Town, Cape Town, South Africa, 30 May - 2 June, 1998

Ombe, Z.A. (2003). Indigenous Land Use Management in Lower Changane Chibuto: Sacred and Profane Desacralisation and Recovery. *Indilinga African Journal of Indigenous Knowledge Systems* Vol.2, No.1, pp. 1-9, ISSN 1683-0296

Ribot, J.C. (2002). *Democratic Decentralization of Natural Resources: Institutionalizing Popular Participation.* World Resources Institute, ISBN 1-56973-522-0, Washington, D.C.

Robinson, J.G., & Redford, K.H. (1994). Community-Based Approaches to Wildlife Conservation in Neotropical Forests. In: *Natural Connections: Perspectives in Community Based Conservation,* eds. D. Western & R.M. Wright, pp. 300-322, Island Press, ISBN 1559633468, Washington, D.C.

Seragaldin, I., & Taboroff, J. (1992). *Culture and Development in Africa.* Proceedings of Culture and Development in Africa - International Conference, ISBN 0-8213-2780-1, World Bank, Washington, D.C., 2-3 April 1992

Smith, E.A. (2001). On the Coevolution of Cultural, Linguistic, and Biological Diversity. In: *On Biocultural Diversity: Linking Language, Knowledge, and the Environment,* ed. L. Maffi, pp. 95-117, Smithsonian Institution Press, ISBN 1560989300, Washington, D.C.

Statistics South Africa (2003). *Census 2001.* Retrieved from http://www.statssa.gov.za/census01/html/default.asp

Stevens, S. (1997). Introduction. In: *Conservation Through Cultural Survival: Indigenous Peoples and Protected Areas,* ed. S. Stevens, pp. 1-8, Island Press, ISBN 9781559634496, Washington, D.C.

Sudman, S. (1983). Applied sampling. In: *Handbook of Survey Research,* eds. P.H. Rossi, J.D. Wright, & A.B. Anderson, pp. 145-194, Academic Press, ISBN 0125982275, San Diego, CA

Terblanche, H.P. (1994). Geselekteerde Tegniese Skeppinge van die Tsongavrou met spesifieke verwysing ma die Tsongakraal opelug Museum [Selected Technical Creations of Tsonga Women with Specific Reference to the Tsonga Village Open Air Museum]. Unpublished Masters thesis, University of Pretoria, Pretoria, South Africa.

Twine, W., Siphugu, V., & Moshe, D. (2003). Harvesting of Communal Resources by 'Outsiders' in Rural South Africa: A Case of Xenophobia or a Real Threat to Sustainability? *International Journal of Sustainable Development and World Ecology* Vol.10, No. 3, (September 2003), pp. 263-274, ISSN 1350-4509

Veech, J.A. (2003). Incorporating Socioeconomic Factors into the Analysis of Biodiversity Hotspots. *Applied Geography* Vol.23, No.1, (January 2003), pp. 73-88, ISSN 0143-6228

Weisberg, H.F., Krosnick, J.A., & Bowen, B.D. (1996). *An Introduction to Survey Research, Polling, and Data Analysis* (3rd ed.), Sage, ISBN 0803974027, Thousand Oaks, CA

Wolpert, L. (1993). *The Unnatural Nature of Science,* Harvard University Press, ISBN 0571169724, Cambridge, MA

Woodroffe, R., & Ginsberg, J.R. (1998). Edge Effects and the Extinction of Populations Inside Protected Areas. *Science* Vol.280, No.5372, (June 1998), pp. 2126-2128, ISSN 0036-8075

2

A Study on Biodiversity Mechanism by the Creativity Theory of Ecosystem

Yong Liu, Guolei Li, Haiqun Yu and Ruiheng Lv
Key Laboratory for Silviculture and Conservation of Ministry of Education, Beijing Forestry University, Beijing China

1. Introduction

The mechanisms of biodiversity have been intensively studied in recent decades. Significant attention has been given to finding those mechanisms that explain the patterns of species richness found changing with latitudinal gradients (Hubbell, 1979; Jablonski, 2006; Lyons, 1999; Root, 1988). A large number of these species richness hypotheses have been proposed, and new ones continue to appear, with the total now exceeding thirty (Hawkins et al., 2003; Huston, 1979; Ritchie & Olff 1999). Yet there remains considerable controversy about the hypotheses that underlie the observed patterns of biodiversity (Kerswell, 2006; Willing et al., 2003).

The theories of local determinism generally try to find a few key environmental factors and establish their simple relationships with species richness in that distinct environment (Ricklefs, 2006). By doing so, the species richness could hopefully be predicted by measuring these environmental factors and their results could then become the principles of biodiversity conservation. Initially a single prominent factor is regressed against species richness, for example, the *species-energy hypothesis*, *species-area hypothesis*, or *species-productivity hypothesis* treats the single factor of energy, area, and productivity, respectively, as the most important factor to influence species richness (Allen et al., 2002; Mittelbach et al., 2001; Turner et al., 1988). Later on multiple factors are used to explain the causes of biodiversity, such as the *hypothesis of water-energy dynamics* that suggests the link between water-energy and species richness is widespread and generally strong (Hawkins et al., 2003). Ironically, more and more environmental factors are found to be important, and the relationships between these factors and species richness are variants according to different locations and scales.

The theories of community explain the forces that maintain species diversity from the aspect of community ecology, for example the *niche-assembly theory* asserts that species co-occur in a community only when they differ from one another in resource use. But this theory has some difficulties to explaining the diversity often observed in species-rich communities such as tropical forests (Zhou & Zhang, 2006). The *neutral theory*, on the other hand, assumes that all individuals of all species in a trophically similar community are ecologically equivalent. The number of species in a community is controlled by species extinction and immigration, and speciation of new species (Hubbell, 2001). Based on the fundamental processes of birth, death, dispersal and speciation, neutral theory presented a mechanism that generates species abundance distributions remarkably similar to those observed in nature, however controversy persists (McGill, 2003). Some ecologist believe that the most important task ahead is to integrate niche and neutral theories, that is to add more processes in neutral theories and more stochasticity in niche theories (Alonso & McKane, 2004; Chave, 2004). This demonstrates that there is a strong wish for the ecologists to search for a general principle about the mechanisms of biodiversity (Tilman, 1999). Nevertheless such integration only within the community ecology may not solve the existing problems and identify the general principle embraced. Ricklefs (2006) pointed out that to assess the relative roles of local ecological constraint compared to regional and historical unfolding of diversity-environment relationships, we must abandon localized concepts of the community and adopt historical (particularly phylogenetic) and geographic methods to evaluate the evolution of diversity within large regions and its influence on diversity at local scales.

Although many hypotheses have been proposed and ecologists have amassed a wealth of detail to explain global patterns in species richness, there is no integrated hypothesis of how the ecosystems work as a whole evolved entity. For example, local determinism explains the biodiversity gradients from a physical environment perspective, and the theories of community consider more about ecological processes or population dynamics. Moreover, it is inappropriate to run a regression of species richness against environmental factors by treating the species richness as a dependent variable and the environmental factors as independent variables. This is because from the aspect of ecosystems, both species richness and physical environments are independent variables and their interactions contribute to the properties of the ecosystems. Therefore species richness itself through interactions among species also contributes to further species richness. In addition, we all agree that biodiversity is the emergent property of ecosystems through interactions of physical environments and organisms after a long evolutionary history. In contrast, there have been few studies on such indicators to reflect the ecosystem's potential of emerging properties, which undoubtedly influences our effort to understand the mechanisms of biodiversity thoroughly. Obviously, it is necessary to establish an integrated theory to study the causes of biodiversity by embracing *system thinking*, and regarding the ecosystem as an entity and treating biodiversity as the emergent property of the ecosystems.

The objective of this paper is to study the mechanisms of biodiversity by establishing a new theory from the aspect of ecosystem creativity. Here we develop an integrated theory, which we call Creativity Theory of Ecosystems (CTE), to study the mechanisms of biodiversity with a different perspective compared to more conventional approaches. Basing our theory on system thinking, the CTE establishes a model according to four concepts of creativity, energy, environment diversity and adaptability and the relationships among them. Chiefly by introducing the adaptability as one of the independent variables, the CTE model not only integrates biotic and abiotic factors but also combines spatial and temporal scales needed to predict plant species richness. This new approach is a very general theory and can be applied to any ecosystem because it is premised on system thinking, and is not tied to any specific scale or particular experimental design. A quantitative test of CTE was also conducted with statistical methods according to data obtained for species richness and environmental factors from 27 provinces of China.

2. Creativity Theory of Ecosystems (CTE)

An ecosystem is one of the complex systems whose properties are not fully explained by an understanding of its component parts (Gallagher & Appenzeller, 1999). It is an essential approach for an ecologist to view the ecosystem as an evolved entity, by doing so we find that biodiversity patterns actually are the self-creations of the global ecosystems after a long history of interactions among organisms and physical environments in different scales. It is clear that biodiversity is one of the most prominent emergent properties for the ecosystems, and creativity can naturally be employed as an indicator for the potential of yielding emergent features. Thus, the CTE is based on four main concepts:

1. Ecosystem creativity is a measure of an ecosystem's potential to yield emergent properties. There are three indispensable factors influencing ecosystem creativity. These include energy, environmental diversity, and adaptability within the ecosystem.
2. Energy is the most fundamental of the ecosystems and can be regarded as the capacity for doing work being associated with material bodies or for motion with systems, thus there cannot be creation without energy involved. The higher the energy input, the more emergent properties would be produced within a system. This quality has been proved by species richness distribution of birds (Hawkins, 2003; Root, 1988; Turner, 1988), Lizard (Scheibe, 1987), vascular plants (Mittelbach et al., 2001), benthic marine algae (Kerswell, 2006), etc. So energy is generally positively related to ecosystem creativity.
3. Environmental diversity is defined as environmental complexity of the ecosystems, which at least will include spatial heterogeneity and climatic variability. Theoretically the more heterogeneous and complex the physical environments, the more complex the plant and animal communities will be, and the higher the species diversity (Krebs, 2001). Ecological studies have also shown positive relationships between environmental complexity and species diversity for many groups of organisms, including mammals, lizards, plankton, marine gastropods, reef fish, algae and plants (Manuel, 2002). Analysis using 85 data sets ranging from plants to vertebrates and invertebrates on publications, Hawkins et al (2003) found that climatic variables were the strongest predictors of richness in 83 of the 85 cases. This finding offers widespread support for the hypothesis that climate in general has a major influence on diversity gradients across large spatial extents. It is obvious that environmental diversity has a positive relationship with ecosystem creativity.
4. Adaptation is an ordinary phenomenon within the biotic kingdom and has been considered as a primary force in evolution (Zhang, 1998). Global biodiversity today can also be interpreted as the result of organism adaptation, because organisms change their material environment as well as adapt to it (Lovelock, 2003). Naturally, using adaptability as a measure of adaptation for the ecosystems should be a reasonable indicator to understand biodiversity mechanisms from the aspect of an organism's contribution. Adaptability can be defined as the quality or state of being adaptable, where adaptable means capable of being adapted or suitable without change (Gove, 1976). With this concept, it is clear that adaptability includes a continuum of states from adaptable (without change) to not adaptable (maximum change). Because creation must involve changes, and great creation means great changes, adaptability therefore is negatively related to creativity. This relationship has been demonstrated by Buckling et al. (2003) with the bacterium *Pseudomonas fluoresceus,* that adaptation itself is likely to limit a population's ability to diversify. In general, local adaptation to source habitats can limit local adaptation in sinks and restrict the use of alternate niches (Urban, 2006).

Based on the above definitions we find that ecosystem creativity is positively correlated with energy (e) and environmental diversity (d), but negatively correlated with adaptability (a) of the ecosystems. We then introduce creativity index (CI) as an indicator to reflect the creativity of ecosystems. We can write CI as

$$CI = f(e, d, a) \tag{1}$$

Generally speaking, in this equation energy can be easily understood and measured, however environmental diversity and adaptability are problematic. Because there is no single parameter to completely express environmental diversity, it is probably wise to consider it according to spatial scales. Freestone & Inouye (2006) found that the mechanisms driving species coexistence and diversity in serpentine seeps appear scale-dependent. Willis & Whittaker (2002) classified the spatial scales in five categories as local, landscape, regional, continental, and global. It is becoming increasingly apparent that the factor best accounting for patterns of biodiversity seems to be delimited by scale. Consequently, d is a variable of scale-dependent in Eq. (1).

According to the definitions, an ecosystem is completely adaptable to its environmental conditions if there are no changes occurring, but the ecosystem displays some non-adaptable quality if there are emergent properties occurring after a certain period of evolutionary time. Thus the more emergent properties that appear within the ecosystem the bigger the changes become and the lower the level of adaptability. We then can calculate the a value by the reciprocal of the ecosystem's changed rate with the following equation:

$$a = 1/(N-n)/n = n/(N-n) \tag{2}$$

Where, n is the number of original properties at time t_1, N is the number of properties at time t_2, Δt ($\Delta t = t_2 - t_1$) is the evolutionary time of an ecosystem. Thus, $N-n$ is the number of emergent properties and $(N-n)/n$ is the changed rate of ecosystem properties during evolutionary time period Δt.

Apparently we measure the adaptability of the ecosystem from historical and evolutionary aspects, because in the biological sense current adaptations are the result of selection that was in progress at some time in the past (Ridley, 2004). We actually judge the adaptability from differences between current properties of the ecosystem and its properties in the past. This then implies that our a value is also a variable that is time-scale-dependent. In addition, though the ecological processes of competition and predation (Bush, 2003; Fine et al., 2006; Schmitz, 2006; Straub, 2006) are greatly different within evolutionary time periods, the Bible teaches us to judge a tree by its fruit. Correspondingly, we judge the perfection of an organism by its power to survive and multiply (Egbert Giles Leigb, 1971) or in our hypotheses by the final emergent properties of ecosystems.

If the a variable in Eq. (1) is replaced by Eq. (2), the model yields: $CI = f(e, d, a) = f[e, d, n/(N-n)]$, Since a is negatively correlated with creativity, the above equation should be:

$$CI = f[e, d, (N-n)/n] \tag{3}$$

Hence, CTE treats the ecosystems as a consolidated entity, and biodiversity is its emergent property through interactions among organisms and environments after evolving through a certain spatial and temporal scale. For an ecosystem the higher the energy input the more diversified environments become and the less adaptability. Subsequently, the greater the creativity also means the higher the biodiversity.

However, due to the combinations among the variables of *e, d, and a* could be various forms such as plus, multiplication, power, etc., and the *CI* model (3) is only a functional equation that cannot be calculated directly. This is a reflection of complexity of the ecosystems, i.e., one cannot predict the creativity of every ecosystem with a single combination of these variables because the variables of *d* and *a* are spatial and temporal dependent, respectively. Thus to calculate the *CI* value quantitatively we need to specify ecosystems in spatial and temporal scales and find the relationship among variables for certain scale ecosystems. We then test the CTE with the methods of Principal Component Analysis (PCA) and regression for ecosystems on a regional scale.

3. Test of the CTE

3.1 Material and methods

China, with a vast area of 9.6 million km^2, is an ideal region to test CTE. Its territorial distance from south to north is 5500km, including tropic, subtropic, temperate warm and temperate cool zones. From China's east to west is 5200km, including a great expanse of land from the Pacific Ocean to Mt. Everest with the terrain rising gradually. Due to the various geographic and climatic conditions, China is one of the regions with the most abundant displays of biodiversity in the world (Shi, 1991).

We collected the data from 27 provinces (22 provinces and 5 autonomous regions) in China (Table 1 in Appendix 1) Three municipalities directly under the Central Government (Beijing, Tianjing, Shanghai) were excluded because their relative small areas have been strongly influenced by urbanization. The data from the Chongqing municipality was included in Sichuan province. The areas of Hong Kong, Macao, and Taiwan were not included because we were unable to get relevant data.

The variables were collected and determined as follows:

1. Animal species richness. The number of terrestrial vertebrate species (including birds, reptiles, amphibians, and mammals) was used to indicate the animal species richness. Data were taken from Editorial board for series of natural resources in China (1995) and Editorial board for the complete series of Chinese agriculture (1998).
2. Plant species richness. The number of vascular plant species (including pteridophyte, gymnosperm, and angiosperm) was used to indicate the plant species richness. Data were taken from publications related to the flora and vegetation for every selected province (see Appendix 1).
3. Energy (*e*). Energy input was estimated by annual mean temperature. Data were available online from the Scientific Database of the Chinese Academy of Science (http://www.sdb.ac.cn/).
4. Environmental diversity (*d*). We used annual precipitation, altitude difference (maximum minus minimum altitudes), and land area to estimate environmental diversity. Water availability is a critical factor to constrain species distribution and altitude difference is the most prominent geographic feature in China. Area is also employed as a variable to influence environmental diversity because the areas were greatly different among provinces. The combination of these three factors can predict the *d* variable in regional scale very well. Data were taken from the publication Editorial board for series of natural resources in China (1995) and Editorial board for the complete series of Chinese agriculture (1998) and online from (http://www.sdb.ac.cn/).
5. Reciprocal of adaptability (a^{-1}). According to Eq. (2): $a^{-1} = 1/a = 1/\,n/(N-n) = (N-n)/n$, this actually is the changed rate of ecosystem properties during evolutionary time period Δt. Because we treated every province as a consolidated entity, and it is impossible to

measure the complete properties of the ecosystems, we used the animal species richness at present-day (N) to estimate the ecosystem properties. We then assumed that the number of original properties (animal species richness) at the processes occurring was 1, that is $n = 1$. The evolutionary time period (Δt) might be over the last 10000 years, i.e., since the end of the last glacial period (Willis 2002).

CI values for every province were calculated from Eq. (3) by the following steps and methods: First, the data (from Table 1 in Appendix 1) were normalized using the standard deviation method (Xu, 2002) in order to eliminate influences caused by different units and dimensions (Table 2 in Appendix 1). Second, a Principal Component Analysis (PCA) was performed on the e, d and a^{-1} in an effort to reduce the dimensionality of the data sets. The varimax rotation was used to simplify the interpretation of the results. Two components accounted for 87.63% of variance in e, d and a^{-1} depending on the eigenvalues, percent of variance, and cumulative percent (Table 3 in Appendix 1). The major components of the first factor are temperature, precipitation, and the reciprocal of adaptability (Table 4 in Appendix 1). This is consistent with the natural situation in China where there are two most prominent climatic features of temperature increasing from north to south and precipitation decreasing from east to west. Factor 2 is composed of altitude difference and area (Table 4 in Appendix 1), each of which reflects the geographic characteristics in China. From the rotated component matrix (Table 4 in Appendix 1) and component score coefficient matrix (Table 5 in Appendix 1), we find the component score coefficient rotated matrix for all the provinces, which were F1 and F2 (Table 6 in Appendix 1). Finally, the *CI* value (Table 6 in Appendix 1) was estimated by the integrated factor ($\sum F$) according to the percent of variance for F1 and F2:

$$CI = \sum F = \%\text{ of variance for component }1 \times F1 + \%\text{ of variance for component }2 \times F2$$
$$= 0.55569 \times F1 + 0.32065 \times F2$$

Therefore, *CI* represents integrated levels of energy, reciprocal of adaptability, and environmental diversity in terms of annual precipitation, altitude difference, and area. *CI* value is above average if it is positive, and below average if negative.

In order to test the CTE we took the plant species richness as the emergent property of the ecosystem for each province and as the dependent variable, where CI was the independent variable. We ran regressions on the normalized data using the equation: plant species richness = a + b(CI) + c(CI)^2, in linear, quadratic and cubic models. Only the best model was shown in Table 1. The results were compared with that of conventional methods, i.e., *energy-hypothesis, spatial heterogeneity hypothesis, area hypothesis*, and *water-energy dynamic hypothesis*, with a multiple regression equation of: plant species richness = a + b (energy or area…) + c (energy or area…)^2 also in linear, quadratic and cubic models. In order to be as liberal as possible in discovering patterns, relationships were considered significant if $P < 0.05$. The majority of relationships considered significant had $P < 0.01$.

3.2 Results

Regression analysis shows (Table 1) relationships that are almost exactly those predicted by the CTE hypotheses. We find the CI value is the best predictor with the cubic model and accounted for 94.0% of the variation in plant species richness (F=137.516, P<0.0001) , whereas model with an integrated environmental factor excluding adaptability can only account for 42.3% of the variation in plant species richness (F=20.054, P<0.0001). The models with a single parameter of temperature, altitude difference, or land area are not ideal and account for 12.3%,17.6%, and 28.0% of the variation in plant species richness respectively, with less F value and $P<0.05$. The model with multiple parameters of temperature, precipitation, and altitude difference shows much better than that of models with a single parameter, and account for 59.3% of the variation in plant species richness ($F=19.915$, $P<0.001$), and is far behind the CI model.

3.3 Discussion

The CTE successfully predicts the plant species richness distribution in provinces of China. This success comes mainly from our new system thinking approach of study. First, we treat every province as a consolidated ecosystem (though their land areas may differ greatly), and that plant species richness is one of its emergent properties through the interactions of biotic and abiotic factors. Conventional approaches using equal and small area quadrat as the basic studying unit actually divide the ecosystems into many small component parts. This approach is probably reasonable for a small scale evaluation within a community or landscape, but is inappropriate for a large spatial scale like China with its 9.6 million km^2 land area, and distances exceeding 5000 km from south to north and east to west. Adding up all of the component parts is not equal to a whole ecosystem. This also explains why Willis and Whittaker (2002) concluded that variables that best account for species richness on a local spatial scale may not be the same as those accounting for richness at regional spatial scales.

Secondly, we use an integrated factor from the results of PCA to calculate *CI*, which represents the contribution weights of every independent variable to the whole system. This approach is much better than the single variable model of temperature, precipitation, or area, because species richness is the emergent property of ecosystems through interactions of multiple factors. So a single factor model cannot explain the ecosystem property, especially on a large scale. On the other hand, though model established by multiple variables with stepwise regression fits better than a single variable model, it has a major drawback from the aspect of system thinking. This is because it selects variables only considering the correlations between dependent and independent variables, but neglects the interactions among variables. For instance, the last model in Table 1 only selects altitude difference and precipitation and rejects temperature. But we well know that nothing will happen in the real world without energy input. Thus, Hawkins et al. (2003) concluded that the interaction between water and energy provides a strong explanation for globally extensive plant and animal diversity gradients. Those analyses that do not include water-energy variables are missing a key component for explaining the broad-scale patterns of diversity. In our theory we not only consider correlation, but also pay great attention to the indispensable components of the emergent properties of ecosystems.

Finally, and most importantly, we introduce adaptability into the independent variables. This is estimated by the changed rate of animal species richness in an evolutionary time scale. In this way, we not only integrate biotic and abiotic factors but also combine spatial and temporal scales to predict plant species richness. This is a brand new approach when compared to the conventional hypotheses. Our results show that the changed rate of animal species richness has a great influence on plant species richness. Not only do theoretical works support this, but also empirical studies at the population and community levels have documented that herbivores can reduce a plants' potential distribution, restricting them to a subset of habitats that they might physiologically tolerate (Harley, 2003). More studies demonstrated that higher trophic-levels can have important effects on plant diversity and ecosystem properties (Fine, 2006; Schmitz, 2006).

Variable	Regression	Degrees of freedom	F value	P
Plant species richness – *CI*, in cubic				
Coefficient of determination (R^2)	0.940	3/23	137.516	0.0000
Constant	-0.184			
CI	0.818			
CI^2	0.293			
CI^3	0.432			
Plant species richness – integrated environmental factor of temperature, precipitation, altitude difference, and area, in linear				
Coefficient of determination (R^2)	0.423	1/25	20.054	0.0001
Constant	$-6.377E^{-07}$			
Integrated factor	0.935			
Plant species richness – temperature, in linear				
Coefficient of determination (R^2)	0.123	1/25	4.656	0.0407
Constant	$-7.407E^{-11}$			
Temperature	0.396			
Plant species richness – altitude difference, in linear				
Coefficient of determination (R^2)	0.176	1/25	6.553	0.0169
Constant	$-1.247E^{-10}$			
Altitude difference	0.456			
Plant species richness –area, in inverse				
Coefficient of determination (R^2)	0.280	1/25	11.132	0.0027
Constant	0.132			
Land area	0.201			
Plant species richness – temperature, precipitation and altitude difference, in linear				
Coefficient of determination (R^2)	0.593	2/24	19.915	0.000
Constant	$-2.11E^{-10}$			
Altitude difference	0.761			
Precipitation	0.714			

Table 1. Regression analysis of plant species richness against *CI* value and other environmental variables

	Area	Altitude difference	Annual precipitation	Annual Mean air temperature	Animal species richness
Plant species richness	0.0648	0.4557	0.3886	0.3962	0.9239
P	0.748	0.017	0.045	0.041	0.000

Table 2. Partial Correlation Coefficients between plant species richness and area, altitude difference, precipitation, temperature, and animal species richness

In addition, we find an interesting phenomenon that the geographical area has a very weak partial correlation coefficient with plant species richness, and the best model is an inverse one. This finding is counter to the *area hypothesis* (Table 2, Table 1). We postulate that the *area hypothesis* mainly considers a small scale, and that the pattern of species richness increasing with area will not exist if the area exceeds a critical size. Lyons & Willing (1999) also found that area effects on species richness for bats and marsupials are a minor importance at the area scales of 1000-25000km^2. We believe that the smallest area in our study (33900km^2 Hainan Province) may have been sufficiently large to have sampled most taxa as a regional species pool within China.

4. Conclusions

The mechanism of biodiversity is a complex issue that needs additional study from both system-specific models, and a more general theoretical framework that subsumes system-specific models as special cases (Fox, 2006). By embracing system thinking and regarding an ecosystem as an entity, and by treating biodiversity as the emergent property of the ecosystem, the Creativity Theory of Ecosystems integrates biotic and abiotic factors and combines spatial and temporal scales into a single model. Among the three variables of the model, the introduction of an adaptability variable is a unique and most important innovation. This enables our model to embrace both biotic and temporal factors. Thus we believe that the CTE provides a new approach to the study of the mechanisms of biodiversity from the aspect of a general theoretical framework. In addition, using the methods of PCA, the CI can be quantitatively calculated and will successfully predict plant species richness distribution on a regional scale within China. This demonstrates that the Creativity Theory of Ecosystems is feasible and promising.

5. Acknowledgements

We thank Z. H. Li for providing climatic data assistance Q. Q. Zhang for data collecting and analyzing assistance, and Mr. Richard R. Faltonson for editing in English. This work was supported by the National Natural Science Foundation of China (NSFC) grant 30972353 and by the Doctoral Discipline Special Foundation of High Educational Universities in China grant 20090014110011.

6. Appendix 1

6.1 Tables of basic data and steps for CI value calculation by PCA

Provinces (or Autonomous region)	Area ($\times 10^4 Km^2$)	Altitude difference (m)	Annual Precipitation (mm)	Annual Mean temperature (°C)	Animal species richness (No. of species)	Plant species richness (No. of species)
Anhui	13.98	1860	1192.1	15.06	535	3644
Fujian	12.14	2148	1588.3	18.21	809	4709
Gansu	45.4	5258	292.48	6.88	821	4164
Guangdong	17.8	1922	1762.5	21.53	829	6621
Guangxi	23.67	1941	1596.7	21.13	878	7148
Guizhou	17.61	2763	1125.5	15.19	910	6665
Hainan	3.39	1863.1	1670.5	24.9	561	3585
Hebei	18.77	2879	518.31	10.31	540	2888
Henan	16.7	2123	739.38	13.5	428	3779
Heilongjiang	45.46	1366	518.62	1.97	496	1846
Hubei	18.59	3105.4	1216.3	15.45	546	4295
Hunan	21.17	2076	1438.4	16.82	578	4705
Jilin	18.74	2686	644.19	4.33	410	2516
Jiangsu	10.26	624.7	1025.8	14.9	480	2492
Jiangxi	16.69	2138	1665.8	17.54	531	4552
Liaoning	14.59	1348	666.54	8.58	477	1358
Inner Mongolia	118.34	3474.4	284.13	4.84	506	2781
Ningxia	5.18	2756	279.56	8.22	384	1647
Qinghai	72.12	5210	339.37	1.17	398	2703
Shandong	15.7	1530	676.2	12.35	450	1616
Shanxi	15.6	2878	491.29	8.64	405	2751
Shanxi(Xi'an)	20.56	3647	631.29	11.5	564	3813
Sichuan	56.71	7476	935.72	12.68	1006	9249
Tibet	120.1	7348.13	506.78	3.91	730	5780
Xinjiang	166.49	8765	132.81	7.26	560	3500
Yunnan	39.4	6663.6	1133	16.46	1314	15444
Zhejiang	10.53	1933	1441.7	15.99	638	4579

Table 1. The data on climate, environmental diversity, and species richness for the 27 provinces in China.

Provinces (or Autonomous region)	Area	Altitude difference	Annual Precipi-tation	Annual Mean temperature	Animal species richness	Plant species richness	a-1§
Anhui	-0.547	-0.668	0.571	0.472	-0.398	-0.267	-0.398*
Fujian	-0.594	-0.530	1.367	0.991	0.860	0.109	0.860
Gansu	0.256	0.964	-1.236	-0.877	0.915	-0.084	0.915
Guangdong	-0.449	-0.638	1.717	1.539	0.952	0.783	0.952
Guangxi	-0.300	-0.629	1.384	1.473	1.177	0.969	1.177
Guizhou	-0.454	-0.234	0.437	0.494	1.323	0.799	1.323
Hainan	-0.818	-0.667	1.532	2.095	-0.278	-0.288	-0.278
Hebei	-0.425	-0.179	-0.783	-0.311	-0.375	-0.534	-0.375
Henan	-0.478	-0.542	-0.339	0.215	-0.889	-0.219	-0.889
Heilongjiang	0.257	-0.905	-0.782	-1.686	-0.577	-0.902	-0.577
Hubei	-0.429	-0.070	0.620	0.536	-0.347	-0.037	-0.347
Hunan	-0.363	-0.564	1.066	0.762	-0.200	0.107	-0.200
Jilin	-0.425	-0.271	-0.530	-1.297	-0.971	-0.665	-0.971
Jiangsu	-0.642	-1.261	0.237	0.446	-0.650	-0.674	-0.650
Jiangxi	-0.478	-0.535	1.523	0.881	-0.416	0.053	-0.416
Liaoning	-0.531	-0.914	-0.485	-0.596	-0.664	-1.074	-0.664
Inner Mongolia	2.119	0.107	-1.253	-1.213	-0.531	-0.572	-0.531
Ningxia	-0.772	-0.238	-1.262	-0.656	-1.091	-0.972	-1.091
Qinghai	0.938	0.941	-1.142	-1.818	-1.026	-0.599	-1.026
Shandong	-0.503	-0.827	-0.465	0.025	-0.788	-0.983	-0.788
Shanxi	-0.506	-0.179	-0.837	-0.587	-0.994	-0.582	-0.994
Shanxi(Xi'an)	-0.379	0.190	-0.556	-0.115	-0.264	-0.208	-0.264
Sichuan	0.544	2.029	0.056	0.080	1.764	1.710	1.764
Tibet	2.164	1.967	-0.806	-1.366	0.497	0.487	0.497
Xinjiang	3.349	2.648	-1.557	-0.814	-0.283	-0.318	-0.283
Yunnan	0.102	1.639	0.452	0.703	3.178	3.896	3.178
Zhejiang	-0.635	-0.633	1.072	0.625	0.075	0.063	0.075

Notes: § $a^{-1} = (N-n)/n$, where N is the animal species richness, n is 1, first calculated it from table 1, then normalized it.

* The data are normalized by standard deviation method with the following equations:

$$x'_{ij} = \frac{x_{ij} - \overline{x_j}}{s_j} (i = 1,2,...m; j = 1,2,...,n)$$

where,
$$\overline{x_j} = \frac{1}{m}\sum_{i=1}^{m} x_{ij} \qquad s_j = \sqrt{\frac{1}{m}\sum_{i=1}^{m}(x_{ij} - \overline{x_j})^2}$$

x_{ij} are the original data from table 1, x_{ij}' are the normalized data

Table 2. The normalized data from Table 1

Component	Initial Eigenvalues			Extraction Sums of Squared Loadings		
	Total	% of Variance	Cumulative %	Total	% of Variance	Cumulative %
1	2.778	55.569	55.569	2.778	55.569	55.569
2	1.603	32.065	87.634	1.603	32.065	87.634
3	0.373	7.453	95.087			
4	0.159	3.181	98.269			
5	0.087	1.731	100.000			

Component	Rotation Sums of Squared Loadings		
	Total	% of Variance	Cumulative %
1	2.284	45.681	45.681
2	2.098	41.953	87.634

Note: Component 1 and 2 accounted for 87.63% of variance in *e*, *d* and a^{-1} depending on the initial eigenvalues in percent of variance and cumulative percent, which fulfill the requirements of cumulative percent for the two components larger than 80.00%, and of total initial eigenvalues larger than 1. So, we need only calculate the two components, F1 and F2, and they can represent the integrated level of the total variables.

Table 3. Total Variance Explained

	Component	
	1	2
Annual Mean Temperature	0.895	-0.305
Annual Precipitation	0.884	-0.343
a^{-1}	0.730	0.578
Altitude difference	-0.113	0.953
Area	-0.396	0.803

Note: The major components of the first factor (F1) include temperature, precipitation, and the reciprocal of adaptability (a^{-1}); F2 is composed of altitude difference and area.

Table 4. Rotated Component Matrix

	Component	
	1	2
Annual Mean Temperature	0.371	-0.0609
Annual Precipitation	0.382	-0.0399
a^{-1}	0.419	0.391
Altitude difference	0.0709	0.474
Area	-0.0820	0.360

Table 5. Component Score Coefficient Matrix

Provinces (or Autonomous region)	F1§	F2※	% of Variance for component 1	% of Variance for component 2	*CI*∗
Ningxia	-1.10848	-0.70104	0.55569	0.32065	-0.84075973
Heilongjiang	-1.23733	-0.43886	0.55569	0.32065	-0.82829237
Jilin	-1.06291	-0.56709	0.55569	0.32065	-0.77248587
Shanxi	-0.90525	-0.57085	0.55569	0.32065	-0.68608143
Qinghai	-1.52923	0.51429	0.55569	0.32065	-0.68487073
Liaoning	-0.69384	-0.81544	0.55569	0.32065	-0.64703079
Shandong	-0.50111	-0.83787	0.55569	0.32065	-0.54712483
Inner Mongolia	-1.29225	0.71707	0.55569	0.32065	-0.48816191
Henan	-0.4076	-0.75018	0.55569	0.32065	-0.46704446
Hebei	-0.53405	-0.31812	0.55569	0.32065	-0.39877142
Jiangsu	-0.05018	-1.09471	0.55569	0.32065	-0.37890329
Shanxi (Xi'an)	-0.31057	-0.10938	0.55569	0.32065	-0.20765334
Anhui	0.2189	-0.70929	0.55569	0.32065	-0.1057933
Hubei	0.31361	-0.37556	0.55569	0.32065	0.053846627
Gansu	-0.35612	0.99807	0.55569	0.32065	0.122138823
Hunan	0.58149	-0.56129	0.55569	0.32065	0.14315054
Jiangxi	0.71499	-0.70264	0.55569	0.32065	0.172011277
Xinjiang	-1.07383	2.43101	0.55569	0.32065	0.182786764
Zhejiang	0.66299	-0.57856	0.55569	0.32065	0.182901649
Tibet	-0.63809	1.97223	0.55569	0.32065	0.277815317
Hainan	1.24772	-0.87933	0.55569	0.32065	0.411388362
Guizhou	0.90862	0.19321	0.55569	0.32065	0.566863834
Fujian	1.23396	-0.24656	0.55569	0.32065	0.606639768
Guangdong	1.58504	-0.25305	0.55569	0.32065	0.799650395
Guangxi	1.52001	-0.08699	0.55569	0.32065	0.816761013
Sichuan	0.87298	1.80694	0.55569	0.32065	1.064501567
Yunnan	1.84052	1.96399	0.55569	0.32065	1.652511952

Notes: § F1=-0.0820×area+0.0709×altitude difference + 0.371 × precipitation + 0.382 × temperature + 0.419 × a^{-1}

※ F2=0.360×area+0.474×altitude difference -0.0609 × precipitation -0.0399 × temperature + 0.391 × a^{-1}

∗ $CI = \sum F$ = % of variance for component 1 × F1 + % of variance for component 2 × F2 = 0.55569 × F1 + 0.32065 × F2

Table 6. Component score coefficient rotated matrix and *CI* values for all the provinces

6.2 References to data of plant species richness for 27 Provinces of China

Chen, H. B. (1992). *Flora of Shandong (Vol.1)*. Qingdao Publishing House, Qingdao, China, pp13-151

Chen, H. Y. (1964). *Flora of Hainan (Vol.1)*. Science Press, Beijing, China, pp1-222

Cooperating Group for Flora of Anhui. (1985, 1987, 1991, 1990, 1992). *Flora of Anhui (Vol.1-5)*. Anhui Science and Technology Publisher, Hefei, China

Cooperating Group for Vegetation of Sichuan. (1980). *Vegetation of Sichuan*. Sichuan People's Publishing House, Chengdu, China

Ding, B. Z.; Wang, S. Y.& Gao, Z. Y. (1981). *Flora of Henan (Vol.1)*. Henan People's Publishing House, Zhengzhou, China, pp1-123

Editorial Board for Flora of Guizhou. (1982, 1986, 1986, 1989, 1988, 1989, 1989, 1988, 1989). *Flora of Guizhou (Vol.1-10)*. Guizhou People's Publishing House, Guiyang, China

Editorial Board for Flora of Hebei. (1986, 1988, 1991). *Flora of Hebei (Vol.1-3)*. Hebei Scientific and Technical Publisher, Shijiazhuang

Editorial Board for Flora of Hunan. (2004). *Flora of Hunan (Vol.1, Vol.2)*. Hunan Scientific and Technical Publisher, Changsha, China, pp1-50

Editorial Board for Flora of Inner Mongolia. (1998, 1991, 1990, 1992, 1994). *Flora of Inner Mongolia (Vol.1-5)*. Inner Mongolia People's Publishing House, Huhhot, China

Editorial Board for Flora of Shanxi. (1992). *Flora of Shanxi (Vol.1)*. Scientific and Technical Publisher of China, Beijing

Editorial Board for Flora of Zhejiang. (1993, 1992, 1993, 1993, 1989, 1993, 1993). *Flora of Zhejiang (Vol.1-7)*. Zhejiang Scientific and Technical Publisher, Hangzhou, China

Editorial Board for Series of Natural Resources in China. (1995). *The Series of Natural Resources in China (Heilongjiang Volume)*. China Environmental Science Press, Beijing, China, pp. 469

Editorial Board for Series of Natural Resources in China. (1995)..*The Series of Natural Resources in China (Xinjiang Volume)*. China Environmental Science Press, Beijing, China, pp319

Editorial Board for the Complete Series of Chinese Agriculture. (1999). *The Complete Series of Chinese Agriculture (Henan Volume)*. Chinese Agriculture Publisher, Beijing, China, pp. 41

Editorial Board for the Complete Series of Chinese Agriculture. (2001). *The Complete Series of Chinese Agriculture (Jiangxi Volume)*. Chinese Agriculture Publisher, Beijing, China, pp22-23

Fu. S. X. (2001). *Flora of Hubei (Vol.1)*. Hubei Scientific and Technical Publisher, Wuhan, China, pp1

Guangdong Institute of Botany. (1976). *Vegetation of Guangdong*. Science Press, Beijing, China, pp18

Huang, D.X. (1997). *Flora of Gansu*. Gansu Science and Technology Press, Lanzhou, China, pp36-37

Institute of Botany in Jiangsu Province. (1982, 1997). *Flora of Jiangsu (vol.1-2)*. Jiangsu People's Publishing House, Nanjing, China

Kunming Institute of Botany, Chinese Academy of Science. (1977, 1995, 2006). *Flora of Yunnan (Vol.1, 16, 20)*. Science Press, Beijing, China

Lei, M. D. (1999). *Vegetation of Shanxi*. Science Press, Beijing, China, pp52

Li, S. X. (1988). *Flora of Liaoning (Vol.1)*. Liaoning Scientific and Technical Publisher, Dalian, China

Ma, D. Z. & Liu, H. L. (1986, 1990). *Flora of Ningxia (Vol.1, Vol. 2)*. Ningxia People's Publishing House, Yinchuan, China

Ma, Z. Q. (2001). *Vegetation of Shanxi*. Scientific and Technical Publisher of China, Beijing, China, pp41

Ni, H.W. & Tang, S.B. (1995). Studies on the Floristic Components and Characteristics of Heilongjiang Province, China. *Territory & Natural Resources Study* 64:57-61

Northwest Plateau Institute of Biology, Academia Sinica. (1997, 1999, 1996, 1999). *Flora of Qinghai (Vol.1-4)*. Qinghai People's Publishing House, Xining, China

Qi, C. J. (1990). *Vegetation of Hunan*. Hunan Scientific and Technical Publisher, Changsha, China, pp40

Scientific and Technical Committee of Fujian Province and Editorial Board for Flora of Fujian. (1982, 1995). *Flora of Fujian (Vol.1, Vol.6)*. Fujian Science and Technology Publisher, Fuzhou, China

South China Institute of Botany, Academia Sinica. (1964, 1965, 1974, 1977). *Flora of Guangdong (Vol.1-4)*. Science Press, Beijing, China

South China Institute of Botany, Academia Sinica. (2000). *Flora of Guangdong (Vol.4)*. Guangzhou Scientific and Technical Publisher, Guangzhou, China

Su, Z. Y. ; Liao, W. B. & Zhang, H. D. (1996). Studies on the Floristic Components and Characteristics of Guangxi Province, China: I. The Floristic Components. *The Journal of Sun Yatsen University* 2 (Suppl.):81-86

The Comprehensive Scientific Expedition to the Qinghai-Xizang Plateau, Academia Sinica. (1983). *Flora of Tibet (Vol.1)*. Science Press, Beijing, China, pp. 1, 356, 406

Wang, P. S. & Wang, X. Y. (2001). *Ferns of Guizhou*. Guizhou Scientific and Technical Publisher, Guiyang, China

Wang, W.Y. & Pei, L. H. (1998). The Variety and Protection of Gymnospermae in Jiangxi, China. *Jiangxi Forestry Science and Technology (Suppl.)*:13-15,34

Wuhan Institute of Botany, Academia Sinica. (2001, 2001, 2002, 2002). *Flora of Hubei (Vol.1-4)*. Hubei Scientific and Technical Publisher，Wuhan, China

Yang, C. Y. (1993). *Flora of Xinjiang (Vol.1)*. Xinjiang Science and Technology and Hygiene Publishing House, Wulumuqi, China, pp1-51

Zhang, W. & Zhao, S. L. (2001). The Evolution and Sources of the Flora in Shandong Province. *Journal of Wuhan Botanical Research* 19:57-64

Zhang, Y. J. (1989). The phytogeographical distribution and floristic characteristics of pteridophytes of Gansu Province. *Journal of Lanzhou University* 25:87-92

Zheng, J. H. (1987). Research on the basic elements and main features of pteridophytes flora of Hubei province, China. *Journal of Wuhan Botanical Research* 5:227-233

Zhou, H.G. et al. (2004). Study on pteridophyte flora of Guangxi, China. *Guihaia* 24:311-316

7. References

Allen, A. P.; Brown, J. H. & Gillooly, J. F. (2002). Global biodiversity, biochemical kinetics, and the energetic-equivalence rule. *Science* 297:1545-1548

Alonso, D. & McKane, A. J. (2004). Sampling Hubbell's neutral theory of biodiversity. *Ecology letters* 7:901-910

Buckling, A.; Wills, M. A. & Colegrave, N. et al. (2003). Adaptation limits diversification of experimental bacterial populations. *Science* 302:2107-2109

Bush, M. B. (2003). *Ecology of a Changing Planet (Third edition).* Pearson Education Asia Limited and Tsinghua University Press, Beijing

Chave, J. (2004). Neutral theory and community ecology. *Ecology letters* 7:241-253

Editorial Board for Series of Natural Resources in China. (1995). *The Series of Natural Resources in China (in Chinese).* China Environmental Science Press, Beijing, China

Editorial Board for the Complete Series of Chinese Agriculture. (1998). *The Complete Series of Chinese Agriculture.* China Agriculture Press, Beijing

Egbert Giles Leigb J.r. (1971). *Adaptation and diversity: natural history and the mathematics of evolution.* Freeman, Cooper & Company, San Francisco, California

Fine, P. V.; Miller Z.J. & Mesones, I. et al. (2006). The growth-defense trade-off and habitat speciation by plants in Amazonian forests. *Ecology* 87 (Suppl.): S150-S162

Fox, J. W. (2006). Using the price equation to partition the effects of biodiversity loss on ecosystem function. *Ecology* 87:2687-2696

Freestone, A. L. & Inouye B. D. (2006). Dispersal limitation and environmental heterogeneity shape scale-dependent diversity patterns in plant communities. *Ecology* 87:2452-2432

Gallagher, R. & Appenzeller, T. (1999). Beyond reductionism. *Science* 284:79

Gove, P. B. & the Merriam-Webster editorial staff. (1976). *Webster's Third New International Dictionary of the English Language Unabridged.* G. & C. MERRIAM COMPANY, publishers

Harley, C. D. G. (2003). Abiotic stress and herbivory interact to set range limits across a two-dimensional stress gradient. *Ecology* 84:1477-1488

Hawkins, B. A.; Porter, E. E. & Diniz-Filho, J. A. F. (2003). Productivity and history as predictors of the latitudinal diversity gradient of terrestrial birds. *Ecology* 84:1608-1623

Hawkins, B. A.; Field, R. & Cornell, H. V. et al. (2003) .Energy, water, and broad-scale geographic patterns of species richness. *Ecology* 84:3105-3117

Hubbell, S. P. (1979). Tree dispersion, abundance, and diversity in a tropical dry forest. *Science* 203:1299-1309

Hubbell, S. P. (2001). *A Unified Neutral Theory of Biodiversity and Biogeography.* Princeton University Press, Princeton, NJ

Huston, M. A. (1979) .A general hypothesis of species diversity. *American Naturalist* 113:81-101

Jablonski, D.; Roy, K. & Valentine J. W. (2006). Out of the tropics: evolutionary dynamics of the latitudinal diversity gradient. *Science* 314:102-106

Kerswell, A. P. (2006). Global biodiversity patterns of benthic marine algae. *Ecology* 87:2479-2488

Krebs, C.J. (2001) Ecology: *The Experimental Analysis of Distribution and Abundance (5 edition)*. Benjamin Cummings.

Lovelock, J. (2003). The living earth. *Nature* 426:769~770

Lyons, S. K. & Willig, M. R. (1999). A hemispheric assessment of scale dependence in latitudinal gradients of species richness. *Ecology* 80:2483-2491

Manuel, C. & Molles, J. r. (2002). *Ecology: Concepts and Applications (second edition)*, Beijing: Higher Education Press and The McGraw-Hill Book Co. Singapore

McGill, B.J, (2003). A test of the unified neutral theory of biodiversity. *Nature* 422:881-885

Mittelbach, G. G.; Steiner, C.F. & Scheiner, S. M. et al. (2001). What is the observed relationship between species richness and productivity? *Ecology* 82:2381-2396

Ricklefs, R. E. (2006). Evolutionary diversification and the origin of the diversity-environment relationship. *Ecology* 87 (Suppl.): S3-S13

Ridley, M. (2004). *Evolution (Second edition)*. Oxford University Press Inc., New York, USA

Ritchie, M. E. &Olff, H. (1999). Spatial scaling laws yield a synthetic theory of biodiversity. *Nature* 400:527-557

Root, T. (1988). Energy constraints on avian distributions and abundances. *Ecology* 69:330-339

Scheibe, J. S. (1987). Climate, competition, and the structure of temperate zone lizard communities. *Ecology* 68:1424-1436

Schmitz, O. J. (2006). Predators have large effects on ecosystem properties by changing plant diversity, not plant biomass. *Ecology* 87:1432-1437

Shi, K.; Deng, J. & Ding, Y. (1991). *Development of Forestry Science and Technology in China*. China Science and Technology Press, Beijing

Straub, C. S. & Snyder, W. E. (2006). Species identity dominates the relationship between predator biodiversity and herbivore suppression. *Ecology* 87:277-282

Tilman, D. (1999) The ecological consequences of changes in biodiversity: a search for general principles. *Ecology* 80:1455-1474

Turner, J. R. G.; Lennon, J. J. & Lawrenson, J. A. (1988) .British bird species distributions and the energy theory. *Nature* 335:539-541

Urban, M. C. & Skelly, D. K. (2006). Evolving metacommunities: toward an evolutionary perspective on metacommunities. *Ecology* 87:1616-1626

Willig, M. R.; Kaufman, D. M. & Stevens, R. D. (2003). Latitudinal gradients of biodiversity: pattern, process, scale, and synthesis. *Annual Review of Ecology, Evolution, and Systematics* 34:273-309

Willis, K. J. & Whittaker, R. J. (2002). Species diversity - scale matters. *Science* 295:1245-1248

Xu, J. H. (2002). *Mathematical Methods in Contemporary Geography*. High Education Press, Beijing

Zhang, Y. (1998). *Biological Evolution*. Beijing University Press, Beijing

Zhou, S. R. & Zhang, D. Y. (2006). Neutral theory in community ecology. *Journal of Plant Ecology*. 30:868-877

3

Ecological Niches and Diversity Maintenance

Peter Chesson
Department of Ecology and Evolutionary Biology,
The University of Arizona, Tucson,
USA

1. Introduction

Why do some places have much higher diversity than others? Explaining patterns of species diversity on the earth is a problem of long-standing for ecologists. Numerous diversity patterns have been documented [1], but the mechanisms behind these patterns remain poorly understood. If we knew the mechanisms, surer decisions in conservation and management would be possible. Understanding can be sought in many different ways. A dominant approach in community ecology is to search for explanations through the study of species interactions [2, 3]. This approach is motivated by the competitive exclusion principle, which suggests that there are limits to how similar species can be in their ecology while coexisting with one another in a stable way [2-6]. However, the competitive exclusion principle is challenged by the existence of many highly diverse communities of seemingly similar species. A response to this challenge is neutral theory, which postulates that many coexisting species are ecologically identical in respects that matter, and do not coexist stably. Instead, they undergo random walks to extinction, with diversity replenished by speciation, and immigration of species from other areas [7]. The neutral theory has been successful in predicting some diversity patterns in nature [8], but fails in other areas [9-12], and is limited in the nature of the predictions that it can produce [13].

Other responses to the diversity challenge take the competitive exclusion principle seriously, although with a broad definition of stability [5]. These responses seek to understand the various ways that species differ from one another, how these differences structure species interactions and ultimately contribute to diversity maintenance in terms of species coexistence mechanisms [3, 5, 14, 15]. Traditional approaches focus on the differences between similar species in the ways they exploit resources, with some attention to the role of natural enemies in mediating or modifying interactions. New developments emphasize that the relationships that species have with their natural enemies are potentially just as important for diversity maintenance as their relationships with resources, and can have very similar outcomes [16]. Other directions focus on structuring and variation in the physical environment, emphasizing that the abilities of similar species to coexist with one another may have much to do with physical environmental structure [15, 17-19].

Testing diversity maintenance hypotheses in nature is a vexing challenge [3]. Data are often limited and manipulating systems experimentally for testing diversity maintenance questions poses serious difficulties. Moreover, devising definitive tests of mechanisms has been problematic [20, 21]. However, work on coexistence mechanisms in variable environments has led to methods of quantifying the strength of coexistence mechanisms, and these quantifications have suggested definitive tests of mechanisms in nature [22]. Implementing these new methods comes with all the usual difficulties of ecological field manipulation, but the ability to focus on surer methods [20] may ultimately lead to the kind of focus in experimental technique that leads to breakthroughs.

In spite of the difficulty of understanding which coexistence mechanisms are active in a given system, enough commonalities exist between different mechanisms of stable coexistence to allow some general advice for conservation and management. Although we are a long way from truly understanding diversity maintenance, we have enough knowledge to suggest a number of areas where caution is needed. Species should not be managed in isolation. Factors improving the situation for one set of species may degrade it for others. Maintaining trophic structure, physical environmental structure, disturbance regimes, and spatial connectivity are all common sense ideas that receive support from existing understanding of diversity maintenance mechanisms.

2. Fitness differences, niches and coexistence

When we consider how species interactions limit diversity, an important concept is the average fitness of a species as a whole, for this average fitness determines a species' ability to dominate other species with which it interacts. This concept is not to be confused with the fitness of a genotype or individual organism. The most common use of the idea of fitness is for genotypes or individual alleles of a gene, where its use is often relative, to predict the survival of a genotype or allele compared with others [23, 24]. Ecologists, however, often give fitness an absolute meaning as the performance of an individual organism, and this individual-level fitness is often measured as the total number of offspring that an individual leaves in its life time [25]. An alternative and more pragmatic meaning is its contribution to the population over a defined period of time through its own survival and reproduction [23]. Fitness at the species level applies in a similar relative context to fitness of a genotype or allele, but in this usage fitness specifies the relative degree of adaptedness of a species compared to others having a similar way of life and living in the same area, i.e. this fitness measure is to relative other species in the same guild, living in the same area [20].

When this species-level fitness measure is applied to problems of species coexistence, it assumes that the area of land in question is large enough for populations to be closed on an ecological timescale [5]. Then the fitness measure, which we denote κ, determines the degree of adaptedness of a species to the conditions applicable on that area of land. If the species in a guild cannot coexist with one another, then it is the species with the largest value of κ that persists, excluding others. More generally, the κ values rank the species in terms of their adaptation to the environment, and in essence rank the species in terms of how secure their persistence is when interacting with the others species in the guild.

2.1 The MacArthur consumer-resource model

A program to measure average fitness measures in nature has been proposed [26], but at the present time, these measures are easier to define in models. We consider the model of MacArthur [27] where a guild of animal species consume common biological resources in a lower trophic level, as reworked by Chesson [5, 28]. This model has had a key role in the development of ideas on resource partitioning for animal guilds (Table 1). Fig. 1 diagrams the foodweb being modeled, and Table 1 specifies the equations.

MacArthur Consumer-resource Equations	$\frac{1}{N_j}\frac{dN_j}{dt} = \sum_l c_{jl} R_l v_l - \mu_j$ (consumer equation) $\frac{1}{R_l}\frac{dR_l}{dt} = r_l^R\left(1 - \alpha_l^R R_l\right) - \sum_j N_j c_{jl}$ (resource equation)
Average fitness measure	$\kappa_j =$ $\frac{1}{s_j}\Big(\underbrace{\sum_l c_{jl} K_l^R v_l}_{\text{Harvesting ability}} - \underbrace{\mu_j}_{\text{Maintenance requirement}}\Big)$ $s_j = \underbrace{\sqrt{\left(\sum_l c_{jl}^2 v_l K_l^R / r_l^R\right)}}_{\text{Sensitivity}}$
Overlap measure	$\rho = \sum_l \frac{c_{jl} v_l K_l^R c_{kl}}{r_l^R} \Big/ s_j s_k$
Derived Lotka-Volterra Equations	$\frac{1}{N_1}\frac{dN_1}{dt} = r_1\left(1 - \alpha_{11} N_1 - \alpha_{12} N_2\right)$ $\frac{1}{N_2}\frac{dN_2}{dt} = r_2\left(1 - \alpha_{21} N_1 - \alpha_{22} N_2\right)$
Notation	N_j : Density of (consumer) species j. R_l : Density of resource l. c_{jl} : Consumption rate of resource l by species j. v_l : unit value of resource l. μ_j: resource maintenance requirement of species j. r_l^R : Maximum growth rate of resource l. α_l^R : Intraspecific competition for resource l $1/K_l^R$, the carrying capacity of the resource. α_{ij}: Coefficient for competition for the effect of species j on species i

Table 1. MacArthur's consumer-resource equations

The fitness measure κ derived from this model involves three things (Table 1). The first is the ability of a species to harvest resources. This is the total resource intake of a species when its resources are at their carrying capacities.

This quantity, therefore, is a maximum resource harvesting rate. Subtracted from it is the resource maintenance requirement, which is the level of resource intake needed for a per capita growth rate of exactly zero, i.e. it is the resource intake that gives an individual organism a fitness of 1, which means that that one individual is replaced by exactly one individual on average in one unit of time. The maximum resource intake, less the maintenance requirement, measures the ability of a species to meet its needs, and is called the net maximum harvesting rate. Finally, the net maximum harvesting rate is divided by a quantity s which measures the sensitivity of the growth rate of the species to changes in resource availability, as discussed in detail by Chesson [20]. The resulting quantity is the average fitness measure κ (Table 1).

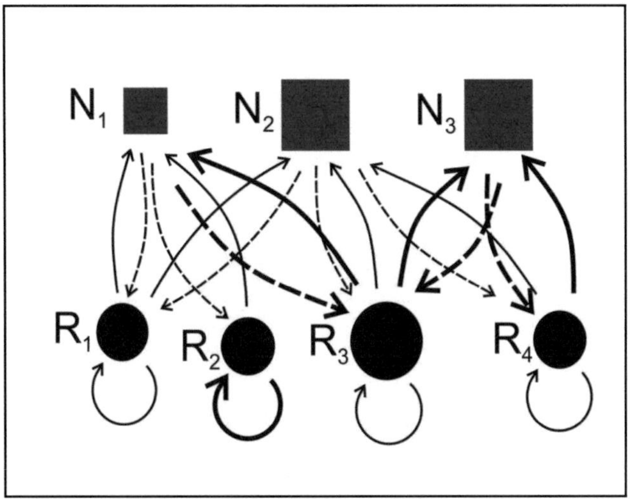

Fig. 1. Foodweb in the MacArthur model. Squares are the focal guild members and circles are resources. Arrows indicate the directions of effects of one species on another, with thickness differences indicating differences in strengths of these effects. The feedback loops for guild members come from links from a square to a circle back to a square. Circular loops for resources indicate resource self-limitation.

Although this fitness measure κ may seem intuitively reasonable, the real test is its ability to predict the outcome of the interactions of this species with others in the same guild. The measure κ has this desired property because it is related to competition coefficients derived from the MacArthur consumer-resource model. These competition coefficients are denoted α_{ij} giving the effect of interspecific competition from species j onto species i, and α_{jj} for intraspecific competition within species j. They measure the feedback loops illustrated in Fig. 1 from one species to another, and from one species back to itself. Moreover, when resource dynamics are fast relative to consumer dynamics, e.g. because resource species have much shorter generation times, the MacArthur consumer-resource model reduces to the Lotka-Volterra competition model defined by these competition coefficients [5] – see Table 1 for the two-species case. Of most importance, the ratio of interspecific competition to intraspecific competition is related to the ratio of the κ values:

$$\frac{\kappa_j}{\kappa_i}\rho = \frac{\alpha_{ij}}{\alpha_{jj}}, \qquad (1)$$

where the quantity ρ measures overlap in resource use between and species j [5]. Fundamentally, we see that the ratio of the fitness measures determines how much one species affects another, compared with how much it affects itself. Understandably, however, relative fitness alone does not do this. It is most important to know how much the species interact with one another also. Thus, the fitness ratio is multiplied by the overlap measure, ρ, which is on a scale from 0 to 1, with 1 meaning complete overlap and 0 meaning no overlap.

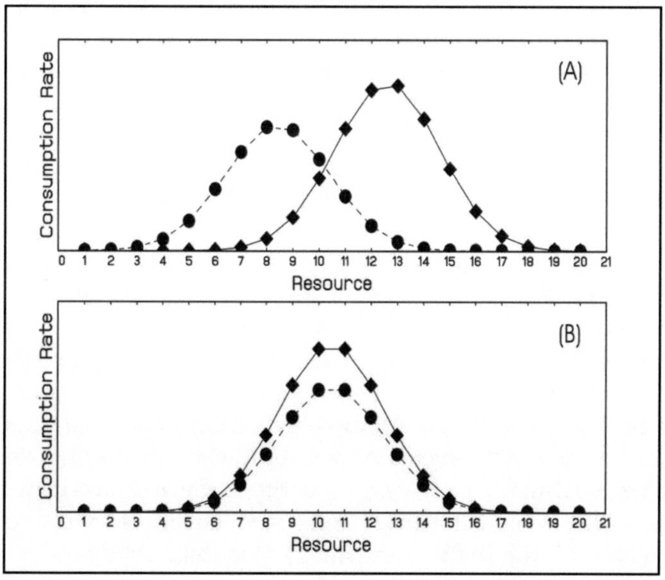

Fig. 2. Niche relationships of two species, in terms of the consumption rates on limiting resources. (A) Partial overlap in resource use, $\rho = 0.33$. (B) Complete overlap, $\rho = 1$.

Fig. 2 shows the niche relationships of two species as specified by their consumption rate curves for both partial overlap and complete overlap, while Table 1 shows how this overlap measure is calculated. The difference in the heights of the consumption rate curves does not affect the overlap, but is a factor explaining the difference between the fitnesses, κ.

Interpreting this result is simplest in the case of two species competing with one another. If expression (1) is greater than 1, then species j excludes i from the system. In terms of competition coefficients, this means that species j places less limitation on its own growth as it increases in abundance than it places on the growth of the species i. Thus, species j continues increasing in abundance beyond the point that the growth of species i becomes negative.

In terms of the fitness ratio and overlap measure, if the species overlap completely in resource use, then species j excludes i if it simply has a higher fitness than species i. However, with less than complete overlap, this fitness ratio needs to be discounted by the degree of overlap to determine if species j can exclude i. For example, if $\rho = 2/3$, then the fitness ratio needs to be more than 3/2, rather than just more than 1. Said another way, if one species is fitter than another, it only harms that other species to the extent that they overlap in resource use. As we shall see below, this statement generalizes to all forms of overlap associated with interactions between the species.

In the case where neither of two competing species can exclude the other, they coexist. For two species labeled 1 and 2, this means that expression (1) must be less than 1 for (i, j) = (1, 2) and (2, 1), discounting the unlikely boundary case of equality with 1, discussed below. This condition can be rearranged to

$$\rho < \frac{\kappa_1}{\kappa_2} < \frac{1}{\rho}, \tag{2}$$

and shows that overlap, ρ, places constraints on how different the average fitness measures for the two species can be and still allow coexistence. Simply put, the more similar the species are in resource use (the closer ρ is to 1), the more similar they have to be in average fitness for coexistence to occur. In opposite terms, the more different the species are in resource use, the more different they must be in average fitness before one excludes the other.

This analysis of MacArthur's model yields some important general lessons. The competitive exclusion principle holds that species that are very similar in their ecology should have difficulty coexisting. Similarity in ecology within the MacArthur model can be interpreted as meaning a ρ value near 1, i.e. high overlap in resource use. Difficulty in coexisting can now be interpreted as meaning that the species have to be evenly matched in average fitness if they are to coexist, i.e. if one species gets a small edge in average fitness over the other, then exclusion will occur.

These findings for the MacArthur model generalize to others [5, 29], and they highlight two different sorts of similarity between species. The first is similarity in average fitness, i.e. having κ values near to each other. As discussed above, the κ values allow ranking of overall adaptation to the environment. Almost by definition, having similar κ values is favorable to species coexistence. However, similarity in κ values is a very different concept from similarity in way of life, or similarity in niches, when niches are defined in terms of the way of life of a species [5, 30]. In the MacArthur model, similarity in way of life reduces to similarity in resource use, which then determines how much the species interact with one another. Thus, similarity in niches constrains differences in average fitness compatible with coexistence, by conditons (2).

Looked at another way, we can think of these similarities and differences as relating to average performance compared with performance under specific conditions. Species 1 and 2 in Fig. 2(A) have very different performance under specific conditions; for example, species 1 derives very little benefit from resources 13 to 17, instead gaining most benefit from resources 5 to 12 while species 2 has an opposite pattern. The average performances of the species, which depend on the heights of the curves in the figure, are nowhere near as different as their performances for most specific resources.

Differences in performance under specific conditions lead to tolerance of average performance differences. For example, in the MacArthur model, specific-condition differences give a small value of ρ, and thus wide tolerance in the κ ratio. More generally, the nature of these specific differences is very important. They must relate directly or indirectly to separation of feedback loops because it is separation of these feedback loops that makes it possible for intraspecific competition to exceed interspecific competition, the key to coexistence. With the MacArthur model, the measure ρ is a specific measure of the degree of separation of those feedback loops. Multiplication by the fitness ratio adjusts the degree of separation of those loops for the relative degrees of resource shortage that the two species can tolerate due to their overall adaptation to the environment, to give the competition coefficient ratio (1).

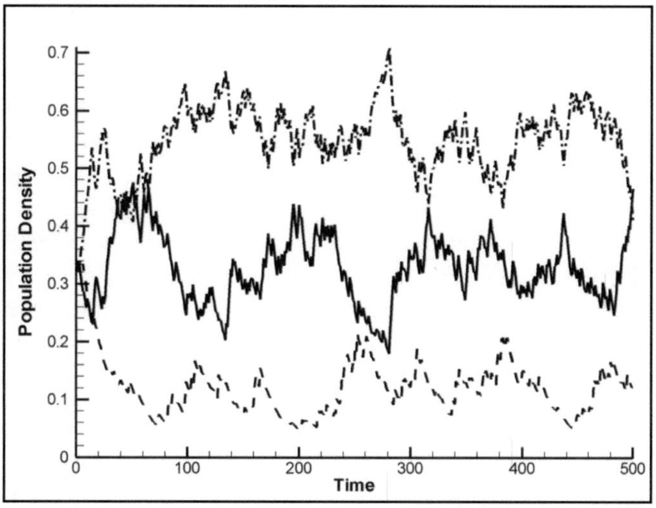

Fig. 3. Simulation showing species coexisting stably according to the lottery competition in a temporally varying environment. Note that each species shows a strong tendency to recover from fluctuations to low density.

2.2 Stable versus neutral coexistence

When the conditions (2) above are satisfied in the MacArthur model, the result is stable coexistence in the sense that if either species is perturbed to low density, it recovers from that low density state. In general, the MacArthur model leads to a stable equilibrium point. However, by stable coexistence, a stable equilibrium is not implied. Instead, the requirement is recovery of each species from low density, which is fully compatible with fluctuating coexistence, as illustrated in Fig. 3. Indeed, Fig. 3 illustrates stable coexistence in the lottery model [31], which requires temporal fluctuations for stable coexistence to occur. i.e. it is an example of fluctuation-dependent coexistence [32], as discussed below under variation in time and space. Neutral models, on the other hand, do not allow stable coexistence. Instead, they are characterized by having average fitnesses the same for all species. Thus, all species have the same κ values, and the niche overlap measures ρ between all pairs of species are all 1. In general, the key assumption of neutral models is stronger: it is that the species identity of an individual has no bearing on how it is affected by any other individual or how it affects other individuals [33]. However, this stronger assumption is no particular concern here. With equal average fitnesses and complete niche overlap, in the MacArthur model expression (1) is exactly equal to 1, and the inequalities (2) are not satisfied. Stable coexistence does not occur, but the model is neutrally stable. For example, in the two-species case, population densities approach the line $\alpha_{11}N_1 + \alpha_{22}N_2 = 1$, but can remain at any point on this line, until perturbed. In particular, there is no tendency for either species to increase from low density if perturbed there, and so there is no stable coexistence.

The significance of neutral models is their ability to predict some patterns of diversity seen in nature, for example in very diverse communities, neutral models have successfully reproduced the observed frequencies of species with different degrees of commonness and rarity [8]. To produce any pattern at all, however, they need to go beyond the strict deterministic description of equations like those in Table 1, and generally include demographic stochasticity. They thus recognize the finiteness of population size, and include independent chance events in the lives of individuals that determine individual fates [7]. For a model like the MacArthur consumer-resource equations, this would mean that the abundances of the species would be constantly randomly perturbed up and down the line $\alpha_{11}N_1 + \alpha_{22}N_2 = 1$, or with more than two species, over the hyperspace $\sum \alpha_{jj}N_j = 1$. Extinctions do occur in such models, and so in modern versions they include stochastic speciation and immigration to balance species losses [7].
The predictions of neutral models have often not held up [9-12], and neutral models do not give predictions about the differential effects that environmental change often has on the species in a guild [13], which are critical for conservation and management questions. On the other hand, models in which individuals of different species are not identical in their ecology can reproduce at least some of the patterns of multispecies diversity predicted by neutral models [34]. Neutral models provide a simpler mathematical route to some predictions, and in this way have highlighted some of the processes that lead to some observed patterns in nature, but these same processes are able to produce these patterns in nonneutral models as well.

2.3 Multitrophic diversity maintenance

Similar species can interact with one another by feedback loops through resources, or by feedback loops through predators. Naturally, they may also interact directly by interfering [35] with or facilitating each other's activities [36], or through intraguild predation [37]. However, the focus here is on the much misunderstood and critical area of feedback loops through predators [38]. Fig. 4 expands the foodweb model of Fig. 1 to include predators of the species in our guild of interest. Much emphasis has focused on idea that competition limits diversity of guild members, while predation modifies what competition does. For example, the keystone species idea was developed around the concept that predation on a competitive dominant would prevent competitive exclusion [39, 40].
The keystone species idea can be view directly within the competition framework above if it is assumed that the predator, in causing mortality, increases the resource maintenance requirement of the dominant species. This is a natural expectation because if a species has higher mortality, to persist it will need higher reproduction, and higher resource consumption to fuel that reproduction. The increased maintenance requirement for the dominant then decreases its κ value making it more comparable to that of other species, potentially enabling coexistence. For example, in the two-species case, the coexistence conditions (2) might become satisfied [6].
The above keystone species discussion focuses just on the mortality that the predators cause, and neglects the feedback loops associated with them. In causing mortality, the predator benefits and can build up in density, with the potential then of inflicting greater mortality. The feedback loops from guild members (the prey) to guild members arise because higher densities allow predator numbers to increase, increasing mortality on guild members. As Holt [38, 41] pointed out many years ago, these feedback loops through predators can have

very similar effects to feedback loops through resources. As consequence Holt coined the term "apparent competition" [41]. While previously we focused on species differences based on their use of resources, equally we can focus on species differences based on their patterns of susceptibility to the various predator species of the guild in question. Thus, the idea of a species' niche can be expanded to include susceptibility to predator species [3, 5].

Within this expanded concept of a niche, the niche overlap measure ρ is extended to predator overlap too. Indeed, the MacArthur consumer-resource model extends to a consumer-resource-predator model, with ρ depending on both resource and predator overlap, measuring resource-sensitivity and predator-sensitivity similarities between species [16]. For example, ρ would be lowered if one species were more sensitive to changes in resource abundance while the other species were more sensitive to changes in predator-abundance. The extended fitness measure, κ, accounts for overall predation susceptibility as well as resource harvesting ability.

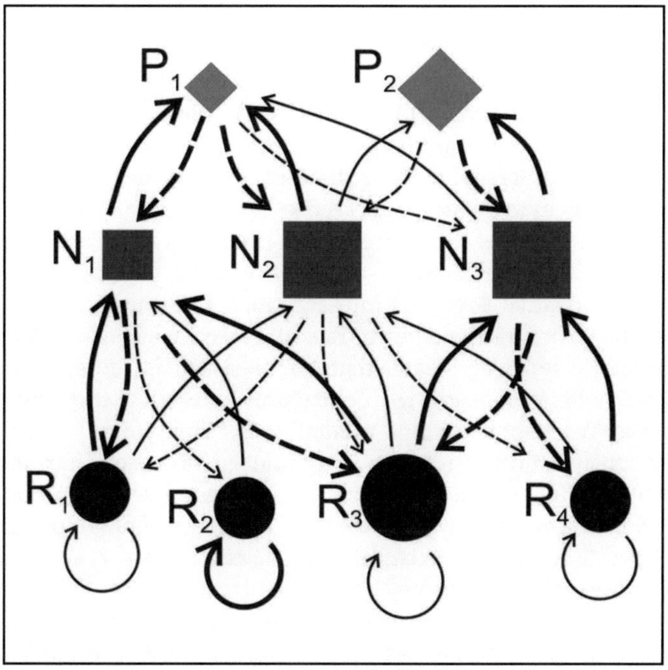

Fig. 4. Foodweb for the consumer-resource-predator model. As for Fig. 1, but with predators signified by the diamond symbols.

In this fuller concept of the niche, competition coefficients α_{ij}, and α_{jj}, become coefficients of interspecific and intraspecific density dependence, and thus include both resource competition and apparent competition. Of most importance, the relationship (1) between competition coefficient ratios, fitness ratios and niche overlap still holds, and the coexistence conditions (2) remain valid. Thus, no longer can competition and predation be regarded as having very different effects on species coexistence. We should not think of competition as the primary factor limiting coexistence, with predation modifying what competition does. Instead, the view emerging corresponds to Holt's [38] advice that what is possible for competition is possible for predation too.

Relationships with resources and with predators can equally promote or limit diversity depending on the circumstances. Coexistence is promoted if different guild members have different relationships with their resources or different relationships with their predators, and exclusion is promoted if these relationships tend to be similar. Differences with respect to either resource sensitivity or predator susceptibility lower ρ, while similarities in either respect increase ρ. As noted above, an extra complication is that complementary relationships between species between sensitivity to resources and susceptibility to predators also lowers ρ, and is referred to in the literature as a competition-predation tradeoff [42].

Feedback loops through resources and through predators naturally do have some differences. For example, predators fitting the most common notion as species that hunt and kill prey species, are often larger than their prey, or of comparable size, and are normally less species rich than the prey they focus on, while the resources of their prey may be more species rich. Opportunities for niche distinctions between species due to their relationships with their predators would therefore seem less than the opportunities due their relationships with their resources. However, predatory behavior can be complex [43], and when predators have frequency-dependent behavior, a single predator has a similar effect in terms of the coexistence of prey species to several predators that do not have this frequency-dependent behavior, but are instead specialists on particular species. Moreover, interpreting "predators" as natural enemies more generally, such as diseases and parasites, makes predators appear every bit as able as resources to define distinct niches for their prey species.

The potential symmetry between resources and predators in promoting coexistence or exclusion depends also on the relative strengths of these processes. The niche overlap measure ρ takes these relative strengths into account. This can be seen in the formula in Table 1, where the reciprocal of the renewal rate of a resource weights its importance in the overlap measure. Slow renewal means that the measure is more strongly affected by consumption, and therefore more strongly contributes to resource competition. Similarly, in the extended overlap measure to include predation, the ability of a predator to build up in response to prey consumption weights its importance, and determines its contribution to apparent competition [16]. Of most significance, if resource separation is strong, but resource competition as a whole is weak relative to apparent competition, then that resource separation will have little effect on coexistence. Whether the species coexist depends on whether their niches are distinguished by predation. If their niches are not distinguished by predation, then they will have difficulty coexisting despite substantial separation at the level of resources. Naturally, the opposite conclusion is reached if it is resource competition that is strong relative to apparent competition.

2.4 Variation in time and space

All natural environments vary substantially in time or space in at least some major ecologically significant ways. Environmental variables and population densities nearly always vary substantially in both time and space. Empirical studies show this, but at great variance with the realities of nature, attention of theory to these facts historically was not a mainstream endeavor [44], and sadly that remains true today. Despite the absence of mainstream attention, considerable theory on the role of variation in time and space has been developed [14, 17, 19, 20, 38, 45-52]. Although the influence of this theory on empirical studies has not yet matched its potential importance, there is now a growing body of related empirical studies [53-56]

Some ideas about spatial and temporal variation, especially those under the heading of disturbance, have viewed variation in the same way that has often coloured thinking about predation: competition has been seen as limiting diversity, with variation in time and space as modifying or even nullifying what competition does [57-60]. Many of these views are influenced by the idea that variation in time and space make ecological communities "nonequilibrial," while niche ideas are presumed to depend on the concept of equilibrium [61]. Nothing could be further from the truth [3, 5].

Theoretical studies converge on the conclusion that variation in time and space are best viewed as providing more opportunities for the niches of species to be distinguished, defined by the responses of species to the changing conditions they encounter [3, 5]. This is most easily understood with respect to spatial variation [50, 62-64]. If species favor different habitats, and their resources are consumed and regenerated locally in those habitats, the habitats provide separate feedback loops, just like separate resources. Separate feedback loops with respect to predators are less likely, because predators are often wider ranging than their prey and so prey densities in one habitat can affect predation in other habitats, but this case has been developed theoretically [38]. The same would be true of resources too if resources moved between habitats, or if habitats changed their status over time on a shorter time scale than resource changes, i.e. if there were rapid spatio-temporal environmental variation [65]. Similar issues apply to temporal variation [5].

Niche theory for spatial and temporal variation has developed techniques for assessing the extent to which spatial and temporal variation can separate the niches of species [20, 22]. This is the concept of covariance between environment and competition and its generalization to include apparent competition [51, 52]. Such covariances assess whether density-dependent feedback loops change with the environmental conditions under consideration, and can thus separate the niches of species. Based on them, powerful techniques for definitively testing diversity maintenance hypotheses based on spatial and temporal have been proposed [20] and implemented in some cases [22, 66]. Extension of these ideas for other niche-based diversity maintenance mechanisms should be possible, and might well lead to much progress on deciding between the various diversity maintenance ideas that have been proposed.

3. Conservation and management

Although it is not possible at the present time to say with confidence which of the many possible diversity maintenance ideas applies in any particular system, these ideas share enough features in common that some general advice is possible. Our principal example has been MacArthur's consumer-resource model, and its extension to a consumer-resource-predator model, for which average fitness and overlap measures are easy to define. Per capita growth rates in these models are linear functions of the densities of the various species. Models similar in spirit, but having nonlinear relationships, have been studied extensively by Tilman [17, 67], especially for plants limited by nonbiological resources, and compared with others by Chase and Leibold [3]. Although these simple measures of average fitness and niche overlap are not available, the underlying concepts remain applicable [3]. Other models of resource and apparent competition in both constant and variable environments have yielded measures of fitness and niche overlap, when they have been sought explicitly [5, 29, 47, 51, 52, 68], clearly demonstrating how these ideas extend beyond MacArthur's [27] beginnings. Moreover, Shigesada [69] provides an explicit representation of spatial competition models in the MacArthur framework.

The fact that different species do differ from one another ecologically means that changes in the environment do affect different species differently or to different degrees. Under the individual-species approach to conservation and management, this would simply mean that different species are at risk to different degrees from some change in the environment due to human activities. Under the interactive model given here, changes in the environment need not directly affect a species for it to be affected indirectly through its linkages with other species. The fitness measures κ reveal this most simply. A mortality rate of a particular species might be changed by hunting, elimination of a predator, or changes in physical stress in the environment. That change would have a direct effect on that species, either increasing or decreasing its abundance and perhaps putting it at risk, but that change would also have effects on a species' relationships with others. In particular, in the MacArthur model, we see that the ratio of the κ values would be affected. An increase in the κ ratio in one species favor disadvantages another, potentially leading to its extinction.

One example of this κ-ratio analysis is its application to understanding invasive species of large effect [30, 70]. A large κ value in a particular system would allow a species to invade that system, but would have also have the effect of depressing native species and potentially driving them extinct. There are a number of ways that an invasive species might achieve a larger κ value including low susceptibility to natural enemies in the invaded system [71], advantages in resource harvesting [72], and lower sensitivity to competition than the native species [73]. Alternatively, an invasive species might have a larger κ value than native species because habitat degradation has increased stresses on native species, lowering their κ values, and rendering the system vulnerable to invasion [30, 74]. Viewed from another perspective, a native species might become endangered not because some environmental change has directly affected it, but because other guild members have had their κ values elevated by changes that directly benefited them.

Naturally, changes in resource supply also have important effects [75]. Increasing the carrying capacity of a resource that one species depends on directly increases that species' κ value, and so can increase its ability to compete with another species that does not depend on that resource. This other species would thus be disadvantaged, potentially driving it extinct. However, not just average fitness, κ, but niche overlap, ρ, would be affected by the increase in supply of a particular resource because ρ depends on resource carrying capacities (see Table 1). In particular, ρ would decrease with an increase in the carrying capacity of a resource that only one species uses. Nevertheless, the change in the κ value has the larger effect confirming the conclusion that changes directly benefiting only one species will harm its competitors indirectly. Effects like this are most clearly seen for invasive species where an increase in resource supply, for example nitrogen deposition for an invasive plant species [76, 77], gives advantages to invaders, depressing native species. However, these concerns apply not only between invaders and native species but between native species where factors helping just some species may well negatively affect others in the same guild [75].

These various effects stemming from changes in resource richness have their counterparts in habitat availabilities because habitat variation in space can have effects similar to resource diversity, as explained above in the section on variation in time and space. Thus, increasing the availability of a habitat type specific to one species, can negatively affect another species that does not use that habitat type, but shares other habitat types with the species that does.

This outcome can be understood intuitively: one species increases in abundance when a valuable habitat type becomes more available, and, due to dispersal between habitat types, its abundance everywhere is increased. It thus has greater competitive effects on other species in all habitats. From the perspective of the theory above, average fitnesses and niche overlap change in the same way as when resource supply is changed. Similar affects apply to temporal environmental variation, an issue of particular concern with climate change [78]. For example, desert annual plants are believed to coexist with one another because different species are favored in different years, depending on the weather [19, 55]. Changes in weather patterns can change the relative κ values of the species if the weather becomes on average more favorable for some than for others. Changes in disturbance regimes can have similar effects through changes in the patterns of spatio-temporal variation [79].

Although systems where coexistence relies on disturbance are sometimes regarded as nonequilibrial [80], they do not escape these general priniciples. Individual species may be adapted to survive and even take advantage of temporal change, but there is every reason to expect that individual species, and the system as a whole, are sensitive to changes in the average frequency and intensity of environmental events, such as disturbances [81, 82]. Surviving and taking advantage of temporal change may rely on life-history attributes such as dormancy, but may also depend on dispersal to and from refuge habitats that escape change either permanently or at given time (e.g when a fire sweeps through) [83]. Maintaining landscape connectivity, for example through maintenance of dispersal corridors, is especially important in such circumstances [84].

Finally, we come to the important question of trophic structure. Recognition that predation and competition can have similar roles in diversity maintenance, and that the effects that each has can be undermined by the other, depending on their relative strengths, means that maintenance of trophic structure is vital for conservation [85]. There is much concern about loss of large predators in many systems [86-89]. These predators are often valued for their own sake. Conservation measures focused on them might aim merely for their persistence in nature, rather than for maintenance of the roles that they have in systems [90, 91]. For example, practices in the United States relating to wolf restoration face contentious arguments over the mere persistence of wolves, versus maintaining their numbers so that the ecosystem roles are restored as well [92, 93]. Discussion of trophic cascades, which are increasingly seen in marine systems, as the larger species are eliminated, focus much on the overall abundance of a given trophic level [94, 95]. However, the theory discussed here emphasizes that decimating one trophic level will affect the diversity maintenance roles that it has in other trophic levels for which is either a predator trophic level or a resource trophic level, as has been found empirically in some systems [96].

4. Conclusions

Taken together, these principles and examples reemphasize that species should not be managed in isolation. Factors improving the situation for one set of species may degrade it for others. Maintaining trophic structure, physical environmental structure, disturbance regimes, and spatial connectivity are all common sense ideas that are backed up by the theory of diversity maintenance mechanisms.

5. Acknowledgements

This research was supported by NSF grant DEB-0542991.

6. References

[1] M. L. Rosenzweig, *Species Diversity in Space and Time*. Cambridge: Cambridge University Press, 1995.
[2] J. Roughgarden, "Community structure and assembly," in *Perspectives in Ecological Theory*, J. Roughgarden, R. M. May, and S. A. Levin, Eds. Princeton, N.J.: Princeton University Press, 1989, pp. 203-226.
[3] J. M. Chase and M. A. Leibold, *Ecological Niches: linking classical and contemporary approaches*. Chicago: The University of Chicago Press, 2003.
[4] S. A. Levin, "Community equilibria and stability, and an extension of the competitive exclusion principle," *American Naturalist*, vol. 104, pp. 413-423, 1970.
[5] P. Chesson, "Mechanisms of maintenance of species diversity," *Annual Review of Ecology and Systematics*, vol. 31, pp. 343–66, 2000.
[6] [P. Chesson and N. Huntly, "The roles of harsh and fluctuating conditions in the dynamics of ecological communities," *The American Naturalist*, vol. 150, pp. 519-553, 1997.
[7] S. P. Hubbell, *The unified neutral theory of biodiversity and biogeography*. Princeton, NJ: Princeton University Press, 2001.
[8] I. Volkov, J. R. Banavar, S. P. Hubbell, and A. Maritan, "Patterns of relative species abundance in rainforests and coral reefs," *Nature*, vol. 450, pp. 45-49, 2007.
[9] C. J. Chu, Y. S. Wang, G. Z. Du, F. T. Maestre, Y. J. Luo, and G. Wang, "On the balance between niche and neutral processes as drivers of community structure along a successional gradient: Insights from alpine and sub-alpine meadow communities," *Annals of Botany*, vol. 100, pp. 807-812, 2007.
[10] P. B. Adler, "Neutral models fail to reproduce observed species-area and species-time relationships in Kansas grasslands," *Ecology*, vol. 85, pp. 1265-1272, 2004.
[11] C. K. Kelly, M. G. Bowler, O. Pybus, and P. H. Harvey, "Phylogeny, niches, and relative abundance in natural communities," *Ecology*, vol. 89, pp. 962-970, 2008.
[12] T. Wiegand, S. Gunatilleke, and N. Gunatilleke, "Species associations in a heterogeneous Sri lankan dipterocarp forest," *American Naturalist*, vol. 170, pp. E67-E95, 2007.
[13] E. G. Leigh, "Neutral theory: a historical perspective," *Journal of Evolutionary Biology*, vol. 20, pp. 2075-2091, 2007.
[14] D. Tilman, "Niche tradeoffs, neutrality, and community structure: A stochastic theory of resource competition, invasion, and community assembly," *Proceedings of the National Academy of Science USA*, 2004.
[15] P. Amarasekare, "The geometry of coexistence," *Biological Journal of the Linnean Society*, vol. 71, pp. 1-31, 2000.
[16] P. Chesson and J. J. Kuang, "The interaction between predation and competition" *Nature*, Vol 456, pp 235-238, 2008.
[17] D. Tilman, *Resource Competition and Community Structure*. Princeton, New Jersey: Princeton University Press, 1982.
[18] P. Amarasekare, M. F. Hoopes, N. Mouquet, and M. Holyoak, "Mechanisms of coexistence in competitive metacommunities," *The American Naturalist*, vol. 164, 2004.
[19] P. Chesson, R. L. E. Gebauer, S. Schwinning, N. Huntly, K. Wiegand, M. S. K. Ernest, A. Sher, A. Novoplansky, and J. F. Weltzin, "Resource pulses, species interactions and diversity maintenance in arid and semi-arid environments," *Oecologia*, vol. 141, pp. 236-257, 2004.
[20] P. Chesson, "Quantifying and testing species coexistence mechanisms," in *Unity in Diversity: Reflections on Ecology after the Legacy of Ramon Margalef*, F. Valladares, A.

Camacho, A. Elosegui, C. Gracia, M. Estrada, J. C. Senar, and J. M. Gili, Eds. Bilbao: Fundacion BBVA, 2008, pp. 119-164.
[21] A. M. Siepielski and M. A. McPeek, "On the evidence for species coexistence: a critique of the coexistence program," *Ecology,* vol. 91, pp. 3153-3164, 2010.
[22] A. L. W. Sears and P. Chesson, "New methods for quantifying the spatial storage effect: an illustration with desert annuals," *Ecology,* vol. 88, pp. 2240-2247, 2007.
[23] J. A. Endler, *Natural Selection in the Wild.* Princeton, New Jersey: Princeton University Press, 1986.
[24] E. D. Brodie, A. J. Moore, and F. J. Janzen, "Visualizing and quantifying natural selection," *Trends in Ecology and Evolution,* vol. 10, pp. 313-318, 1995.
[25] E. L. Charnov, Life History Invariants: some explorations of symmetry in evolutionary ecology. Oxford: Oxford University Press, 1993.
[26] P. B. Adler, J. HilleRisLambers, and J. M. Levine, "A niche for neutrality," *Ecology Letters,* vol. 10, pp. 95-104, 2007.
[27] R. MacArthur, "Species packing and competitive equilibrium for many species," *Theoretical Population Biology,* vol. 1, pp. 1-11, 1970.
[28] P. Chesson, "MacArthur's consumer-resource model.," *Theoretical Population Biology,* vol. 37, pp. 26-38, 1990.
[29] R. Haygood, "Coexistence in MacArthur-style consumer–resource models," *Theoretical Populaiton Biology,* vol. 61, pp. 215-223, 2002.
[30] K. Shea and P. Chesson, "Community ecology theory as a framework for biological invasions," *Trends in Ecology and Evolution,* vol. 17, pp. 170-176, 2002.
[31] P. L. Chesson and R. R. Warner, "Environmental variability promotes coexistence in lottery competitive systems," *The American Naturalist,* vol. 117, pp. 923-943, 1981.
[32] P. Chesson, "Multispecies competition in variable environments," *Theoretical Population Biology,* vol. 45, pp. 227-276, 1994.
[33] P. Chesson and M. Rees, "Commentary on Clark et al . (2 0 0 7): Resolving the biodiversity paradox," *Ecology Letters,* vol. 10, pp. 659-661, 2007.
[34] D. W. Purves and S. W. Pacala, "Ecological drift in niche-structured communities: neutral pattern does not imply neutral process," in *Biotic Interactions in the Tropics,* B. D., M. Pinard, and S. Hartley, Eds. Cambridge, UK: Cambridge University Press, 2005, pp. 103-138.
[35] P. Amarasekare, "Interference competition and species coexistence," *Proceedings of the Royal Society of London Series B-Biological Sciences,* vol. 269, pp. 2541-2550, 2002.
[36] R. M. Callaway, Positive Interactions and Interdependence in Plant Communities. Dordrecht :: Springer,, 2007.
[37] R. D. Holt and G. A. Polis, "A theoretical framework for intraguild predation," *The American Naturalist,* vol. 149, pp. 745-764, 1997.
[38] R. D. Holt, "Spatial heterogeneity, indirect interactions, and the coexistence of prey species," *American Naturalist,* vol. 124, pp. 377-406, 1984.
[39] R. T. Paine, "Food web complexity and species diversity," *American Naturalist,* vol. 100, pp. 65-75, 1966.
[40] J. Lubchenco, "Plant species diversity in a marine intertidal community: importance of herbivore food preference and algal competitive abilities," *American Naturalist,* vol. 112, pp. 23-39, 1978.
[41] R. D. Holt, "Predation, apparent competition, and the structure of prey communities," *Theoretical Population Biology,* vol. 12, pp. 197-229, 1977.

[42] J. P. Grover and R. D. Holt, "Disentangling resource and apparent competition: realistic models for plant-herbivore communities," *Journal of Theoretical biology,* vol. 191, pp. 353-376, 1998.
[43] D. W. Stephens and J. R. Krebs, *Foraging Theory*. Princeton, N. J.: The Princeton University Press, 1986.
[44] D. R. Strong, "Density vagueness: abiding the variance in the demography of real populations.," in *Community Ecology,* J. Diamond and T. e. Case, Eds. New York: Harper & Row Publishing, 1986.
[45] Y. Iwasa and J. Roughgarden, "Interspecific competition among metapopulations with space-limited subpopulations," *Theoretical Population Biology,* vol. 30, pp. 194-214, 1986.
[46] D. Tilman, "Competition and biodiversity in spatially structured habitats," *Ecology,* vol. 75, pp. 2-16, 1994.
[47] R. E. Snyder and P. Chesson, "How the spatial scales of dispersal, competition, and environmental heterogeneity interact to affect coexistence," *American Naturalist,* vol. 164, pp. 633-650, 2004.
[48] B. M. Bolker, S. W. Pacala, and C. Neuhauser, "Spatial dynamics in model plant communities: What do we really know?," *American Naturalist,* vol. 162, pp. 135-148, 2003.
[49] P. Amarasekare, "Competitive coexistence in spatially structured environments: a synthesis," *Ecology Letters,* vol. 6, pp. 1109-1122, 2003.
[50] S. Muko and Y. Iwasa, "Incomplete mixing promotes species coexistence in a lottery model with permanent spatial heterogeneity," *Theoretical Population Biology,* vol. 64, pp. 359-368, 2003.
[51] J. J. Kuang and P. Chesson, "Interacting coexistence mechanisms in annual plant communities: frequency-dependent predation and the storage effect " *Theoretical Population Biology,* vol. 77, pp. 56-70, 2010.
[52] P. Chesson and J. J. Kuang, "The storage effect due to frequency-dependent predation in multispecies plant communities," *Theoretical Population Biology,* vol. 78, pp. 148-164, 2010.
[53] P. B. Adler, S. P. Ellner, and J. M. Levine, "Coexistence of perennial plants: an embarrassment of niches," *Ecology Letters,* vol. 13, pp. 1019-1029, 2010.
[54] P. B. Adler, J. HilleRisLambers, P. C. Kyriakidis, Q. F. Guan, and J. M. Levine, "Climate variability has a stabilizing effect on the coexistence of prairie grasses," *Proceedings of the National Academy of Sciences of the United States of America,* vol. 103, pp. 12793-12798, 2006.
[55] A. L. Angert, T. E. Huxman, P. Chesson, and D. L. Venable, "Functional tradeoffs determine species coexistence via the storage effect," *Proceedings of the National Academy of Sciences of the United States of America,* vol. 106, pp. 11641-11645, 2009.
[56] P. Amarasekare, "Spatial dynamics in a host-multiparasitoid community," *Journal of Animal Ecology,* vol. 69, pp. 201-213, 2000.
[57] J. H. Connell, "Diversity and the coevolution of competitors, or the ghost of competition past," *Oikos,* vol. 35, pp. 131-138, 1980.
[58] G. E. Hutchinson, "The paradox of the plankton," *The American Naturalist,* vol. 95, pp. 137-145, 1961.
[59] J. H. Connell, "Diversity in tropical rain forests and coral reefs," *Science,* vol. 199, pp. 1302-1310, 1978.
[60] P. L. Chesson and T. J. Case, "Overview: nonequilibrium community theories: chance, variability, history, and coexistence," in *Community Ecology,* J. Diamond and T. J. Case, Eds. New York: Harper and Row, 1986, pp. 229-239.

[61] M. Huston, "A general hypothesis of species diversity," *The American Naturalist,* vol. 113, pp. 81-101, 1979.
[62] P. Amarasekare and R. M. Nisbet, "Spatial heterogeneity, source-sink dynamics, and the local coexistence of competing species," *The American Naturalist,* vol. 158, pp. 572-584, 2001.
[63] P. L. Chesson, "Coexistence of competitors in spatially and temporally varying environments: a look at the combined effects of different sorts of variability," *Theoretical Population Biology,* vol. 28, pp. 263-287, 1985.
[64] P. Chesson, "General theory of competitive coexistence in spatially-varying environments," *Theoretical Population Biology,* vol. 58, pp. 211-237, 2000.
[65] R. E. Snyder, "Spatiotemporal population distributions and their implications for species coexistence in a variable environment," *Theoretical Population Biology,* vol. 72, pp. 7-20, 2007.
[66] A. L. W. Sears, "Quantifying the effects of spatial environmental variation on dynamics of natural plant populations: field tests for covariance between response to the environment and competition," in *Population biology* Davis: University of California, Davis, 2004, 155 pp.
[67] D. Tilman, *Plant Strategies and the Dynamics and Structure of Plant Communities.* Princeton, New Jersey: Princeton University Press, 1988.
[68] R. E. Snyder, E. T. Borer, and P. Chesson, "Examining the relative importance of spatial and nonspatial coexistence mechanisms," *American Naturalist,* vol. 166, pp. E75-E94, 2005.
[69] N. Shigesada, "Spatial distribution of rapidly dispersing animals in heterogeneous environments.," *Lect. Notes Biomath.,* vol. 54, pp. 478-501., 1984.
[70] A. S. MacDougall, B. Gilbert, and J. M. Levine, "Plant invasions and the niche," *Journal of Ecology,* vol. 97, pp. 609-615, 2009.
[71] R. M. Keane and M. J. Crawley, "Exotic plant invasions and the enemy release hypothesis," *Trends in Ecology and Evolution,* p. this issue, 2002.
[72] K. Petren and T. Case, "An experimental demonstration of exploitation competition in an ongoing invasion," *Ecology,* vol. 77, pp. 118-132, 1996.
[73] R. M. Callaway and E. T. Aschehout, "Invasive plants versus their new and old neighbors: a mechanism for exotic invasion," *Science,* vol. 290, pp. 521-523, 2000.
[74] A. A. Sher and L. A. Hyatt, "The disturbed resource-flux invasion matrix: a new framework for patterns of plant invasion," *Biological Invasions,* vol. 1, pp. 107-114, 1999.
[75] C. M. Clark, E. E. Cleland, S. L. Collins, J. E. Fargione, L. Gough, K. L. Gross, S. C. Pennings, K. N. Suding, and J. B. Grace, "Environmental and plant community determinants of species loss following nitrogen enrichment," *Ecology Letters,* vol. 10, pp. 596-607, 2007.
[76] M. L. Brooks, "Effects of increased soil nitrogen on the dominance of alien annual plants in the Mojave Desert," *Journal of Applied Ecology,* vol. 40, pp. 344-353, 2003.
[77] D. A. Wedin and D. Tilman, "Influence of nitrogen loading and species composition on the carbon balance of grasslands," *Science,* vol. 274, pp. 1720-1723, 1996.
[78] J. R. Malcolm, C. R. Liu, R. P. Neilson, L. Hansen, and L. Hannah, "Global warming and extinctions of endemic species from biodiversity hotspots," *Conservation Biology,* vol. 20, pp. 538-548, 2006.
[79] S. Vuilleumier, C. Wilcox, B. J. Cairns, and H. P. Possingham, "How patch configuration affects the impact of disturbances on metapopulation persistence," *Theoretical Population Biology,* vol. 72, pp. 77-85, 2007.
[80] J. H. Connell, M. D. Lowman, and I. R. Noble, "Subcanopy gaps in temperate and tropical forests," *Austalian Journal of Ecology,* vol. 22, pp. 163-168, 1997.

[81] M. C. Wimberly and T. A. Spies, "Influences of environment and distrubance on forest patterns in coastal Oregon watersheds," *Ecology (Washington D C),* vol. 82, pp. 1443-1459, 2001.
[82] E. Nonaka, T. A. Spies, M. C. Wimberly, and J. L. Ohmann, "Historical range of variability in live and dead wood biomass: a regional-scale simulation study," *Canadian Journal of Forest Research-Revue Canadienne De Recherche Forestiere,* vol. 37, pp. 2349-2364, 2007.
[83] P. L. Chesson, "Geometry, heterogeneity and competition in variable environments," *Philosophical Transactions of the Royal Society of London, Series B,* vol. 330, pp. 165-173, 1990.
[84] R. L. Pressey, M. Cabeza, M. E. Watts, R. M. Cowling, and K. A. Wilson, "Conservation planning in a changing world," *Trends in Ecology & Evolution,* vol. 22, pp. 583-592, 2007.
[85] P. Chesson and J. J. Kuang, "The interaction between predation and competition," *Nature,* vol. 456, pp. 235-238, 2008.
[86] A. S. Glen, C. R. Dickman, M. E. Soule, and B. G. Mackey, "Evaluating the role of the dingo as a trophic regulator in Australian ecosystems," *Austral Ecology,* vol. 32, pp. 492-501, 2007.
[87] C. N. Johnson, J. L. Isaac, and D. O. Fisher, "Rarity of a top predator triggers continent-wide collapse of mammal prey: dingoes and marsupials in Australia," *Proceedings of the Royal Society B-Biological Sciences,* vol. 274, pp. 341-346, 2007.
[88] J. Terborgh, L. Lopez, P. Nunez, M. Rao, G. Shahabuddin, G. Orihuela, M. Riveros, R. Ascanio, G. H. Adler, T. D. Lambert, and L. Balbas, "Ecological meltdown in predator-free forest fragments," *Science,* vol. 294, pp. 1923-1926, 2001.
[89] F. Sergio, T. Caro, D. Brown, B. Clucas, J. Hunter, J. Ketchum, K. McHugh, and F. Hiraldo, "Top Predators as Conservation Tools: Ecological Rationale, Assumptions, and Efficacy," *Annual Review of Ecology, Evolution, and Systematics,* vol. 39, pp. 1-19, 2008.
[90] M. R. Whiles, K. R. LIps, C. M. Pringle, S. S. Kilham, R. J. Bixby, R. Brenes, S. Connelly, J. C. Colon-Gaud, M. Hunte-Brown, A. D. Huryn, C. Montgomery, and S. Peterson, "The effects of amphibian population declines on the structure and function of Neotropical stream ecosystems," *Frontiers in Ecology and the Environment,* vol. 4, pp. 27-34, 2006.
[91] J. Terborgh, K. Feeley, M. Silman, P. Nunez, and B. Balukjian, "Vegetation dynamics of predator-free land-bridge islands," *Ecology,* vol. 94, pp. 253-263, 2006.
[92] C. C. Wilmers, R. L. Crabtree, D. W. Smith, K. M. Murphy, and W. M. Getz, "Trophic facilitation by introduced top predators: grey wolf subsidies to scavengers in Yellowstone National Park," *Journal of Animal Ecology,* vol. 72, pp. 909-916, 2003.
[93] C. C. Wilmers, E. Post, R. O. Peterson, and J. A. Vucetich, "Predator disease out-break modulates top-down, bottom-up and climatic effects on herbivore population dynamics," *Ecology Letters,* vol. 9, pp. 383-389, 2006.
[94] R. A. Myers, J. K. Baum, T. D. Shepherd, S. P. Powers, and C. H. Peterson, "Cascading effects of the loss of apex predatory sharks from a coastal ocean," *Science,* vol. 315, pp. 1846-1850, 2007.
[95] M. R. Heithaus, A. Frid, A. J. Wirsing, and B. Worm, "Predicting ecological consequences of marine top predator declines," *Trends in Ecology & Evolution,* vol. 23, pp. 202-210, 2008.
[96] O. J. Schmitz, "Predators have large effects on ecosystem properties by changing plant diversity, not plant biomass," *Ecology,* vol. 87, pp. 1432-1437, 2006.

4

Using Multiple Linear Regression Models to Identify Factors Underlying Avian Species Imperilment in Sub-Saharan Africa and Europe

R. Eliot Crafton and Brandon P. Anthony
Department of Environmental Sciences & Policy,
Central European University, Budapest
Hungary

1. Introduction

Determining what factors influence the threats faced by the world's flora and fauna is of key importance to conservation biologists (Cardillo et al., 2008; Davies et al., 2006; Smith et al., 2003; see Spangenberg, 2002). A plethora of research has been directed at this effort and has looked extensively at biological and anthropogenic factors, including social and socio-economic conditions (e.g. Holland et al., 2009; Huby et al., 2006; Kerr & Currie, 1995; Lenzen et al., 2009; McKee et al., 2003). This chapter intends to supplement the existing literature by utilizing updated data to address this issue from a primarily socio-economic perspective for birds in a selection of sub-Saharan African and European countries. We generate several models using multiple linear regression to test the explanatory power of a host of variables, including human population density (HPD) per km^2, Corruption Perception Index (CPI) score (as a proxy for governance), GDP per capita, and the average degrees from the equator. In addition, the results are considered in light of projected changes to HPD levels for the year 2050 (United Nations [UN], 2008).

Bird species are currently being impacted by several threats, resulting in the need for conservationists to address a wide range of issues (Brooke et al., 2008). These include land-use change, habitat destruction, invasive species, unsustainable exploitation, climate change, and insufficient governance (Brooke et al., 2008; Butchart, 2008; Lemoine et al., 2007; Lenzen et al., 2009; Reif et al., 2008; Smith et al., 2003; Thomas et al., 2004). These threats have the potential to impact the quality or quantity of available resources, directly impact the population, or change the conditions that a species may face. In addition, while these threats exist in isolation, many are correlated or exist in a cascading fashion (for example climate change can lead to habitat destruction). While these threats, and others, can be enumerated, the underlying drivers of these pressures are often rooted in socio-economic conditions, including corruption, HPD, and poverty level (Kerr & Currie, 1995; Pandit & Laband, 2009; Pellegrini & Gerlagh, 2004).

Understanding how these socioeconomic factors influence conservation agendas has been the focus of several authors, occasionally with contradictory results. While correlation can often be observed, the exact causal influence of these factors on the species of interest is typically a matter for interpretation. Corruption has been speculated to influence the availability of resources that are necessary to implement and enforce conservation efforts (Smith & Walpole, 2005). While corruption may potentially positively affect conservation efforts by minimizing the economic development of a country, it may at the same time jeapordize conservation efforts by compromising the support structure necessary for conservation efforts to be successful (Laurance, 2004). HPD has also been shown to be an essential factor to consider when implementing biodiversity conservation strategies (Kerr & Currie, 1995). While a relationship between HPD and species richness clearly exists, the mechanism(s) by which HPD influences biodiversity is less clear; the extent to which HPD influences biodiversity through habitat loss or use patterns is unresolved (Chown et al., 2003, Kerr & Currie, 1995). Similarly, while it is accepted that biodiversity loss and poverty are linked, the relationship between the two is not universally agreed upon (Adams et al., 2004). However, it has been suggested that poverty may hamper efforts by encouraging both violation of protected areas and minimizing ability to dedicate funds for conservation.

These threats are not unique to bird species and, according to the Millennium Ecosystem Assessment [MEA] (2005), biodiversity and the ecosystems that support it are in many cases imperiled. However, specific to birds, it has been suggested that extinctions per million species per year (E/MSY; a unit used to describe relative extinction rates [Pimm et al., 1995]) are occurring at 1000 - 10 000 times above background rates; more conservative estimates are 100 - 1000 times above background rates (see Brooke et al., 2008; Pimm et al., 2006).

While current research suggests that conservation efforts are successful at reducing the amount of extinctions being experienced, existing efforts are still lacking to successfully protect the world's biodiversity (Brooke et al., 2008; Pimm et al., 2006). In order for these conservation methods to be successful, they must not only deal with the flora and fauna that are being threatened but must address the underlying drivers for the pressures exerting the threats; it is essential that conservation measures are considered for management of people in addition to nature (Luck, 2007). According to the Millennium Ecosystem Assessment, effective ecosystem management "requires substantial changes in institutions and governance, economic policies and incentives, social and behavior factors, technology, and knowledge" (MEA, 2005).

2. Methods

2.1 Country selection

We first considered all countries from sub-Saharan Africa and Europe for our study. However, countries of either region were omitted from the analyses if data was not available for at least one of the variables under consideration. Our final list includes 73 countries, 42 from sub-Saharan Africa and 31 from Europe (see *Fig. 1 & 2*).

2.2 Data acquisition

Data was acquired for the variables utilized in the model through an internet search. Data for each of the variables was freely available from online datasets provided through the respective sources. As the dependent variable, the proportion of threatened bird species as a percentage of total bird species per country was utilized. This value was derived by dividing the number of threatened bird species (International Union for Conservation of Nature and Natural Resources [IUCN] designations critically endangered [CR], endangered

sub-Saharan Africa (n = 42; *Fig. 1*): Angola, Benin, Botswana, Burkina Faso, Burundi, Cameroon, Central African Republic, Chad, Congo, Cote d'Ivoire, Democratic Republic of the Congo, Djibouti, Equatorial Guinea, Eritrea, Ethiopia, Gabon, Gambia, Ghana, Guinea, Guinea-Bissau, Kenya, Lesotho, Liberia, Malawi, Mali, Mauritania, Mozambique, Namibia, Niger, Nigeria, Rwanda, Senegal, Sierra Leone, Somalia, South Africa, Sudan, Swaziland, Togo, Uganda, United Republic of Tanzania, Zambia, Zimbabwe.

Fig. 1. Map of sub-Saharan countries included in regression analyses (dark gray shaded countries) (Base map source: CIA, n.d.).

[EN], and vulnerable [VU]) in each country as listed in the 2009.1 IUCN Red List database (http://www.iucnredlist.org; International Union for Conservation of Nature and Natural Resources [IUCN], 2009) by the total number of bird species recorded in each country (United Nations Environment Programme-World Conservation Monitoring Centre, 2008) and multiplying by 100%.

Europe (n = 31; *Fig. 2*): Albania, Austria, Belgium, Bosnia and Herzegovina, Bulgaria, Croatia, Czech Republic, Denmark, Estonia, Finland, France, Germany, Greece, Hungary, Iceland, Ireland, Italy, Latvia, Lithuania, Netherlands, Norway, Poland, Portugal, Romania, Slovakia, Slovenia, Spain, Sweden, Switzerland, The former Yugoslav Republic of Macedonia, United Kingdom.

Fig. 2. Map of European countries included in regression analyses (dark gray shaded countries) (Base map source: CIA, n.d.).

Four independent variables were considered in our study. These included HPD per km^2 (UN, 2005), CPI score (Transparency International, 2007), GDP per capita (medium variant in Current International Dollars [CID]) (UN, 2006), and average distance from the equator based on the approximate geographic center for each country (see www.cia.gov for an explanation of how geographic center is determined; Central Intelligence Agency (CIA) n.d.). This last latitudinal variable was added to control for the trend towards higher species richness closer to the equator and associated differences in variance between the size of country bird lists. While HPD and GDP per capita (as a proxy for poverty level) are utilized in other similar research, CPI score is intended to illustrate the relative national level of governance (e.g. Kerr & Currie, 1995; McKee et al., 2003; Smith et al., 2003). CPI score is based on a ten point scale with 10 being the highest score, indicating the presence of less corrupt governance (for more information, consult http://transparency.org).

2.3 Model generation and manipulation

SPSS (ver. 16.0) software was utilized to generate the models under consideration. Within the program, the built in Regression-Linear option was performed. This function output an equation of the form:

Proportion of threatened bird species (%) = constant + a (HPD per km²) + b (CPI score) + c (GDP per capita) + d (degrees from the equator)

Equations were derived for each model performed using a different set of explanatory variables.

Two datasets, comprised of the countries listed for sub-Saharan Africa and Europe, were run through the regression function utilizing the four independent variables. In addition to the original models containing all independent variables, a second set was generated excluding the GDP per capita variable for both country groups, as it showed strong co-linearity to CPI score. A final set of models were run for Africa and Europe, which included only CPI score and CPI score and HPD per km^2, respectively.

The results from the models run with all variables except GDP per capita were used to illustrate the effects of a theoretical shift in the HPD per km^2 independent variable to reflect projected HPD per km^2 in the year 2050. Projections were taken from the 2008 Revision of the World Population Prospects generated by the Population Division of the Department of Economic and Social Affairs of the United Nations Secretariat (UN, 2008). The projections include values for three variants (low, medium, and high) and current fertility rate to reflect the range of plausible population scenarios for both sub-Saharan Africa and Europe. For more information consult http://esa.un.org/unpp.

3. Results

The mean proportion of threatened bird species for the sub-Saharan African countries considered is significantly lower than that for the European countries (t = -6.26; P = 0.000; mean = 1.88 ± 0.86 and 3.10 ± 0.76, respectively). Models generated using all four independent variables for Europe and sub-Saharan Africa showed a disparity in level of significance and explanatory power (P = 0.002 and 0.129; adjusted R^2 = 0.384 and 0.082; F = 5.673 and 1.912, respectively). Despite the difference in level of significance, there was a high degree of correlation between CPI score and GDP per capita for both Europe and sub-Saharan Africa (Pearson correlations of 0.914 and 0.550, respectively) (*Fig. 3*). Removing GDP per capita from the independent variables considered improved the significance and explanatory power of the Europe and Africa models (P = 0.001 and 0.069; adjusted R^2 = 0.397 and 0.103; F = 7.571 and 2.563, respectively). Using a paired down set of independent variables returned an improvement in the Africa model (CPI score only; P = 0.009; adjusted R^2 = 0.136; F = 7.454) and an impairment in the European model (HPD per km^2 and CPI score; P = 0.006; adjusted R^2 = 0.258; F = 6.221).

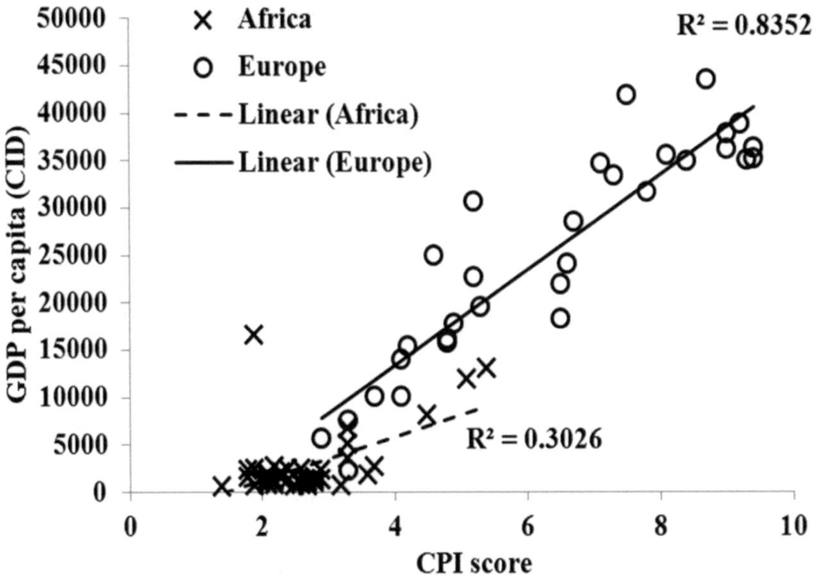

Fig. 3. Scatterplot of GDP per capita and CPI Score with linear trend line. sub-Saharan Africa n = 42; Europe n = 31.

Within the models, not all independent variables considered were significant (*Table 1*). For the African models, no single variable was significant for either the model containing all four independent variables or the model with GDP per capita removed. In both instances, CPI score was the closest to being significant (P = 0.075 and 0.057, respectively). In the African model considering only CPI score, CPI score became a significant predictor (P = 0.009). For the European models containing all four and all but GDP per capita, the only significant individual variable was degrees from the equator (P = 0.010 and 0.011, respectively). In the model generated using just HPD per km^2 and CPI score, CPI score was significant and HPD was nearly significant (P = 0.004 and 0.053, respectively).

Figure 4 shows the relationships between HPD per km^2, CPI score, and degrees from the equator and proportion of threatened bird species when plotted individually against one another. For all three independent variables, the trends are opposite for the two country groups *(Fig. 4)*.

For Europe, increasing HPD per km^2 and decreasing CPI score and degrees from equator correspond to an increasing proportion of threatened bird species. For Africa, the opposite is true; decreasing HPD per km^2 and increasing CPI score and degrees from equator correspond to an increasing proportion of threatened species. Of these trends, degrees from equator is the variable with the strongest relationship for Europe followed by CPI score (R^2 = 0.44 and 0.21, respectively). For Africa, CPI score is the strongest followed by degrees from the equator (R^2 = 0.16 and 0.08, respectively). For Europe and Africa, HPD per km^2 has the weakest correlation to proportion of threatened species (R^2 = 0.06 and 0.004, respectively).

Variable	sub-Saharan Africa				Europe			
	Coeff.	SE	*P*	Mean	Coeff.	SE	*P*	Mean
Model 1								
Constant	0.769	0.465	0.106		5.996	0.849	0	
HPD	-5.21E^{-4}	0.002	0.798	58.75	1.31E^{-3}	1.35 E^{-3}	0.343	113.08
CPI	0.393	0.214	0.075	2.67	0.030	0.145	0.839	6.35
GDP	-0.17E^{-4}	0.46E^{-4}	0.717	2772.45	-0.16E^{-4}	0.24E^{-4}	0.511	25231.84
Degrees	0.012	0.021	0.580	11.63	-0.057	0.021	0.010	49.96
Model 2								
Constant	0.813	0.443	0.075		6.024	0.840	0	
HPD	-3.16E^{-4}	1.92E^{-3}	0.870	58.75	1.22E^{-3}	1.33E^{-3}	0.368	113.08
CPI	0.351	0.179	0.057	2.67	-0.054	0.070	0.442	6.35
Degrees	0.012	0.021	0.547	11.63	-0.054	0.020	0.011	49.96
Model 3								
Constant	0.785	0.419	0.068		3.945	0.388	0.000	
HPD	-	-	-	-	0.003	0.001	0.053	113.08
CPI	0.409	0.150	0.009	2.67	-0.182	0.057	0.004	6.35

Table 1. Model coefficients (Coeff.) with standard error (SE) and significance values *(P)* generated using the Regression-Linear function and associated current mean values for independent variables considered. Model 1: all four independent variables; Model 2: all but GDP; Model 3: CPI score only for sub-Saharan Africa and HPD per km^2 and CPI score for Europe. HPD = HPD per km^2, CPI = CPI score, GDP = GDP per capita (CID), and Degrees = average degrees from equator. sub-Saharan Africa n = 42; Europe n = 31.

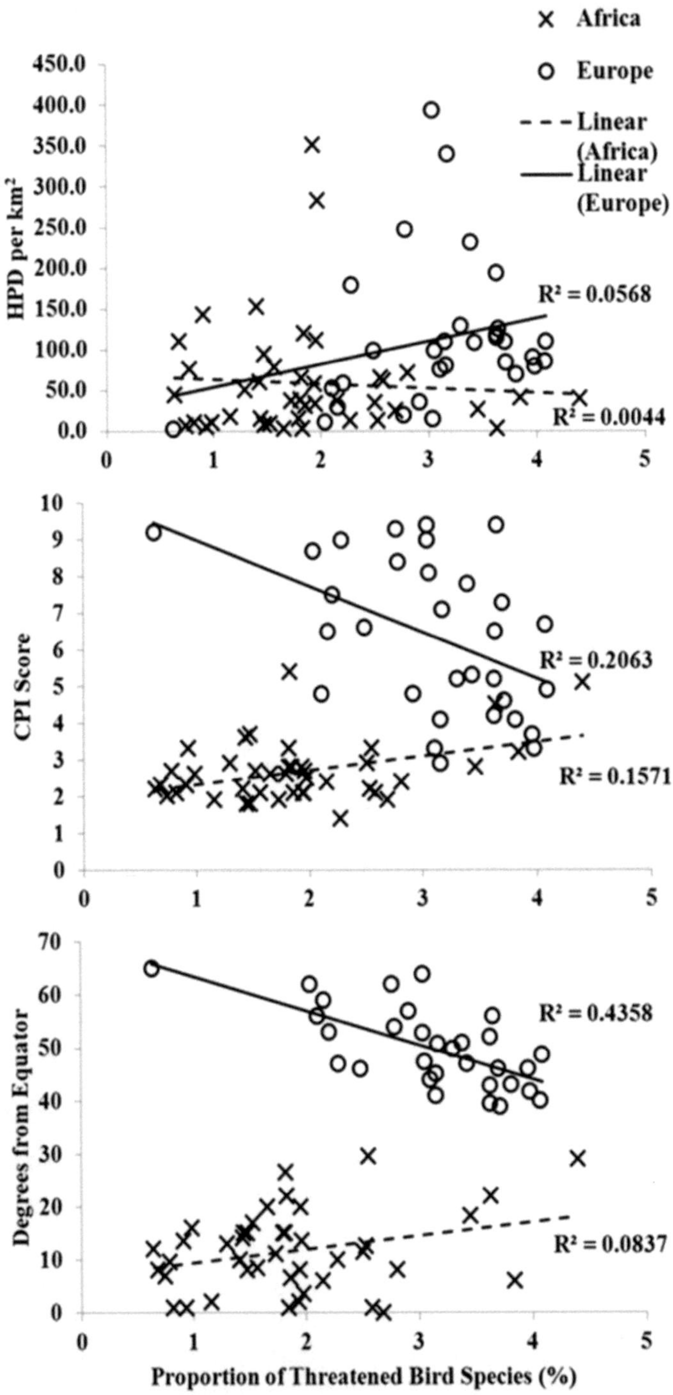

Fig. 4. Scatterplots of HPD per km², CPI score, and degrees from equator plotted against proportion of threatened bird species (%). sub-Saharan Africa n = 42; Europe n = 31.

Using the results from the models generated with all variables except GDP per capita, HPD per km^2 projections for the year 2050 (UN, 2008; *Table 2*) were utilized to predict the impact of future population projections on the proportion of threatened bird species. For both the European and African models, the HPD coefficients were small and not significant (*Table 1*). However, the African model resulted in a negative relationship, while the European model contained a positive one (*Fig. 4*). As a result of the small coefficient values, these manipulations resulted in modest to no changes to the proportion of threatened bird species for all scenarios despite the difference between projected HPD changes in Africa and Europe (*Table 2*).

Scenario	sub-Saharan Africa				Europe			
	Value	Change	Prop.	Change	Value	Change	Prop.	Change
Current	58.75	-	1.88	-	113.08	-	3.09	-
Low	116.31	1.98	1.86	0.99	100.61	0.89	3.08	1.00
Medium	132.90	2.26	1.85	0.99	113.94	1.01	3.10	1.00
High	150.60	2.56	1.85	0.98	128.55	1.14	3.11	1.01
Constant	199.71	3.40	1.83	0.98	109.71	0.97	3.09	1.00

Table 2. Values for current and projected HPD per km^2 levels and corresponding modeled proportion of threatened bird species (Prop.) (%) based on the Model 2 scenario (without GDP per capita independent variable included). Projected values are based on manipulation of HPD per km^2 to reflect 2050 projections for Low Variant (Low), Medium Variant (Medium), High Variant (high), and current fertility rate (constant) scenarios (derived from UN, 2008 data). Change is categorized in relation to current values. sub-Saharan Africa n = 42; Europe n = 31.

4. Discussion

It is clear that conservation efforts are criticially important to protect biodiversity from the increasing multitude of threats that species are presented with. These threats are both direct and indirect and primarily driven by anthropogenic activities. However, the threats that these species face (e.g. land-use change/habitat destruction, overexploitation, invasive species, climate change) are in many cases symptoms of underlying social conditions rooted in politics and economics. The research conducted within this project helps to further the analysis of what factors influence the threatened status of bird species. Without suggesting causation, these models represent a plausible set of proxies which can significantly explain up to 13.6% and 39.7% of the variation in the sub-Saharan African and European countries examined, respectively. While it was possible to develop a significant model for both country groups, there were disparities in the amount of variation explained within each model and the relative impact and significance of the individual parameters considered. In terms of significance, the only parameter to register as significant (at $P < 0.05$) for both country sets was CPI score in the Model 3 scenario. In the other two model scenarios, sub-

Saharan Africa had no significant individual parameters, and for Europe, only degrees from the equator was significant.

The lack of significance in the Model 1 and Model 2 scenarios for the three socio-economic variables (HPD per km^2, CPI score, and GDP per capita) for both Europe and sub-Saharan Africa provides an interesting perspective on their relative importance within these models. However, more interesting is the opposite trends observed between the two country groups. In the Model 2 scenario, all three variables considered had opposite influence between country groups. This can also be seen in the scatterplots in Figure 4, which show that the trends in the existing data are inverted between country groups. While the significance of these variables in the Model 2 scenario makes comparison difficult, Model 3 suggests that CPI score is an important, significant factor for both the African and European country groups. However, this model suggests that improving governance in Europe decreases the proportion of threatened species, while in Africa improving governance increases the proportion of threatened species. While this model does not suggest reasons to why this is the case, it highlights the importance of creating models that are specific to the region within question. For similar models to be useful in providing insight into conservation issues and strategies, it is important that regionally specific conditions are considered in model generation.

Furthermore, comparison between the sub-Saharan African and European data highlights the fact that these models are reliant on abundant and accurate data for creating reasonable forecasts, which can be a problem in areas with insufficient data (problems of this nature were encountered when searching for historic data for sub-Saharan Africa). Indeed, national level data may also be too coarse to tease out confounding factors contributing to species imperilment and, for example, may ignore potentially vast differences in HPD *within* countries, which may have significantly greater effects on birds than what our models indicated. Brown & Laband (2006), for example, utilized state-level data to evaluate correlations between species imperilment and the level and spatial distribution of human settlement and infrastructure development in the United States. It was only at this scale that they were able to identify that the number of people and households, incidence of roads, and intensity of nighttime lights were all significantly correlated with the ecological imperilment of species. Additionally, Pandit & Laband (2007) point out that modeling the determinants of threats to species using country-level data may also be complicated by the fact that factors that influence species imperilment may extend or operate beyond arbitrary political boundaries. Therefore, they advise controlling for spatial autocorrelation in models focusing on imperilment of flora and fauna. As data becomes more abundant at finer resolutions, and more easily accessible, we expect that better, more functional models are likely to be produced.

HPD has been shown to have very little impact on the proportion of threatened species in both the African and European countries considered due to the extremely small coefficients within the models. This outcome is surprising given the potential increase in HPD values in the 2050 projections and contradictory to assumptions based on existing theory, which suggest that increasing HPD should have a negative impact on species. The violation of the exisitng assumption is especially true in regard to the sub-Saharan Africa countries correlation, which has an inverse relationship between HPD and proportion of threatened bird species. While HPD and the other variables are not significant individually in all models, the trends observed in the models provide an interesting insight into potential conservation issues. These trends suggest that in order to develop effective conservation strategies in Europe, a set of goals must be pursued including decreasing HPD and increasing CPI and GDP. In Africa, the exact opposite is the case.

Developing an understanding of the relationships between socioeconomic drivers and the threats faced by avian and other species worldwide is essential for developing a functional conservation strategy. Understanding the regional specific trends allows for a more focused application of effort to maxamize the conservation outcome. However, these trends are purely correlational, and understanding the cause of these relationships requires interpretation of the data and a detailed understanding of the sociopolitical dynamics of the region in question. To this end, the trends observed in this study are perplexing and counter intuitive to what would be expected from existing theory. In particular, CPI score as a metric of governance suggests that in Europe strong governance and lack of corruption are beneficial while in sub-Saharan Africa the opposite relationship exists. This model suggests that corruption may function to protect bird species under certain conditions. It may be the case that in under-developed regions, corruption and poor governance obstruct development, with the effect that even while human populations remain mired in poverty, wildlife habitat is left intact. Conversely, in highly developed areas a great deal of governance effort is required to protect what habitat remains. While this is an interesting idea, given that CPI and GDP are colinear, suggesting that lower CPI will benefit birds has significant social implications for the nations concerned. These postulates concerning the disparate effects of corruption and governance across our regions were not tested in our analyses, but they are interesting and would be worthy of continued research effort, on both regional and national scales.

5. Conclusion

The determination of what factors influence the threat status of the world's flora and fauna is of key interest to conservation biologists worldwide. While these models do not suggest causation and it is likely that the independent variables suggested represent proxies for the pressures rather than the actual pressures themselves, the models generated within this research have significant explanatory power. Given that these models are built on a combination of anthropogenic and non-anthropogenic variables, they highlight the interdependent nature of humans and the environment for conservation purposes. Finally, given the disparities in the results between geographic regions, these models suggest that accurate data reflecting regionally specific differences must be taken into account when considering which pressures and conservation strategies are most applicable to the area in question.

It is evident that these models, though significant, explain only a fraction of the variability in threat status faced by bird species in these regions. Additionally, these models result in a plethora of questions that are of criticial importance for conservation strategies. While correlations are presented, these models do not present either a mechanism or answers to why regional differences exist. There is no reason why any of these factors independent of their causal effects should threaten biodiversity. Therefore it is essential to develop a better understanding of what the cascading effects of these socio-economic factors might be, and it is equally important to recognize that the same socioeconomic condition may result in a very different outcome in different regions as demonstrated by these models. Furthermore, what factors are influencing the remaining variability is left unresolved.

6. References

Adams, W., Aveling, R., Brockington, D., Dickson, B., Elliott, J., Hutton, J., Roe, D., Vira, B. & Wolmer, W. (2004). Biodiversity conservation and the eradication of poverty. *Science*, Vol.306, No.5699, (November 2004), pp. 1146-1149, ISSN 1095-9203

Brooke, M., Butchart, S., Garnett, S., Crowley, G., Mantilla-Beniers, N. & Stattersfield, A. (2008). Rates of movement of threatened bird species between IUCN Red List categories and toward extinction. *Conservation Biology*, Vol.22, No.2, (April 2008), pp. 417-427, ISSN 1523-1739

Brown, R.M., & Laband, D.N. (2006). Species imperilment and spatial patterns of development in the United States. *Conservation Biology*, Vol. 20, No.1, (February 2006), pp. 239-244, ISSN 1523-1739

Butchart, S. (2008). Red List indices to measure the sustainability of species use and impacts of invasive alien species. *Bird Conservation International*, Vol.18, No.S1, (August 2008), pp. S245-S262, ISSN 1474-0001

Cardillo, M., Mace, G., Gittleman, J., Jones, K., Bielby, J. & Purvis, A. (2008), The predictability of extinction: biological and external correlates of decline in mammals. *Proceedings of the Royal Society B*, Vol.275, No.1641, (June 2008), pp. 1441-1448, ISSN 1471-2954

Central Intelligence Agency, United States of America (CIA). (n.d.). The world factbook, field listings: geographic coordinates, In: *Publications*, 20.2.2010, Available from: https://www.cia.gov/library/publications/the-world-factbook/index.html

Chown, S., van Rensburg, B., Gaston, K., Rodrigues, A., & van Jaarsveld, A. (2003). Energy, species richness, and human population size: conservation implications at a national scale. *Ecological Applications*, Vol.13, No.5, (October 2003), pp. 1233-1241, ISSN 1051-0761

Davies, R., Orme, C., Olson, V., Thomas, G., Ross, S., Ding, T., Rasmussen, P., Stattersfield, A., Bennett, P., Blackburn, T., Owens, I. & Gaston, K. (2006). Human impacts and the global distribution of extinction risk. *Proceedings of the Royal Society B*, Vol.273, No.1598, (September 2006), pp. 2127-2133, ISSN 1471-2954

Holland, T., Peterson, G. & Gonzalez, A. (2009). A cross-national analysis of how economic inequality predicts biodiversity loss. *Conservation Biology*, Vol.23, No.5, (October 2009), pp. 1304-1313, ISSN 1523-1739

Huby, M., Cinderby, S., Crowe, A., Gillings, S., McClean, C., Moran, D., Owen, A. & White, P. (2006). The association of natural, social and economic factors with bird species richness in rural England. *Journal of Agricultural Economics*, Vol.57, No.2, (July 2006), pp. 295-312, ISSN 1477-9552

International Union for Conservation of Nature and Natural Resources. (2008). IUCN Redlist Database 2008, 20.2.2010, Available from: http://www.iucnredlist.org

Kerr, J. & Currie, D. (1995). Effects of human activity on global extinction risk. *Conservation Biology*, Vol.9, No.6, (December 1995), pp. 1528-1538, ISSN 1523-1739

Laurance, W. (2004). The perils of payoff: corruption as a threat to global biodiversity. *Trends in Ecology and Evolution*, Vol.19, No.8, (August 2004), pp. 399-401, ISSN 0169-5347

Lemoine, N., Bauer, H., Peintinger, M. & Böhning-Gaese, K. (2007). Effects of climate and land-use change on species abundance in a central European bird community. *Conservation Biology*, Vol.21, No.2, (April 2007), pp. 495-503, ISSN 1523-1739

Lenzen, M., Lane, A., Widmer-Cooper, A. & Williams, M. (2009). Effects of land use on threatened species. *Conservation Biology*, Vol.23, No.2, (April 2009), pp. 294-306, ISSN 1523-1739

Luck, G.W. (2007), A review of the relationships between human population density and biodiversity. *Biological Reviews*, Vol.82, No.4, (November 2007), pp. 607-645, ISSN 1469-185X

McKee, J., Sciulli, P., Fooce, C. & Waite, T. (2003). Forecasting global biodiversity threats associated with human population growth. *Biological Conservation*, Vol.115, No.1 , (January 2004), pp. 161-164, ISSN 0006-3207

Millennium Ecosystem Assessment. (2005). *Ecosystems and Human Well-Being: Synthesis*, Island Press, ISBN 1-59726-040-1,Washington, DC

Pandit, R. & Laband, D.N. (2007). Spatial autocorrelation in country-level models of species imperilment. *Ecological Economics*, Vol.60, No.13, (January 2007), pp. 526-532, ISSN 0921-8009

Pandit, R. & Laband, D.N. (2009). Economic freedom, corruption, and species imperilment: across-country analysis. Society & Natural Resources, Vol.22, No.9, (September2009), pp. 805-823, ISSN 1521-0723

Pellegrini, L. & Gerlagh, R. (2004). Corruption's effect on growth and its transmission channels. *Kyklos* Vol. 57, No.3, pp. 429-456, ISSN 0023-5962

Pimm, S., Russell, G., Gittleman, J. & Brooks, T. (1995). The future of biodiversity. *Science*, Vol.269, No.5222, (July 1995), pp. 347-350, ISSN 1095-9203

Pimm, S., Raven, P., Peterson, A., Şekercioğlu, C. & Ehrlich, P. (2006). Human impacts on the rates of recent, present, and future bird extinctions. *Proceedings of the National Academy of Sciences of the United States of America*, Vol.103, No.29, (July 2006), pp. 10941-10946, ISSN 1091-6490

Reif, J., Storch, D., Voříšek, P., Štatsný, K. & Bejček, V. (2008). Bird-habitat associations predict population trends in central European forest and farmland birds. *Biodiversity and Conservation*, Vol.17, No.13 , (December 2008), pp. 3307-3319, ISSN 1572-9710

Smith, R. & Walpole, M. (2005). Should conservationists pay more attention to corruption? *Oryx*, Vol.39, No.3, pp. 251-256, ISSN 1365-3008

Smith, R., Muir, R., Walpole, M., Balmford, A. & Leader-Williams, N. (2003). Governance and the loss of biodiversity. *Nature*, Vol.426, No.6962, (November 2003), pp. 67-70, ISSN 1476-4687

Spangenberg, J. (2002). Environmental space and the prism of sustainability: frameworks for indicators measuring sustainable development. *Ecological Indicators*, Vol.2, No.3, (December 2002), pp. 295–309, ISSN 1470-160X

Thomas, C., Cameron, A., Green, R., Bakkenes, M., Beaumont, L., Collingham, Y., Erasmus, B., de Siqueira, M., Grainger, A., Hannah, L., Hughes, L., Huntley, B., van Jaarsveld, A., Midgely, G., Miles, L., Ortega-Huerta, M., Peterson, A., Phillips, O. & Williams, S. (2004). Extinction risk from climate change. *Nature*, Vol.427, No.6970, (January 2004), pp. 145-148, ISSN 1476-4687

Transparency International. (2007). Corruption perception index 2007. 20.2.2010, Available from: http://www.transparency.org/policy_research/surveys_indices/cpi/2007

United Nations. (2005). United Nations Population Information Network. 20.2.2010, Available from: http://www.un.org/popin/data.html#Global Data

United Nations (2006). UN data. 20.2.2010, Available from: http://data.un.org/Data.aspx?d=CDB&f=srID%3A29922

United Nations. (2008). Population Division of the Department of Economic and Social Affairs of the United Nations Secretariat, world population prospects: the 2008 revision. 20.2.2010, Available from: http://esa.un.org/unpp

United Nations Environment Programme-World Conservation Monitoring Centre. (2008). UNEP-WCMC Species Database. 20.2.2010, Available from: http://www.unep-wcmc.org/isdb/Taxonomy/

5

Towards a Better Understanding of Beta Diversity: Deconstructing Composition Patterns of Saproxylic Beetles Breeding in Recently Burnt Boreal Forest

Ermias T. Azeria[1], Jacques Ibarzabal[2], Jonathan Boucher[2] and Christian Hébert[1]
[1]*Natural Resources Canada, Canadian Forest Service, Laurentian Forestry Centre, Québec,*
[2]*Université du Québec à Chicoutimi, Département des sciences fondamentales, Québec, Canada*

1. Introduction

Biodiversity patterns vary across space and time, and this variation is thought to be driven by several ecological/biogeographical processes that act upon species distributions and leave their imprint at various spatial and temporal scales (Huston, 1994; Ricklefs, 2004). The regional or geographical (gamma diversity) diversity patterns, in addition to local diversity (alpha diversity), are consequent upon the extent of "spatial" variation in species composition among sites (beta diversity) (Whittaker, 1960; 1972). Arrays of ecological hypotheses have been proposed as determinants of beta diversity patterns. It can arise from variation in environmental/habitat heterogeneity among sites, spatial constraints, disturbance regimes, and corresponding species-level variations in life history traits (e.g., dispersal, niche-breadth) (Harrison *et al.*, 1992; Nekola & White, 1999; Chase, 2007; Veech & Crist, 2007; Baselga, 2008) as well as neutral and/or stochastic processes (Hubbell, 2001; Chase, 2010). Nevertheless, beta diversity patterns and the mechanisms by which ensembles of local communities maintain their variations and, consequently, influence diversity at regional scale remains a central challenge in ecological and conservation studies (Wilson & Shmida, 1984; Veech & Crist, 2007; Baselga, 2010; Chase, 2010; Tuomisto, 2010; Anderson *et al.*, 2011).

Despite its apparent conceptual simplicity, the empirical assessment of beta diversity has been mired with extensive debates over a perplexing array of indices, analytical approaches, or scales of analysis (Wilson & Shmida, 1984; Lande, 1996; Loreau, 2000; Koleff *et al.*, 2003; Tuomisto & Ruokolainen, 2006; Legendre *et al.*, 2008; Veech & Crist, 2010; for a detailed outline of the issues, see recent review by Anderson *et al.*, 2011). One of the most fundamental and recurrent challenges in beta diversity studies has been distinguishing and quantifying the pure "spatial" turnover component of beta diversity from that caused by variation in species richness between local communities (Wilson & Shmida, 1984; Koleff *et al.*, 2003; Veech & Crist,

2010; Chase *et al.*, 2011). Various indices and frameworks have been proposed to estimate beta diversity patterns independent of richness gradients (Wilson & Shmida, 1984; Koleff *et al.*, 2003). In this context, the recently proposed framework by Baselga (2010) appears to provide an improved means to systematically distinguish between the two independent components of beta diversity - the pure "spatial" turnover of species (due to species replacement) from the nestedness-driven pattern due to "species loss or gain" associated to variation in species richness. These two distinct patterns might also have different underlying causes.

The nestedness component of beta diversity essentially reflects differences in the composition of nested assemblages, i.e., when communities in species-poor sites are subsets of those found in species-rich sites (Patterson & Atmar, 1986). In other words, the difference in composition between species-poor and species-rich sites arises because the latter contains species not present in species-poor sites and not the converse. Nestedness patterns can result from mechanisms/factors that sort species hierarchically such as differential extinction or colonization rates, nested distributions or suitability of habitat gradients (Atmar & Patterson, 1993; Hylander *et al.*, 2005; Azeria *et al.*, 2006; Ulrich & Gotelli, 2007; Azeria & Kolasa, 2008). It should be emphasized that the nestedness component of beta diversity is the dissimilarity due to a nestedness effect or richness difference and not a measure of nestedness itself (Baselga, 2010).

The "spatial" turnover component of beta diversity, on the other hand, reflects distinct community ensembles and their corresponding underlying causal factors or processes , which depart notably from hierarchical effects underlying nestedness (idiosyncartic communities; Atmar & Patterson, 1993; Azeria *et al.*, 2006; 2009b). Potential causes include distinct habitats sustaining distinct communities or spatial segregations due to interspecific interactions (within same habitat template) (Loreau, 2000; Baselga, 2010; Chase *et al.*, 2011). It is clear that distinguishing between the two components is important for our understanding of biodiversity patterns. For instance, a recent study by Baselga (2010) has demonstrated that beta diversity of longhorn beetles in southern Europe was primarily caused by spatial turnover (and associated endemics due to historical effects) while in the northern Europe it was caused by spatial turnover but also by nestedness (ordered loss of species towards the north) (Baselga, 2008; Baselga, 2010). This crucial information was not evident in the un-partitioned beta diversity, which indicated a similar pattern in northern and southern Europe.

Another promising approach to decipher beta diversity patterns beyond richness gradient effects is by using null model analysis (Chase, 2007; Vellend *et al.*, 2007; Anderson *et al.*, 2011; Chase *et al.*, 2011). Null model tests have, however, been remarkably neglected in beta diversity studies (but see Chase, 2007; Chase *et al.*, 2011) despite their wide applicability in many areas of ecology and biogeography (Connor & Simberloff, 1983; Gotelli & Graves, 1996; Gotelli, 2001). Null models quantify and assess whether observed patterns depart from random expectations by comparing them against patterns emerging by randomization of observation data. This approach has been helpful in establishing non-random species co-occurrence patterns beyond random expectations (Connor & Simberloff, 1983; Gotelli & Graves, 1996; Azeria *et al.*, 2009a) and has recently been extended in "biodiversity deconstruction" for identifying species groups that are distinct in some ecological sense (Azeria *et al.*, 2009a; 2011). Similarly, null models can be useful in deciphering beta diversity patterns that are beyond (higher or lower) random chance expectations (given variation in species richness). Such beyond chance patterns may imply a deterministic causal processes; for example, Chase (2007) demonstrated that communities in drought ponds were more similar than expected by random chance, probably due to a deterministic filtering effect caused by a harsh environment (also see Chase *et al.*, 2011).

Clearly, a common ground between the partitioning framework and null model analysis of beta diversity needs to be established in order to promote the state of the art and practical understanding of beta diversity patterns. The idea behind null model analysis is to disentangle beta diversity patterns beyond richness gradients, which should be reflected in the turnover component of beta diversity. The nestedness component of beta diversity, on the other hand, is primarily driven by richness differences, which may be related to beta diversity expected under null distribution given richness gradients.

The present study examines the overall beta diversity components as well as the turnover and nestedness-driven components of beta diversity of saproxylic beetles emerging from tree boles following forest fire. In the present study, we have two major goals. First, we will establish the relationship of the overall beta diversity as well as its two components, turnover and nestedness, with beta diversity expected under and beyond random assembly of communities. We partition or deconstruct the overall beta diversity pattern following the partitioning framework of Baselga (2010). We perform null model analyses for beta diversity with a similar methodological basis as in Chase *et al.* (2011) to estimate the extent of beta diversity expected under and beyond random assembly, given variations in species richness among sites and the regional frequency of species. Second, we will demonstrate if the two components of beta diversity have different underlying causes, here habitat attributes defined primarily by tree-species, burn-severity, and tree-size classes.

Our study system is located in the western spruce–moss bioclimatic subdomain (northwestern Quebec, Canada) where fire is an important natural disturbance generating mosaics of stands with high structural and compositional heterogeneity in the forest landscape (Bergeron *et al.*, 2004). The post-fire environment is characterised by huge amounts of freshly killed and stressed trees, which are very important habitat attributes that sustain saproxylic beetles that feed directly on the bark/wood of dead and dying trees, saprophagous, mycophagous and their predators (McCullough *et al.*, 1998; Grove, 2002; Saint-Germain *et al.*, 2004; Boulanger *et al.*, 2010). The early post-fire environments are characterised by a diversity/abundance of saproxylic beetles that rapidly attack or colonize burned forests. Yet, the influence of the more-local, post-fire habitat legacies such as host-tree species, tree-size, and burn-severity gradients on beta diversity patterns of saproxylic beetles is poorly known. In this study, we examine their effect on the overall, turnover and nestedness components of beta diversity by using distribution data of saproxylic beetles that actually breed in the burned forest, i.e., beetles that develop in, and emerge from, naturally fire-killed tree boles. Understanding the importance of local factors is undoubtedly crucial in conserving these essential groups, whose feeding activity may facilitate the decay of dead trees and, consequently, could have a substantial role in nutrient cycling of disturbed forests (Grove, 2002; Cobb *et al.*, 2010). On the other hand, post-fire salvage logging is currently increasing to maintain timber supply and the negative ecological consequences of this practice has become an emerging ecological issue in forest management (Lindenmayer *et al.*, 2008).

2. Material and methods

2.1 Study area and bole sampling

The study was conducted in 72 sites within four forest burns (burned in 2005) within the western spruce–moss bioclimatic subdomain of northwestern Quebec, Canada (49°15′-50 °

40′N and 75°00′W-73°45′W). This forest subdomain is typically dominated by black spruce (*Picea mariana*) and, to a lesser extent, jack pine (*Pinus banksiana*). The forest landscape also contains various combinations of Balsam fir (*Abies balsamea*), trembling aspen (*Populus tremuloides*) and paper birch (*Betula papyrifera*). In this forest subdomain, forest fire is an important disturbance agent and occurs in a relatively short cycle (120-180 years) and, as consequence, the landscape is dominated by even-aged forest stands (Bergeron *et al.*, 2004).

We sampled sites using a systematic factorial design: 2 tree species X 3 burn levels X 4 tree size categories X 3 replicates (Total = 72 forest sites). Our focus was to examine saproxylic beetles that actually develop in, and emerge from, fire-killed trees following forest fire. Therefore, from each sampling site, we retrieved a 50-cm bole section from five fire-killed trees that we felled (Total 360 trees). This was carried in early June 2006, one year after fire, and when initial colonization or attack by saproxylic beetles was achieved naturally (Boulanger and Sirois 2007). The retrieved bole sections were then enclosed in rearing (emergence) cages placed in a field insectarium, where they were suspended while respecting their natural vertical orientation. The offspring were bred out, and the emerging adults/larvae were collected monthly from June to November in 2006 and 2007 in a vial (with preservative) placed under the bole section. Vouchers are conserved in the insect collection of the Laurentian Forestry Centre. All collected specimens were identified to the lowest taxonomic level (species or genus level) whenever possible, otherwise they were identified only to the family level. The "species" lists of boles were pooled per-site in subsequent beta diversity analysis.

Our sampling protocol thus enabled us to investigate simultaneously the effects of variation in tree species (black spruce and jack pine), burn severity (low, moderate and high) and tree size (db1=8-12 cm; db2=12-16cm; db3=16-20 cm; db4=20-24 cm) for beta diversity of these saproxylic beetles. The tree size classes were based on diameter at breast height (dbh); and burn severity classes were visually defined following criteria adopted by *Ministère des Ressources naturelles de la Faune du Québec* (MRNF) but with modifications of the classes. These habitat variables have been shown to differentially influence the distribution of saproxylic beetles in burned forests (Saint-Germain *et al.*, 2004; Boulanger *et al.*, 2010); and we expect them to also differentially contribute to the turnover and nestedness component of beta diversity. For example, different tree species might attract or host different beetle species and, consequently, might contribute more to the turnover than nestedness component of beta diversity. The effects of burn severity and tree size class is to generally create a gradient of habitat suitability, the lowest suitability being in small trees and/or severe-burns and highest suitability being in large trees of low severity classes (Boulanger *et al.*, 2010). Such a hierarchical suitability gradient would potentially contribute to the nestedness component of beta diversity. On the other hand, the within habitat class contribution for beta diversity components might differ; for example, there could be more dispersion or turnover within-small than within-large tree size classes due to intensive competition in smaller trees that cause segregated species co-occurrences, or some species being biased towards species-poor sites, termed as idiosyncratic species (Azeria *et al.*, 2006; 2009b).

2.2 Saproxylic beetle (emerging) distribution data

We compiled a presence–absence matrix (sites x species) of all saproxylic beetles in each sampling site by pooling the respective data from five fire-killed tree boles. The total number of "species" recorded was 62 species. Nearly half of these species (30 species) were very rare recorded from only one or two sites, and many were primarily common in dead trees in unburnt forest such as those generated by gap dynamics (Boucher, 2011). We omitted them as such species often tend to have an unduly large influence in multivariate analysis and, consequently, distort the overall pattern (e.g. in ordinations). In addition, it is advisable that species that are uncharacteristic of the species pool (here the species might be unlikely to colonize fire-killed tress in burned forests) are omitted to minimize their effect in null model analysis of beta diversity (see Chase et al 2011 for discussion related to the latter issue). We thus restrict our analysis to 32 species that occurred in more than two sites. These species accounted for 94% of the presence–absence matrix and for 97% of the total abundance (1639 individuals). We also omitted one site that only had a single species, which was recorded in nearly all sites (71 of 72 sites, thus its composition dissimilarity was always zero with respect to this site) as it was not possible to generate a "null community" and consequently "null" beta diversity values for the site given the constraints imposed in the null model.

3. Statistical analysis

3.1 Deconstruction or partitioning of beta diversity

Our first goal was to partition the overall beta diversity pattern of the saproxylic beetles into two components: the "spatial" turnover and nestedness component. We also wanted to examine their relationship to beta diversity values expected *under* and *beyond* random community assembly given a null model. We emphasize that throughout the paper, beta diversity refers to site-to-site composition dissimilarity. We partitioned beta diversity of saproxylic beetles following the framework of Baselga (2010) as: $\beta_{sor} = \beta_{sim} + \beta_{nes}$; where β_{sor} (Sorensen dissimilarity) represents the total difference in species composition between two sites, and β_{sim} (Simpson dissimilarity) and β_{nes} (nestedness-driven dissimilarity) are its "turnover" and 'nestedness" components, respectively. We computed these components using the function provided by Baselga (2010), as implemented in R version 2.11 (R-Development-Team, 2010).

3.2 Null model analysis of beta diversity

We used null models to disentangle beta diversity values expected under null distributions and beyond random assembly given constraints set by null models ("delta" beta diversity; also see Chase et al 2011). In principle, the null model analysis for site-to-site beta diversity is methodologically the same as that used for species-to-species co-occurrence patterns in the context of a "biodiversity deconstruction" framework (Azeria *et al.*, 2009a).

First, we computed the beta diversity for the observed data using the Sorensen dissimilarity index (β_{sor}), consistent with beta diversity partitioning framework. Second, null models are applied to randomize the observation data to generate "null" communities (n=1000) for which beta diversity will be calculated. We used two null models: the Fixed-Fixed (FF) and Fixed-Equiprobable (FE) null models. Both null models maintain species richness of sites from the observation matrix. The FF null model also maintains species frequency as in the observation data, while FE null model sample species from the regional species pool equiprobably. We used the function *permatfull*, a wrapper for *commsimulator*, in the R-package Vegan (Oksanen *et al.*, 2010) to generate 1000 null matrices according each null model. For the FF null model, we used the quasi swap algorithm (Miklós & Podani, 2004), which generates matrices that are independent of each other and different from the original matrix.

We computed beta diversity (using the Sorensen dissimilarity index) for each of the 1000 null matrices ($\beta_{sor\text{-}null.mat}$), from which the beta diversity expected by "random" chance was estimated by computing their mean ($\beta_{sor\text{-}null}$) and standard deviation ($\beta_{sor\text{-}null.sd}$). These values effectively represent the beta diversity expected by random chance, given differences in richness among sites (and species frequencies for FF null model). Third, we estimated beta diversity independent of and beyond random chance ($\beta_{sor\text{-}diff}$) by computing the difference in beta diversity values between the observation data (β_{sor}) and null communities ($\beta_{sor\text{-}null}$), i.e. $\beta_{sor\text{-}diff} = \beta_{sor} - \beta_{sor\text{-}null}$. The difference could also be expressed by effect sizes (or standard deviation units) as $\beta_{sor\text{-}SES} = \beta_{sor\text{-}diff} / \beta_{sor\text{-}null.sd}$. The $\beta_{sor\text{-}SES}$ measures the number of standard deviations that the observed dissimilarity β_{sor} is above or below the mean index of the dissimilarity obtained in the null distribution ($\beta_{sor\text{-}null}$). The value of $\beta_{sor\text{-}diff}$ (and $\beta_{sor\text{-}SES}$) will be positive for sites pairs that are more dissimilar than expected and negative for those less dissimilar (more similar) than expected. For ordinations or other graphical representations, the $\beta_{sor\text{-}SES}$ can be rescaled using the ranging formula as $(\beta_{sor\text{-}SES} - \beta_{sor\text{-}SES.min})/(\beta_{sor\text{-}SES.max} - \beta_{sor\text{-}SES.min})$, where $\beta_{sor\text{-}SES.min}$ and $\beta_{sor\text{-}SES.max}$ are the minimum and maximum $\beta_{sor\text{-}SES}$ values. This will rescale the $\beta_{sor\text{-}SES}$ between 0 (for sites that are more similar) and 1 (for sites that are more dissimilar) in the same manner as applied for species-pairs in Azeria et al. (2009a; 2011).

Our main goal was to link the beta diversity components with that expected under null models. It is important that all the terms are expressed in the same units; therefore we examined how the overall (β_{sor}), and its partition to "spatial" turnover (β_{sim}) and nestedness (β_{nes}) components were related to that expected under the null model ($\beta_{sor\text{-}null}$) and beyond ($\beta_{sor\text{-}diff}$).

We also applied the null model recently proposed by Chase et al (2011; also see Chase, 2007) which is a modification of the Raup-Crick metric (β_{RC}) (Raup & Crick, 1979). The null model for computing β_{RC} maintains species richness of site pairs as in the observation data, and species are sampled *proportional* to their frequency, rarity and commonness. The original implementation of the metric (Raup & Crick, 1979) also maintains for species richness, but it does not constrain for species incidence, i.e., it assumes species would be sampled equiprobably. The latter option is also available if desired. We computed β_{RC} using the R-function provided by Chase et al (2011). The program computes a β_{RC} that shows the probability that the observed site-to-site dissimilarity is by chance by counting the number of null matrices where observed shared species is less than simulated. The computed probabilities (between 0 and 1) are rescaled by subtracting -0.5 and multiplying by 2 into values between -1 (less dissimilar) and 1 (more dissimilar). Values around 0 are not different from expected by chance (for other details see Chase *et al.*, 2011).

It is worth mentioning the similarities and differences in our implementation of the null model from that of Chase et al. (2011). Our implementation of the FF null model maintains the species frequency *exactly* as in the observed data, while the null model associated for β_{RC} samples species *proportional* to their incidence in the data set. On the other hand, our implementation of the FE null model will be effectively similar to that of β_{RC} when species are sampled equiprobably, as in the original formulation of Raup-Crick metric (Raup & Crick, 1979). The other difference is, the $\beta_{sor\text{-}diff}$ indicates actual differences in dissimilarity values between observed and null communities in beta diversity-units, while β_{RC} expresses the difference (for the corresponding null model) in terms of a probability index. The probability associated with $\beta_{sor\text{-}diff}$ can be estimated by applying inverse-logit (R function *plogis*) or normal (R function *pnorm*) transformations on its standardized form, the $\beta_{sor\text{-}SES}$. Transformed values will be closer to zero (negative $\beta_{sor\text{-}SES}$) for site pairs that are less dissimilar (more similar) than expected by random chance (one tail), and the converse is

true for more dissimilar site pairs. Note that the index is based on dissimilarity; the tests for statistical significance of "more dissimilar than expected" are made by taking the complement of the transformed values (for index values 0.5 to 1, subtract them from 1). Alternatively, to test statistical significance, one may use the proportion of null matrices in which $\beta_{sor\text{-}null.mat}$ is the same or larger (or lower) β_{sor} for obtaining higher (lower) dissimilarity between sites. In the latter case, a comparable index with high β_{RC} values (closer to 1, which was based on shared species) will be the number of matrices for which β_{sor} was higher than $\beta_{sor\text{-}null.mat}$.
To assess the relationships between the different algorithms, we examined the probability computed for $\beta_{sor\text{-}SES}$ (based on the FF and FE null models) and that of β_{RC} metric.

3.3 Analysis of habitat effects on overall beta diversity and components of beta diversity

Our second goal was to examine the contribution of a set of habitat variables (tree species, burn severity classes, tree size classes) as well as their interactions for the overall (β_{sor}), turnover (β_{sim}) and nestedness-driven (β_{nes}) beta diversity of saproxylic assemblages. In addition, we examined the spatial effect on composition dissimilarity by considering its correlation with site-to-site geographical distance, and then by considering the dissimilarity between and within the forest burns (four burns). We used a nonparametric, Permutatitional Multivariate Analysis of Variance (PERMANOVA; Anderson, 2001) to test whether categories for each habitat factor differed in their variability in species composition or beta diversity (based on β_{sor}, β_{sim} and β_{nes}). The method computes a pseudo F-ratio, which is the ratio of composition dissimilarity within a treatment (habitat class) to that of between treatments, and then tests its significance by permutation (9999 replicates). Accordingly, significant results might indicate differences in dissimilarity among the classes (difference between treatment's centroid in multivariate space), due to differences in the within-class dispersion (i.e., mean distances of members to their group centroid) or both. The potential role of each is distinguished by running a complementary analysis, the Permutational Analysis of Multivariate Dispersions (PERMDISP; Anderson *et al.*, 2006), which tests whether classes (treatments) differed in their within-treatment dispersion or beta diversity. As an example, consider tree species, Black spruce (BSP) and Jack pine (JPI), as sources of variability. A significant result by PERMANOVA and a non-significant difference by PERMDISP would suggest differences in their across class dissimilarity (BSP ≠ JPI), i.e., the treatments differ in their centroid in multivariate space and not in the within-treatment dispersion (BSP-BSP = JPI-JPI dispersion). When significant results are also obtained by PERMDISP, one may run pairwise tests to examine which of the classes had higher dispersion (particularly when dealing with more than two classes). We performed PERMANOVA and PERMDISP using the functions *adonis* and *betadisper*, respectively, in the R-package Vegan (Oksanen *et al.*, 2010).
We also applied multivariate regression analysis on distance matrices (MRM) (Zapala & Schork, 2006; Lichstein, 2007), which might help model the beta diversity variation of within- and between habitat classes, as well as the difference between them, simultaneously. MRM is essentially an ANOVA-like analysis performed on the site-to-site dissimilarity matrix expanded into a vector and modelled against the corresponding contrast of classes of habitat (for each habitat variable) changed into vector. For example, when considering the effect of tree species, the contrasts [within black spruce (BSP-BSP), within Jack pine (JPI-JPI) and between the two species (BSP-JPI)] are unfolded into vector. Note that according to our

construction, the within comparisons (BSP-BSP and JPI-JPI) are regarded as distinct values or factors and indexed accordingly in the explanatory variable. This re-indexing of the pairwise factors is done for each explanatory variable considered here, i.e., tree species, burn severity, tress size classes and forest burns where the sampling sites were located. The significance of MRM is tested by permutation of the distance vector. Note that this provides a computationally efficient way to compute an equivalent permutation test if the raw data of species composition were permuted and distance was computed afterwards (Hayden *et al.*, 2009). We will focus mainly on the results of pairwise tests among the "new classes" (e.g., BSP-BSP, BSP-JPI and JPI-JPI) of the MRM with that of PERMANOVA and PERMDISP to further unravel the source of variation for the overall beta diversity and beta diversity components.

4 Results and discussion

4.1 The relationship of beta diversity components and expectations under and beyond null models

We present results (Fig. 1 and Fig. 2) that illustrate the relationship between overall, observed beta diversity (β_{sor}), and more specifically its turnover (β_{sim}) and nestedness (β_{nes}) components (sensu Baselga, 2010) with beta a diversity pattern expected by *"random"* ($\beta_{sor\text{-}null}$) and with *deviations beyond* ($\beta_{sor\text{-}diff}$) random distribution of species among sites using two null models. The first null model preserves both species richness and species incidence (FF null model; Fig. 1), while the second null model preserves only species richness (FE null model; Fig. 2). While the overall beta diversity (β_{sor}) value was generally correlated to that expected by *random* ($\beta_{sor\text{-}null}$) under FF (Fig.1a) and FE (Fig.1a) null models, a stronger relationship was evident with that of *deviations beyond* null model expectations, i.e., $\beta_{sor\text{-}diff}$ (FF: Fig.1b; FE: Fig. 2b). This indicates that beta diversity patterns of saproxylic beetles show a strong signal of deterministic underlying processes that deviates from random expectations given variations in species richness and species incidence.

A more explicit examination of the beta diversity components pattern via partitioning into its turnover (β_{sim}) and nestedness (β_{nes}) components (sensu Baselga, 2010) revealed that the two components exhibited distinct relationships to beta diversity values expected under ($\beta_{sor\text{-}null}$) and beyond ($\beta_{sor\text{-}diff}$) null models. Thus, the turnover component was linearly related to beta diversity deviating *beyond* "random" expectations (Figs 1d and 2d), while the nestedness component of beta diversity was related to the *null* expectations (Figs. 1e and 2e). The converse relationships, i.e., turnover with expectations under a null model (Figs. 1c and 2c) and that of nestedness components with deviations beyond "random" (Figs. 1f and 2f) were not significant. Taken together our results demonstrate quantitatively (via null models) that the beta diversity components (sensu Baselga , 2010) do indeed reflect different extents of dependence on richness gradients. More specifically, we show the independence of the turnover and the dependence of the nestedness component on richness variations. In addition, our result emphasizes that the nestedness-driven component (sensu Baselga, 2010) should be interpreted as a general effect of richness difference on beta diversity.

Moreover, our results indicate that the relationship of turnover and nestedness components with respective null model expectations, i.e., $\beta_{sor\text{-}diff}$ and $\beta_{sor\text{-}null}$, were stronger when the null model also preserved the species incidence (FF-null model: Fig. 1d, r=0.84; Fig.1e, r=0.73) than when species were assumed to be sampled equiprobably (FE-null model: Fig. 2d, r=0.68; Fig.2e, r=0.44). Thus, while the nestedness and turnover components would reflect the beta diversity consequent upon and beyond richness gradients respectively, our results suggest that both components might be conditioned by species frequency (commonness and rarity) patterns. This also implies that the appropriate null model for beta diversity analysis *sensu lato* should preserve not only observed species richness and but also the species frequency. The importance for null models to also control for species commonness and rarity has been emphasized in the modified Raup-Crick metric (β_{RC}) proposed by Chase *et al.* (2011). They modified the original Raup-Crick probabilistic index that samples species equiprobably (Raup & Crick, 1979) by implementing a null model that samples species *proportional* to their regional frequency in the observed data.

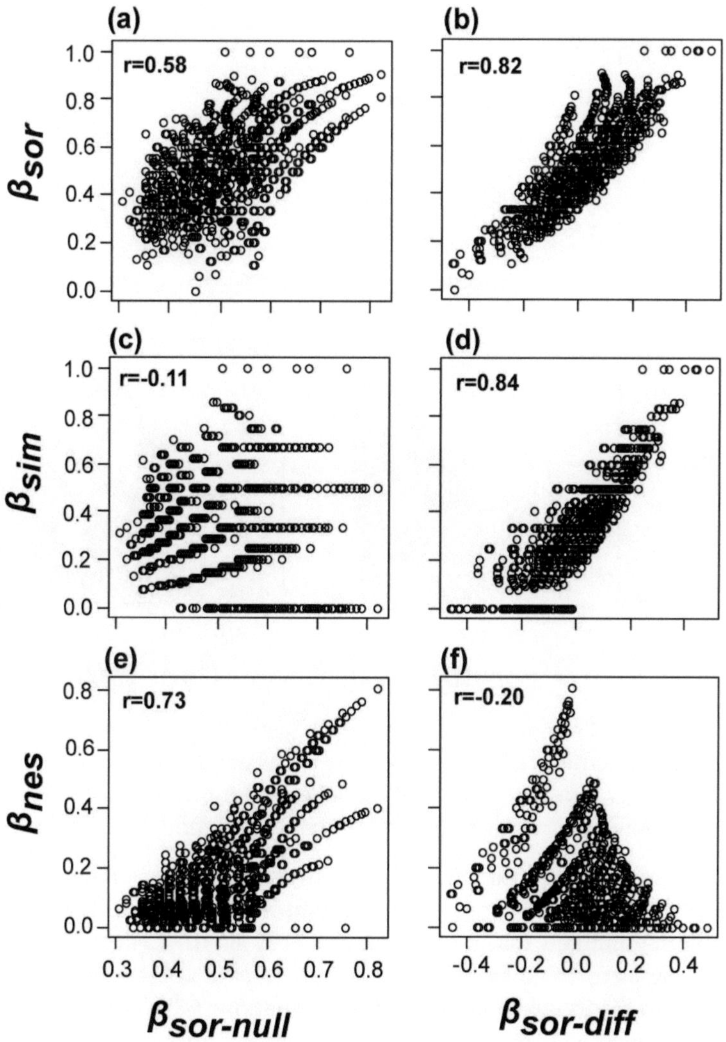

Fig. 1. The relationship of overall betadiversity (β_{sor}), and its partitions, the "spatial" turnover (β_{sim}) and nestedness (β_{nes}) components, with beta diversity values expected under null distributions ($\beta_{sor\text{-}null}$) and beyond-null model expectations ($\beta_{sor\text{-}diff}$). The null communities were generated using fixed-fixed null model (FF null model).

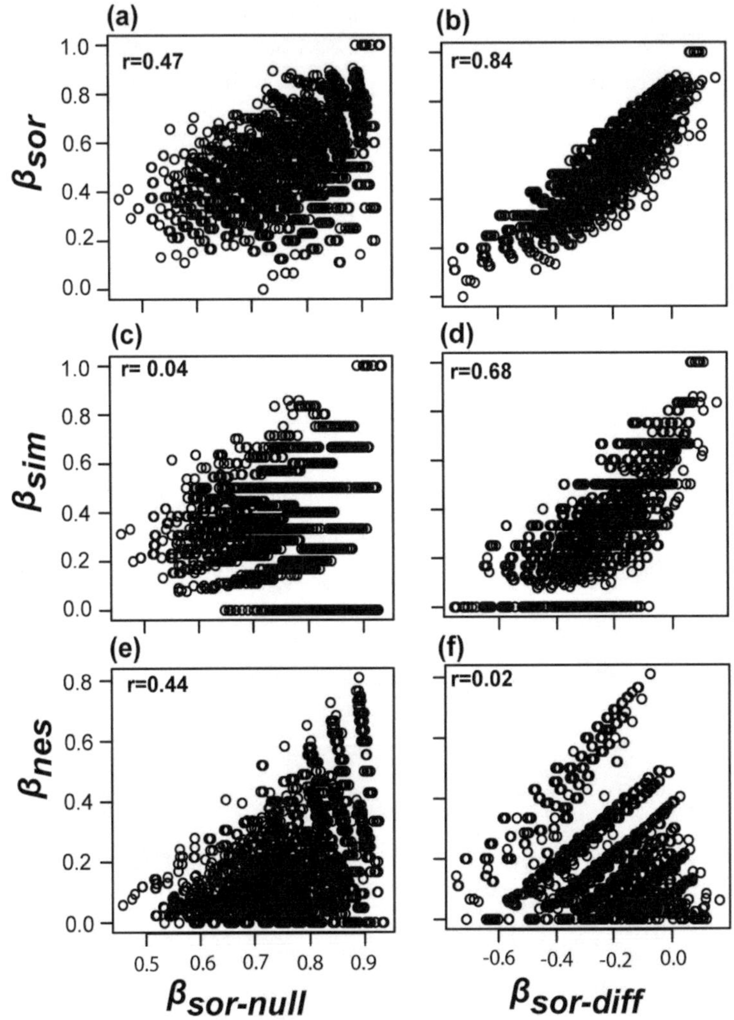

Fig. 2. The relationship of overall betadiversity (β_{sor}), and its partitions, the "spatial" turnover (β_{sim}) and nestedness (β_{nes}) components, with beta diversity values expected under null distributions ($\beta_{sor\text{-}null}$) and beyond-null model expectations ($\beta_{sor\text{-}diff}$). The null communities were generated using fixed-equiprobable null model (FE null model).

The utility of null models for disentangling beta diversity patterns independent of richness variation has only recently been emphasized (Anderson *et al.*, 2011; Chase *et al.*, 2011). Chase *et al.* (2011) have highlighted that the β_{RC} dissimilarity metric (as expressed in probability between 0 and 1, or as rescaled from -1 to 1; see Chase et al 2011) reflects the beta diversity or dissimilarity patterns independent of richness variations (also see Chase, 2007). We also found that β_{RC} was generally correlated withthe turnover component of beta diversity β_{sim} (r =0.73) as well as to our computation of beta diversity beyond null expectations $\beta_{sor\text{-}diff}$ (r =0.90). However, it should be emphasized that the β_{RC} dissimilarity metric is a *probability* index rather than a direct measure of the departure from random chance in terms of "beta diversity units" sensu stricto (β_{sim} or $\beta_{sor\text{-}diff}$). To obtain a dissimilarity measure comparable to β_{RC} metric, we estimated the probability associated with $\beta_{sor\text{-}diff}$ values by applying a inverse-logit transformation on the standard effect size of their deviations (standard deviation units), i.e., $\beta_{sor\text{-}SES}$ (also see Azeria *et al.*, 2009a; 2011). Accordingly, the $\beta_{sor\text{-}SES}$ computed using the FF-null model was linearly related to β_{RC}, albeit the data points for β_{RC} were slightly lower from the one-to-one line (Fig. 3a). This slight deviation is expected given that our FF-null model *preserves* observed species frequency while the null model for β_{RC} samples species *proportional* to their respective observed frequencies. On the other hand, values of β_{RC} were higher than the values computed for $\beta_{sor\text{-}SES}$ under FE-null model, which samples species equiprobably (Fig 3b). When the two metrics were computed using a comparable null models that sampled species equiprobably, the $\beta_{sor\text{-}SES}$ and β_{RC} were highly correlated (Fig. 3c). The same results were obtained when $\beta_{sor\text{-}SES}$ was transformed by normal distribution.

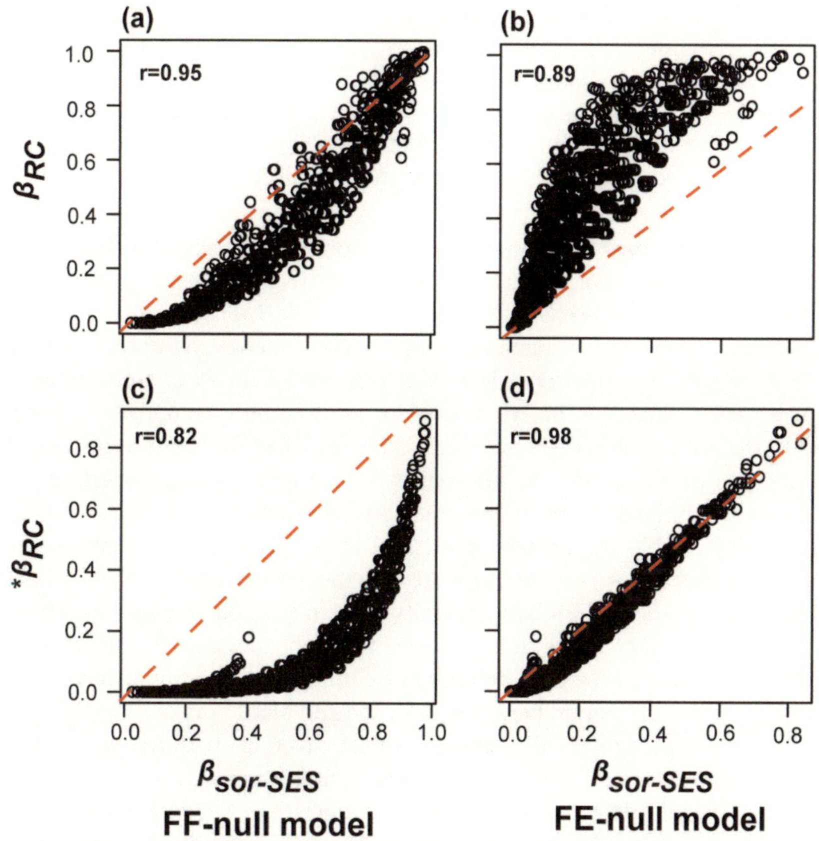

Fig. 3. Relationship between dissimilarity/beta diversity values beyond null expectations as expressed in terms of standard effect size ($\beta_{sor\text{-}SES}$) and the Raup-Crick metric (β_{RC}). The $\beta_{sor\text{-}SES}$ values were transformed using logistic function. In all null models, the species richness of sites is maintained as in the observed. The species frequency was preserved in computing the $\beta_{sor\text{-}SES}$ (FF-null model: a,c), or equiprobably (FE-null model: b,d) (see method). The null models for computing the β_{RC} sampled species proportional to observed frequencies (β_{RC}: a,b) or equiprobably ($^*\beta_{RC}$: c,d) (for details of the method see Chase et al., 2001).

These results underscore that a clear distinction should be made between the *"actual"* values of turnover beyond null model (β_{sim} and $\beta_{sor\text{-}diff}$) and the *probabilities* associated with these values as estimated from null model, using the $\beta_{sor\text{-}SES}$ and β_{RC} metrics. As demonstrated above, a more direct link to the turnover component of beta diversity (β_{sim}) is obtained by beta diversity beyond "null" distribution $\beta_{sor\text{-}diff}$. It is noteworthy that $\beta_{sor\text{-}diff}$ will be negative when observed dissimilarity is less than expected (sites were more similar), and a proper rescaling should be made (e.g., subtract the minimum value) in subsequent analysis that require positive values (e.g. ordinations or PERMANOVA). Certainly, the null model derived $\beta_{sor\text{-}SES}$ (this study, also see Azeria *et al.*, 2009a) and β_{RC} metrics are applicable and important measures for studying beta diversity independent of richness variation (Chase, 2007; Chase *et al.*, 2011). However, caution should be exercised as the values can become biased for site pairs that have extremely low species richness relative to the regional species pool. For example, although the difference between the observed and null expectation $\beta_{sor\text{-}diff}$ is small, the variation of the null expectations $\beta_{sor\text{-}sd}$ might be so small that $\beta_{sor\text{-}SES}$ become inflated (for related caveats with β_{RC} and other issue see Chase *et al.*, 2011). Although, the inverse-logit/normal transformation should minimize the bias (see Azeria *et al*, 2009a), caution should still be exercised when using the value in subsequent analysis. In addition, our results offer interesting qualitative comparisons among the null model based indices of beta diversity and how constraints imposed on species incidence would influence the values.

4.2 Effect of habitat factors for overall and components of beta diversity

We found a significant effect of tree species, burn severity, and tree-size class on overall beta diversity (β_{sor}) of saproxylic beetles (PERMANOVA Table 1, Fig. 4a, d, and g). In addition, we found a marginal effect of interaction terms between tree species and burn severity. The effect of tree species and tree-size class was primarily due to compositional difference between treatments (location of treatment in multivariate response) but not due to differences in the within-class dispersion (PERMDISP, Table 2). In contrast, the influence of burn severity was primarily due to differences in the within-class dispersion, which was lower for low-severity burn (more homogeneous) than that of moderate- and high-severity burns (PERMDISP, Table 2). Results from multivariate regression analysis on distance matrices (MRM) provide a concise summary of the simultaneous effect of within- and between-treatment on the overall beta diversity pattern (Fig. 5a, b, c and d). The MRM also showed some trends that were not evident through the use of PERMANOVA and PERMIDISP. For example, the beta diversity or composition dissimilarity within jack pine (JPI-JPI) was similar to that found between jack pine and black spruce (BSP-JPI) (Fig 5a).

Overall, the effect of geographical distance on beta diversity patterns was only marginal, and when detected it was due to differences in the within-burns dissimilarity (lower for F1, forest burn in the north, than the others, Fig 5d) rather than between-burns differences. In other words, there was no increase in composition dissimilarity of saproxylic beetles in burned forests with increasing site-to-site geographical distance (β_{sor}: r= 0.032; β_{sim}: r= 0.040; β_{nes}: r= -0.017). Thus our results do not provide support for the "distance decay of similarity" hypothesis (Nekola & White, 1999). It seems that saproxylic beetles might be good dispersers due to the ephemeral nature of their habitats (Boulanger *et al.*, 2010) and thus may not be strongly limited by dispersal, at least at the scale of our study (up to 200 km). Baselga (2010) has shown that composition dissimilarity of longhorn beetles increase with geographical distance measured across larger scales (up to 3000 km) across Europe.

		Total (β_{sor})		Turnover (β_{sim})		Nestedness (β_{nes})	
Source of variability	df	*SS*	*F- value*	*SS*	*F- value*	*SS*	*F- value*
Tree species (Ts)	1	0.911	**8.546*****	0.706	**11.327*****	-0.030	-1.835
Burn severity (Bs)	2	0.717	**3.362*****	-0.058	-0.466	0.486	**14.956*****
Tree size/dbh class (Dc)	3	1.502	**4.696*****	0.334	1.789	0.578	**11.869*****
Forest burns "zone"	3	0.484	***1.513§***	0.394	***2.109§***	0.035	0.726
Ts x Bs	2	0.369	***1.729§***	0.281	***2.253§***	-0.033	-1.026
Ts x Dc	3	0.317	0.992	0.181	0.969	0.082	1.683
Bs x Dc	6	0.555	0.868	0.240	0.642	0.213	***2.184§***
Ts x Bs x Dc	6	0.684	1.069	0.300	0.804	0.050	0.510
Residuals	44	4.691		2.741		0.714	
Total	70	10.230		5.119		2.094	

***p <0.001; ** <0.01; * <0.05; § <0.10

Table 1. PERMANOVA table of saproxylic assemblages indicating the effect of habitat variables on the overall beta diversity (β_{sor}) and its "spatial" turnover (β_{sim}) and nestedness-driven (β_{nes}) components. Note that $\beta_{sor} = \beta_{sim} + \beta_{nes}$. Significance of the pseudo F-ratio was tested using a permutation test (9999 permutations); significant results Pr (>F-value) are indicated by bold typeface, and those indicated in italics are marginal.

Taken together our results suggest that the total beta diversity (βsor) pattern of saproxylic beetles was driven by differences between tree species and between tree size classes, as well as variation in within-treatment dissimilarity among burn severity classes and to some extent among sites within forest burns. The effects of these habitat attributes on overall beta diversity may be through influences on its "spatial" turnover (dissimilarity by species replacement) and/or nestedness (dissimilarity by richness variation) components. It is crucial that the two components of beta diversity are disentangled for a proper understanding of the most likely distinct underlying mechanisms (Baselga, 2008; 2010).

Our results indicate that the turnover (β_{sim}) and nestedness (β_{nes}) components of beta diversity of saproxylic beetles are indeed dependent upon different habitat attributes (Tables 1 &2; Figs. 4 & 5). The turnover component of beta diversity was primarily driven by tree species, which showed a significant composition differentiation between black spruce and jack pine (PERMANOVA, Table 1; Fig. 4b). In addition, there was a significant difference in the within-treatment turnover among tree species: turnover was higher within jack pine than within black spruce (PERMDISP, Table 2; Fig. 4b; Fig. 5e). In fact, the within-treatment species turnover for jack pine was to the same extent as that found between jack pine and black spruce (Fig. 5e). The within forest-burn turnover was also lower for the two north burns (F1, F2) than for the southern (F3, F4) forest burns (Tables 2; Fig. 5h), but composition differentiation between burns were not significant (Tables 1; Fig. 5h).

The influence of tree species on the turnover component of beta diversity indicates that there is some level of differentiation in community ensembles between black spruce and jack pine. This pattern is expected among saproxylics that exhibit host-tree specificity (Allison *et al.*,2004; Janssen et al. in press) Such distinct habitat preferences or suitability of trees for component species can lead to segregated species distributions (Azeria *et al.*, 2010). It was intriguing that the within-treatment dissimilarity was higher for jack pine than black spruce;

this might be related to jack pine being less common (although widely distributed) than black spruce in the landscape. The high turnover within the southern forest burns compared to the northern forest burns may be related to their higher heterogeneity in terms of composition and structure. Notably, the dissimilarity within and between the southern burns (F3 and F4) was of the same extent as that observed with respect to the northern forest burns (F1 and F2).

			Total (β_{sor})		Turnover (β_{sim})		Nestedness (β_{nes})	
Source of variability		df	*SS*	*F value*	*SS*	*F value*	*SS*	*F value*
Tree species (Ts)	Groups	1	0.029	2.697	0.065	**5.802***	0.014	1.594
	Residual	69	0.751		0.770		0.587	
	Pairwise tests		BSP=JPI		BSP <JPI		BSP=JPI	
Burn severity (Bs)	Groups	2	0.085	**4.583***	0.033	1.496	0.077	**5.991****
	Residual	68	0.634		0.755		0.438	
	Pairwise tests		L < (M=H)		L =M=H		L < (M§=H)	
Tree size/dbh class (Dc)	Groups	3	0.068	***2.185*§**	0.047	1.486	0.043	***2.168*§**
	Residual	67	0.694		0.712		0.443	
	Pairwise tests		Db1=Db2=Db3= Db4		Db1=Db2=Db3= Db4		Db1=Db2=Db3=Db4	
Forest burns "zone"	Groups	3	0.111	**4.279****	0.112	**3.702***	0.053	**2.377§**
	Residual	67	0.579		0.674		0.496	
	Pairwise tests		F1<(F3=F4)=F2		(F1=F2) § < F3=F4		F1=F2=F3=F4	

** P<0.01; * <0.05; § <0.10

Table 2. Summary of PERMDISP analysis examining differences in the within-treatment dispersion/dissimilarity based on metrics of overall, turnover and nestedness components of beta diversity. Significant differences are indicated in bold letters; while bold-italics indicate marginally significant results (9999 permutations). Treatments or classes of habitat attributes considered are- Tree species: BSP=Black spruce and JPI=Jack pine; Burn severity: L=Low, M=Moderate and H=High; Tree size/dbh (diameter at breast height in cm) classes: Db1=8-12, Db2=12-16, Db3=16-20 and Db4=20-24. Forest burns were located north to south: F1-F2-F3/F4.

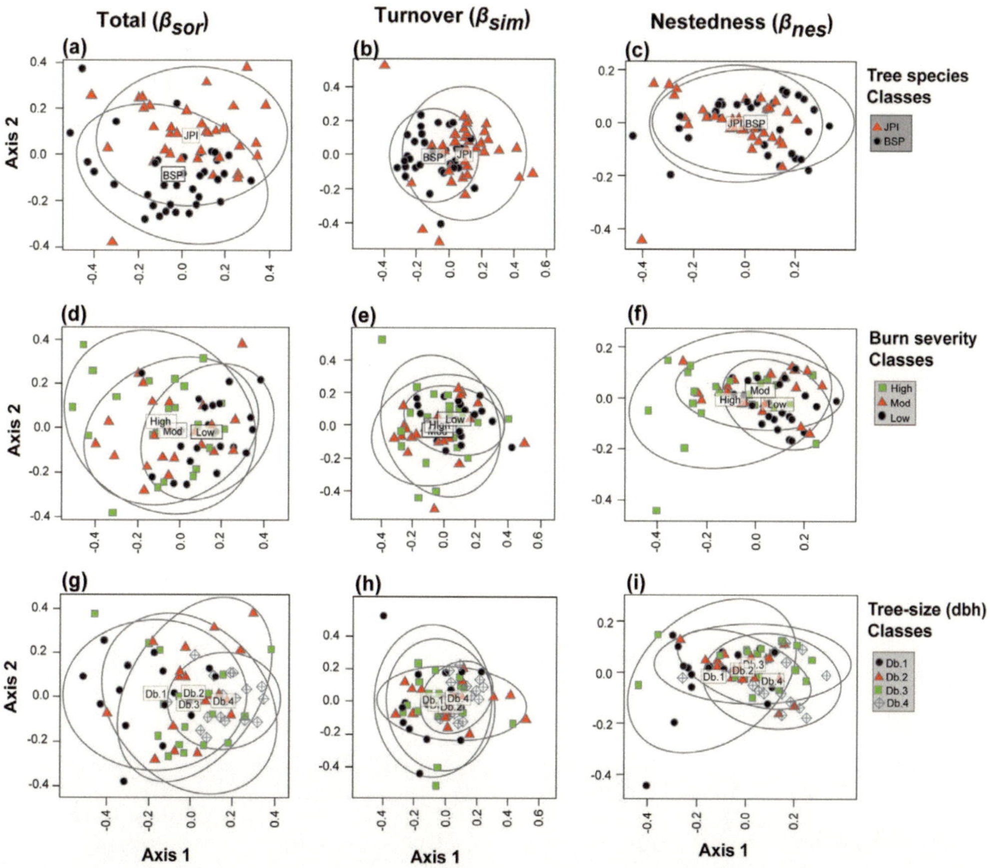

Fig. 4. A two-dimensional MDS representation of site-to-site dissimilarity showing the effect of tree species (a-c), burn severity (d-f) and tree-size class (g-i) on saproxylic beetle's total beta diversity(a,d,g), and when partitioned into "spatial" turnover (b,e,h) and nestedness-driven (c,f,i) components. The ellipse encloses 80% of the dispersion of habitat classes from their respective group centroid as accounted by the first two axes. Habitat classes are Tree species: BSP=Black spruce and JPI=Jack pine; Burn severity: L=Low, M=Moderate and H=High; Tree size/dbh (diameter at breast height in cm) classes: db1=8-12, db2=12-16, db3=16-20 and db4=20-24.

The nestedness-driven beta diversity (β_{nes}), however, was influenced by gradients of burn severity and tree size/dbh, but not by tree species (PERMANOVA, Table 1, Fig 4f, i). There was significant nestedness-driven composition dissimilarity between high severity burns and both low- and moderate-severity burns (Fig 4f and Fig. 5j); and between small-sized trees (Db1) and large-sized trees (Db3 and Db4) (Fig 4i; Fig 5k). In addition, there was a significant difference in the within-class dissimilarity for burn severity; β_{nes} was higher for the high-severity burns and moderate-severity burns than in the low-severity burn class (PERMDISP; Table 2, Fig 4f and Fig 5j). Finally, although PERMIDISP indicated a non-significant result for tree species, the MRM analysis indicated that the nestedness-driven dissimilarity was higher within black spruce than jack pine (Fig. 5i), which was opposite to that observed for turnover component (Fig. 5e).

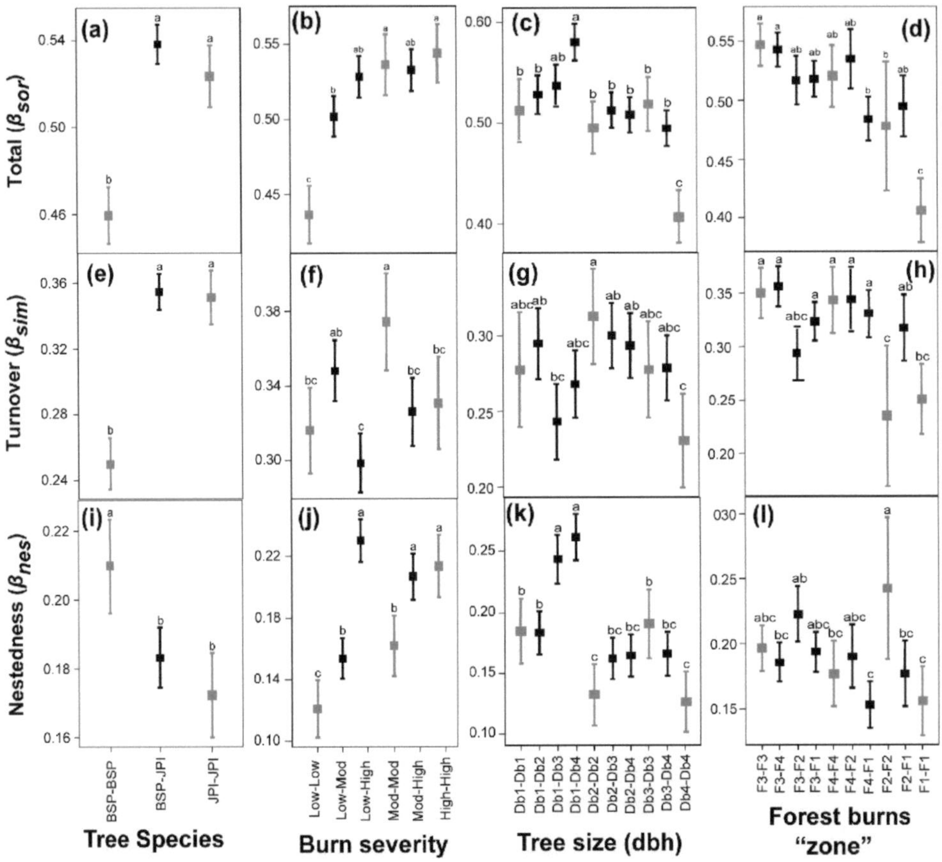

Fig. 5. A simultaneous analysis of beta diversity patterns (β_{sor}, β_{sim}, β_{nes}) differences within- (gray bars) and between-habitat (black bars) classes according tree species (a,e,i), burn severity (b,f,j), tree-size class (c, g, k) and burns (d, h, i) on saproxylic beetle's total beta diversity(a-d), and when partitioned into "spatial" turnover (e-h) and nestedness-driven (i-l) components. On the y-axis are mean values of dissimilarity (beta diversity) for the corresponding habitat contrasts. Habitat codes are as in Fig. 4.

The effects of burn severity and tree size gradients on the nestedness component of beta diversity indicate that hierarchical suitability of the attributes is causing richness-driven composition differences between respective treatments. Generally, large trees of low burn severity are more suitable to saproxylic beetles than small trees of high burn severity, because they provide suitable oviposition conditions and high quality food resources (i.e., subcortical tissues) for larvae feeding directly on the bark and wood of trees (Allison *et al.*, 2004; Saint-Germain *et al.*, 2004). Indirectly, these trees may also be more suitable to predators of these larvae (Kenis *et al.*, 2004). Indeed, large-sized and/or less severly burned treess have been shown to support more saproxylic beetle species than small-sized and/or very severely burned trees (Azeria *et al.*, 2010; Boulanger *et al.*, 2010). It is interesting, and counter intuitive, that a similar trend was also observed in the within-class nestedness-driven composition dissimilarity among burn severity classes; thus, nestedness-driven dissimilarity was higher for high-severity than low-severity burns. This might be explained by the high variance (relative to average) in species richness exhibited within severe burns. We suspect that an interaction effect between burn severity and tree-size might play a role, but this is difficult to isolate statistically from their independent effects in PERMDISP analysis.

The above results clearly underscore that the turnover and nestedness components of saproxylic beetles are driven by different post-fire habitat legacies. Disentangling causal factors influencing beta diversity patterns of saproxylic beetles can have important conservation implications, such as in post-fire salvage logging operations where maintaining saproxylic diversity can concern management plans. Indeed, concern over ecological consequence of post-fire salvage logging on biodiversity and ecosystem function is a pressing management issue (Lindenmayer *et al.*, 2008), and saproxylic beetles constitute key component of burned forest ecosystems (Grove, 2002; Cobb *et al.*, 2010). Knowledge about the factors influencing beta divesity components can be crucial in setting salvage logging practices that will conserve also these essential groups. For example, given the species composition turnover exhibited between jack pine and black spruce, conserving the totality of saproxylic beetle species will require a management approach that maintains a mosaic of both tree species in the landscape. In addition, higher turnover within-jack pine than black spruce (and within moderate severity burns) may require special considerations to conserve all species occupying that habitat. On the other hand, the nestedness driven component will perhaps require emphasis on habitat attributes that increase species richness, e.g., large trees of lower severity burns. These decisions could also have implications for the structure of forest stands that are to be set aside for protective purposes. For example, in old-growth stands with numerous large-diameter trees (e.g., Db4=20-24), a smaller number of trees might suffice to capture the totality of species given the low differentiation of both turnover and nestedness component of beta diversity within large trees (given all other attributes are considered). However, if the available forest stands contain only moderately sized trees (e.g., Db2=12-16), then protecting more trees during salvage-logging might be required (given high turnover), although variation of nestedness-driven beta diversity was to the same extent as that of larger trees. These examples are just to illustrate the implication of disentangling the turnover and nestedness components and underlying factors for management (also see Azeria *et al.*, 2006; 2009b; Baselga, 2010).

5. Conclusion

Ecologists face a continuing challenge to disentangle and explain the species 'turnover' from composition dissimilarity that is driven by variation in species richness among sites. The recently proposed beta diversity partitioning framework and null model approaches make a significant step forward in meeting this challenge. We demonstrate the explicit quantitative relationship of the beta diversity components as depicted in the partitioning framework with that expected *under* and *beyond* random assembly of communities by using null models. Our results indicate that the turnover component indeed reflects the beta diversity deviating *beyond* that expected from "random" assembly given species richness variations, while the nestedness component of beta diversity was related to the *null* expectations. In addition, the beta diversity components were conditioned by variation in species frequency; this also implies that null models for beta diversity studies should perhaps preserve both the observed richness and species frequency patterns. We concur with others (Baselga 2010, Anderson et al. 2011) that the distinction between the two beta diversity components is important, because they were driven by different underlying causes (habitat factors) and this has implications for post-fire management in the boreal forest. Additional studies that directly incorporate habitat and spatial effects into null model analysis such as using habitat- and spatially-constrained null models are needed in order to increase our understanding of the factors that control beta diversity patterns across space (also see Chase *et al.*, 2011).

6. Acknowledgements

We thank Chantiers Chibougamau and Barrette Chapais Ltée for logistical support, and all of our field assistants that helped in conducting the field work and in sample sorting. We are indebted to Y. Dubuc and G. Pelletier for specimen identification. We thank J. Hodson for his comments and revising the English. This study was supported by the *Fonds de recherche sur la nature et les technologies* and the *Ministère des Ressources naturelles et de la Faune* through the *Programme de recherche en partenariat sur l'aménagement et l'environnement forestiers*-II, by the iFor consortium (Université Laval), by Environment Canada, by the Canadian Forest Service, and by *la Fondation de l'Université du Québec à Chicoutimi.*

7. References

Allison, J., Borden, J. & Seybold, S. (2004) A review of the chemical ecology of the Cerambycidae (Coleoptera). *Chemoecology*, 14, 123-150

Anderson, M., Crist, T., Chase, J., Vellend, M., Inouye, B., Freestone, A., Sanders, N., Cornell, H., Comita, L. & Davies, K. (2011) Navigating the multiple meanings of diversity: a roadmap for the practicing ecologist. *Ecology Letters*, 14, 19-28

Anderson, M. J. (2001) A new method for non parametric multivariate analysis of variance. *Austral Ecology*, 26, 32-46

Anderson, M. J., Ellingsen, K. E. & Mcardle, B. H. (2006) Multivariate dispersion as a measure of beta diversity. *Ecology Letters*, 9, 683-693

Atmar, W. & Patterson, B. D. (1993) The measure of order and disorder in the distribution of species in fragmented habitat. *Oecologia*, 96, 373-382

Azeria, E. T., Bouchard, M., Pothier, D., Fortin, D. & Hébert, C. (2011) Using biodiversity deconstruction to disentangle assembly and diversity dynamics of understorey plants along post fire succession in boreal forest. *Global Ecology and Biogeography*, 20, 119-133

Azeria, E. T., Carlson, A., Pärt, T. & Wiklund, C. G. (2006) Temporal dynamics and nestedness of an oceanic island bird fauna. *Global Ecology and Biogeography*, 15, 328-338

Azeria, E. T., Fortin, D., Hébert, C., Peres-Neto, P., Pothier, D. & Ruel, J. C. (2009a) Using null model analysis of species co-occurrences to deconstruct biodiversity patterns and select indicator species. *Diversity and Distributions*, 15, 958-971

Azeria, E. T., Fortin, D., Lemaître, J., Janssen, P., Hébert, C., Darveau, M. & Cumming, S. G. (2009b) Fine scale structure and cross taxon congruence of bird and beetle assemblages in an old growth boreal forest mosaic. *Global Ecology and Biogeography*, 18, 333-345

Azeria, E. T., Ibarzabal, J. & Hebért, C. (2010) Importance of habitat and biotic interactions for assembly of Saproxylic beetles emerging from logs: A test using habitat-constrained null models. *Ecological Society of America (ESA) 95th Annual Meeting*, Pittsburgh, PA

Azeria, E. T. & Kolasa, J. (2008) Nestedness, niche metrics and temporal dynamics of a metacommunity in a dynamic natural model system. *Oikos*, 117, 1006-1019

Baselga, A. (2008) Determinants of species richness, endemism and turnover in European longhorn beetles. *Ecography*, 31, 263-271.

Baselga, A. (2010) Partitioning the turnover and nestedness components of beta diversity. *Global Ecology and Biogeography*, 19, 134-143

Bergeron, Y., Gauthier, S., Flannigan, M. & Kafka, V. (2004) Fire regimes at the transition between mixedwood and coniferous boreal forest in northwestern Quebec. *Ecology*, 85, 1916-1932

Boucher, J. (2011) *Impacts de la coupe de récupération après feu sur les coléoptéres associés aux brûlis en forêt boréale: une dynamque temporaelle*. Université du Québec à Chicoutim, Chicoutimi, Canada.

Boulanger, Y., Sirois, L. & Hébert, C. (2010) Distribution of saproxylic beetles in a recently burnt landscape of the northern boreal forest of Québec. *Forest Ecology and Management*, 260, 1114-1123

Chase, J. (2007) Drought mediates the importance of stochastic community assembly. *Proceedings of the National Academy of Sciences*, 104, 17430-17434

Chase, J. M. (2010) Stochastic community assembly causes higher biodiversity in more productive environments. *science*, 328, 1388-1391

Chase, J. M., Kraft, N. J. B., Smith, K. G., Vellend, M. & Inouye, B. D. (2011) Using null models to disentangle variation in community dissimilarity from variation in α-diversity. *Ecosphere*, 2, 19-28

Cobb, T. P., Hannam, K. D., Kishchuk, B. E., Langor, D. W., Quideau, S. A. & Spence, J. R. (2010) Wood feeding beetles and soil nutrient cycling in burned forests: implications of post fire salvage logging. *Agricultural and Forest Entomology*, 12, 9-18

Connor, E. F. & Simberloff, D. (1983) Interspecific competition and species co-occurrence patterns on islands: null models and the evaluation of evidence. *Oikos*, 41, 455-465

Gotelli, N. J. (2001) Research frontiers in null model analysis. *Global Ecology and Biogeography*, 10, 337-343

Gotelli, N. J. & Graves, G. R. (1996) *Null Models in Ecology*, edn. Smithsonian Institution, Washington, DC

Grove, S. J. (2002) Saproxylic insect ecology and the sustainable management of forests. *Annual Review of Ecology and Systematics*, 33, 1-23

Harrison, S., Ross, S. J. & Lawton, J. H. (1992) Beta diversity on geographic gradients in Britain. *Journal of Animal Ecology*, 61, 151-158

Hayden, D., Lazar, P. & Schoenfeld, D. (2009) Assessing statistical significance in microarray experiments using the distance between microarrays. *PLoS One*, 4.

Hubbell, S. P. (2001) *The unified neutral theory of biodiversity and biogeography*, edn. Princeton University, Princiton, NJ.

Huston, M. A. (1994) *Biological diversity: the coexistence of species on changing landscapes.* Cambridge University Press, Cambridge, UK

Hylander, K., Nilsson, C., Gunnar Jonsson, B. & Göthner, T. (2005) Differences in habitat quality explain nestedness in a land snail meta community. *Oikos*, 108, 351-361.

Janssen, P., Hébert, C. & Fortin, D. (in press). Biodiversity conservation in old-growth boreal forest: black spruce and balsam fir snags harbour distinct assemblages of saproxylic beetles. Biodiversity and Conservation.

Kenis, M., Wermelinger, B. & Grégoire, J. (2004) Research on parasitoids and predators of Scolytidae–a review. *Bark and Wood Boring Insects in Living Trees in Europe, a Synthesis* (ed. by F. Lieutier & K.R. Day & A. Battisti & J.-C. Grégoire & H.F. Evans), pp 237-290. Kluwer Academic Publishers, Dordrecht

Koleff, P., Gaston, K. & Lennon, J. (2003) Measuring beta diversity for presence–absence data. *Journal of Animal Ecology*, 72, 367-382

Lande, R. (1996) Statistics and partitioning of species diversity, and similarity among multiple communities. *Oikos*, 76, 5-13

Legendre, P., Borcard, D. & Peres-Neto, P. R. (2008) Analyzing or explaining beta diversity? Comment. *Ecology*, 89, 3238-3244

Lichstein, J. (2007) Multiple regression on distance matrices: a multivariate spatial analysis tool. *Plant Ecology*, 188, 117-131

Lindenmayer, D., Burton, P. J. & Franklin, J. F. (2008) *Salvage logging and its ecological consequences*. Island Press, Washington, DC

Loreau, M. (2000) Are communities saturated? On the relationship between α, β and γ diversity. *Ecology Letters*, 3, 73-76

McCullough, D. G., Werner, R. A. & Neumann, D. (1998) Fire and insects in northern and boreal forest ecosystems on North America. *Annual Review of Entomology*, 43, 107-127

Miklós, I. & Podani, J. (2004) Randomization of presence-absence matrices: comments and new algorithms. *Ecology*, 85, 86-92.

Nekola, J. C. & White, P. S. (1999) The distance decay of similarity in biogeography and ecology. *Journal of Biogeography*, 26, 867-878

Oksanen, J., Kindt, R., Legendre, P., O'hara, B., Stevens, M. & Oksanen, M. (2010) The vegan package. *vegan: Community Ecology Package. http://CRAN.R-project.org/package/vega* (Accessed on 2010-03-03)

Patterson, B. D. & Atmar, W. (1986) Nested subsets and the structure of insular mammalian faunas and archipelagos. *Biological Journal of the Linnean Society*, 28, 65-82

R-Development-Team (2010) R: A language and environment for statistical computing. *R Foundation for Statistical Computing (Vienna, Austria) http://cran.r-project.org*

Raup, D. & Crick, R. (1979) Measurement of faunal similarity in paleontology. *Journal of Paleontology*, 53, 1213-1227

Ricklefs, R. E. (2004) A comprehensive framework for global patterns in biodiversity. *Ecology Letters*, 7, 1-15

Saint-Germain, M., Drapeau, P. & Hébert, C. (2004) Xylophagous insect species composition and patterns of substratum use on fire-killed black spruce in central Quebec. *Canadian Journal of Forest Research*, 34, 677-685

Tuomisto, H. (2010) A diversity of beta diversities: straightening up a concept gone awry. Part 2. Quantifying beta diversity and related phenomena. *Ecography*, 33, 23-45

Tuomisto, H. & Ruokolainen, K. (2006) Analyzing or explaining beta diversity? Understanding the targets of different methods of analysis. *Ecology*, 87, 2697-2708

Ulrich, W. & Gotelli, N. J. (2007) Disentangling community patterns of nestedness and species co occurrence. *Oikos*, 116, 2053-2061

Veech, J. & Crist, T. (2010) Diversity partitioning without statistical independence of alpha and beta. *Ecology*, 91, 1964-1969

Veech, J. A. & Crist, T. O. (2007) Habitat and climate heterogeneity maintain beta diversity of birds among landscapes within ecoregions. *Global Ecology and Biogeography*, 16, 650-656

Vellend, M., Verheyen, K., Flinn, K., Jacquemyn, H., Kolb, A., Van Calster, H., Peterken, G., Graae, B., Bellemare, J. & Honnay, O. (2007) Homogenization of forest plant communities and weakening of species–environment relationships via agricultural land use. *Journal of Ecology*, 95, 565-573

Whittaker, R. (1960) Vegetation of the Siskiyou mountains, Oregon and California. *Ecological Monographs*, 30, 279-338

Whittaker, R. (1972) Evolution and measurement of species diversity. *Taxon*, 21, 213-251

Wilson, M. & Shmida, A. (1984) Measuring beta diversity with presence-absence data. *The Journal of Ecology*, 72, 1055-1064

Zapala, M. & Schork, N. (2006) Multivariate regression analysis of distance matrices for testing associations between gene expression patterns and related variables. *Proceedings of the National Academy of Sciences*, 103, 19430

Part 2

History of Biodiversity

6

Twenty Landmark Papers in Biodiversity Conservation

Corey J. A. Bradshaw[1,2], Navjot S. Sodhi[3†], William F. Laurance[4,5] and Barry W. Brook[1]

[1]*The Environment Institute and School of Earth and Environmental Sciences, University of Adelaide, Adelaide,*
[2]*South Australian Research and Development Institute,*
[3]*Department of Biological Sciences, National University of Singapore,*
[4]*School of Marine and Tropical Biology, James Cook University, Cairns,*
[5]*Smithsonian Tropical Research Institute, Balboa, Ancón*
[1,2,4]*Australia*
[3]*Republic of Singapore*
[5]*Panama*

1. Introduction

Biodiversity conservation was first defined as a science less than three decades ago (Meine, 2010), but is now a well-developed, multidisciplinary research endeavour (Sodhi & Ehrlich, 2010). The consolidation of this 'crisis' field (Soulé, 1985) is inextricably tied to mounting global environmental degradation as the human enterprise now threatens most of the world's biodiversity and the ecosystems of which they are part (Ehrlich & Pringle, 2008). Although the history of the field is complex and its maturation gradual (Meine, 2010; Meine et al., 2006), a modest number of key ideas have subsequently sparked enormous progress in our understanding of biodiversity's response to human impacts, and how such knowledge might help avoid extinctions.

As the planet's biotic crisis escalates, we reflect on some of the most important research discoveries in biodiversity conservation science and its progenitor disciplines. Although this is a subjective list, our rationale was to select 20 papers that either built new paradigms or tore down old ones, and set thinking along new and interesting pathways towards biodiversity conservation. Other authors would no doubt list different papers, or challenge the true origin of certain ideas we highlight. Our goal here is simply to stimulate biodiversity scientists to think about what serious innovation looks like—with the benefit of perfect hindsight—and to use this retrospective to help guide future thinking.

In the remainder of this chapter, we briefly assess these 20 papers, but make no attempt to rank their relative importance - apart from a citation analysis (Fig. 1). Although we do not consider scientific citations alone reflect a paper's value sufficiently, it does provide a simple indication of its influence on research directions. Indeed, many of the highlighted papers have proposed ideas that have subsequently been discredited or rendered obsolete. This is a natural part of the progression of science. There is nonetheless little doubt that each of these papers has inspired new research paths and directed conservation interventions that have arguably benefited biodiversity.

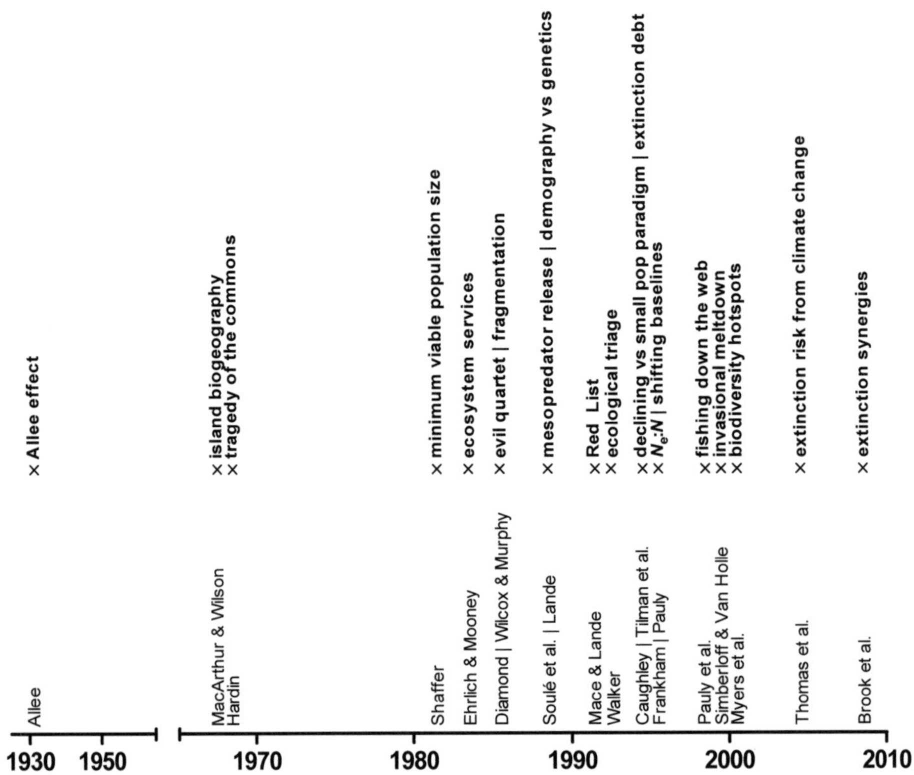

Fig. 1. Time line of the 20 landmark biodiversity conservation papers.

2. Landmark papers

2.1 Allee effect (Allee, 1931)

The 'Allee effect' can be broadly defined as a "…positive relationship between any component of individual fitness and either numbers or density of conspecifics" (Stephens et al., 1999). The idea is attributed to Warder Clyde Allee, an American ecologist from the early half of the 20th Century, although Odum (1953) first named it "Allee's principle". We consider Allee's 1931 book (Allee, 1931) to be the classic source (Fig. 2). Allee discussed the evidence for the effects of crowding on demographic and life history traits of populations, which he subsequently redefined as "inverse density dependence" (Allee, 1941).

Broadly speaking, when populations become small, a range of positive feedbacks can reduce a population's average fitness (measured in many ways, such as survival probability, reproductive rate, or growth rate). The many types of Allee effects (see Berec et al., 2007) can be mutually reinforcing (synergistic), and so drive populations even faster toward extinction than expected by their additive effects (Brook et al., 2008). Thus, ignoring potential Allee effects can compromise everything from estimates of minimum viable population size (Paper 4) to restoration attempts and predictions of extinction risk (Gregory et al., 2010).

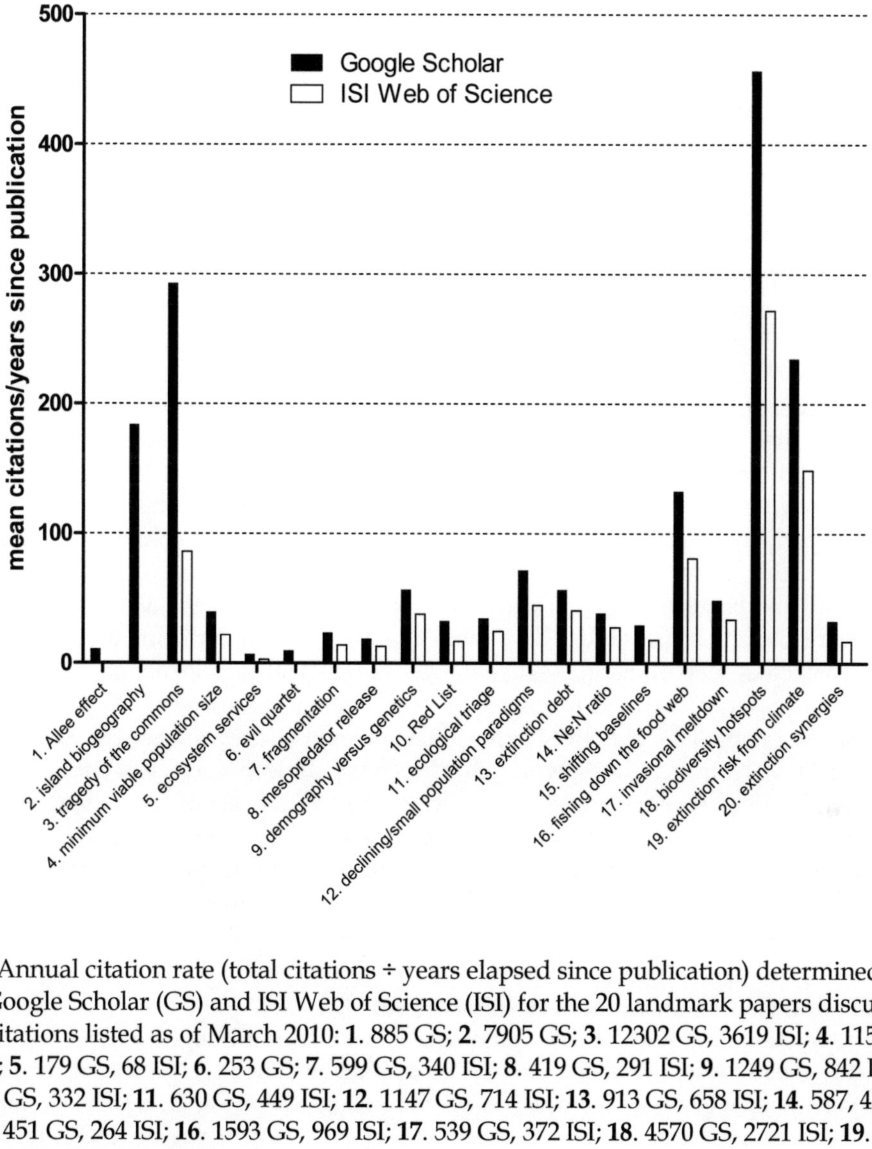

Fig. 2. Annual citation rate (total citations ÷ years elapsed since publication) determined from Google Scholar (GS) and ISI Web of Science (ISI) for the 20 landmark papers discussed. Total citations listed as of March 2010: **1**. 885 GS; **2**. 7905 GS; **3**. 12302 GS, 3619 ISI; **4**. 1152 GS, 624 ISI; **5**. 179 GS, 68 ISI; **6**. 253 GS; **7**. 599 GS, 340 ISI; **8**. 419 GS, 291 ISI; **9**. 1249 GS, 842 ISI; **10**. 625 GS, 332 ISI; **11**. 630 GS, 449 ISI; **12**. 1147 GS, 714 ISI; **13**. 913 GS, 658 ISI; **14**. 587, 426 ISI; **15**. 451 GS, 264 ISI; **16**. 1593 GS, 969 ISI; **17**. 539 GS, 372 ISI; **18**. 4570 GS, 2721 ISI; **19**. 1411 GS, 896 ISI; **20**. 65 GS, 34 ISI.

2.2 Island biogeography (MacArthur & Wilson, 1967)

This classic monograph (MacArthur & Wilson, 1967) could be considered the principal progenitor of modern conservation science considering its influence on reserve design and the estimation of extinction rates. The original theory proposed that the number of species on islands resulted from a dynamic equilibrium between the opposing forces of colonization (a function of island isolation) and extinction (a function of island size). Put simply: the bigger the island and the closer it is to the mainland (source), the more species it should contain. This ultimately led to the branch of biogeography and conservation biology that applied species-area relationships (Giam et al., 2010; Guilhaumon et al., 2008; Koh & Ghazoul, 2010; Koh et al., 2010) to habitat fragments to extrapolate total species number and—more importantly in the context of the modern extinction crisis—to estimate rates of species loss (Koh & Ghazoul, 2010; Koh et al., 2010; Laurance, 2008). The species-area relationship literature is a hot-bed of critique and polemic (Guilhaumon et al., 2008), yet no one can deny that this seminal work spawned the idea that reduced, fragmented, and isolated areas are bad for biodiversity (e.g., Mellin et al., 2010). We would not have as many nature reserves today were it not for this simple, yet foundational, piece of science.

2.3 Tragedy of the commons (Hardin, 1968)

Although not a conservation biology paper *per se*, Hardin's classic essay (Hardin, 1968) changed the way we think about managing natural resources that lack definitive ownership. The thesis of the "tragedy of the commons" is that individuals are inherently selfish and usually place their own interests first in using commonly owned resources, thereby resulting in their depletion. Hardin used a hypothetical and simplified situation based on medieval land tenure in Europe (herders sharing a common parcel of land) on which each herder was entitled to graze his cattle. Each herder maximized his gains by putting additional cattle onto the land, even if the carrying capacity of the common was exceeded and overgrazing ensued. The herder, by making an "individually rational decision," received all the benefits from his cattle, but could in the process deplete the common resource for the entire group. If all herders make such selfish decisions then the common will be depleted, jeopardizing the livelihoods of all. Hardin's paper is now a central paradigm in natural resources management (e.g., fisheries); however, his work has been criticized most notably by Elinor Ostrom - the first woman to be awarded the Noble Prize in economics in 2009. In her classic work, Ostrom (1990) showed that when communities are given the freedom to self-govern, they are, under certain conditions, able to use the commons sustainably. Another controversial theme of Hardin's paper is that an expanding human population is a detriment to the planet and its ability to support human existence, and thus he implies that humanity needs to be educated to relinquish the freedom to breed without limit (for more recent discussion see Ehrlich & Pringle, 2008).

2.4 Minimum viable population size (Shaffer, 1981)

Small populations are particularly vulnerable to extinction through random variation in birth and death rates, varying resource or habitat availability, predation, competitive interactions, single-event catastrophes and inbreeding (Gilpin & Soulé, 1986). Enter the concept of the minimum viable population (MVP) size, which was originally defined by Shaffer (1981) as the smallest number of individuals required for an isolated population to persist (at some predefined 'high' probability) for some 'long' time into the future. In other words, the MVP size is the number of individuals needed to withstand normal (expected) variation in all the processes that affect persistence.

This simple, yet fundamental, concept provides a quantitative target for species recovery, setting reserve sizes and sustainable harvest, if calculated correctly. A danger is that it encourages 'threshold'-based decision making that is not scientifically or ethically supportable (Gilpin, 1996). MVP size served as the basis for the infamous '50/500' rule—the notion that 50 breeding individuals are needed for short-term persistence of a population, and 500 for the maintenance of long-term genetic variability—which in turn has recently been updated to the '500/5000' rule (Traill et al., 2010a). Indeed, MVP is a concept implicitly underlying threatened species lists worldwide, including the IUCN Red List (see Paper 10). While various methodological issues, genetic considerations and policy implementation problems remain (Clements et al., 2011; Traill et al., 2010a), Shaffer's original paper (Shaffer, 1981) spawned an entire generation of research into quantitative risk assessment in conservation biology, and set the stage for deriving tangible, mathematically based conservation targets (Bradshaw & Brook, 2010).

2.5 Ecosystem services (Ehrlich & Mooney, 1983)

The concept embodied in Ehrlich and Mooney (1983) is that intact biological communities and functioning species interactions provide humanity with a host of 'services' that support or improve our quality of life. The ongoing assault on species and habitats around the globe are, in their words, "accompanied by severe degradation of the public service functions of the systems". But what are ecosystem services? The list is long, but still largely unquantified. It includes processes such as carbon storage, soil maintenance, crop pollination (Potts et al., 2010), seed dispersal, food, fuel, fibre and timber provision, water purification, flood regulation (Bradshaw et al., 2009; Bradshaw et al., 2007), disease prevention, waste decomposition and detoxification, nutrient cycling, pharmaceuticals, and cultural appreciation.

Valuing ecosystem services in any conventional sense is controversial because of the difficulty of measuring the link between ecosystem function and the services they provide, but also by the decision of whether to include direct or indirect costs of providing artificial replacement. Attempts to put them into financial terms (e.g., Costanza et al., 1997; Naidoo et al., 2008) have inevitably attracted criticism (e.g., Bockstael et al., 2000; Turner et al., 2003). Nonetheless, estimates of the gross value of ecosystem services are often remarkably high—totalling some trillions of dollars globally (European Communities, 2008)—with individual services such as insect pollination valued at more than $US200 billion per year (Gallai et al., 2009).

2.6 The evil quartet (Diamond, 1984)

The 'evil quartet', or 'four horsemen of the ecological apocalypse', was probably the first treatment of extinction dynamics as a biological discipline in its own right. Diamond (Diamond, 1984) took a sweeping historical and contemporary view of extinction, then simplified the problem to four principal mechanisms: 1. overhunting (or overexploitation), 2. introduced species, 3. habitat destruction and 4. chains of linked extinctions (trophic cascades, or co-extinctions). Far from a mere review or list of unrelated mechanisms, Diamond's evil quartet crystallized conservation biologists' thinking about key mechanisms and, more importantly, directed attention towards those factors likely to drive extinctions in the future. The unique combination of prehistorical through to modern examples gave conservation biologists a holistic view of extinction dynamics and helped spawn many of the papers described hereafter.

2.7 Habitat fragmentation (Wilcox & Murphy, 1985)

Despite the long-recognized notion that the more habitat you remove, the fewer species (and individuals) will persist (see Paper 2), it took decades to develop a strong consensus about the harmful consequences of fragmentation to biodiversity (Wilcox & Murphy, 1985). Put simply, it occurs when a contiguous expanse of habitat is broken up into a number of smaller, isolated patches. The surrounding area is typically defined a less-suitable 'matrix', and in the case of forest fragmentation, this generally means degraded habitat, fewer native species and urban, rural or agricultural development. Fragmentation is bad for many reasons: it 1. reduces patch area, 2. increases isolation among populations associated with fragments, 3. creates artificial 'edges' where unmodified habitat abuts matrix habitat (Saunders et al., 1991), and fundamentally alters the community and population dynamics for surviving species (Laurance, 2002). Each of these changes impacts habitat suitability, for we now know that 1. the smaller an area, the fewer individuals and species it can contain, 2. the more isolated a population, the less chance that immigrants will rescue it from catastrophes (Brown & Kodric-Brown, 1977), and 3. abrupt habitat edges allow the invasion of alien species, alter microclimatic conditions, increase access to people (e.g. hunters, loggers), and can lead to cascading ecological events, such as the penetration of fire (Cochrane & Laurance, 2002). The more fragmented an environment, the worse will be the extinction rates of species therein (Saunders et al., 1991).

2.8 Mesopredator release (Soulé et al., 1988)

Although popularized by Crooks and Soulé (1999), Soulé et al. (1988) first coined the term that described how ecosystems can be unbalanced by a reduction of a dominant predator that formerly exerted so-called 'top-down' control on the abundance of species at lower trophic levels. The idea had theoretical support (Litvaitis & Villafuerte, 1996; Wright et al., 1994) but it was Soulé and colleagues (1988) who first demonstrated the phenomenon. Crooks and Soulé (1999) described an example where the decline in coyotes (*Canis latrans*), in combination with urbanization-driven habitat fragmentation, leads to an increase in cat (*Felis catus*) densities and the subsequent decline in scrub-breeding birds. More recent examples attest to the broad importance of mesopredator release: Myers et al. (2007) linked the decline of coastal shark species to mesopredator release of cownose rays (*Rhinoptera bonasus*), thereby driving a reduction in commercially important shellfish densities; and Johnson et al. (2006) showed how dingoes (*Canis lupus dingo*) in Australia suppress populations of exotic predators such as cats and foxes, leading to more locally abundant populations of native marsupials.

2.9 Demography versus genetics (Lande, 1988)

In an influential review, Lande (1988) argued that "demography may usually be of more immediate importance than population genetics in determining the minimum viable size of wild populations". It was a well-reasoned case, and was widely interpreted to mean that demographic and ecological threats would provide the 'killer blow' to threatened species before genetic factors such as inbreeding and fitness effects of loss of genetic diversity had time to exert a major influence on small population dynamics. Lande's paper ignited a fire under the belly of conservation geneticists, and led to a concerted effort to come up with stronger evidence for the role of genetics in elevating extinction risk. This in turn resulted in innovative field experiments (Saccheri et al., 1998), meta-analyses on genetically effective population size (Paper 14), studies on the enhanced effect of inbreeding on wild populations (Crnokrak & Roff, 1999), and a pairwise comparison of 170 threatened taxa showing that the majority had indeed suffered from a reduction genetic diversity compared to their non-threatened congeners (Spielman et al., 2004). The net conclusion is that demographic and genetic changes can work in concert in small populations to threaten their viability and survival (Mills & Smouse, 1994).

2.10 Red List (Mace & Lande, 1991)

The IUCN Red List (www.iucnredlist.org) has been one of the most influential conservation policy tools constructed. For over 40 000 listed species, the Red List is the main instrument used to judge their status, extinction risk, and recovery potential worldwide. The Red List is focused predominantly on vertebrates and is still far from complete (with < 2 % of described species), but it is an evolving and improving expert assessment. It has become much more than a mere list - it is often used as a proxy for extinction risk (e.g., Bradshaw et al., 2008) and in cross-cutting analyses of threats to biodiversity (e.g., Laurance & Useche, 2009). From extinction theory to conservation policy, Mace and Lande's (1991) original definition of 'red list criteria' was a major step forward for biodiversity conservation.

2.11 Ecological triage (Walker, 1992)

Ecological triage (Walker, 1992) is taken from the medical term *triage* used in emergency or wartime situations. Ecological triage refers to the conservation prioritization of species that provide unique or necessary functions to ecosystems, and little focus on those that do not have unique ecosystem roles or that face almost certain extinction given that they fall well below their MVP size (Walker, 1992). Financial resources such as investment in recovery programs, purchase of land for reserves, and habitat restoration are allocated accordingly; the species that contribute most to ecosystem function and have the highest probability of persisting are earmarked for conservation, whereas others are largely ignored (Hobbs & Kristjanson, 2003). This pragmatic approach to conservation (labelled 'emotionally empty' or 'accounting-type' by some) is controversial because public favourites such as pandas, kakapo parrots and whales cannot be easily justified in terms of their intrinsically high costs and potentially limited benefits. According to the triage viewpoint, however, it makes no long-term conservation or economic sense to waste money on the doomed and ecologically redundant (Bottrill et al., 2008). Many in the conservation business apply ecological triage without being fully aware of it. Finite pools of money for conservation mean that we have to set priorities. Reserve design is just one example of this sacrifice-the-doomed-for-the good-of-the-ecosystem approach.

2.12 Declining and small population paradigms (Caughley, 1994)

Caughley's (1994) classic paper argued forcefully that two major paradigms had arisen in conservation biology, the 'declining' population paradigm and the 'small' population paradigm, yet only the former was truly scientific in allowing for the explicit testing of hypotheses. The declining population paradigm involves identifying and managing the processes that depress the demographic rates (e.g., survival and reproduction) of a species and cause its populations to decline deterministically, whereas the small population paradigm is the study of the dynamics of small populations that have declined from some (deterministic) perturbation and which are more susceptible to extinction via chance (stochastic) events. Caughley essentially argued that instead of focussing on theoretical aspects of extinction processes operating in small populations, we should instead focus our efforts on testing hypotheses exposing the drivers of population decline and arresting these. This proposition led to much debate (e.g., Hedrick et al., 1996) yet also fostered progress on more considered unification of these ideas and their applications to effective management goals such as minimum viable population size targets (Traill et al., 2010a). For instance, we now more clearly recognize that the forces that send populations into decline are not necessarily those that drive the final nail into a species' coffin. Hence, we must manage for both types of processes simultaneously, and the synergies between them (Brook et al., 2008; Laurance & Useche, 2009), if we want to reduce the likelihood of species going extinct.

2.13 Extinction debt (Tilman et al., 1994)

This paper (Tilman et al., 1994) shattered optimistic notions that the full consequences of human activity are obvious and amenable to quick fixes. 'Extinction debt' is the concept that, as habitats become increasingly fragmented, long-lived species that are reproductively isolated from conspecifics can take generations to perish (e.g., large trees in forest fragments). This gives rise to a higher number of species than would be otherwise expected for the size of the fragment, and the false impression that many species can persist in habitat patches that are too small to sustain MVPs (Clements et al., 2011; Traill et al., 2010a). These 'living dead' or 'zombie' species are therefore committed to extinction regardless of whether habitat loss is arrested or reversed. Only assisted dispersal (Hoegh-Guldberg et al., 2008) or artificially enhanced reproduction is likely to save such species. Neglecting the extinction debt is one reason why some people have over-estimated the value of fragmented and secondary forests in guarding species against extinction (see Brook et al., 2006). It means that data on species presence collected shortly after the main habitat degradation or destruction event could give misleading signals to conservation managers about the health of ecosystems. Appreciating that the extent of expected extinctions can take generations (potentially many centuries for long-lived organisms) to come to light, Tilman and colleagues' (1994) concept forced deeper thinking in the quest to minimize habitat loss and fragmentation.

2.14 N_e:N ratio (Frankham, 1995b)

Although it had long been recognized that inbreeding and loss of genetic diversity were accentuated in small, isolated populations (Charlesworth & Charlesworth, 1987), genetic hazards were generally considered to be of less consequence to extinction risk than demographic and environmental stochasticity (Paper 9). Frankham (1995b) helped overturn this viewpoint, using a meta-analysis to draw together comprehensive evidence on the ratio of genetically effective to actual population size (N_e:N). He assessed the effects of fluctuating population size, variance in family size and uneven sex ratios from 102 species. He found that the N_e:N ratio averaged only 10 % in wild populations, signalling that even apparently large and 'secure' populations might, in reality, already be suffering from inbreeding depression and thereby require genetic management to lower their extinction risk and retain evolutionary potential.

2.15 Shifting baselines (Pauly, 1995)

Pauly's (1995) concept describes the way that changes to a system are measured against baselines which themselves are often degraded versions of the original state of the system. Pauly (1995) originally meant it in a fisheries context, where "… fisheries scientists sometimes fail to identify the correct 'baseline' population size (e.g., how abundant a fish species population was before human exploitation) and thus work with a shifted baseline". It is now considered a mantra in fisheries and marine science (Jackson et al., 2001), but it has been extended to many other conservation issues. Yet, quantifying shifting baselines in conservation is difficult, with little empirical evidence (but see Jackson et al., 2001), despite the logic and general acceptance of its ubiquity by conservation scientists. Papworth et al. (2009) have recently addressed this knowledge gap by defining two kinds of shifting baselines: (1) general amnesia ("... individuals setting their perceptions from their own experience, and failing to pass their experience on to future generations") and (2) personal amnesia ("... individuals updating their own perception of normality; so that even those who experienced different previous conditions believe that current conditions are the same as past conditions"). Humans inevitably have short memories when it comes to environmental degradation, suggesting that real-world biodiversity declines are probably far worse than many scientists realize.

2.16 Fishing down the food web (Pauly et al., 1998)

The trend identified by Pauly and colleagues (1998) is simple but telling: data from the United Nations' Food and Agriculture Organization revealed that the average trophic level of harvested fish has declined globally by an overall average of 0.2 units (in the sea, trophic levels range from 1 for phytoplankton to 4.6 for top predators). The trends varied by region and whether or not one takes into account certain overrepresented species, but the average decline was more or less consistent across the dataset. This work revealed that fishing on a global commercial scale has been systematically removing the largest species and then moving to smaller species as each stock is exhausted. Many species not considered palatable even a short while ago are now high-demand market items. This is one of the first high-impact papers to show that the over-fishing problem was global in extent and massive in magnitude.

2.17 Invasional meltdown (Simberloff & Von Holle, 1999)

This seminal paper (Simberloff & Von Holle, 1999) describes the mutually reinforcing ways in which non-indigenous species become invasive, transform habitats, degrade ecosystem services (Paper 5), and reduce biodiversity and thereby constitute one of the greatest threats to ecosystems today—perhaps nearly as important as habitat loss and over-exploitation (Sodhi et al., 2009). Invasional meltdown describes the process by which the negative impacts induced on native ecosystems by one invading species are exacerbated by interactions with another. Although direct information on invasion meltdowns is limited, one good example is the introduction of the yellow crazy ant (*Anoplolepis gracilipes*) on Christmas Island, leading to the decline of native red crabs (*Gecarcoidea natalis*), the dominant terrestrial herbivore. This in turn led to a proliferation of forest undergrowth from reduced crab herbivory (O'Dowd et al., 2003) and altered habitat suitability for native plant species. Invasive species interactions will be reinforced by global climate change through increasing temperatures, rising sea levels and changing rainfall patterns that increase the potential range and spread of new invading species (Brook et al., 2008), so the problem is likely to worsen.

2.18 Biodiversity hotspots (Myers et al., 2000)

Although published only a decade ago, Myers and colleagues' (2000) 'hotspots' concept has already become part of the classic conservation lexicon, defining areas with high species endemism and severe degradation by humans. These are places where, at current rates of habitat loss and exploitation, we stand to lose far more irreplaceable species than in similar habitats elsewhere (the concept was originally applied to terrestrial environments but later extended to the marine realm) (Roberts et al., 2002). The concept has been criticized for its inability to account for all types of threats - indeed, many other prioritization criteria have been proposed (e.g., Brooks et al., 2006; Orme et al., 2005) - but the general concept has helped to set conservation priorities for many countries.

2.19 Extinction risk from climate change (Thomas et al., 2004)

Throughout the 1990s and into the new millennium, ecologists and conservation biologists accumulated many case studies quantifying the extent to which species had shifted in their geographic ranges, phenology and behaviour in response to a rapidly warming planet (Parmesan & Yohe, 2003). However, it was not until the paper by Thomas and colleagues (2004) that the magnitude of potential species extinctions arising from climate change was projected on a global scale. While simplified and controversial (e.g., Akçakaya et al., 2006; Thuiller et al., 2004), the study has helped to catalyse the exponential rise in climate change research in conservation biology (see Parmesan, 2006; Traill et al., 2010b for recent reviews). Indeed, climate change is now seen as one of the major potential drivers of mass extinction over the next few centuries, causing Diamond's evil quartet (Paper 6) to be updated to an 'evil quintet' (Brook et al., 2008).

2.20 Extinction synergies (Brook et al., 2008)

We have elected to finish this list with one of our own papers (Brook et al., 2008). This was the first synthetic compilation of the evidence for extinction 'synergies' or 'multiple' stressors. Although the concept was highlighted over twenty years ago (Myers, 1987), and elaborated later in the context of deep-time extinctions (McKinney, 1997), it seems to have been largely ignored by conservation biologists until only recently (Laurance & Useche, 2009). The concept builds on a number of ideas discussed in the other 19 papers: that separate drivers of extinction, such as habitat loss, over-exploitation, climate change, and invasive species, tend to conspire through mutually reinforcing feedbacks to heighten the extinction probability of the species they affect in a way that is greater than the sum of the individual effects. Our review compiles evidence from field studies, experiments and meta-analyses that demonstrate the amplifying actions of multiple drivers of population decline and extinction. We argued that conservation actions only targeting single drivers will usually be inadequate (and therefore potentially pointless) due to the cascading effects caused by unmanaged synergies. Time will tell if we are right.

3. Conclusion

Our list of 20 landmark papers in biodiversity conservation is but a tiny slice of the huge number of important research papers published in the field over the last 30 years. Although one could dispute some our specific choices, there is no doubt that the papers and ideas listed herein were highly influential and helped shape the advance of conservation biology. We excluded other papers because of the difficulty of attributing the original idea to a single source. Examples include papers discussing ecosystem stability (May, 1974), metapopulation dynamics (Hanski, 1998), inbreeding depression (Charlesworth & Charlesworth, 1987), reserve selection (Pressey et al., 1993), estimating biodiversity (Colwell & Coddington, 1994), the future of biodiversity (Pimm et al., 1995), and conservation genetics (Frankham, 1995a).

We speculate that, ironically, many of the authors we highlighted were probably not genuinely aware of the paradigm-shifting impact of their work at the time of publication. With this in mind, we hope that equally inspiring papers will revolutionize our thinking and actions in the coming decades. Such examinations of past leaps forward, combined with research wish lists (Sutherland et al., 2009) and future horizon-scanning exercises (e.g., Sutherland et al., 2010), will inspire the next generation of conservation-science innovators to take the field forward.

4. Acknowledgements

This article was inspired by CJAB's blog http://ConservationBytes.com. Funded in part by an Australian Research Council Discovery Grant (DP0878582) to CJAB and NSS. Dedicated to the memory of Navjot S. Sodhi who died on 12 June 2011.

5. References

Akçakaya, H.R.; Butchart, S.H.M.; Mace, G.M.; Stuart, S.N. & Hilton-Taylor, C. (2006). Use and misuse of the IUCN Red List Criteria in projecting climate change impacts on biodiversity. *Global Change Biology*, Vol.12, No.11, pp. 2037-2043, ISSN 1365-2486

Allee, W.C. (1931). *Animal Aggregations: A Study in General Sociology*, University of Chicago Press, ISBN 0404145019, Chicago, USA

Allee, W.C. (1941). Integration of problems concerning protozoan populations with those of general biology. *American Naturalist*, Vol.75, No.760, pp. 473-487, ISSN 0003-0147

Berec, L.; Angulo, E. & Courchamp, F. (2007). Multiple Allee effects and population management. *Trends in Ecology and Evolution*, Vol.22, No.4, pp. 185-191, ISSN 0169-5347

Bockstael, N.E.; Freeman, A.M.; Kopp, R.J.; Portney, P.R. & Smith, V.K. (2000). On measuring economic values for nature. *Environmental Science and Technology*, Vol.34, No.8, pp. 1384-1389, ISSN 0013-936X

Bottrill, M.C.; Joseph, L.N.; Carwardine, J.; Bode, M.; Cook, C.; Game, E.T.; Grantham, H.; Kark, S.; Linke, S.; McDonald-Madden, E.; Pressey, R.L.; Walker, S.; Wilson, K.A. & Possingham, H.P. (2008). Is conservation triage just smart decision making? *Trends in Ecology and Evolution*, Vol.23, No.12, pp. 649-654, ISSN 0169-5347

Bradshaw, C.J.A. & Brook, B.W. (2010). The conservation biologist's toolbox – principles for the design and analysis of conservation studies, In: *Conservation Biology for All*, N.S. Sodhi & P.R. Ehrlich (Eds.), 313-334, Oxford University Press, ISBN 978-0199554249, Oxford

Bradshaw, C.J.A.; Brook, B.W.; Peh, K.S.-H. & Sodhi, N.S. (2009). Flooding policy makers with evidence to save forests. *Ambio*, Vol.38, No.2, pp. 125-126, ISSN 0044-7447

Bradshaw, C.J.A.; Giam, X.; Tan, H.T.W.; Brook, B.W. & Sodhi, N.S. (2008). Threat or invasive status in legumes is related to opposite extremes of the same ecological and life-history attributes. *Journal of Ecology*, Vol.96, No.5, pp. 869-883, ISSN 0022-0477

Bradshaw, C.J.A.; Sodhi, N.S.; Peh, K.S.H. & Brook, B.W. (2007). Global evidence that deforestation amplifies flood risk and severity in the developing world. *Global Change Biology*, Vol.13, No.11, pp. 2379-2395, ISSN 1466-8238

Brook, B.W.; Bradshaw, C.J.A.; Koh, L.P. & Sodhi, N.S. (2006). Momentum drives the crash: mass extinction in the tropics. *Biotropica*, Vol.38, No.3, pp. 302-305, ISSN 0006-3606

Brook, B.W.; Sodhi, N.S. & Bradshaw, C.J.A. (2008). Synergies among extinction drivers under global change. *Trends in Ecology and Evolution*, Vol.25, No.8, pp. 453-460, ISSN 0169-5347

Brooks, T.M.; Mittermeier, R.A.; da Fonseca, G.A.B.; Gerlach, J.; Hoffmann, M.; Lamoreux, J.F.; Mittermeier, C.G.; Pilgrim, J.D. & Rodrigues, A.S.L. (2006). Global biodiversity conservation priorities. *Science*, Vol.313, No.5783, pp. 58-61, ISSN 0036-8075

Brown, J.H. & Kodric-Brown, A. (1977). Turnover rates in insular biogeography: effect of immigration on extinction. *Ecology*, Vol.58, No.2, pp. 445-449, ISSN 0012-9658

Caughley, G. (1994). Directions in conservation biology. *Journal of Animal Ecology*, Vol.63, No.2, pp. 215-244, ISSN 0021-8790

Charlesworth, D. & Charlesworth, B. (1987). Inbreeding depression and its evolutionary consequences. *Annual Review of Ecology and Systematics*, Vol.18, No.1, pp. 237-268, ISSN 0066-4162

Clements, G.R.; Bradshaw, C.J.A.; Brook, B.W. & Laurance, W.F. (2011). The SAFE index: using a threshold population target to measure relative species threat. *Frontiers in Ecology and the Environment*, Vol.In press, pp., ISSN 1540-9295

Cochrane, M.A. & Laurance, W.F. (2002). Fire as a large-scale edge effect in Amazonian forests. *Journal of Tropical Ecology*, Vol.18, No.3, pp. 311-325, ISSN 0266-4674

Colwell, R.K. & Coddington, J.A. (1994). Estimating terrestrial biodiversity through extrapolation. *Philosophical Transactions of the Royal Society B*, Vol.345, No.1311, pp. 101-118, ISSN 0080-4622

Costanza, R.; d'Arge, R.; de Groot, R.; Farber, S.; Grasso, M.; Hannon, B.; Limburg, K.; Naeem, S.; O'Neill, R.V.; Paruelo, J.; Raksin, R.G.; Sutton, P. & van den Belt, M. (1997). The value of the world's ecosystem services and natural capital. *Nature*, Vol.387, No.6630, pp. 253-260, ISSN 0028-0836

Crnokrak, P. & Roff, D.A. (1999). Inbreeding depression in the wild. *Heredity*, Vol.83, No.3, pp. 260-270, ISSN 0018-067X

Crooks, K.R. & Soulé, M.E. (1999). Mesopredator release and avifaunal extinctions in a fragmented system. *Nature*, Vol.400, No.6744, pp. 563-566, ISSN 0028-0836

Diamond, J.M. (1984). 'Normal' extinction of isolated populations, In: *Extinctions*, M.H. Nitecki (Ed.), 191-246, Chicago University Press, ISBN 0226586901, Chicago, USA

Ehrlich, P.R. & Mooney, H.A. (1983). Extinction, substitution, and ecosystem services. *BioScience*, Vol.33, No.4, pp. 248-254, ISSN 0006-3568

Ehrlich, P.R. & Pringle, R.M. (2008). Where does biodiversity go from here? A grim business-as-usual forecast and a hopeful portfolio of partial solutions. *Proceedings of the National Academy of Sciences of the USA*, Vol.105, No.Supplement 1, pp. 11579-11586, ISSN 0027-8424

European Communities (2008). The Economics of Biodiversity and Ecosystems. An Interim Report, European Communities, Wesseling, Germany, Available from http://ec.europa.eu/environment/nature/biodiversity/economics/

Frankham, R. (1995a). Conservation genetics. *Annual Review of Genetics*, Vol.29, No.1, pp. 305-327, ISSN 0066-4197

Frankham, R. (1995b). Effective population size/adult population size ratios in wildlife: a review. *Genetical Research*, Vol.66, No.2, pp. 95-107, ISSN 0016-6723

Gallai, N.; Salles, J.-M.; Settele, J. & Vaissière, B.E. (2009). Economic valuation of the vulnerability of world agriculture confronted with pollinator decline. *Ecological Economics*, Vol.68, No.3, pp. 810-821, ISSN 0921-8009

Giam, X.; Sodhi, N.S.; Brook, B.W.; Tan, H.T.W. & Bradshaw, C.J.A. (2010). Relative need for conservation assessment of vascular plant species among ecoregions. *Journal of Biogeography*, Vol.38, No.1, pp. 55-68, ISSN 1365-2699

Gilpin, M. (1996). Forty-eight parrots and the origins of population viability analysis. *Conservation Biology*, Vol.10, No.6, pp. 1491-1493, ISSN 0888-8892

Gilpin, M.E. & Soulé, M.E. (1986). Minimum viable populations: processes of species extinction, In: *Conservation Biology: The Science of Scarcity and Diversity*, M.E. Soulé (Ed.), 19-34, Sinauer, ISBN 0878937951, Sunderland, USA

Gregory, S.; Bradshaw, C.J.A.; Brook, B.W. & Courchamp, F. (2010). Limited evidence for the demographic Allee effect from numerous species across taxa. *Ecology*, Vol.91, No.7, pp. 2151–2161, ISSN 0012-9658

Guilhaumon, F.; Gimenez, O.; Gaston, K.J. & Mouillot, D. (2008). Taxonomic and regional uncertainty in species-area relationships and the identification of richness hotspots. *Proceedings of the National Academy of Sciences of the USA*, Vol.105, No.40, pp. 15458-15463, ISSN 0027-8424

Hanski, I. (1998). Metapopulation dynamics. *Nature*, Vol.396, No.6706, pp. 41-49, ISSN 0028-0836

Hardin, G. (1968). The tragedy of the commons. *Science*, Vol.162, No.3859, pp. 1243-1248, ISSN 0036-8075

Hedrick, P.W.; Lacy, R.C.; Allendorf, F.W. & Soulé, M.E. (1996). Directions in conservation biology: comments on Caughley. *Conservation Biology*, Vol.10, No.5, pp. 1312-1320, ISSN 0888-8892

Hobbs, R.J. & Kristjanson, L.J. (2003). Triage: how do we prioritize health care for landscapes? *Ecological Management and Restoration*, Vol.4, No.Supplement S1, pp. S39-S45, ISSN 1442-8903

Hoegh-Guldberg, O.; Hughes, L.; McIntyre, S.; Lindenmayer, D.B.; Parmesan, C.; Possingham, H.P. & Thomas, C.D. (2008). Assisted colonization and rapid climate change. *Science*, Vol.321, No.5887, pp. 345-346, ISSN 0036-8075

Jackson, J.B.C.; Kirby, M.X.; Berger, W.H.; Bjorndal, K.A.; Botsford, L.W.; Bourque, B.J.; Bradbury, R.H.; Cooke, R.; Erlandson, J.; Estes, J.A.; Hughes, T.P.; Kidwell, S.; Lange, C.B.; Lenihan, H.S.; Pandolfi, J.M.; Peterson, C.H.; Steneck, R.S.; Tegner, M.J. & Warner, R.R. (2001). Historical overfishing and the recent collapse of coastal ecosystems. *Science*, Vol.293, No.5530, pp. 629-638, ISSN 0036-8075

Johnson, C.N.; Isaac, J.L. & Fisher, D.O. (2006). Rarity of a top predator triggers continent-wide collapse of mammal prey: dingoes and marsupials in Australia. *Proceedings of the Royal Society B: Biological Sciences*, Vol.274, No.1608, pp. 341-346, ISSN 1471-2954

Koh, L.P. & Ghazoul, J. (2010). A matrix-calibrated species-area model for predicting biodiversity losses due to land-use change. *Conservation Biology*, Vol.24, No.4, pp. 994-1001, ISSN 0888-8892

Koh, L.P.; Lee, T.M.; Sodhi, N.S. & Ghazoul, J. (2010). An overhaul of the species–area approach for predicting biodiversity loss: incorporating matrix and edge effects. *Journal of Applied Ecology*, Vol.47, No.5, pp. 1063-1070, ISSN 1365-2664

Lande, R. (1988). Genetics and demography in biological conservation. *Science*, Vol.241, No.4872, pp. 1455-1460, ISSN 0036-8075

Laurance, W.F. (2002). Hyperdynamism in fragmented habitats. *Journal of Vegetation Science*, Vol.13, No.4, pp. 595-602, ISSN 1654-1103

Laurance, W.F. (2008). Theory meets reality: how habitat fragmentation research has transcended island biogeographic theory. *Biological Conservation*, Vol.141, No.7, pp. 1731-1744, ISSN 0006-3207

Laurance, W.F. & Useche, D.C. (2009). Environmental synergisms and extinctions of tropical species. *Conservation Biology*, Vol.23, No.6, pp. 1427-1437, ISSN 0888-8892

Litvaitis, J.A. & Villafuerte, R. (1996). Intraguild predation, mesopredator release, and prey stability. *Conservation Biology*, Vol.10, No.2, pp. 676-677, ISSN 0888-8892

MacArthur, R.H. & Wilson, E.O. (1967). *The Theory of Island Biogeography*, Princeton University Press, ISBN 0691088365, Princeton, USA

Mace, G.M. & Lande, R. (1991). Assessing extinction threats: toward a re-evaluation of IUCN threatened species categories. *Conservation Biology*, Vol.5, No.2, pp. 148-157, ISSN 0888-8892

May, R.M. (1974). *Stability and Complexity in Model Ecosystems*, Princeton University Press, ISBN 0691088616, Princeton, USA

McKinney, M.L. (1997). Extinction vulnerability and selectivity: combining ecological and paleontological views. *Annual Review of Ecology and Systematics*, Vol.28, No.1, pp. 495-516, ISSN 0066-4162

Meine, C. (2010). Conservation biology: past and present, In: *Conservation Biology for All*, N.S. Sodhi & P.R. Ehrlich (Eds.), 7-26, Oxford University Press, ISBN 0199554234, Oxford, United Kingdom

Meine, C.; Soulé, M. & Noss, R.F. (2006). "A mission-driven discipline": the growth of conservation biology. *Conservation Biology*, Vol.20, No.3, pp. 631-651, ISSN 0888-8892

Mellin, C.; Huchery, C.; Caley, M.J.; Meekan, M.G. & Bradshaw, C.J.A. (2010). Reef size and isolation determine the temporal stability of coral reef fish populations. *Ecology*, Vol.91, No.11, pp. 3138-3145, ISSN 0012-9658

Mills, L.S. & Smouse, P.E. (1994). Demographic consequences of inbreeding in remnant populations. *American Naturalist*, Vol.144, No.3, pp. 412-431, ISSN 0003-0147

Myers, N. (1987). The extinction spasm impending: synergisms at work. *Conservation Biology*, Vol.1, No.1, pp. 14-21, ISSN 0888-8892

Myers, N.; Mittermeier, R.A.; Mittermeier, C.G.; da Fonseca, G.A.B. & Kent, J. (2000). Biodiversity hotspots for conservation priorities. *Nature*, Vol.403, No.6772, pp. 853-858, ISSN 0028-0836

Myers, R.A.; Baum, J.K.; Shepherd, T.D.; Powers, S.P. & Peterson, C.H. (2007). Cascading effects of the loss of apex predatory sharks from a coastal ocean. *Science*, Vol.315, No.5820, pp. 1846-1850, ISSN 0036-8075

Naidoo, R.; Balmford, A.; Costanza, R.; Fisher, B.; Green, R.E.; Lehner, B.; Malcolm, T.R. & Ricketts, T.H. (2008). Global mapping of ecosystem services and conservation priorities. *Proceedings of the National Academy of Sciences of the USA*, Vol.105, No.28, pp. 9495-9500, ISSN 0027-8424

O'Dowd, D.J.; Green, P.T. & Lake, P.S. (2003). Invasional 'meltdown' on an oceanic island. *Ecology Letters*, Vol.6, No.9, pp. 812-817, ISSN 1461-0248

Odum, E. (1953). *Fundamentals of Ecology*, WB Saunders Co., ISBN 0534420664, Philadelphia, USA

Orme, C.D.L.; Davies, R.G.; Burgess, M.; Eigenbrod, F.; Pickup, N.; Olson, V.A.; Webster, A.J.; Ding, T.-S.; Rasmussen, P.C.; Ridgely, R.S.; Stattersfield, A.J.; Bennett, P.M.; Blackburn, T.M.; Gaston, K.J. & Owens, I.P.F. (2005). Global hotspots of species richness are not congruent with endemism or threat. *Nature*, Vol.436, No.7053, pp. 1016-1019, ISSN 0028-0836

Ostrom, E. (1990). *Governing the Commons: The Evolution of Institutions for Collective Action*, Cambridge University Press, ISBN 0521405998, Cambridge, United Kingdom

Papworth, S.K.; Rist, J.; Coad, L. & Milner-Gulland, E.J. (2009). Evidence for shifting baseline syndrome in conservation. *Conservation Letters*, Vol.2, No.2, pp. 93-100, ISSN 1755-263X

Parmesan, C. (2006). Ecological and evolutionary responses to recent climate change. *Annual Reviews in Ecology, Evolution and Systematics*, Vol.37, No.1, pp. 637-669, ISSN 0066-4162

Parmesan, C. & Yohe, G. (2003). A globally coherent fingerprint of climate change impacts across natural systems. *Nature*, Vol.421, No.6918, pp. 37-42, ISSN 0028-0836

Pauly, D. (1995). Anecdotes and the shifting baseline syndrome of fisheries. *Trends in Ecology and Evolution*, Vol.10, No.10, pp. 430, ISSN 0169-5347

Pauly, D.; Christensen, V.; Dalsgaard, J.; Froese, R. & Torres Jr., F. (1998). Fishing down marine food webs. *Science*, Vol.279, No.5352, pp. 860-863, ISSN 0036-8075

Pimm, S.L.; Russell, G.J.; Gittleman, J.L. & Brooks, T.M. (1995). The future of biodiversity. *Science*, Vol.269, No.5222, pp. 347-350, ISSN 0036-8075

Potts, S.G.; Biesmeijer, J.C.; Kremen, C.; Neumann, P.; Schweiger, O. & Kunin, W.E. (2010). Global pollinator declines: trends, impacts and drivers. *Trends in Ecology and Evolution*, Vol.25, No.6, pp. 345-353, ISSN 0169-5347

Pressey, R.L.; Humphries, C.J.; Margules, C.R.; Vane-Wright, R.I. & Williams, P.H. (1993). Beyond opportunism: key principles for systematic reserve selection. *Trends in Ecology and Evolution*, Vol.8, No.4, pp. 124-128, ISSN 0169-5347

Roberts, C.M.; McClean, C.J.; Veron, J.E.N.; Hawkins, J.P.; Allen, G.R.; McAllister, D.E.; Mittermeier, C.G.; Schueler, F.W.; Spalding, M.; Wells, F.; Vynne, C. & Werner, T.B. (2002). Marine biodiversity hotspots and conservation priorities for tropical reefs. *Science*, Vol.295, No.5558, pp. 1280-1284, ISSN 0036-8075

Saccheri, I.; Kuussaari, M.; Kankare, M.; Vikman, P.; Fortelius, W. & Hanski, I. (1998). Inbreeding and extinction in a butterfly metapopulation. *Nature*, Vol.392, No.6675, pp. 491-494, ISSN 0028-0836

Saunders, D.A.; Hobbs, R.J. & Margules, C.R. (1991). Biological consequences of ecosystem fragmentation: a review. *Conservation Biology*, Vol.5, No.1, pp. 18-32, ISSN 0888-8892

Shaffer, M.L. (1981). Minimum population sizes for species conservation. *BioScience*, Vol.31, No.2, pp. 131-134, ISSN 0006-3568

Simberloff, D. & Von Holle, B. (1999). Positive interactions of nonindigenous species: invasional meltdown? *Biological Invasions*, Vol.1, No.1, pp. 21-32, ISSN 1387-3547

Sodhi, N.S.; Brook, B.W. & Bradshaw, C.J.A. (2009). Causes and consequences of species extinctions, In: *The Princeton Guide to Ecology*, S.A. Levin (Ed.), 514-520, Princeton University Press, ISBN 978-0-691-12839-9, Princeton, USA

Sodhi, N.S. & Ehrlich, P.R. (2010). *Conservation Biology for All*, Oxford University Press, ISBN 0199554234, Oxford, United Kingdom

Soulé, M. (1985). What is conservation biology? *BioScience*, Vol.35, No.11, pp. 727-734, ISSN 0006-3568

Soulé, M.E.; Bolger, D.T.; Alberts, A.C.; Sauvajot, R.S.; Wright, J.; Sorice, M. & Hill, S. (1988). Reconstructed dynamics of rapid extinctions of chaparral-requiring birds in urban habitat islands. *Conservation Biology*, Vol.2, No.1, pp. 75-92, ISSN 0888-8892

Spielman, D.; Brook, B.W. & Frankham, R. (2004). Most species are not driven to extinction before genetic factors impact them. *Proceedings of the National Academy of Sciences of the USA*, Vol.101, No.42, pp. 15261-15264, ISSN 0027-8424

Stephens, D.W.; Sutherland, W.J. & Freckleton, R.P. (1999). What is the Allee effect? *Oikos*, Vol.87, No.1, pp. 185-190, ISSN 0030-1299

Sutherland, W.J.; Adams, W.M.; Aronson, R.B.; Aveling, R.; Blackburn, T.M.; Broad, S.; Ceballos, G.; Côté, I.M.; Cowling, R.M.; Da Fonseca, G.A.B.; Dinerstein, E.; Ferraro, P.J.; Fleishman, E.; Gascon, C.; Hunter Jr., M.; Hutton, J.; Kareiva, P.; Kuria, A.; Macdonald, D.W.; Mackinnon, K.; Madgwick, F.J.; Mascia, M.B.; Mcneely, J.; Milner-Gulland, E.J.; Moon, S.; Morley, C.G.; Nelson, S.; Osborn, D.; Pai, M.; Parsons, E.C.M.; Peck, L.S.; Possingham, H.; Prior, S.V.; Pullin, A.S.; Rands, M.R.W.; Ranganathan, J.; Redford, K.H.; Rodriguez, J.P.; Seymour, F.; Sobel, J.; Sodhi, N.S.; Stott, A.; Vance-Borland, K. & Watkinson, A.R. (2009). One hundred questions of importance to the conservation of global biological diversity. *Conservation Biology*, Vol.23, No.3, pp. 557-567, ISSN 1523-1739

Sutherland, W.J.; Clout, M.; Côté, I.M.; Daszak, P.; Depledge, M.H.; Fellman, L.; Fleishman, E.; Garthwaite, R.; Gibbons, D.W.; De Lurio, J.; Impey, A.J.; Lickorish, F.; Lindenmayer, D.; Madgwick, J.; Margerison, C.; Maynard, T.; Peck, L.S.; Pretty, J.; Prior, S.; Redford, K.H.; Scharlemann, J.P.W.; Spalding, M. & Watkinson, A.R. (2010). A horizon scan of global conservation issues for 2010. *Trends in Ecology and Evolution*, Vol.25, No.1, pp. 1-7, ISSN 0169-5347

Thomas, C.D.; Cameron, A.; Green, R.E.; Bakkenes, M.; Beaumont, L.J.; Collingham, Y.C.; Erasmus, B.F.N.; de Siqueira, M.F.; Grainger, A.; Hannah, L.; Hughes, L.; Huntley, B.; van Jaarsveld, A.S.; Midgley, G.F.; Miles, L.; Ortega-Huerta, M.A.; Peterson, A.T.; Phillips, O.L. & Williams, S.E. (2004). Extinction risk from climate change. *Nature*, Vol.427, No.6970, pp. 145-148, ISSN 0028-0836

Thuiller, W.; Araújo, M.B.; Pearson, R.G.; Whittaker, R.J.; Brotons, L. & Lavorel, S. (2004). Biodiversity conservation: uncertainty in predictions of extinction risk. *Nature*, Vol.430, No.6995, pp. doi:10.1038/nature02716, ISSN 0028-0836

Tilman, D.; May, R.M.; Lehman, C.L. & Nowak, M.A. (1994). Habitat destruction and the extinction debt. *Nature*, Vol.371, No.6492, pp. 65-66, ISSN 0028-0836

Traill, L.W.; Brook, B.W.; Frankham, R. & Bradshaw, C.J.A. (2010a). Pragmatic population viability targets in a rapidly changing world. *Biological Conservation*, Vol.143, pp. 28-34, ISSN 0006-3207

Traill, L.W.; Lim, M.L.M.; Sodhi, N.S. & Bradshaw., C.J.A. (2010b). Mechanisms driving change: altered species interactions and ecosystem functions from global warming. *Journal of Animal Ecology*, Vol.79, No.5, pp. 937-947, ISSN 0021-8790

Turner, R.K.; Paavola, J.; Cooper, P.; Farber, S.; Jessamy, V. & Georgiou, S. (2003). Valuing nature: lessons learned and future research directions. *Ecological Economics*, Vol.46, No.3, pp. 493-510, ISSN 0921-8009

Walker, B.H. (1992). Biodiversity and ecological redundancy. *Conservation Biology*, Vol.6, No.1, pp. 18-23, ISSN 0888-8892

Wilcox, B.A. & Murphy, D.D. (1985). Conservation strategy: the effects of fragmentation on extinction. *American Naturalist*, Vol.125, No.6, pp. 879-887, ISSN 0003-0147

Wright, S.J.; Gompper, M.E. & DeLeon, B. (1994). Are large predators keystone species in Neotropical forests? The evidence from Barro Colorado Island. *Oikos*, Vol.71, No.2, pp. 279-294, ISSN 0030-1299

Part 3

Evaluating Biodiversity

7

Evaluating Biodiversity and Spatial Simulation of Land-Cover Change in the Tropical Region of Western Ghats, India

Giriraj A.[1,2,3], Murthy M.S.R.[2] and Basanta Shrestha[3]
[1]*International Water Management Institute (IWMI), Colombo,*
[2]*National Remote Sensing Centre, Hyderabad,*
[3]*International Centre for Integrated Mountain Development (ICIMOD), Kathmandu,*
[1]*Sri Lanka*
[2]*India*
[3]*Nepal*

1. Introduction

Excessive alterations of the global environment by human activities have led to various changes in global biochemical cycles, and transformation of land, and have increased the mobility of the biota. These anthropogenically-induced changes have triggered the sixth major extinction event in the history of life on earth, and have caused widespread changes in the global distribution of organisms (Sala et al., 2000; Midgley et al., 2002; Parmesan & Yohe, 2003; Root et al., 2003). With an increase in resource requirement, more and more natural areas (virgin forests/landscape) is being encroached upon by humans and their resources are exploited, thus leading to a loss of biological diversity. Much of the attempt is on the study and documentation of forest degradation in the tropics, and is not actually related to a quantitative spatial assessment of species loss and composition. It is therefore necessary to determine accurately the rate of such loss, and its spatial patterns to formulate sustainable strategies for conserving and monitoring relatively undisturbed landscapes.

Conservation approaches primarily focus on targeting vegetation types, economic/ ecological species, habitats and landscape units. In any region, land conversion forces the declining population towards the edge of their species range, where they become increasingly vulnerable, and collapse if exposed to further human impact (Channell & Lomolino, 2000). Temporal and spatial mapping and modeling of potential species distribution (e.g. *Ephedra gerardiana; Porwal* et al., 2003, *Hippophae rhamnoides;* Roy et al., 2001) helps to quantify and understand the current status, and to assess species loss to specific habitats and landscape. Amarnath et al., 2003, have also shown that conjunctive analysis of patch characteristics and species distribution can be used in identifying the areas of priority in terms of eco-restoration and conservation in wet evergreen forests of India. Thus, a systematic planning is necessary to conserve large areas (Margules & Pressey, 2000), to ensure viability and long-term persistence of species in situ. In practice, the management of reserves is inadequately funded, unplanned and often threatened by illegal extraction of forest products and commercial activities in the tropics (James, 1999). Thus conservation management of naturally occurring undisturbed areas should ensure that the natural values are retained in the face of internal natural dynamics, disturbances from the outside, and varied anthropogenic pressures.

Long-term ecological monitoring sites such as large-scale biosphere atmosphere (LBA) experiment, establishment of biological dynamics of forest fragment projects (BDFFP), and minimum critical size of ecosystem projects (MCSEP) to evaluate the species losses and composition are very few in the tropics (Laurance et al., 2004). Such sites coupled with habitat and landscape characterization enable us to understand the processes regulating biological diversity. In the absence of such intensive data in spatial and temporal domain, satellite remote sensing helps to address habitat loss and analyze probable changes in species composition, based on the extrapolation of spatial changes in species pattern between intact and changed habitats. These studies indicate that conservation approaches based on spatial and temporal information derived from satellite-based platforms, can help identify remnant areas of rich biodiversity.

The present study proposes an approach for monitoring the conservation of phytodiversity using remote sensing and GIS in the Kalakad–Mundanthurai Tiger Reserve, Tamil Nadu (India), an ecological hotspot. The focus is on identifying the remnant patches of intact evergreen forest using multi-temporal satellite data for delineation of vegetation types and their likelihood of sustenance in the coming decades, studying the compositional changes in vegetation in these patches through the conjunctive use of satellite data and phytosociology, analysing the processes regulating the changes, and identifying the conservation areas and monitoring parameters in one of the ecologically sensitive biodiversity hotspots.

2. Materials and methods

2.1 Study site secondary heading, left justified

Kalakad–Mundanthurai Tiger Reserve (KMTR) is located at the southern end of the Western Ghats (hereafter 'WG'), Tamil Nadu, India (Fig 1), and lies between $8^0 21'$ - $8^0 52'$ N latitude and $77^0 10'$ - $77^0 33'$ E longitude in the biogeographic provinces (Udvardy, 1975) 4.1.1 (Malabar rainforest) and 4.14.4 (Deccan thorn forest). The area falls in two districts, namely Tirunelveli and Kanya Kumari of Tamil Nadu and is bound in the west by the state of Kerala.

KMTR covers an area of 907km^2, with hills towering to majestic heights ranging from 100 to 1880m (Agasthiar peak). Agasthyamalai hills at the southern end of the WG are known for high species diversity, harbouring about 2000 flowering plant species with 7.5% endemism (Henry et al., 1984). The mid-elevation zone (700 – 1400m) makes up the tropical wet evergreen forest of the Cullenia–Mesua–Palaquium series (Pascal, 1988). The climate of the area is typically wet with a minimum rainfall of 1200mm and a maximum of 5000mm. The annual average temperature ranges from 13.5°C in the evergreen to 23°C in the deciduous forests. The dry period lasts for 3–5 months and the number of rainy days is in the order of 89–92 days (Pascal, 1982). Forest types such as tropical evergreen, tropical semi-evergreen, tropical moist deciduous, tropical dry deciduous, grasslands and secondary succession exist in the study area. The common trees species are Cullenia exarillata, Mesua ferrea, Palaquium elliptium, Myristica dactyloides, Aglaia bourdillonii in the evergreen forest, while the deciduous species includes Anogeissus latifolia, Terminalia chebula and Terminalia bellirica. Major invasive species includes Lantana camara, Ageratum conyzoides and Eupatorium species.

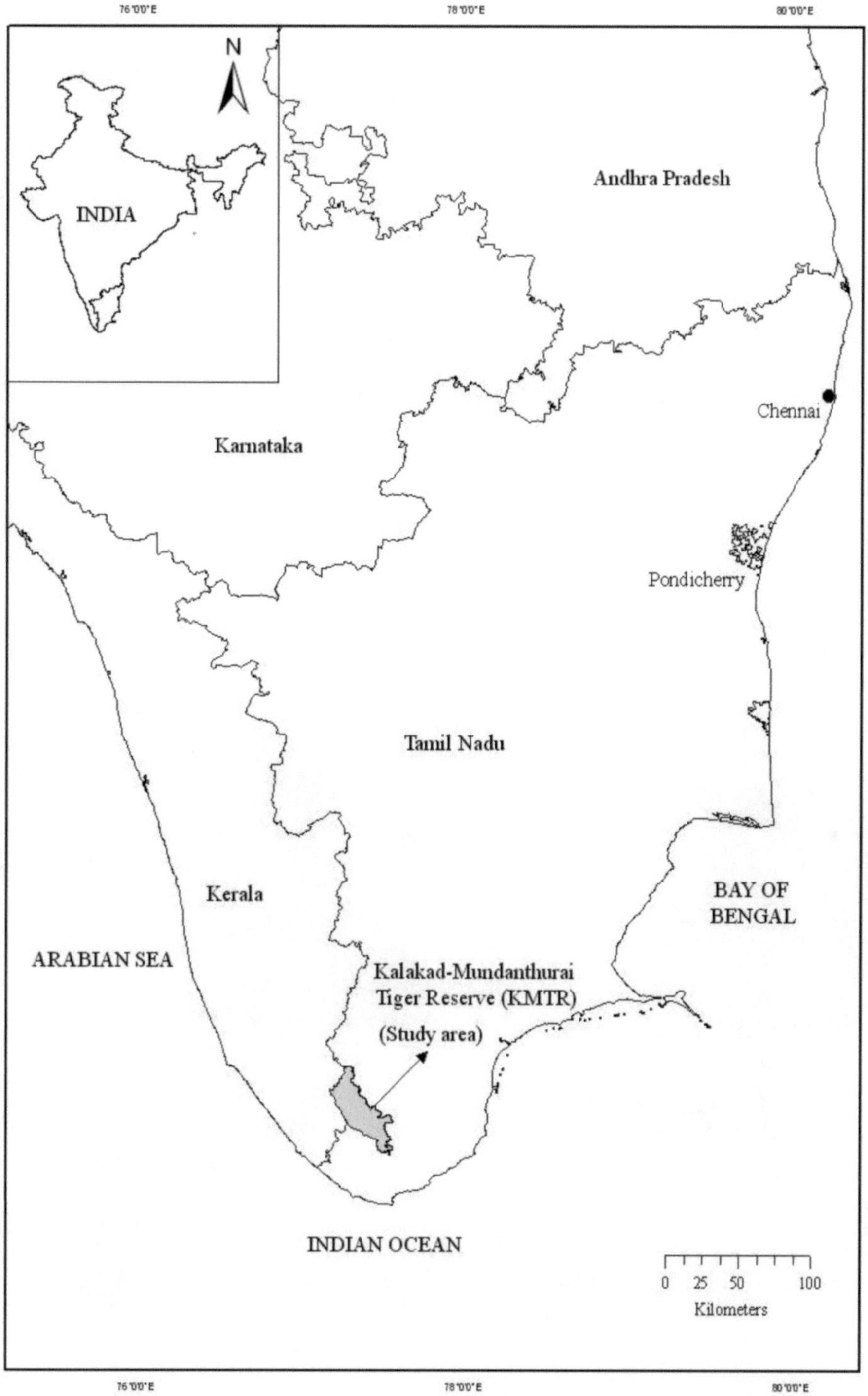

Fig. 1. Map showing the geographic position of Kalakad-Mundanthurai Tiger Reserve, Southern Western Ghats, in the Tirunelveli and Kanykumari districts of Tamil Nadu, India.

2.2 Study site

The evergreen forests of KMTR have a long history of change in the forest cover and land-use type over the last 250 years. Ramesh et al., (1997), have pointed out that there has been a significant loss of biological rich areas between 1960 and 1990 – 85.6km^2 to plantation, 42km^2 to encroachment and 36.4km2 to reservoirs. A total of 28 enclaves have been identified within the KMTR (Ali & Pai, 2001). A total of 189 eco-development villages are in the immediate vicinity of the park in the eastern perimeter in a 5km broad strip. Ecological damage due to forest fires, invasion of reeds and erosion are some of the major causes, which may have led to the change in floristic composition, degeneration and loss of endemics.

2.3 Classification using satellite data and land-cover change

Cloud-free data of LANDSAT Multispectral Scanner (MSS) of March 1973 covering path and row 154/54 was obtained from USGS, EROS Data Center, Sioux Falls, SD; IRS-1B LISS II satellite data of April 1990 and IRS-P6 LISS III satellite data of March 2004 covering path and rows viz. 101/67 and 101/68 from National Remote Sensing Agency, Hyderabad (Fig. 2); LANDSAT-MSS data with a spatial resolution of 80m and the spectral wavelength (B1 0.5 - 0.6, B2 0.6 - 0.7, B3 0.7 - 0.8 and B4 0.8 - 1.1µm); IRS-1B LISS I with a spatial resolution of 72.5m and the spectral wavelength (B1 0.45 - 0.52, B2 0.52 - 0.59, B3 0.62 - 0.68 and B4 0.77 - 0.86µm) and IRS-P6 LISS-III with a spatial resolution of 24m and the spectral wavelength of four bands (B2 0.52 – 0.59, B3 0.62 – 0.68, B4 0.77 – 0.86 and B5 1.55 – 1.70µm) were analysed in the study.

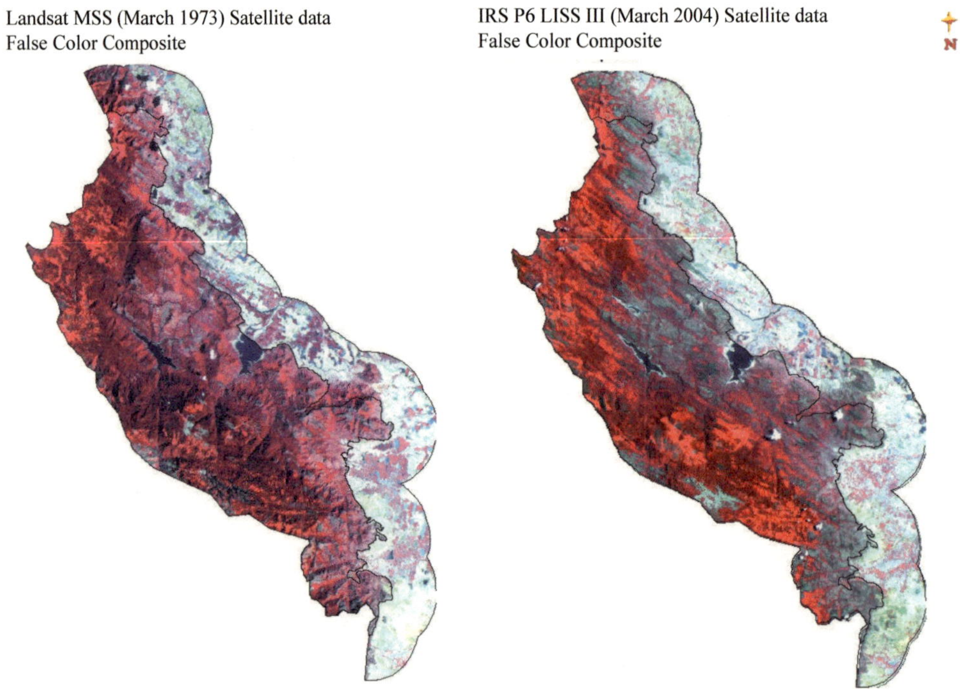

Fig. 2. False color composite for the study region (1973 & 2004) showing different vegetation formations, and also the variation in tone and texture in the Southern Western Ghats, Tamil Nadu, India.

Supervised classification technique was used to classify satellite image of 2004 (LISS-III) based on the information of terrain, topography and species database collected during the landscape-biodiversity characterization program for the Western Ghats region (Anonymous, 2002). Using the 2004 vegetation and land cover map, an area of interest (AOI) was selected for the evergreen patches of 1973 using MSS data assuming that these patches of 2004 had remained unchanged from 1973. The reflectance properties of the evergreen and semi-evergreen patches for the both the satellite images showed similar trends (Fig. 3). Taking into consideration of the above criteria, the 1973 MSS image and LISS-III 2004 image were used to generate the LULC maps of 1973 and 2004. Likewise, all the spectral classes were assigned training sets from the geometrically corrected images and were then classified. The maximum likelihood algorithm was used to classify these scenes (Lillesand & Kiefer, 2000). Major forest types delineated were tropical evergreen, semi-evergreen, moist deciduous, dry deciduous, dry evergreen, grasslands, scrubs, reeds (Ochlandra sp.), and orchards. The tone and textural differences in these forest types can be clearly seen in the satellite imagery (Fig. 2). The classified vegetation map was validated by verification on ground and found to be 85% accurate. Finally, IRS LISS-III dataset were resampled to 80m (equivalent to MSS) to facilitate comparison.

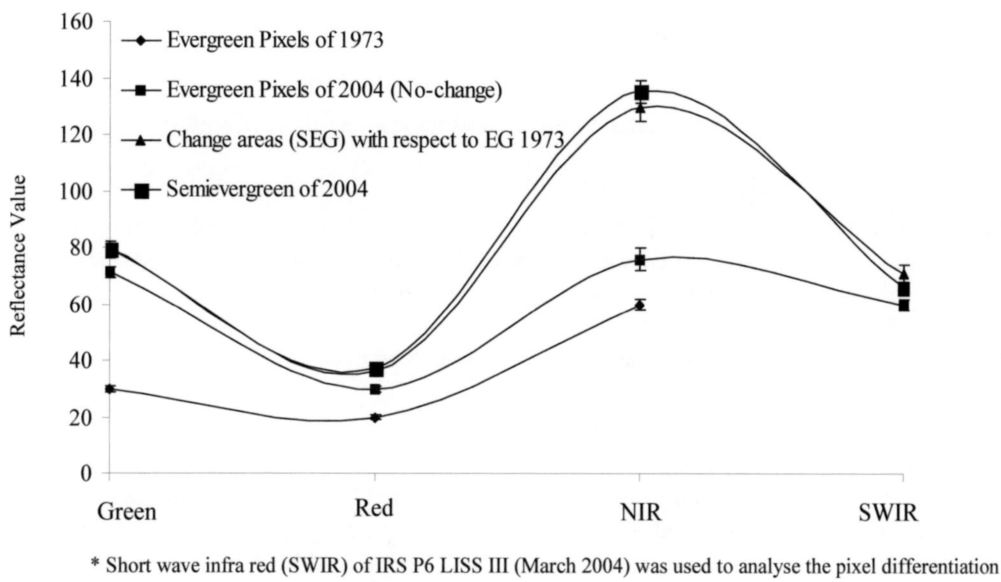

Fig. 3. Spectral reflectance value (X-axis has spectral bands and Y-axis has reflectance value) differentiating evergreen and semi-evergreen forest types of KMTR, Southern Western Ghats of Tamil Nadu, India.

A brief methodology describing land-cover change analysis carried out for the present study is given in Fig 4. It combines information on phytosociology (no-change and change- areas plots) and land-cover information, modelling evergreen change using spatial drivers to identify areas for biodiversity monitoring and prioritization.

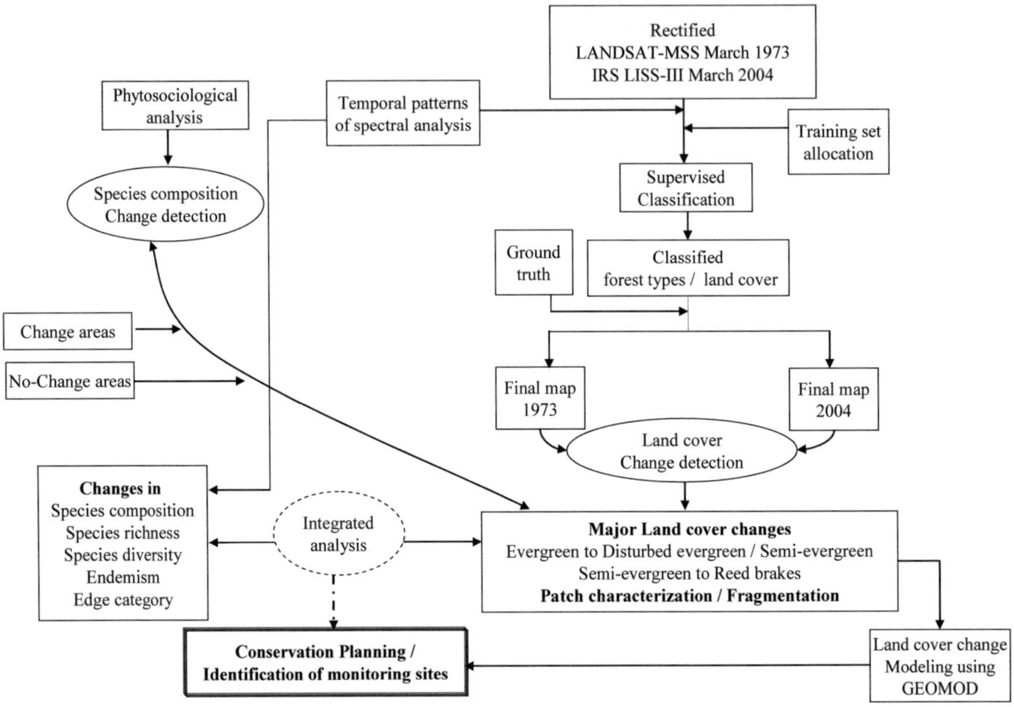

Fig. 4. Methodology used for the study region combining satellite based land-cover and phytosociological information to characterize areas for conservation planning

2.4 Landscape analysis

Landscape is defined as "an aggregation of heterogeneous elements, which interact with each other". Landscape has three intrinsic properties: structure, function and change. These can be explained in terms of porosity, fragmentation, patch density, patchiness, interspersion, juxtaposition, contagion etc (Forman & Godron, 1986). In the present study the following landscape metrics have been studied to assess landscape changes and forest fragmentation patterns.

2.4.1 Fragmentation modeling

Land-cover map with a spatial resolution of 80m was used to characterize the fragmentation levels around the evergreen and semi-evergreen forests pixel. To perform the fragmentation calculations, we used a "moving window" algorithm developed by Riitters et al., (2000). The model was designed to identify patterns of forest fragmentation using coarse and fine scale resolution land-use and land-cover information.

To implement the fragmentation model, the size of the analysis window had to be determined. After considering the resolution of the data, delineation of the forest features having the smallest areas, and practicality assessment of various window sizes, a 5x5 window was found to maintain an adequate representation of the proportion (Pf) of pixels in the window and also to maintain interior forests at an appropriate level.

Using the results from the forest fragmentation model, further research was conducted to produce maps, which identify the state of forest fragmentation in a specified region. The purpose of the forest fragmentation index was to provide a quick means to assess the extent of forest fragmentation within a region, and to track trends in forest fragmentation to identify areas that would benefit from possible reforestation. Different indexes were used to generate forest fragmentation e.g. total forest proportion (TFP), forest continuity (FC) and weighted forest area (WFA). These can be referred from Civco et al., 2002; Vogelmann, 1995; and Wickham et al., 1999.

2.4.2 Patch analysis

Patch Size, Number and Shape: After classifying the forest type using satellite data, it was vectorised in the GIS domain to characterize the patches. The information on patch size and number was extracted from the vectorised classified data. A minimum of 3x3 pixel window was set for patch analysis. A simple measure of patch shape is the perimeter-to-area ratio. This measure is often standardized to the most compact form; either a square or a circle, and is taken as equal to 1. Higher perimeter value indicates increase of edge effect, an ecologically undesirable influence on most species population and communities.

Contagion: Contagion metric was first proposed by O'Neil et al., (1988), and later by several others (Gustafson & Parker, 1992; Herold et al., 2005; Li & Reynolds, 1993). It is a measure of clumping or aggregating the patches. It is used as an indication of the degree of fragmentation of a landscape.

Fractal Dimension: Fractal dimension has been used for measurement, simulation and spatial analysis in the mapping sciences (Despland, 2003; Krummel et al., 1987; Leduc et al., 1994). Changes in the fractal dimension of the remote sensing images, imply changes in the environmental conditions (Lam & Quattrochi, 1992). A number of studies have found that the fractal dimension of the landscape varies according to the type of land-use (Despland, 2003; Krummel et al., 1987).

Patch Per Unit (PPU): PPU is low when the landscape is not fragmented. As the landscape becomes more fragmented, the PPU increases (Frohn, 1998).

Landscape shape complexity: As an alternative to the use of fractal dimension for quantifying patch shape complexity, SqP has been introduced (Frohn, 1998). SqP considers the perimeter area relationship for raster data structures and normalizes the perimeter - area ratio to a value between 0 and 1.

Landscape Shape Index (LSI): Landscape shape index (LSI) provides a standardized measure of the total edge or edge density and adjusts for the size of the landscape. In contrast to total edge, LSI can be interpreted directly because it is standardised, and it is a measure of patch aggregation or disaggregations. An increase in LSI indicates an increase in disaggregation.

2.5 Predictive modelling of evergreen forest

An attempt has been made to model land-use and land-cover change (LULC) using GEOMOD, to predict future changes (Fig. 5). GEOMOD is a simple unidirectional linear change modelling tool (Pontius & Batchu, 2003) that uses suitability image/s, produced by combining a variety of driver images to predict locations of change for a given quantum of change between two time periods. After comparing its past and present performances, using satisfactory suitability image/s, one can actually simulate future change for various scenarios of change between two different time frames. The most interesting part of this type of change modelling is its ability to model location-specific changes for different quantities of change.

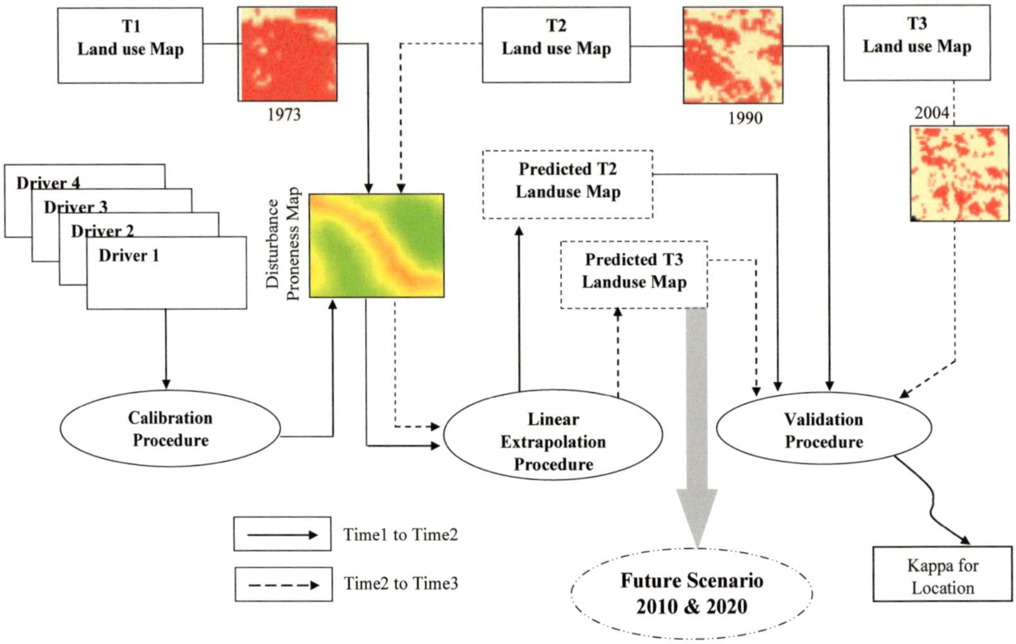

Fig. 5. GEOMOD based modeling of evergreen forest change using spatial drivers, extrapolation and validation procedure for the future evergreen forest scenario in KMTR

2.5.1 Reclassification of land-use types

GEOMOD can evaluate change in two land-use types at a time. Therefore, each of the vegetation and land-cover map was reclassified as evergreen (evergreen) and non-evergreen (semi-evergreen, deciduous and other land-cover) areas. The most common reclassification is to classify all undisturbed forest as type 1, and all other land-use types, which can be characterized as having undergone some human intervention, such as disturbed evergreen, orchards, and Ochlandra areas as type 2. Area estimation is done using reclassified data, to determine how many forest pixels existed for a particular time period. Future rate of change is calculated using the simple subtraction to find the area deforested during the interim period.

2.5.2 Spatial drivers of land cover change

Thirteen carefully chosen drivers (Fig. 6) including altitude, slope, aspect, proximity to protected area (PA) boundary, settlement, tea and coffee plantation, road and footpath, rainfall intensity and existing sites of Ochlandra reeds were integrated using appropriate fuzzy set membership functions into a single suitability image (Eastman, 2003). This suitability image was then used in GEOMOD to model change from 1973 to 1990, 1990 to 2004 and then from 2004 to 2020. The driver maps used for each calibration run are added together to create a (disturbance proneness area) suitability map (Fig. 6). GEOMOD uses this map of ranked potentials, or likelihoods, to simulate deforestation at a third point in time, the results of which are validated against the actual map of that same time period to test how well the drivers did in predicting the spatial pattern of deforestation. This 'test' is called the validation process and is discussed further below.

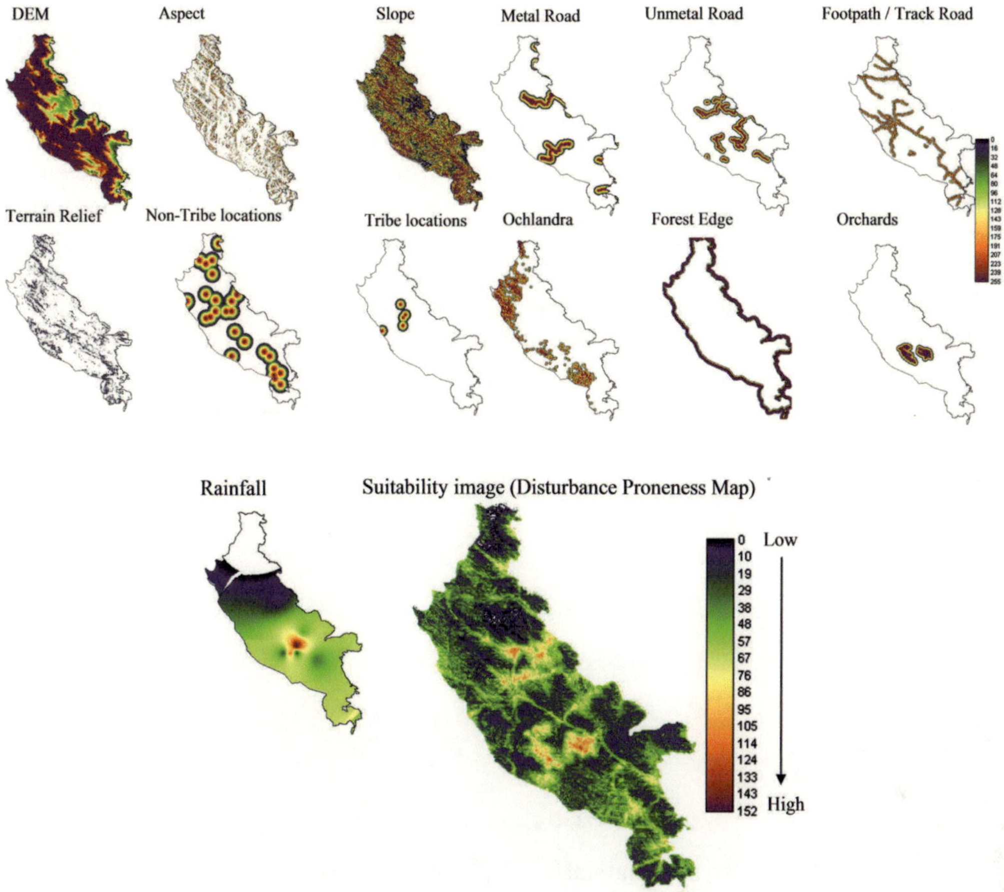

Fig. 6. List of potential spatial drivers generated using multi-criteria decision support to generate suitability image for the evergreen forest change modeling of KMTR, Southern Western Ghats, Tamil Nadu, India.

2.5.3 Validation

To validate the results created by GEOMOD, the actual evergreen map at a known point of time is compared with the predicted evergreen map of that same time, based on analysis of the pattern at an earlier time point using the "Validation" tool available in the IDRISI module. In the past, measures of the 'goodness of fit' commonly were performed by using a simple percent-correct measure or, at best, a multiple-resolution percent-correct measure (Costanza, 1989; Hall et al., 1995), but this provides little assessment of a model's ability to predict the correct quantity of change versus its ability to identify the correct location of change (Pontius & Batchu, 2003; Pontius & Pacheco, 2004). Spatial measures of 'goodness of fit' have been developed, that measures the degree to which a simulated map agrees with a reality map with respect to both location (Kappa-for-location) and quantity of cells correct (Kappa-for-quantity).

3. Results

3.1 Vegetation types

The changes in the vegetation cover based on digital classification of Landsat-MSS and IRS LISS-III data shows a significant decrease of evergreen forest mostly to semi-evergreen forest types (Table 1a). Of the total area covered by natural vegetation (857sq.km) the evergreen and semi-evergreen forest occupied 60% of the area. The map showing multi-temporal vegetation and land-cover classification is given in Fig. 7. In 1973, evergreen forests constituted 316sq.km, followed by semi-evergreen having 194sq.km. In 2004 the evergreen forest had diminished to 188sq.km (i.e. 40% loss of land-cover), whereas semi-evergreen forest gained by 36% to 265.1sq.km. The grassland covering of 73sq.km is largely distributed among the Kodayar, Manjamparai and adjoining areas of the Agasthyamalai region. The area under grassland has shown a significant increase of ca. 166% during the period 1973 – 2004.

S.No	Types	1973	2004
Phenelogical types			
1	Evergreen	316.72	188.49
2	Semi-evergreen	194.40	265.10
3	Moist deciduous	143.59	132.02
4	Dry deciduous	38.95	98.30
5	Dry evergreen	136.14	61.55
6	Grassland	27.36	73.14
	Subtotal	**857.16**	**818.60**
Other landcover types			
7	Shrubs	1.11	16.01
8	Ochlandra	13.56	23.74
9	Orchards	2.33	16.42
10	Fallow/barren	10.04	9.86
11	Water	15.22	16.86
12	Shadow	6.97	2.50
13	Cloud	1.07	3.46
	Grand total	**907.46**	**907.46**

Table 1a. Vegetation and land-cover distribution in KMTR, Southern Western Ghats, Tamil Nadu, India. for the years1973 and 2004 using satellite imagery.

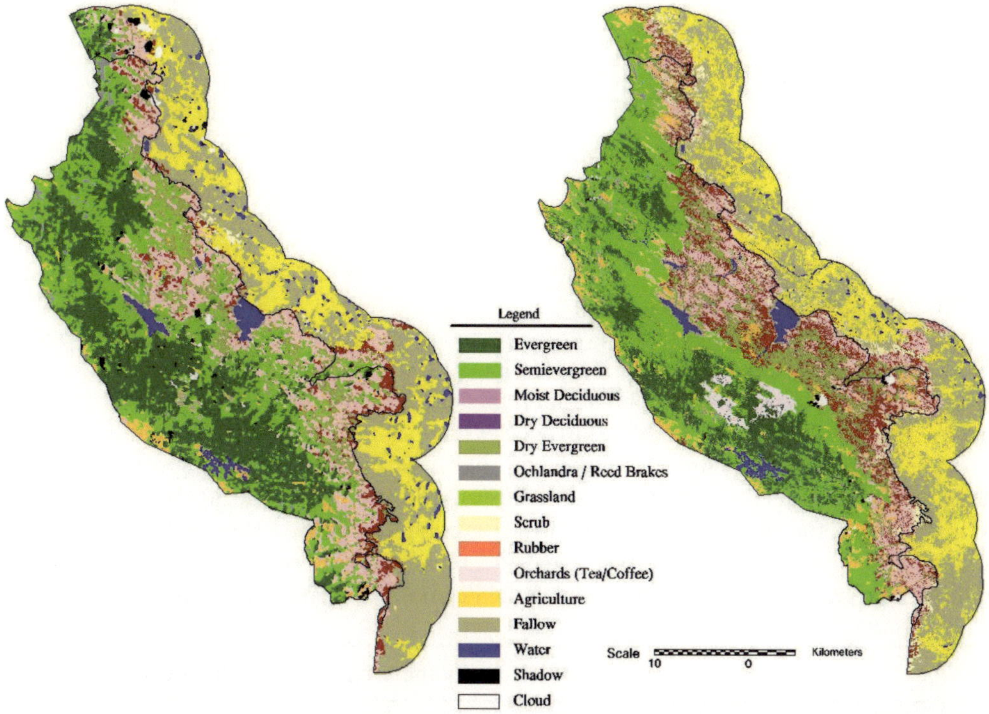

Fig. 7. Vegetation and land cover classification map for the study region (1973 & 2004) of KMTR, Southern Western Ghats, Tamil Nadu, India.

3.2 Spectral analysis

The spectral reflectance of the evergreen forest in the NIR band of 1973 (60±2 SD) is quite similar to the spectral reflectance of the 2004 data having 76±4 SD. The semi-evergreen patch showed a spectral reflectance of 92±3 SD from the 1973 data, while the semi-evergreen reflectance in 2004 was 135±4 SD (Fig. 3). The evergreen patches which had been converted to semi-evergreen patches during 2004 showed a reflectance of 130±5 in the NIR. The SWIR band of IRS P6 LISS-III has shown that the intact evergreen and disturbed evergreen patches can be distinctly differentiated from this reflectance value.

3.3 Land-cover change analysis

Change detection analysis was performed was performed on the data from two time periods - 1973 and 2004 - and change matrix is given in Table 1b. Major changes in the land-cover of KMTR has been the change from evergreen to semi-evergreen forest type (121.74km^2), constituting ca. 38% of the total land-cover changes, evergreen to reeds brake (Ochlandra) (6.64km^2), and evergreen to orchards (4.87km^2). Conspicuous changes in semi-evergreen forest type were to grassland (17.64km^2), Ochlandra (10.72km^2) and orchards (6.76km^2). It is observed that the changed vegetation cover is mostly in patch sizes of either 100ha or more, or in the size of 10 -50ha.

2004 / 1973	Evergreen	Semi-evergreen	Others	**Total**
Evergreen	166.31	121.74	28.67	**316.72**
Semi-evergreen	21.02	119.19	54.20	**194.40**
Others	1.16	24.17	371.02	**396.34**
Total	**188.49**	**265.10**	**453.88**	**907.46**

Table 1b. Change matrix for KMTR between 1973 and 2004, Southern Western Ghats, Tamil Nadu, India.

3.4 Phytodiversity change analysis

To understand the changes of evergreen to semi-evergreen patches in terms of phytosociological data, a total of 95 sample plots (0.1ha each) were grouped into change (40 plots) and no-change areas (55 plots) for identification of predominant species, percentage of endemic and edge species. Differences in species composition based on predominant species for change and no-change area are given in Table 2. Of the total 266 tree species in the no-change area 38% were endemic and 25% were edge species. While in the change areas having 249 tree species 31% were endemic, 31% belonged to the edge species. Among the saplings in the no-change area (170 species) 44% were endemic and 19% were edge species, in the change area (146 species) 34% were endemic and 29% were of the edge species.

	No-Change area	**Change area**
Tree Species ≥ 30 cm		
No. of plots	55	40
No. of Species	266	249
No. of Individuals	3340	1977
Stand density (ha^{-1})	607	494
Basal area (m^2ha^{-1})	54.39	44.11
No. of endemic species / individuals	100 (38) / 1274 (38)*	78 (31) / 737 (37)
No. of edge species / individuals	66 (25) / 294 (9)	78 (31) / 296 (15)
Saplings (Tree and Shrub) ≤ 30 cm		
No. of plots	51	33
No. of Species	170	146
No. of Individuals	863	535
Stand density (ha^{-1})	169	162
Basal area (m^2ha^{-1})	0.47	0.42
No. of endemic species / individuals	75 (44) / 399 (46)	49 (34) / 155 (29)
No. of edge species / individuals	33 (19) / 60 (7)	42 (29) / 142 (27)

* Figures in bracket indicates percentage

Table 2. Consolidation of phytosociological analysis in no-change and change area plots in KMTR, Southern Western Ghats, Tamil Nadu. India.

The first three predominant species composition in the no-change area is largely composed of *Cullenia exarillata* – *Myristica dactyloides* – *Mesua ferrea* when compared to the change areas having *Cullenia exarillata* - *Dimocarpus longan* - *Kingiodendron pinnatum* (Table 3a). The percentage of endemic species in the change area is greater by 7% and edge species lesser by 7% when compared to the no-change area. Interestingly in change areas large levels of secondary successional species were noticed, which includes *Epiprinus mallotiformis, Macaranga peltata, Mallotus philippensis, Scolopia crenata,* and *Schleichera oleosa.* Regeneration trend in no-change areas showed a dominance of *Antidesma menasu, Cinnamomum malabatrum* and *Mesua ferrea* while in change areas *Dimocarpus longan, Mallotus philippensis* and *Mesua ferrea* species are seen to be dominant. Further, the dominants in the semi-evergreen are again *Dimocarpus* longan, *Mallotus philippensis* and *Mesua ferrea* indicating that the evergreen patch is gradually being changed to semi-evergreen (Table 3b). Species similarity between no-change and change areas was found to be at 68%.

Similar trends on endemic and edge species have been noticed in the no-change and change areas. No-change areas of endemic tree species showed dominance in *Mesua ferrea, Aglaia bourdillonii* and *Hopea utilis,* while for saplings it was *Mesua ferrea, Cinnamomum malabatrum* and *Cryptocarya bourdillonii* (Table 4a). In the case of edge tree species the dominant ones were *Macaranga peltata, Mallotus philippensis* and *Memecylon edule* while among saplings *Mallotus philippensis, Macaranga peltata* and *Olea dioica* were dominant (Table 4b).

Species no-change area	**IVI**	**Species change area**	**IVI**	**Semievergreen**	**IVI**
Cullenia exarillata	22.51	*Cullenia exarillata*	20.17	*Dimocarpus longan*	31.58
Myristica dactyloides	9.88	*Dimocarpus longan*	20.10	*Mesua ferrea*	15.00
Mesua ferrea	9.72	*Kingiodendron pinnatum*	8.18	*Canarium strictum*	12.18
Agrostistachys meeboldii	9.64	*Mesua ferrea*	7.83	*Diospyros paniculata*	11.04
Syzygium gardneri	7.74	*Filicium decipens*	6.86	*Filicium decipens*	10.93
Dimocarpus longan	7.11	*Hopea parviflora*	5.88	*Holigarna arnottiana*	10.44
Calophyllum austroindicum	6.41	*Vitex altissima*	5.60	*Garcinia gummi-gutta*	8.40
Aglaia bourdillonii	6.23	*Hopea utilis*	5.26	*Kingiodendron pinnatum*	8.06
Antidesma menasu	5.72	*Myristica dactyloides*	5.22	*Madhuca indica*	7.33
Hopea utilis	5.26	*Diospyros foliolosa*	4.25	*Mallotus philippensis*	6.98
Acronychia pedunculata	5.17	*Macaranga peltata*	4.23	*Hopea parviflora*	6.61
Artocarpus heterophyllus	4.98	*Agrostistachys meeboldii*	4.18	*Gordonia obtusa*	6.44
Xanthophyllum flavescens	4.86	*Mallotus philippensis*	3.75	*Wrightia tinctoria*	6.02
Palaquium ellipticum	4.78	*Syzygium mundagam*	3.52	*Schleichera oleosa*	5.67
Mangifera indica	4.66	*Pterospermum diversifolium*	3.40	*Diospyros sp.*	5.43
Gomphandra coriacea	4.27	*Scolopia crenata*	3.24	*Bischofia javanica*	4.85
Cryptocarya bourdillonii	4.04	*Syzygium gardneri*	3.19	*Pterospermum xylocarpum*	4.81
Epiprinus mallotiformis	4.00	*Cinnamomum malabatrum*	3.00	*Persea macrantha*	4.72
Pterospermum xylocarpum	3.82	*Tricalysia apiocarpa*	2.97	*Myristica dactyloides*	4.72
Cinnamomum malabatrum	3.62	*Diospyros paniculata*	2.95	*Acronychia pedunculata*	4.52

Table 3a. Tree species composition in no-change and changes areas of the evergreen and semi-evergreen forests in KMTR, Southern Western Ghats, Tamil Nadu, India.

Species no-change area	IVI	Species change area	IVI	Semievergreen	IVI
Antidesma menasu	12.09	*Dimocarpus longan*	25.19	*Dimocarpus longan*	13.00
Cinnamomum malabatrum	10.61	*Mallotus philippensis*	13.20	*Mallotus philippensis*	12.01
Mesua ferrea	10.61	*Mesua ferrea*	8.07	*Mesua ferrea*	10.00
Cullenia exarillata	9.62	*Hydnocarpus alpina*	7.89	*Macaranga peltata*	9.84
Cryptocarya bourdillonii	8.79	*Nothopegia beddomei*	6.49	*Hydnocarpus alpina*	9.23
Syzygium gardneri	8.75	*Xanthophyllum flavescens*	5.90	*Pavetta hispidula*	8.74
Myristica dactyloides	7.55	*Macaranga peltata*	5.80	*Epiprinus mallotiformis*	8.41
Agrostistachys meeboldii	6.94	*Meliosma pinnata*	5.72	*Chomelia asiatica*	7.84
Xanthophyllum flavescens	6.81	*Pavetta hispidula*	5.24	*Scolopia crenata*	7.00
Palaquium ellipticum	6.33	*Mallotus stenanthus*	5.21	*Acronychia pedunculata*	6.81
Neolitsea scorbiculata	5.82	*Canthium angustifolium*	5.01	*Kingiodendron pinnatum*	6.01
Nothopegia heyneana	5.17	*Kingiodendron pinnatum*	5.01	*Filicium decipens*	5.71
Gomphandra coriacea	5.02	*Olea dioica*	4.87	*Persea macrantha*	5.11
Mallotus stenanthus	4.77	*Holigarna arnottiana*	4.65	*Olea dioica*	4.86
Epiprinus mallotiformis	3.91	*Filicium decipens*	4.42	*Holigarna arnottiana*	4.15
Hydnocarpus alpina	3.84	*Eugenia thwaitesii*	4.11	*Canthium angustifolium*	3.95
Cinnamomum verum	3.83	*Eugenia calcadensis*	4.03	*Meliosma pinnata*	3.11
Litsea mysorensis	3.67	*Chomelia asiatica*	3.82	*Xanthophyllum flavescens*	2.99
Litsea floribunda	3.67	*Epiprinus mallotiformis*	3.77	*Nothopegia beddomei*	2.11
Octotropis travancorica	3.64	*Scolopia crenata*	3.68	*Diospyros paniculata*	2.00

Table 3b. Species composition in tree and shrub saplings of no-change and changes areas of the evergreen and semi-evergreen forests in KMTR, Southern Western Ghats, Tamil Nadu, India.

Species	Trees ≥ 30 cm		Species	Sapling ≤ 30 cm	
	No-change	Change		No-change	Change
Mesua ferrea	116	53	*Mesua ferrea*	34	12
Aglaia bourdillonii	107	16	*Cinnamomum malabatrum*	34	2
Hopea utilis	80	37	*Cryptocarya bourdillonii*	27	1
Litsea floribunda	50	1	*Nothopegia heyneana*	21	5
Cryptocarya bourdillonii	44	1	*Palaquium ellipticum*	21	0
Palaquium ellipticum	43	15	*Mallotus stenanthus*	15	9
Cinnamomum malabatrum	36	29	*Octotropis travancorica*	11	1
Drypetes confertiflorus	36	1	*Litsea mysorensis*	10	2
Calophyllum austroindicum	34	1	*Nothopegia beddomei*	9	10
Symplocos macrocarpa	33	8	*Aglaia bourdillonii*	9	2

Table 4a. Predominant endemic species and its individuals in no-change and change areas of the evergreen forest in KMTR, Southern Western Ghats, Tamil Nadu, India.

Species	Trees ≥ 30 cm		Species	Sapling ≤ 30 cm	
	No-change	Change		No-change	Change
Macaranga peltata	41	38	*Mallotus philippensis*	2	23
Mallotus philippensis	14	33	*Macaranga peltata*	1	11
Memecylon edule	5	12	*Olea dioica*	5	10
Margaritaria indica	2	12	*Canthium angustifolium*	1	10
Cleistanthus malabaricus	1	12	*Chomelia asiatica*	4	7
Gordonia obtusa	7	11	*Glycosmis mauritiana*	1	7
Olea dioica	7	9	*Ixora lanceolata*	3	5
Canthium dicoccum	6	9	*Oreocnide integrifolia*	2	5
Ixora brachiata	3	8	*Psychotria subintegra*	2	5
Chomelia asiatica	11	7	*Gordonia obtusa*	0	5

Table 4b. Predominant edge species and its individuals in no-change and change areas of the evergreen forest in KMTR, Southern Western Ghats, Tamil Nadu, India.

3.5 Landscape analysis

3.5.1 Forest fragmentation

The area under fragmentation in the evergreen forest type showed varied changes in the different categories from 1973 - 2004 (Table 5a). While the interior category decreased from 113.90sq.km (36%) to 23.27sq.km (12%), the perforated category decreased from 85.0 to 23.7sq.km, whereas the other categories showed significant increase in fragmentation (patch category 14.82 to 25.3sq.km; edge category 66.8 to 80.15sq.km). There appear to be no significant change in the transitional category. In case of semi-evergreen patches, almost all the categories except the perforated and transitional category showed significant increase (Table 5a). Like evergreen forests, in these forests also the transitional category did not show any significant changes from 1973 - 2004. The perforated category showed a decrease from 60.4 to 31.7sq.km. The total forest proportion (TFP) of the evergreen forest has decreased by 40%, while the semi-evergreen patch showed an increase of TFP by 36%. A similar trend was also shown in the weighted forest area (WFA) values in the evergreen and semi-evergreen forests. Evergreen forest continuity (FC) decreased by 88.4% while the semi-evergreen forest continuity was seen to increase by 77% (Table 5b).

3.5.2 Patch analysis

The patch size and distribution for the period of 1973 - 2004 shows a relative decrease in the number of smaller patches and an increase in the number of larger patches in the evergreen as well as the semi-evergreen type (Table 6). In the 1973 patches of < 50ha these constitute 7% (131 patches), and in the 100 - 500ha constitute 6.22% (9 patches) of the total evergreen area. Contrastingly in 2004, < 50 ha constituted 9.77% (110 patches) and 100 to 500ha made up 12.53% (10 patches). Interestingly, >1000ha patches in 1973 hold 90% (3 patches) when compared to 2004 making up only 67% (4 patches). Similarly in semi-evergreen forest of 1973 >1000ha patches showed 23% (3 patches) when compared to the 2004 area of 60% (3 patches). The results revealed that the distribution of patches could be categorized into four different patterns, namely, larger areas covered by lesser number of patches (evergreen forest of 1973), lesser areas covered by fewer patches (evergreen forest of 2004), lesser areas covered by large number of patches (Semi-evergreen forest of 1973), and lesser areas covered by least number of patches (Semi-evergreen forest of 2004) as seen in Table 6.

Sl	Evergreen category	1973 (Sq. km)	Area (%)	2004 (Sq. km)	Area (%)
1	Interior	113.90	35.96	23.27	12.35
2	Patch	14.78	4.67	25.33	13.44
3	Transitional	32.95	10.40	31.60	16.76
4	Edge	66.88	21.12	80.15	42.52
5	Perforated	85.00	26.84	23.73	12.59
6	Undetermined	3.20	1.01	4.40	2.33
	Total	**316.72**	**100.0**	**188.49**	**100.00**

Sl	Semievergreen category	1973 (Sq. km)	Area (%)	2004 (Sq. km)	Area (%)
1	Interior	10.47	5.39	29.35	11.07
2	Patch	29.56	15.21	35.24	13.29
3	Transitional	49.55	25.49	49.42	18.64
4	Edge	44.16	22.72	115.28	43.49
5	Perforated	60.41	31.08	31.69	11.95
6	Undetermined	0.25	0.13	4.12	1.55
	Total	**194.40**	**100.00**	**265.10**	**100.00**

Table 5a. Forest fragmentation categories for the evergreen and semi-evergreen (1973 - 2004) in KMTR, Southern Western Ghats, Tamil Nadu, India.

	Evergreen		Semievergreen	
	1973	2004	1973	2004
TFP	0.355	0.212	0.218	0.298
WFA	274.38	147.06	150.28	207.19
FC	0.043	0.005	0.002	0.009

Table 5b. Forest fragmentation conditions based on TFP and FC for evergreen and semi-evergreen (1973 - 2004) in KMTR, Southern Western Ghats, Tamil Nadu, India.

Patches having perimeter-to-area (PA) ratio of <0.015 were 51% in 1973 (i.e. 80 patches), as compared to only 30% in 2004 (i.e. 42 patches) indicating the contiguity of large patches with lesser perimeter (Table 6). The higher PA ratio of >0.025 was observed in 2004 to be 37.4% (from 52 patches) as compared to 10.8% in 1973 (from 17 patches) indicating more complex shapes due to increasingly higher PA ratios.

For evergreen forests, fractal dimension (FD) increased from 1.32 to 1.35 while in the case of semi-evergreen forests, it was constant at 1.37 (Table 6). The patch per unit (PPU), which is an indication of clumping, showed an increase in the case of evergreen forest type from 1.18E-07 to 2.17E-07, while in the case of semi-evergreen the PPU showed a slight decrease from 2.44E-07 to 2.31E-07. The patch shape complexity (SqP) for evergreen forests showed an increase from 0.956 to 0.965, whereas in the semi-evergreen type it showed a marginal increase from 0.973 to 0.974. The degree of fragmentation indicator based on the contagion matrix showed a slight decrease in the case of evergreen forest (0.452 to 0.450), while in the case of semi-evergreen there was an increase from 0.457 to 0.469.

Sl.	Parameters	Evergreen		Semievergreen	
		1973	2004	1973	2004
1	Area (sq.km)	290.25	150.14	193.45	229.89
2	Number of Patches	157	139	361	199
3	Patch density	0.54	0.93	1.87	0.87
5	Patch size (ha)				
	< 50	7.05⁺ (131)*	9.77 (110)	24.52 (292)	10.54 (171)
	50 to 100	3.18 (14)	6.58 (14)	10.37 (29)	2.81 (10)
	100 to 500	6.22 (9)	12.53 (10)	32.83 (34)	12.14 (11)
	500 to 1000	0.00	3.33 (1)	8.87 (3)	13.90 (4)
	>1000	90.60 (3)	67.79 (4)	23.41 (3)	60.61 (3)
6	Perimeter / area ratio				
	< 0.015	50.96 (80)	30.22 (42)	48.75 (176)	14.57 (29)
	0.016 - 0.020	38.22 (60)	32.37 (45)	41.00 (148)	39.20 (78)
	> 0.025	10.83 (17)	37.41 (52)	10.25 (37)	46.23 (92)

+ Non Bracketed number indicate % of evergreen forest

* Bracketed number indicate number of patches

Type	LSI	Patch Cohesion	Contagion	D	PPU	SqP
	Evergreen					
1973	16997.04	99.9976	0.452	1.32	1.1764E-07	0.965
2004	16373.33	99.9952	0.450	1.36	2.1697E-07	0.966
	Semievergreen					
1973	16997.04	99.9961	0.457	1.38	2.4433E-07	0.973
2004	16373.33	99.9967	0.460	1.38	2.3063E-07	0.974

Table 6. Patch characteristics and other landscape metrics analyzed for the evergreen and semi-evergreen forest in KMTR, Southern Western Ghats, Tamil Nadu, India.

3.6 Predictive modelling of evergreen forest cover

This study revealed considerable type changes from evergreen to semi-evergreen forests within the KMTR in three decades. Ideally we should have a database of 1973 to validate these changes in terms of species composition or its drivers of land-cover changes. However, in the absence of such databases, two approaches are being used to discuss the observed changes: spectral similarity and species similarity. Evergreen ground sample points of 2004 showed similar spectral characteristics when overlaid on the MSS data of 1973. Comparison of the 1973 data based on spectral signature similar to the 2004 evergreen signature showed that the area classified as evergreen was observed to be around 128 sq km more in the former than the 2004 extent. On evaluating the ground sample points of the area shown to be evergreen in the 1973 data but not in the 2004 data, it was observed that the species composition of the latter were of the semi-evergreen type and showed spectral characteristics similar to the semi-evergreen of the 1973 data. Thus these patches are spectrally distinct in terms of their tone and texture. (See Fig. 2). Table 3a&b highlights the evergreen plots of change areas and the intact evergreen plots which differ widely in terms of their species compositions. Similarly, the evergreen changed areas are closer towards semi-evergreen areas. Based on these findings the changed evergreen areas are considered to be semi-evergreen forest types.

Results indicate that between 1973 and 1990 (the protection began in 1987) about 42% of the total non-degraded forest was actually lost to the degraded category. However the rate of degradation substantially reduced after the protection became effective and the results indicate that only 9% of the total forest that remained in 1990 was lost to the degraded category (Fig. 8 & Table 7). Using GEOMOD it is perceived that even if the same level of protection persists; an additional 27% of the non-degraded forest would be lost to the degraded category by 2020. The results were validated using the validation tool of IDRISI software. The pixel location and quantity comparison of the actual and predicted map of 1990 showed 80% accuracy, while for the 2004 maps it was 88% accuracy (Fig. 9). Based on the validation inputs, the predicted maps of 2010 and 2020 were prepared (Fig. 8).

Evergreen Forest	Area (sq.km)
1973	316
1990	182
2004	166

Current Scenario

Category	No. of Pixels
Evergreen forest - 1973	47709
Evergreen forest - 1990	27616
Total change	20093
Total change (Sq.km)	129
Annual rate of change (Sq km)	7.56
Evergreen forest - 2004	24903
Total change	2713
Total change (Sq.km)	17
Annual rate of change (Sq km)	1.2

Future Scenario

Change in Evergreen Forest	Area (Sq.km)
2010	134.72
2020	112.79

Table 7. Current and future scenarios for the evergreen forest changes using GEOMOD approach in KMTR, Southern Western Ghats, Tamil Nadu, India.

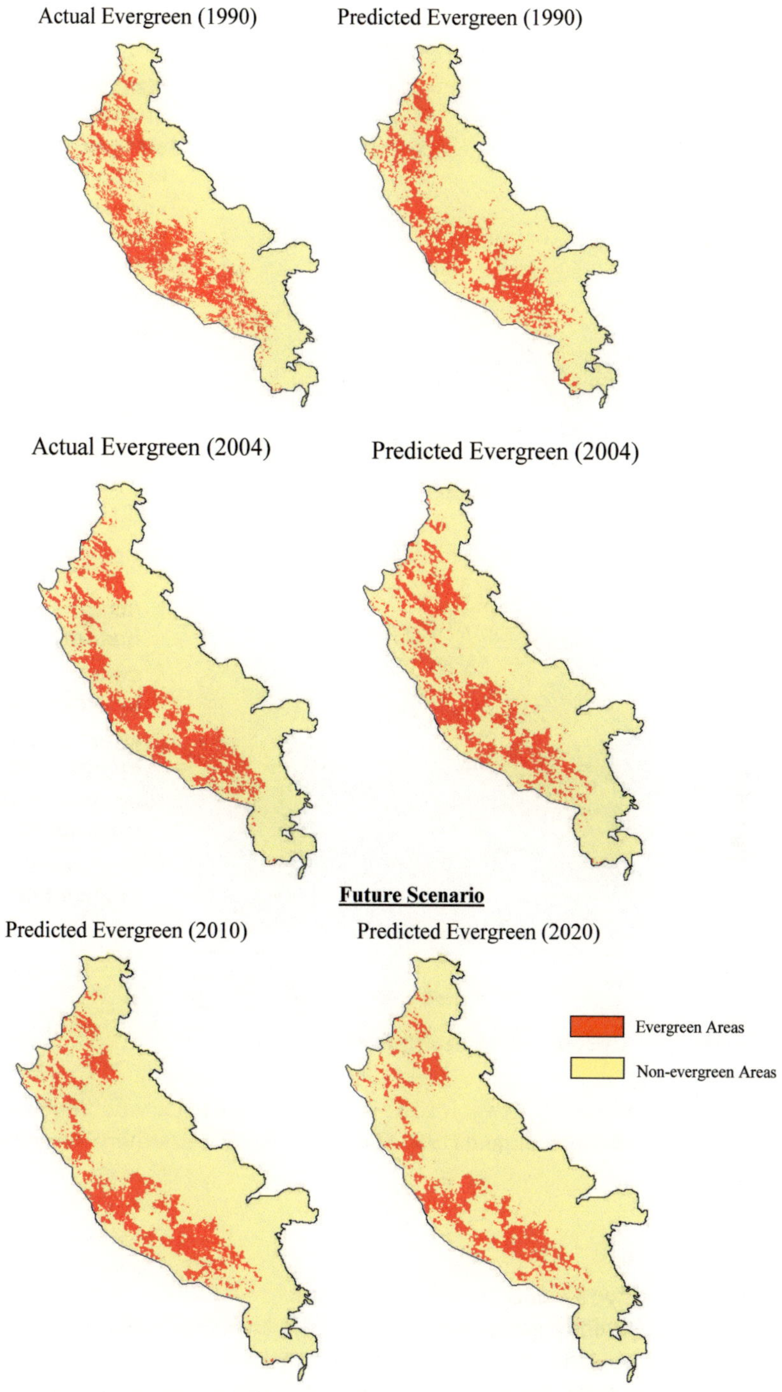

Fig. 8. Comparison of actual and predicted evergreen forest change between 1990 & 2004 using GEOMOD modeling and also the future scenario map of 2010 & 2020 for the study region (KMTR), Southern Western Ghats, Tamil Nadu, India.

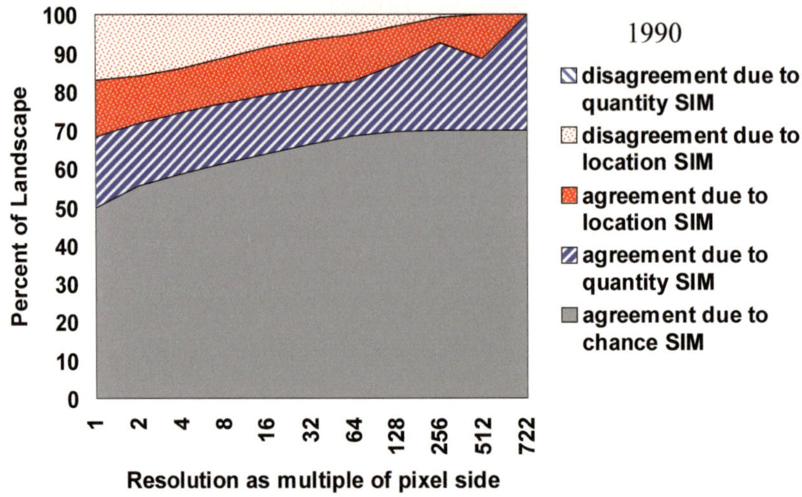

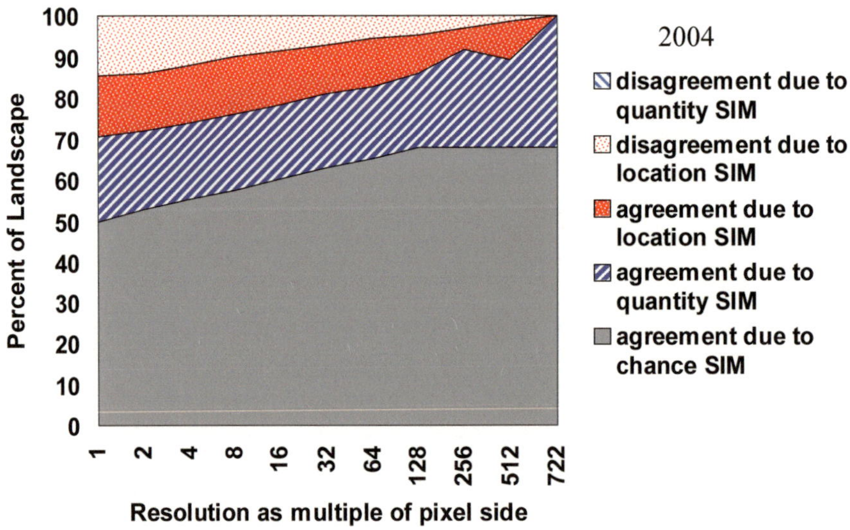

Fig. 9. Evaluation of the evergreen forests change using Kappa validation module (IDRISI) for the actual and reference image (1990 & 2004) of KMTR, Southern Western Ghats, Tamil Nadu, India.

4. Discussion

4.1 Area loss and type transition

The maintenance of biodiversity in a landscape depends on the extrinsic and intrinsic factors regulating the ecosystem and its functioning. Extrinsic factors include area loss (forest to non-forest), forest type transitions, destruction of natural vegetation by overgrazing, encroachment, logging, changes in landscape parameters (i.e. porosity, fragmentation, patchiness, shape and neighbourhood), forest edge influences, forest fires, and developmental activities within forest areas (i.e. road, tourism related pressure, fuel wood extraction, and NTFP extraction). Intrinsic factors which affect these changes are wildlife movements, invasive species, functional change in regeneration, modes and rate of propagule dispersion, decrease in girth-class distribution, and change in species composition, to name a few. These can directly or indirectly influence the process, function and regulators of biodiversity.

In most of the tropical regions, the major threat to forests is not outright deforestation. Rather, forests and their biota are affected by (1) simplification, where structurally rich native forests are converted to simplified secondary stands or other forest types due to harvesting of selective species, (Noss & Cooperrider, 1994; Perry, 1994) and (2) fragmentation, where the remaining tracts of native forests are separated into smaller patches by anthropogenic activities, resulting in a terrain that is hostile to many species and poses barriers to movement (Noss & Csuti, 1997; Saunders et al., 1991; Wilcove et al., 1986).

For studies involving change estimation over a spatial and temporal domain it is important to have ground-based surveys to validate the observed changes, but such surveys are rare and in most cases simply do not exist in the tropics. Since tropical forests are the most stressed among all the forest types and are rich in biodiversity, it is important that the changes in these forests are monitored effectively. For such analysis, we need to find some indirect method to validate the satellite-derived data. The comparative evaluation of the spectral property of NIR, SWIR and species composition has been found to be good indirect evidence and has helped in demarcating the distinct vegetation types which have undergone changes over three decades. The study has clearly identified the areas of secondary formation. This approach provides a reliable means of monitoring the landscape level transformation over a temporal frame.

The evergreen forests of the KMTR have undergone extensive degradation during the two and half decades with 16% of the evergreen forest being degraded primarily in the form of selective logging and clear felling to raise plantations of coffee, tea and cardamom resulting in secondary succession stages such as semi-evergreen among the major phenological types. Ramesh et al., (1997) have pointed out that there is significant loss of biologically rich areas between 1960 and 1990 (85.6km^2 to plantation, 42.0km^2 to encroachment and 36.4km^2 to reservoirs). The present study quantified an 8% increase of Ochlandra patches since 1972. The increase is further continuing due to anthropogenic pressure on the intact forest and can lead to secondary succession or degradation process. One of the most important causes for the growth of these reed brakes is the encroachment of cardamom and coffee plantation into evergreen forests over vast areas, including Uttu, Kakachi and Sengaltheri. Abandoned plantations are invaded by Ochlandra undergrowths, which prevent all other regeneration, as this reed is highly combustible during the dry season. Additionally recurring fires gradually enlarge the openings in the forest and gradually are replaced by Ochlandra reeds.

The loss of evergreen forest in KMTR has resulted in decreased spatial extent of interior or core forest. In case of semi-evergreen forests the spatial extent of interior category was increased due to conversion from evergreen forest. The absence of any significant change in the area under the edge category points to the fact that the process of transition of the evergreen to the semi-evergreen is still continuing. Most of this forest is actually lost to areas that are closer to settlements, roads and plantations. It is also observed that the reed growth is more prolific (Ochlandra reeds) in the vicinity of settlements and roads. Further, the rainfall intensity has significant impact over the natural degradation of the forests in the east-facing slopes where the intensity of erosion and runoff is more due to heavy downpour.

Temporal change analysis using GEOMOD using the three time scale data of years 1973, 1990 and 2004 has shown that an additional 27% of the non-degraded forest would be lost to the degraded category by 2020.

4.2 Forest fragmentation

Fragmentation in KMTR is mainly a result of anthropogenic activities like encroachment by plantations and orchards or natural changes like invasion of reed brakes. The initial invasion of reed brakes is due to the creation of canopy gaps in the intact forest patch, either due to felling or the death of mature trees. Once reed brakes establish themselves in the cleared patch, they start competing with the seedlings of the primary species for nutrients leading to the removal of the next generation of primary species. Apart from competing with the seedlings of the primary species, the reed brakes are also prone to fire during dry seasons. This leads to a further opening up of the forest region and results in to more fragmentation (Pascal, 1988). The evergreen forests are poor in soil nutrients due to increased assimilation by mature trees and can lead to irreversible damages resulting from the conversion of intact evergreen patches into degraded successions such as disturbed evergreen, or the invasion of deciduous and exotic species (Artaxo et al., 2002; Lewis et al., 2004; Lugo & Zimmerman, 2003).

We have observed an increase in fragmentation over time, which may lead to an increase in the isolation of evergreen patches in the KMTR. Decreasing patch size may lead to a reduction in the size of populations and to an increased risk of extinction of the remnant populations. Furthermore, colonization rates may be reduced in isolated patches (Joshi et al., 2006; Soons & Heil, 2002). Intuitively, the forests fragmented by anthropogenic sources are at higher risk of further fragmentation or removal than forest fragmented by natural causes. This isolation of the forest patches leads to a negative impact on the stand regeneration and also adversely affects the survival of species, which require contiguous forest patches for their survival and regeneration (Laurance et al., 1998, 2002; Niemi et al., 1998). Most of the disturbance to the forest patches is as a result of indirect anthropogenic pressure rather than direct encroachment or clear felling. This is indicated by a significant increase in the fractal dimension, which is a representation of the patch shape complexity (Krummel et al., 1987; Díaz-Delgadol et al., 2004). Identifying the forest fragmentation solely attributable to human activity may be a useful tool for policy and decision makers, allowing for improved risk assessments and better targeting of areas for protection or remedial action. The method presented produces data that may be summarized and displayed in a myriad of ways, each of which may be useful in the decision process.

4.3 Patch characterization

Patch characteristics of the 1973 and 2004 data in terms of size, proportion, shape and context were significantly different because of type transitions like evergreen to semi-evergreen, and expansion of Ochlandra and orchards. The variation in the physical environment viz., climate, soil, topography, and other landform features may lead to heterogeneous spatial distribution of resources like water, nutrients, and light resulting in the formation of vegetation patches of different characters (Burnett et al., 1998; Kolasa & Pickett, 1991; Nichols et al., 1998; Peters & Goslee, 2001). In addition, naturally occurring and man-made disturbances also play a vital role in patch formation and characteristics (Fuller et al., 1998; Platt, 1975). Factors such as wildlife grazing, wildlife movement, fuel wood extraction, fire, collection of honey and non-timber forest products by local people are also prevalent at different degrees of biotic pressure. The extent of forest and commercial plantation activity, and protection and conservation measures followed in the area also vary in degree and kind.

Remarkably the contiguous patches of >1000ha in 1973 had 90% of evergreen forest with less porosity due to a lesser degree of anthropogenic pressure compared to 2004, when there was 67% of evergreen forest with high level of porosity. The reduction in the patch area might be due to Ochlandra spread and increased plantations and habitation, resulting in the loss of such contiguous patches. These patches may be large enough to allow the natural disturbance regime to operate, maintain characteristic species composition, support mosaics of community formations, and sustain succession patterns and system functions (Pickett & Thompson, 1978). Further, the increase in the fractal dimension also point out to the fact that the KMTR is constantly under indirect pressure from the surrounding area's biotic or climatic/edaphic conditions.

Patch analysis thus provides a simple framework for goal-oriented monitoring and management in a forest landscape that has experienced several degenerative trends: primary evergreen forest have been replaced by semi-evergreen forests and plantation; structurally complex forests of all ages have been replaced by simplified stands; large, well connected patches have been replaced by smaller, more isolated patches; infrastructure has been developed in undisturbed landscapes; and natural fires have been suppressed. Several forests in Western Ghats and elsewhere in tropics have experienced these kinds of changes, with a concomitant loss of native biodiversity and ecological integrity (Dutt et al., 2002; Muthuramkumar et al., 2006). It is commonly accepted that species richness reduces with the fragmentation of tropical forests (Benitez-Malvido & Martinez-Ramos, 2003; Bierregaard, 1992; Chittibabu & Lovejoy, 1986; Laurance & Luizão, 2007). The smaller the fragments are, the lesser species richness the fragments display (Laurance, 1994; Leigh et al., 1993; Newmark, 1991; Pither & Kellman, 2002). Research articles published elsewhere related to forest fragments mostly focused on various animal groups. Studies on plants related to tropical forest fragments are relatively fewer, although there have been some important ones (Benitez-Malvido, 1998; Benitez-Malvido & Martinez-Ramos, 2003; Cadotte et al., 2002; Laurance & Luizão, 2007; Laurance et al., 1998; Leigh et al., 1993; Muthuramkumar et al., 2006; Oliveira-Filho et al., 1997; Turner & Corlett 1996).

Despite legal protection from major human activities, the region is subjected to various processes that ultimately prove detrimental to the sustenance of the native forest system. The focused priority on conservation of these patches may be helpful to sustain the biological diversity as these patches of evergreen forests provide unique habitats for various endemic plant species and wildlife. In this context the moderate spatial and high spectral resolution data from wide field sensors, can be used for generation of extensive information regarding vegetation area, patch shape and size, fragmentation patterns and porosity which are the major indicators of the disturbance and land-use change in a region.

Natural disturbances like fire, wildlife, proliferation of reeds and introduced exotic species, associated changes in geomorphology and soil, along with human pressure on the forests over the decades are the reasons for changes in the overall composition and the resulting establishment of habitat generalist species in KMTR. It is disturbing to note that the change areas have very high species diversity compared to no-change areas indicating that the ecosystem is undergoing changes in species composition and that the probability of further invasion of these species leading to a replacement of the habitat specialist species is high.

Temporal variation in species diversity can be strongly associated to the external disturbance (Chesson et al., 2004; Holt & Lawton, 1994). Disturbances like deforestation, fire, and harvest of selective species can strongly influence patterns of species diversity. A buffer of 100m was analyzed on the selected patches to understand the biotic pressure (e.g. orchards, secondary formations like reeds and semi-evergreen) and exchanges among patches, which determine the ecosystem structure and function. It reveals that 6.38% of evergreen and 5.13% of semi-evergreen forest have undergone several changes in the period between 1973 and 2004. Similar attempts were carried out to understand the patch dynamics and biotic exchanges among patches, and to determine the ecosystem structure and function (Lewin, 1984; Nagendra, 2001). On similar lines as the previous study, high diversity relationship was related to the characteristics of the community, the habitat, the disturbance and the sampling designs, as explained by Huston (1994); Mackey & Currie (2000, 2001); Noss (1996); and Reice (1985). In the present study, the disturbance showed the impact of strong local interactions and thus increased the relative importance of regional-scale processes (dispersal among the patches) in controlling the number of species within a patch. In this connection, the spatial organization of the patches as identified in the study can form a baseline for continuous monitoring and assessment of the changes in habitat conditions. Other similar case studies include Caswell & Cohen (1993); He et al., (2005); Mouquet et al., (2003); and Mouquet & Loreau (2003).

4.4 Trends in species richness and diversity pattern

The density of 607 trees ha^{-1} found in the no-change areas with a comparable girth threshold of ≥ 30cm, is greater than that of 419 trees ha^{-1} reported as mean tree density for WG closed-canopy evergreen forest (Ghate et al., 1998), 482ha^{-1} reported at the Mylodai site in Courtallum RF (Parthasarathy & Karthikeyan, 1997), Kakachi in the Southern WG (583ha^{-1}, Ganesh et al., 1996) and that of Uppangala, central WG (635ha^{-1}, Pascal & Pelissier, 1996). It is interesting to note that changed areas having stand density of 494 ha^{-1} are comparable with the frequently disturbed area having 575 trees ha^{-1} (Parthasarathy, 1999). The tree basal area having 54.39m^2ha^{-1} is well within the range of tropical wet forest: 55.34m^2ha^{-1} values in Sengaltheri mid-elevation forest, KMTR (Parthasarathy, 2001), and 66.87m^2ha^{-1} in Kalakad National Park, Tirunelveli (Parthasarathy, 1999). The nearby Nelliampathy wet evergreen forests of WG have also been reported to be in the range of 61.9m^2ha^{-1} (Chandrashekara & Ramakrishnan, 1994). In the tropical wet evergreen forests the stand basal area ranged from 25.5m^2ha^{-1} in Rio Xingu, Brazil (Campbell et al., 1992) to 82.67m^2ha^{-1} in the tropical rainforest of Reunion islands (Strasberg, 1996).

The study reveals that there is significant change in terms of species composition, stand density, basal area and percentage of endemic and edge species in the change and no-change areas. In the no-change area, the predominant species *Cullenia exarillata* – *Mesua ferrea* – *Myristica dactyloides* are well compared in the work carried out by Pascal (1988), Ganesh et al., (1996) and Parthasarathy (2001). In change areas, the floristic structure of *Cullenia exarillata* – *Dimocarpus longan* – *Kingiodendron pinnatum* has undergone several changes. This might be because of selective logging for establishment of tea and cardamom plantation. Interestingly the saplings of the trees and shrubs have also shown significant difference in the no-change and change areas. In the no-change area the species include *Antidesma menasu* – *Cinnamomum* malabatrum – *Mesua ferrea* whereas in the change area it consists of *Dimocarpus longan* – *Mallotus philippensis* – *Mesua ferrea* (Table 4a&b). Bresee et al., (2004), quantified these changes and rates of change in vegetative composition and structure within the Washburn Ranger

District in northern Wisconsin using Landsat images. The predominant families occurring in no-change areas include Euphorbiaceae, Lauraceae and Bombacaceae. These differ from those occurring in the change areas such as Euphorbiaceae, Sapindaceae and Lauraceae. Species composition, especially the abundance of some species and the dominant ranks of some families have changed with fragmentation (Zhu et al., 2004).

5. Conclusions

The study has demonstrated the capacity of remote sensing and GIS in detecting the land-cover change with data from different sensors in spite of the absence of past ground data, with appreciable level of accuracy. The different disturbance regimes mainly due to land-use and land-cover changes have led to the alteration in the patch size and shape in KMTR. Temporal change analysis using GEOMOD has shown that an additional 27% of the non-degraded forest would be lost to the degraded category by 2020 in the region. Furthermore, the creation of corridors has led to major changes in the biological diversity with bigger patches retaining higher species richness while smaller patches are dominated mainly by the edge species. The rarity of woody species and a greater number of singletons in the site underline the need to preserve the vast area of this forest, as a single large reserve. The establishment of plantations involves alternations of the rainforest habitat. Habitat alteration includes clear felling (for tea estates) and removal of many large trees, climbers and under storey vegetation (in coffee and cardamom plantations). Such a habitat change poses severe threats to wildlife (for e.g. lion tailed macaque, malabar spiny dormouse, nilgiri langur). The KMTR, besides being a biologically important area with unique flora and fauna, is also a habitat for wild relatives of cultivated plants, such as mango, jackfruit and cardamom. It is also a catchment area for more than 6 major river systems. Hence, the protection of this reserve is crucial for the biological conservation of species and human welfare.

6. Acknowledgements

We thank Dr. V. Jayaraman, Director and Dr. P. S. Roy, Deputy Director (RS and GIS-AA), National Remote Sensing Centre, Hyderabad for their encouragement. Thanks are due to DOS-DBT and Rufford Small Grant for their funding support to first author. We are also thankful to BR Ramesh (IFP), KS Bawa and M Irfan-Ullah (ATREE) for their necessary support. The authors are also grateful to the Tamil Nadu Forest Department for providing them the necessary field support.

7. References

Ali, R. and PAI, A., (2001), Human use areas in the Kalakad-Mundanthurai Tiger Reserve. Current Science, 80, pp. 448-452.

Amarnath, G., Murthy, M.S.R., Britto, S.J., Rajashekar, G. and DUTT, C.B.S., 2003, Diagnostic analysis of conservation zones using remote sensing and GIS techniques in wet evergreen forests of Western Ghats - An ecological hotspot, Tamil Nadu, India. Biodiversity and Conservation, 12, pp. 2331-2359.

Anonymous, History of Singampatti, A report of Bombay Burma Trading Company, Manjolai, Tamil Nadu, India

Artaxo, P., Martins, J.V., Yamasoe, M.A., Procopio, A.S. et al., 2002, Physical and chemical properties of aerosols in the wet and dry seasons in Rondonia, Amazonia. Journal of Geophysical Research. 107, (D20), 8081, doi: 10.1029/2001JD0000666.

Benitez-Malvido, J., 1998, Impact of forest fragmentation on seedling abundance in a tropical rain forest. Conservation Biology, 12, pp. 380–389.

Benitez-Malvido, J., and Martinez-Ramos, M., 2003, Impact of Forest Fragmentation on Understory Plant Species Richness in Amazonia. Conservation Biology, 17, pp. 389–400.

Bierregaard, R.O., Lovejoy, T.E., Kapos, V., Santos, A.A., and Hutchings, R.W., 1992, The biological dynamics of tropical rainforest fragments. Bioscience, 42, pp. 859–866.

Bresee, M.K., Moine, J.L., Mather, S., Brosofske, K.D., et al., 2004, Disturbance and landscape dynamics in the Chequamegon National Forest Wisconsin, USA, from 1972 to 2001. Landscape Ecology, 19, pp. 291–309.

Burnett, M.R., August, P.V., Brown, J.H. Jr, and Killingbeck, K.T., 1998, The influence of geomorphological heterogeneity on biodiversity. I. Patch-scale perspective. Conservation Biology, 12, pp. 363–370.

Cadotte, M.W., Franck, R., REZA L., and Lovett-Doust, J., 2002, Tree and shrub diversity and abundance in fragmented littoral forest of south-eastern Madagascar. Biodiversity and Conservation, 11, pp. 1417–1436.

Campbell, D.G., Stone, J.L., and Rosas, Jr. A., 1992, A comparison of the phytosociology and dynamics of three floodplain (Várzea) forests of known ages, Rio Juruá, western Brazilian Amazon. Botanical Journal of the Linnean Society, 108, pp. 213–237.

Caswell, H., and Cohen, J.E., 1993, Local and regional regulation of species–area relations: a patch-occupancy model, pp 99–107 in Ricklefs R.E. and Schluter D., (eds.). Species diversity in ecological communities. The University of Chicago Press, Chicago, Illinois, USA.

Chandrashekara, U.M., and Ramakrishnan, P.S., 1994, Vegetation and gap dynamics of a tropical wet evergreen forest in the Western Ghats of Kerala, India. Journal of Tropical Ecology, 10, pp. 337–354.

Channell, R., and Lomolino, M.V., 2000, Dynamic biogeography and conservation of endangered species. Nature, 403, pp. 84–86.

Chesson, P., Gebauer, R.L.E., Schwinning, S., Huntly, N. et al., 2004, Resource pulses, species interactions, and diversity maintenance in arid and semi-arid environments. Oecologia, 141, pp. 236–253.

Chittibabu, C.V., and Parthasarathy, N., 2000, Attenuated tree species diversity in human-impacted tropical evergreen forest sites in Kolli hills, Eastern Ghats, India. Biodiversity and Conservation, 9, pp. 1493–1519.

Civco, D.L., Hurd, J.D., Wilson, E.H., Arnold, C.L., and Prisloe, S., 2002, Quantifying and Describing Urbanizing Landscapes in the Northeast United States. Photogrammetric Engineering and Remote Sensing 68, pp. 1083–1090.

Costanza, R., 1989, Model goodness of fit: a multiple resolution procedure. Ecological Modeling, 47, pp. 199–215.

Despland, E., 2003, Fractal index captures the role of vegetation clumping in locust swarming, Functional Ecology, 17, pp. 315–322.

Díaz-Delgado R., Lloret, F., and PONS, X., 2004, Spatial patterns of fire occurrence in Catalonia, NE, Spain. Landscape Ecology, 19, pp. 731–745.

Dutt, C.B.S., Srivastava, V.K., Ranganath, B.K., Murthy, M.S.R., et al., 2002, Western Ghats biodiversity assessment at landscape level using remote sensing and GIS:A special issue on biodiversity. NNRMS Bulletin, 27, pp. 49-55.

Eastman, J.R., 2003, IDRISI 14.0 Kilimanjaro Manual. Guide to GIS and Image processing. http://www.clarklabs.org

Forman, R.T.T., and GODRON, M., 1986, Landscape Ecology. John Wiley and Sons, New York.

Frohn, R.C., 1998, Remote Sensing of Landscape Ecology, CRC Press, Boca Raton, Florida.

Fuller, J.L., Foster, D.R., Mclachlan, J.S., and DRAKE, N., 1998, Impact of human activity on regional forest composition and dynamics in central New England. Ecosystems, 1, pp. 76-95.

Ganesh, T., Ganesan, M., Devy, S., Davidar, P., and BAWA, K.S., 1996, Assessment of plant biodiversity at a mid-elevation evergreen forest of Kalakad-Mundanthurai Tiger Reserve, Western Ghats, India. Current Science, 71, pp. 379-392.

Ghate, U., Joshi, N.V., and GADGIL, M., 1998, On the patterns of tree diversity in the Western Ghats of India. Current Science, 75, pp. 594-603.

Gustafson, E.J. and Parker, G.R., 1992, Relationship between land cover proportion and indices of landscape spatial pattern. Landscape Ecology, 7, pp. 101-110.

Hall, C.A.S., Tian, H., Qi, Y., Pontius, Jr. R.G., et al., 1995, Modeling spatial and temporal patterns of tropical land use change. Journal of Biogeography, 22, pp. 753-757.

He, F., Gaston, K.J., Connor, E.F., and SRIVASTAVA, D.S., 2005, The local regional relationship: Immigration, Extinction and Scale. Ecology, 86, pp. 360-365.

Herold M., Couclelis, H., and Clarke, K.C., 2005, The role of spatial metrics in the anlaysis and modeling of urban land use change. Computers, Environment and Urban System, 29, pp. 369-399.

Holt, R.D. and Lawton, J.H., 1994, The ecological consequences of shared natural enemies. Annual Review of Ecological Systematics, 25, pp. 495-520.

Huston, M.A., 1994, Biological diversity: the coexistence of species on changing landscapes. Cambridge University Press, Cambridge, UK.

James, A.N., 1999, Institutional constraints to protected area funding. Parks, 9, pp. 15-26.

Kolasa, J., and Pickett, S.T.A., 1991, Ecological Heterogeneity. Ecological Studies 86. Springer-Verlag, New York.

Krummel, J.R., Gardner, R.H., Sugihara, G., O'Neil, R.V., and Coleman, P.R., 1987, Landscape patterns in disturbed environment. Oikos, 48: pp. 321-324.

Lam, N.S.N., and Ouattrochi, D.A., 1992, On issues of scale analysis and fractal analysis in mapping sciences. Professional Geographer, 44, pp. 88-98.

Laurance, W.F., 1994, Rainforest fragmentation and the structure of small mammal communities in tropical Queensland. Biological Conservation, 69, pp. 23-32.

Laurance, W.F., and Luizão, R.C.C., 2007, Driving a wedge into the Amazon. Nature, 448, pp. 409-410.

Laurance, W.F., Ferreira, L.V., Rankin-De Merona, J.M., and Laurance, S.G., 1998, Rain forest fragmentation and the dynamics of Amazonian tree communities. Ecology, 79, pp. 2032-2040.

Laurance, W.F., Lovejoy, T.E., Vasconcelos, H.L., Burna, E.M. et al., 2002, Ecosystem decay of Amazonian forest fragments: a 22-year investigation. Conservation Ecology, 16, pp. 605-618.

Laurance, W.F., Mesquita R., Luizão R., and Pinto F., 2004, The Biological Dynamics of Forest Fragments Project: 25 Years of Research in the Brazilian Amazon. Tropinet, 15, pp. 1-3.

Leduc, A., Prairie Y.T., and Bergeron, Y., 1994, Fractal dimension estimates of a fragmented landscape: sources of variability. Landscape Ecology, 9, pp. 279-286.

Leigh, E.G.J.,Wright, J., Herre, E.A., and Putz, F.E., 1993, The decline of tree diversity on newly isolated tropical islands: a test of a null hypothesis and some implications. Evolutionary Ecology, 7, pp. 76-102.

Lewin, R., 1984, Parks: how big is big enough? Science, 225, pp. 611-612.

Lewis, S.L., Malhi, Y., and Phillips, O.L., 2004, Fingerprinting the impacts of global change on tropical forests. Philosophical Transactions of The Royal Society of London Series B, 359, pp. 437-462

LI, H., and Reynolds, J.F., 1993, A new contagion index to quantify spatial patterns of landscapes. Landscape Ecology, 8, pp. 155-162.

Lillesand, T.M., and Kiefer, R.W., 2000, Remote sensing and image interpretation (4th edition), John Wiley and Sons, New York

Lugo, A.E., and Zimmerman, J.K., 2003, Ecological life histories of tropical trees with emphasis on disturbance effects, pp. 191-213 in: Vosso J. (eds.), Tropical tree seed manual. USDA Forest Service.

Mackey, R.L., and Currie, D.J., 2000, A re-examination of the expected effects of disturbance on diversity. Oikos, 88, pp. 483-493.

Mackey, R.L., and Currie, D.J., 2001, The diversity-disturbance relationship: Is it generally strong and peaked? Ecology, 82, pp. 3479-3492.

Margules, C.R. and Pressey, R.L., 2000, Systematic Conservation Planning. Nature, 405, pp. 243-53.

Midgley, G.F., Hannah, L., Millar, D., Rutherford, M.C., and Powrie, L.W., 2002, Assessing the vulnerability of species richness to anthropogenic climate change in a biodiversity hotspot. Global Ecology and Biogeography, 11, pp. 445-451.

Mouquet, N., and Loreau, M., 2003, Community patterns in source-sink metacommunities. American Naturalist, 162, pp. 544-557.

Mouquet, N., Munguia, P., Kneitel, J.M., and Miller, T.E., 2003, Community assembly time and the relationship between local and regional richness. Oikos, 103, pp. 618-626.

Nagendra, H., 2001, Incorporating landscape transformation into local conservation prioritization: a case study in the Western Ghats, India. Biodiversity and Conservation, 10, pp. 353-365.

Newmark, W.D., 1991, Tropical forest fragmentation and the local extinction of understory birds in the Eastern Usambara Mountains, Tanzania. Conservation Biology, 5, pp. 67-78.

Nichols, W.F., Killingbeck, K.T., and August, P.V., 1998, The influence of geomorphological heterogeneity on biodiversity. II. A landscape perspective. Conservation Biology, 12, pp. 371-379.

Niemi, G., Hanowski, J., Helle, P., Howe, R. et al., 1998, Ecological sustainability of birds in boreal forests. Conservation Ecology, [online] 2(2), pp. 17.

Noss, R.F., and Cooperrider, A.Y., 1994, Saving Nature's Legacy: Protecting and restoring biodiversity. Island Press. Washington, D.C., pp. 416.

Noss, R.F., and Csuti, B., 1997, Habitat Fragmentation, pp 269-304 in Principles of Conservation Biology, Second Edition (Meffe G.K. and Carroll, C.R. eds.). Sinauer Associates, Inc., Sunderland, Massachusetts.

Noss, R.S., 1996, Conservation of biodiversity at the landscape scale, pp 574-589 in Szaro, R.C. and Johnston, D.W., (eds). Biodiversity in managed landscapes: theory and practice. Oxford, New York, New York, USA.

O'Neil, R.V., Krummel, J.R., Gardner, R.H., Sugihara, G. et al., 1988, Indices of Landscape Pattern. Landscape Ecology, 1, pp. 153–162.

Oliveira-Filho A.T., Mello J.M., and De Scolforo J.R.S., 1997, Effects of past disturbance and edges on tree community structure and dynamics within a fragment of tropical semi deciduous forest in south-eastern Brazil over a five year period. Plant Ecology, 131, pp. 45–66.

Parmesan, C., and YOHE, G., 2003, A globally coherent fingerprint of climate change impacts across natural systems. Nature, 421, pp. 37–42.

Parthasarathy, N., 1999, Tree diversity and distribution in undisturbed and human-impacted sites of tropical wet evergreen forest in southern Western Ghats, India. Biodiversity and Conservation, 8, pp. 1365–1381.

Parthasarathy, N., 2001, Changes in forest composition and structure in three sites of tropical evergreen forest around Sengaltheri, Western Ghats. Current Science, 80, pp. 389–393.

Parthasarathy, N., and Karthikeyan, R., 1997, Biodiversity and population density of woody species in a tropical evergreen forest in Courtallum reserve forest, Western Ghats, India. Tropical Ecology, 38, pp. 297–306.

Pascal, J.P., 1982, Bioclimates of the Western Ghats at 1/500,000 (2 sheets). Institut Français, Pondichérry.

Pascal, J.P., 1988, Wet evergreen forests of the Western Ghats of India. Ecology, structure, floristic composition and succession. Institut Français De Pondichérry, pp. 345.

Pascal, J.P., and Pelissier, R., 1996, Structure and floristic composition of a tropical evergreen forest in south-west India. Journal of Tropical Ecology, 12, pp. 191–214.

Perry, D.A., 1994, Forest Ecosystems. The Johns Hopkins University Press. Baltimore, MD.

Peters, D.P.C., and Goslee, S.C., 2001, Landscape Diversity. In Levin SA (eds) Encyclopedia of Biodiversity 3: pp. 645–658 Academic Press, New York.

Pickett, S.T.A., and Thompson, J.N., 1978, Patch dynamics and the size of natural reserves. Biological Conservation, 13, pp. 27–37.

Pither, R., and Kellmann, M., 2002, Tree species diversity in small, tropical riparian forest fragments in Belize, Central America. Biodiversity and Conservation, 11, pp. 1623–1636.

Platt, W.J., 1975, The colonization and formation of equilibrium plant species associations on badger disturbances in a tall-grass prairie. Ecological Monographs, 45, pp. 285–305.

Pontius, Jr. R.G., and BATCHU, K., 2003, Using the relative operating characteristic to quantify certainty in prediction of location of land cover change in India. Transactions in GIS, 7, pp. 467–484.

Pontius, Jr. R.G., and Pacheco, P., 2004, Calibration and validation of a model of forest disturbance in the Western Ghats, India 1920-1990. GeoJournal, 61, pp. 325–334

Porwal, M.C., Sharma, L., and Roy, P.S., 2003, Stratification and mapping of Ephedra gerardiana Wall. in Poh (Lahul and Spiti) using remote sensing and GIS. Current Science, 84, pp. 208-212.

Ramesh, B.R., Menon, S., and Bawa, K.S., 1997, A vegetated-based approach to biodiversity gap analysis in the Agastyamalai region, Western Ghats, India. Ambio, 26, pp. 529–536.

Reice, S.R., 1985, Experimental disturbance and the maintenance of species diversity in a stream community. Oecologia, 67, pp. 90–97.

Riitters, K., Wickham, J., O'Neill, R.V., Jones, B., and Smith, E., 2000, Global-scale patterns of forest fragmentation. Conservation Ecology 4, pp. 3.

Root, T.L., Price, J.T., Hall, K.R., Schneider, S.H., Rosenzweig, C., and POUNDS, J.A., 2003, Fingerprints of global warming on wild animals and plants. Nature, 421, pp. 57–60.

Roy, P.S., Porwal, M.C., and Sharma, L., 2001, Mapping of Hippophae rhamnoides Linn. in the adjoining areas of Kaza in Lahul and Spiti using remote sensing and GIS. Current Science, 80, pp. 1107–1111.

Sala, O.E., Chapin, F.S.I, Armesto, J.J. Berlow, E., Bloomfield, J., Dirzo, R., et al., 2000, Global biodiversity scenarios for the year 2100. Science, 287, pp. 1770–1773.

Saunders, D.A., HOBBS, R.J., and MARGULES, C.R., 1991, Biological consequences of ecosystem fragmentation, a review. Conservation Biology, 5, pp. 18–32.

Soons, M.B., and Heil, G.W., 2002, Reduced colonization capacity in fragmented populations of wind-dispersed grassland forbs. Journal of Ecology, 90, pp. 1033–1043.

Strasberg, D., 1996, Diversity, size composition and spatial aggregation among trees on 1-Ha rain forest plot at La Réunion. Biodiversity and Conservation, 5, pp. 825–840.

Turner, I.M., and Corlett, R.T., 1996, The conservation value of small isolated fragmentation of lowland tropical rainforest. Tree, 11, pp. 330–333.

Udvardy, M.D.F., 1975, A classification of the biogeographical provinces of the world. (Occasional Paper no. 18). Gland, Switzerland: International Union for the Conservation of Nature and Natural Resources.

Vogelmann, J.E., 1995, Assessment of forest fragmentation in southern New England using remote sensing and geographic information system technology. Conservation Biology, 9, pp. 439–449.

Wickham, J.D., Jones, K.B., Riitters, K.H., Wade, T.G. et al., 1999, Transitions in forest fragmentation: implications for restoration opportunities at regional scales. Landscape Ecology, 14, pp. 137–145.

Wilcove, D.S., Mclellan, C.H., and Dobson, A.P., 1986, Habitat fragmentation in the temperate zone, pp. 237–256 in SOULÉ, M. E., editor. Conservation biology: the science of scarcity and diversity. Sinauer, Sunderland, Massachusetts, USA.

Zhu, H., Xu, Z.F., Wang, H., and LI, B.G., 2004, Tropical rain forest fragmentation and its ecological and species diversity changes in southern Yunnan. Biodiversity and Conservation, 13, pp. 1355–1372.

Part 4

Regional Biodiversity

8

An Agenda for Austrian Biodiversity Research at the Long-Term Ecosystem Research Network (LTER)

Stefan Schindler[1,5], Thomas Dirnböck[2], Franz Essl[2], Richard Zink[3], Stefan Dullinger[1,4], Thomas Wrbka[1] and Michael Mirtl[2]

[1]*Dptm. of Conservation Biology, Vegetation & Landscape Ecology, University of Vienna*

[2]*Environment Agency Austria*

[3]*Research Institute of Wildlife Ecology*

[4]*Vienna Institute for Nature Conservation & Analyses*

[5]*Centro de Investigação em Biodiversidade e Recursos Genéticos, Universidade do Porto*

[1,2,3,4]*Austria*

[5]*Portugal*

1. Introduction

Natural ecosystems provide a wealth of services that are useful, or even critical to humans (Daily, 1997; Millennium Ecosystem Assessment [MEA], 2003). Biodiversity, while being of intrinsic value per se, is meant to be a system property crucial to the provision of many of these services (Kremen, 2005; Luck et al., 2003). However, the link between diversity, ecosystem function and ecosystem services is still poorly understood (Hooper et al., 2005). Given the many threats to the future of biodiversity (Ehrlich & Pringle 2008), our limited knowledge of how human uses depend on and influence biodiversity is particularly alarming. Developing an agenda that links biodiversity research to socio-ecology in general, and to the study of ecosystem service provision and resource management in particular is hence an urgent issue.

In this book chapter, we present a research framework for Austrian biodiversity research under the umbrella of the Long-term Ecosystem Research (LTER) network (Mirtl, 2010; Mirtl et al., 2010). We elaborate research recommendations for the topics natural resources, resource use, energy production, climate change and pollutants, structural abiotic and biotic change, and the development of new methodological approaches. We further discuss institutional requirements for achieving a successful, efficient and competitive biodiversity research in Austria. We address the products of such research and their users as well as interlinks with the other thematic areas of LTER, namely process-oriented ecosystem research and socio-ecological research.

2. Long-Term Ecosystem Research (LTER)

The European Long-term Ecosystem Research (LTER-Europe) is a network linking 400 research sites, 100 institutions and thousands of research projects in 21 national networks, conducting research into the broad range of European terrestrial and aquatic ecosystems from arctic to Mediterranean areas and covering all major longitudinal and altitudinal gradients (Mirtl, 2010). LTER is currently going through a major restructuring (cf. www.ilternet.edu). This involves the design of infrastructure and the development of approaches focused on coupled socio-ecological systems that emerge through continuous interaction of human societies with ecosystems (Haberl et al., 2006, Singh et al., in press). Sponsored by the European Union, the European LTER infrastructure has been designed and implemented based on existing sites (Mirtl, 2010). One component of LTER-Europe, the Long-term Socio-ecological Research (LTSER) aims at the integration of natural science biodiversity research with socio-economic research (www.lter-europe.net). Referring to this umbrella, the LTER research strategy in Austria was formulated and published as the LTER-Austria White Paper (Mirtl et al., 2010). The White Paper covers basic ecosystem research, biodiversity research and conservation biology and LTSER. In Europe, LTSER will be carried out on so called LTSER Platforms which represent geo-political regions where the interaction of nature and human society can be studied. LTSER mainly investigates ecological and societal pressures on ecosystems, their driving forces, the social and economic consequences of changes in ecosystems including the development, monitoring and evaluation of biodiversity management and policies.

3. Priority research themes

Biodiversity research in the context of LTER is conducted over long periods of time, considers the full range of relevant scales, and/or relies on the LTER in situ infrastructure (Dirnböck et al., in press). The biodiversity research priorities presented here are based on several strategic documents targeting the Austrian and the European level. We used only strategic documents which had been compiled by a wide range of scientists and stakeholders to guarantee the integration of the breadth of the national research communities' priorities. The Austrian perspective is provided by documents compiled at the national level, such as the Declaration "Hardegger Erklärung", which was elaborated at the kick-off meeting of the Austrian Platform for Biodiversity Research (Plattform Biodiversität Forschung Austria - BDFA) and signed by 172 Austrians active in the field of biodiversity. We also considered a survey on the prioritization of issues in Austrian biodiversity research, which was conducted by the BDFA (Platform for Biodiversity Research in Austria, 2008), and was based on a British shortlist of the 100 most politically relevant ecological questions (Sutherland et al., 2006). In addition, the members of the conservation platform at the Federal Environment Agency - mainly including representatives of administrative bodies, NGOs, and businesses - were questioned. We focussed on research that is of outmost importance taking the Austrian biophysical conditions and land use patterns into account, i.e. high importance of mountains, forests, freshwater and agricultural ecosystems. As the Austrian biodiversity research priorities are strongly linked to the European research agenda, we included the European perspective which is provided by several strategic documents elaborated by the European Platform for Biodiversity Strategies (EPBRS); this especially applies to "Mountain Biodiversity" (EPBRS, 2006), "Biodiversity in the Wider

Countryside" (EPBRS, 2007a), "Biodiversity and Ecosystem Services" (EPBRS 2007b) and "Freshwater Biodiversity" (EPBRS, 2008), being of particular relevance for the most important Austrian ecosystems. Consideration was also given to the very recently adopted EPBRS recommendations regarding ecosystem services (EPBRS, 2011) and to the "EPBRS Biodiversity Research Strategy 2010-2020" (EPBRS, 2010), which calls for a strong focus on research areas that generate the knowledge necessary to fulfil the following political goals:

- ensuring the long-term survival of species, their genetic diversity, and the ecological integrity and functionality of habitats and ecosystems
- ensuring the long-term provision of ecosystem services
- adapting to global change (including climate change)
- contributing to meeting other Grand Challenges (water, food, energy supply; population growth; human health)

The result of this survey led to three subject areas: resources and resource use, energy production, climate change and pollutants, and structural abiotic and biotic change.

3.1 Resources and resource use

This category includes the study of one or more species, of habitats and of ecosystem processes across guilds and trophic levels. LTER allows for a close alignment of biodiversity research and traditional ecosystem research, which primarily focuses on energy and material flows. Hence, the focus here is on the interaction between organisms and ecosystem processes. LTSER platforms can be used to extrapolate the gained knowledge from LTER based research to the regional, geopolitical scale. Studies about the utilization and conservation of biodiversity as well as the consequences of changes in utilization and their conservation impact are of particular importance. LTER Austria is an optimal frame to provide answers to research questions such as: To what extent do Austria's nature reserves meet a given set of goals (e.g. halting the loss of species, protecting endangered populations as well as endemic, demanding, rare or migratory species, etc.)? What are the consequences of the various (EU-guided) forms of agricultural land use on the conservation of biodiversity (Wrbka et al., 2008)? To what extent do individual forms of land management, such as hunting, fishing, forestry and farming, affect endangered populations?

Several topics that were given priority by the Austrian biodiversity research and conservation community were related to resources and resource use. These prioritized topics mainly dealt with the species themselves (taxonomy, distribution and abundance of species, population ecology, protection of species in situ), but also studies on the impact of organic farming and more investigation related to wetlands are required (Platform for Biodiversity Research in Austria, 2008). Of the research recommendations made by EPBRS, those relating to mountain and freshwater biodiversity (cf. EPBRS, 2006, 2008) are most relevant for biodiversity research at LTER Austria. Of particular interest is:

- a better understanding of the impact of human activities on the long-term sustainability of biodiversity;
- a better understanding of the role of genetic and species diversity for ecosystem dynamics, functions, and services;
- the coupling of research and long-term monitoring to assess the status, patterns and drivers of European mountain biodiversity at various scales of space and time;
- The definition of favourable states for mountain habitats and populations, as well as the identification of reference states for mountain ecosystems evaluating and taking into account ecosystem services;

- the definition of criteria, indicators, methods and processes for efficient conservation and sustainable management of mountain biodiversity;
- increased assessment of status and distribution of little-studied, ecologically important, or endangered freshwater taxa, habitats, and ecosystems;
- further development of tools to effectively conserve and sustainably use freshwater ecosystems, taking into account their specific characteristics, such as spatial and temporal dynamics and connectivity; and
- a better understanding of the functioning and role of soil biodiversity and subterranean freshwater biodiversity, especially as they relate to ecosystem services and their indicators.

3.2 Energy production, climate change and pollutants

The interactions between organisms, biotic communities, and the main driving forces of global change are of central interest here. The related knowledge is still very scarce and more targeted research is necessary to guide effective conservation measures. The following topics were given priority by the Austrian biodiversity researchers: climate change, climate policy, biofuels and hydropower (Platform for Biodiversity Research in Austria, 2008). Studies on ecosystem functioning are the core of LTER. Ideally, experimental and observational studies should be nested in the long-term monitoring schemes, which document changes of biodiversity and the environment over longer timeframes. This is especially true when it comes to climate change, climate policy and climate change mitigation and adaptation measures, which are currently implemented in numerous sectors such as agriculture, forestry, energy production and tourism. In view of the potentially severe effects of climate change in high mountain ecosystems (Engler et al., 2011), research in high-alpine territory is especially important (Dirnböck et al., 2011; Gottfried et al., 2011; Pauli et al., 2007). Studies on the impacts of climate change and its interaction with human land use on mountain biodiversity should constitute a core field in European research (EPBRS, 2006). The effects of fossil fuel emissions and agriculture on biodiversity (e.g. CO_2 effects, excess of reactive nitrogen, toxic substances, etc.) as well as the role of biodiversity for the functioning of ecosystems (e.g. carbon sequestration) are other highly relevant research topics.

3.3 Structural abiotic and biotic change

Structural changes of ecosystems have been massively accelerated by industrialization, land use change, habitat loss and fragmentation, and increased human mobility. The latter factor is the main driver of the invasive spread of non-native species (Pyšek et al., 2010).
The progressive loss of traditional landscape structures drives a massive crisis of farmland biodiversity that will probably not be completely realized until several decades into the future (Kuussaari et al., 2009). This opens a window of opportunity for rapid rethinking and the development of sustainable forms of utilization. Higher altitudes in the Alps still harbour many natural habitats. In the lowlands, natural and semi-natural habitats, which are important for biodiversity conservation (e.g. meadows, pastures, old-growth deciduous forests, and riverine areas) occur currently mainly as fragmented remnants of often an unfavourable status. The following topics related to the "wider countryside" (EPBRS, 2007a) and "freshwater biodiversity" (EPBRS, 2008) were recommended as research themes by EPBRS and should be included within the framework of LTER Austria:

- the importance of landscape structures, patterns and gradients for biodiversity, applied across different scales;
- effects of demographic, social, and economic trends as well as EU policies (including their national implementation) on biodiversity;
- indirect effects of climate changes (e.g. biofuel production);
- improving Agri-Environmental Schemes so that they deliver more measurable positive impacts for biodiversity; and
- the role of refugia in maintaining the long-term adaptive and evolutionary capacities.

Thus, studies related to cultural landscapes, landscape fragmentation and ecological corridors are required. Core research areas should include the effects of agriculture policies and changes in land use (e.g. land abandonment and subsequent afforestation of traditional cultural landscapes) on the species richness and composition of ecological communities (cf. Wrbka et al., 2008), the soil, and the vegetation structure. A special focus should also be given to the easily overlooked long-term effects of changing land use practices on biodiversity ("extinction debt", "invasion debt", cf. Essl et al., 2011; Kuussaari et al., 2009) which represent both a hidden threat and an opportunity for timely countermeasures. The use of genetically modified organisms and associated risks for the ecosystem will also be an essential focus of future research (e.g. Pascher & Gollmann, 1999; Pascher et al., 2011).

Transdisciplinary approaches that include stakeholders (farmers, foresters, hunters, people seeking recreation etc.) are indispensable for the restoration of the ecological integrity of cultural landscapes, traditional landscape patterns, and the ecosystem services associated therewith. While LTSER platforms provide ideal infrastructure for regional case studies, particularly in the context of transdisciplinary research (Singh et al., in press), LTER sites may serve as a pool for long-term monitoring data and sites for experimental approaches.

4. Approaches and methods

Within the framework of the "Hardegger Erklärung zur österreichischen Biodiversitätsforschung" 2008 (Platform for Biodiversity Research in Austria, 2008), the following three research questions were prioritised (compare also EPBRS, 2010):

- How do methods for evaluating the function of biodiversity in ecosystems need to be improved to capture its importance in supporting ecosystem services crucial for human wellbeing?
- How do biodiversity indicators and monitoring systems need to be improved to identify and prospectively assess the interaction between biological diversity and the drivers of global change?
- What are the most effective strategies and methods to assess, conserve, restore and sustainably use biological diversity?

4.1 Ecosystem functions and services

The concept of ecosystem functions and services (Boyd & Banzhaf, 2007; Costanza et al., 1997; Daily, 1997; De Groot et al., 2002) has been increasingly employed during recent years, since it facilitates an approach to evaluating the importance of intact ecosystems for humans. In the "Millennium Ecosystem Assessment" (MEA, 2003) and "The Economics of Ecosystems and Biodiversity" (TEEB, 2009), the importance of biodiversity and the corresponding ecosystem services was analysed and evaluated. 23 ecosystem functions were determined, based on an even larger set of ecosystem goods and services (De Groot et al., 2002, see also Hermann et al., 2011 for a recent review). The contribution of biodiversity to ecosystem services and the influence of drivers and pressures on conservation and use of ecosystems are research aspects of particular importance (Kremen, 2005; EPBRS, 2007b, 2011). In the frame of a recent meeting under the Hungarian EU presidency that took place 27-29 of April 2011, the EPBRS (2011) adopted research recommendations regarding ecosystem services with the following ones being specifically relevant in the context of Austrian biodiversity research in the frame of LTER and LTSER:

- Develop standardized methods and criteria for the measurements, mapping and monitoring of biodiversity and ecosystem services at various temporal and spatial scales;
- Understand the ecological, economic and social aspects of the multiplicity of ecosystem services, identify trade-offs and synergies occurring between services, and develop management mechanisms and innovative uses;
- Identify and characterize linear and non-linear social and ecological dynamics (including tipping points) and their interactions, to foster ecosystem service resilience;
- Improve existing and develop innovative management techniques to reduce or eliminate drivers of dangerous change in ecosystem services or disservices such as biological invasions, chemical pollution including pharmaceuticals, and eutrophication;
- Assess the impacts on ecosystem services of novel or emerging pressures, such as alternative energy production, abrupt changes in management regimes in an oilconstrained world, and pollution by light and noise, nano-particles and micro-plastics;
- Better understand the disruption of ecosystem services, at various scales in time and space, caused by natural and anthropogenic drivers operating through phenomena such as mismatch in processes related to phenology, trophic interactions, and migration;
- Take into account uncertainty, complexity, and all relevant knowledge including local and traditional knowledge, in developing tools and methods to support the integration of ecosystem services into management and decision making in public and private sectors;
- Take into account the potential for changes in values under future scenarios, and the variability of values in various spatial, temporal and cultural contexts;
- Understand and evaluate ecosystem services provided by poorly known ecosystems such as glaciers, groundwater, and aquatic microbial communities;
- Identify the main threats to soil biodiversity (including to specific functional groups) and quantify their impacts on ecosystem processes and services;

4.2 Indicators

Indicators simplify, quantify, and communicate information on ecosystem processes that are too complex to be measured directly (Hammond et al., 1995). Biodiversity and sustainability in their entirety require very complex methods of measurement, which is why indicators are usually applied (Walpole et al., 2009). The indicators that are most relevant in terms of environmental policy are those that are easy to survey, efficient, cost-effective, sensitive to processes of change and robust against other influences (e.g. EEA, 2007; Gregory et al., 2009; Kati et al., 2010; Pauli et al., 2007; Renetzeder et al., 2010; Schindler et al., 2008; Tasser et al.,

2008). Frequently, environmental indicators are related to habitat and species diversity, land use and land cover, and invasive species. Biodiversity indication is a difficult task and the development of standardized methods to harmonize and supplement indicators for biodiversity as well as for its driving forces and the causes of endangerment is a European biodiversity research focus (EPBRS, 2007a). Well established indicators, such as the IUCN Red List Index, can undermine their own indicator performance as conservation actions become targeted towards Red List species (Newton, 2011). To ensure that naturally species-poor habitats (e.g. mires or acidic beech forests) are adequately represented, the contribution of such areas to overall biodiversity must be considered. Current indicators of species diversity have to be expanded towards genetic diversity and ecosystem diversity (Walpole et al., 2009), and multi-taxa approaches must be applied more frequently in conservation practise (Edenius & Mikuszinski, 2006; Poirazidis et al., 2010). Increasing the taxonomic, geographic and temporal area of biodiversity indicators has to be a paramount goal of biodiversity research. Due to long time series, simultaneous in-situ data of environmental and human pressures and its effects and integrative approaches, LTER Austria provides an outstanding opportunity for testing and improving indicators for biodiversity, sustainability, and climate change. In particular the LTSER platforms provide the possibility to relate such indicators to socioeconomics and ecosystem services, which constitutes another important research topic (Sachs et al., 2009).

4.3 Approaches for conservation and sustainable use of biodiversity

To conserve rare natural goods in the long term, research today increasingly has to address not only autecological problems but also synecological aspects on population and metapopulation levels. In this context, the methodological question of choosing the "right" spatial and temporal scale is of crucial importance for the design of new concepts of evidence based conservation and sustainability (Dirnböck et al. in press). The larger the areas designated for research, the more feasible it is to conduct studies on the level of the (meta¬)population (e.g. gene flow). At larger spatial scales, it is normally not feasible to gather field data across the whole investigation area, and ecological modelling is used instead (Elith et al., 2006; Guisan & Thuiller, 2005). Long time series of in-situ data are necessary to increase the precision of models that aim for instance at detecting changes of the composition of communities and population trends. The importance of indicators and modeling is also increasing, as a growing number of research questions is met with ever decreasing budgets, making it more important than ever to use funds economically. Ecological modeling, however, is not only a means of reducing cost, but is actually a field of research in itself. Further methods that, until recently, were still in their infancy regarding their application in biodiversity research (e.g. genetics, remote sensing) are now valuable options, opening up new fields of research (Avise, 2008; Gillespie et al., 2008; Grill et al., 2007; Schindler et al., 2010).

The human use of ecosystems is omnipresent. The socioeconomic component of LTER, namely LTSER, and relevant biodiversity research has gained tremendously in importance over the last two decades (Mirtl et al., 2010; Singh et al., 2010). LTSER platforms provide an optimal infrastructure to meet this new requirement, enabling research that links biophysical processes to governance and communication, consider patterns and processes across several spatial and temporal scales, combines data from in-situ measurements with statistical data, cadastral surveys, and soft knowledge from the humanities (Haberl et al.,

2006). The inclusion of society into the existing research infrastructure facilitates transdisciplinary approaches. These approaches, which include the participation and mutual learning of stakeholders, are crucial when the research focus lies on the indirect drivers of biodiversity loss (Balian et al., 2011; EPBRS, 2010, 2011), or when the gap between science (e.g. conservation planning and research based conservation recommendations) and action (e.g. implementation of conservation actions) should be bridged (Reyes et al., 2010; Schindler et al., 2011). Stakeholder involvement can also be of advantage when defining conservation priorities. For this purpose, transnational conservation initiatives such as the European Habitat and Birds Directives as well as biodiversity-related Multilateral Environmental Agreements have to be innovatively applied (Mauerhofer 2010, 2011) along with local or national assessments (e.g. national red lists, assessment of global conservation responsibilities).

5. Requirements

5.1 Structural requirements

Concerted research efforts are absolutely crucial for developing scientifically substantiated approaches to solving current problems related to biodiversity and ecosystems. Therefore, a research program founded upon a general consensus of the Austrian research community and approved at an international level is of great importance. To further strengthen research efforts, an even more efficient network of existing research facilities, initiatives, nature reserves and conservation programs is needed. A closer connection to European and international ecosystem research (e.g. LTER-Europe) is desirable; education in schools and universities must be encouraged and research institutions such as museums or universities need increased long-term financing. Cooperation and communication between science and the interested public needs to be specifically promoted.

5.2 Institutional requirements

Implementing the above-mentioned structural requirements implies institutional changes. Within the framework of the EPBRS biodiversity research strategy 2010-2020, five fields are presented for developing the research environment that is needed (EPBRS, 2010):

- continuous identification, revision and "horizon scanning" (i.e. wide, interdisciplinary early recognition of future developments; cf. Sutherland et al., 2010, 2011) of research foci;
- support of European and international platforms (e.g. GEO Bon, ILTER, GBIF, Biodiversity-Knowledge);
- increasing capacity through general and advanced education;
- creation of links between research and politics (e.g. via the Intergovernmental Science-Policy Platform on Biodiversity and Ecosystem Services – IPBES); and
- regular evaluation of European biodiversity research with particular reference to its practicability and the applicability of research findings.

From the Austrian research community`s point of view, highest priority should be given to a better access to biodiversity-relevant information and databases (e.g. geodata, biodiversity data, environmental data); the long-term nature and continuity of networks and projects; integration/networking with international biodiversity research and other international initiatives; as well as improved access to research funding (Platform for Biodiversity

Research in Austria, 2008). A central data collection hub that is easily accessible for LTER-researchers, the "Data Center for Biodiversity and Conservation Research", is to function as an infrastructural institution in support of research activities and as such is seen as a vital prerequisite for improving the quality of research. Another key factor is ensuring the long-term support of existing institutions contributing to biodiversity and conservation research (e.g. nature reserves, museums and collections) as well as access to the data stored at these facilities. A consensual approach to the establishment of future research foci also seems to be of particular importance. This is where the concept of LTER comes into play, without which it would be almost impossible for selected LTER sites to bring together manageable amounts of data in a competent way, i.e. linked and made accessible to individual research groups.
The transnational LTER network offers the advantage of access to international data collections related to sites, where a wide range of potential drivers of biodiversity are measured simultaneously. As a first step, it provides meta-information on the existence of data sets and their holders and supports Austrian research teams to present their data and studies to the international research community - a fact that is highly relevant with respect to acquiring European funding. From a present-day perspective, mapping the research foci seems to be imperative and would give funding bodies a better overview of the entire research landscape. Identifying teams worthy of funding could thus be carried out in a balanced way across all sectors, to the benefit of current research foci.
In this context the ESFRI project LifeWatch is of high relevance (www.lifewatch.eu). It links "resources" (elements producing biodiversity related data like LTER Sites or collections) with the scientific users of such resources by supporing data mining, access and workflows related to complex analyses. LTER-Europe represents one of the major in-situ components of LifeWatch. Communities as well as national organisations engaged in LTER-Europe and LifeWatch are highly overlapping in about 50% of all LifeWatch countries, securing efficient lobbying and maximum use of synergies. In Austria a national LifeWatch strategy has been adopted (Mirtl et al., 2011), integrating LTER-Austria, the BDFA and the Austrian Biodiversity Documentation (museums and collections organized as national GBIF consortium).

6. Products and users

The driving forces of global change force public officials and conservation bodies to deal with complex questions, such as "Where do conservation measures make sense from an ecological or economic standpoint?" or "On which spatial scale are they likely to provide positive results?".
The more precisely it is possible to assess future developments, the easier it is to successfully counteract undesirable developments. Reflecting the wide spectrum of expertise involved, the range of results from biodiversity and conservation research is immensely varied. Their products should be made available to the research community, but should also serve policy makers and society as a basis for future planning and decision-making. Precisely because of the many interfaces between them and the various land use sectors, agriculture, forestry and recreational industries, the transdisciplinary results of biodiversity and conservation research provide practical approaches to the sustainable exploitation of traditionally-used resources. Decision-makers and in many cases the custodians of essential goods (e.g. water) are thus direct beneficiaries.

7. Conclusion

The global loss of natural habitats, biodiversity and ecosystem services represent one of the biggest challenges facing mankind. Emerging issues that could have substantial impacts on the conservation of biological diversity may become reality in the near future (Sutherland et al., 2010, 2011). By combining research and long-term monitoring and creating the necessary infrastructure for this, LTER Austria - in cooperation with LTER networks in other countries - can provide science based answers to the problems arising at an ever increasing rate due to global change.

8. Acknowledgment

We are grateful to the other Austrian researchers, who collaborated in the compilation of the LTER Austria White paper (Mirtl et al., 2010) and to Volker Mauerhofer for his helpful comments on this manuscript. This contribution was partly funded by LTER Austria as well as by the project "Bioserve" of the Austrian Academy of Science.

9. References

Avise, J.C. (2008). Three ambitious (and rather unorthodox) assignments for the field of biodiversity genetics. *Proceedings of the National Academy of Sciences of the United States of America,* Vol. 105, Supplement 1, (August, 2008), pp. 11564-11570, ISSN 00278424

Balian, E.V.; Berhault, A.; Rode, J.; Schindler, S. & Sharman, M. (2011). Report of the Positive Visions for Biodiversity Summit: the 2010 European Platform for Biodiversity Research Strategy (EPBRS) meeting under the Belgian Presidency of the European Union. EPBRS, Brussels. 14.05.2011. Available from: http://www.positivevisionsforbiodiversity.org/pg/file/read/2952/report-of-the-positive-visions-for-biodiversity-summit

Boyd, J. & Banzhaf, S. (2007). What Are Ecosystem Services? The Need for Standardized Environmental Accounting Units. *Ecological Economics,* Vol.63, No.2-3, (August 2007), pp. 616-626, ISSN 0921-8009

Costanza, R.; d'Arge, R.; de Groot, R.S.; Farber, S.; Grasso, M.; Hannon, B.; Limburg, K.; Naeem, S.; O'Neill, R.; Paruelo, J.; Raskins, R.; Sutton, B. & van den Belt, M. (1997). The value of the world's ecosystem services and natural capital, *Nature,* Vol.387, No.6630, (May 1997), pp. 253-260, ISSN 0028-0836

Daily, G. (1997). *Nature's Services: Societal Dependence on Natural Ecosystems,* Island Press, ISBN 978-155-9634-76-2, Washington, DC

De Groot, R.S.; Wilson, M.A. & Boumans, R.M.J. (2002). A typology for the classification, description and valuation of ecosystem functions, goods and services. *Ecological Economics,* Vol.41, No.3, (June 2002), pp. 393-408, ISSN 0921-8009

Dirnböck, T.; Bezák, P.; Dullinger, S.; Haberl, H.; Lotze-Campen, H.; Mirtl, M.; Peterseil, J.; Redpath, S.; Singh, S.; Travis, J. & Wijdeven, S.M.J. (in press). Critical scales for integrated biodiversity research, In: *Long term socio-ecological research: Studies in society-nature interactions across spatial and temporal scales.* S.J. Singh, H. Haberl, M. Chertow, M. Mirtl & M. Schmid (Eds.) Springer, ISBN 978-94-007-1176-1

Dirnböck, T.; Essl, F. & Rabitsch, W. (2011). Disproportional extinction risk of high-altitude endemic species under climate change. *Global Change Biology,* Vol.17, No.2 (February 2011), pp. 990-996, ISSN 1365-2486

Edenius, L. & Mikusinski, G. (2006). Utility of habitat suitability models as biodiversity assessment tools in forest management. *Scandinavian Journal of Forest Research,* Vol.21, Suppl.7, (February 2006), pp. 62-72, ISSN 1400-4089

EEA (2007). European Environment Agency Report No. 11. Halting the loss of biodiversity by 2010: proposal for a first set of indicators to monitor progress in Europe. Copenhagen, Denmark, 14.05.2011, Available from http://www.eea.europa.eu/publications/technical_report_2007_11

Ehrlich, P.R. & Pringle, R.M. (2008). Where does biodiverstiy go from here? A grim business-as-usual forecast and a hopeful portfolio of partial solutions. *Proceeedings of the National Academy of Sciences of the USA,* Vol.105, Supplement 1 (August 2008), pp. 11579-11586, ISSN 1091-6490

Elith, J.; Graham, C.H.; Anderson, R.P.; Dudík, M.; Ferrier, S.; Guisan, A.; Hijmans, R.J.; Huettmann, F.; Leathwick, J.R.; Lehmann, A.; Li, J.; Lohmann, L.G.; Loiselle, B.A.; Manion, G.; Moritz, C.; Nakamura, M.; Nakazawa, Y.; Overton, J.McC.; Peterson, A.T.; Phillips, S.J.; Richardson, K.S.; Scachetti-Pereira, R.; Schapire, R.E.; Soberón, J.; Williams, S.; Wisz, M.S. & Zimmermann, N.E. (2006). Novel methods improve prediction of species' distributions from occurrence data. *Ecography,* Vol.29, No.2 (April, 2006), pp. 129-151, ISSN 1600-0587

Engler, R.; Randin, C.; Thuiller, W.; Dullinger, S.; Zimmermann, N.E.; Araújo, M.B.; Pearman, P.B.; Le Lay, G., Piédallu, C.; Albert, C.H.; Choler, P.; Coldea, G.; de Lamo, X.; Dirnböck, T.; Gégout, J.-C.; Gómez-García, D.; Grytnes, J.-A.; Heegaard, E.; Høistad, F.; Nogués-Bravo, D.; Normand, S.; Puşcas, M.; Sebastià, M.-T.; Stanisci, A.; Theurillat, J.-P.; Trivedi, M.; Vittoz, P. & Guisan, A. (2011). 21st century climate change threatens mountain flora unequally across Europe. *Global Change Biology,* Vol.17, No.7 (July, 2011), pp. 2330-2341, ISSN 1365-2486

EPBRS (2006). EPBRS Recommendations on Europe's Mountain Biodiversity: Research, Monitoring, Management. Recommendations of the meeting of the European Platform for Biodiversity Research Strategy held under Austrian Presidency of the EU. EPBRS, Vienna, Austria, 14.05.2011, Available from http://www.epbrs.org/PDF/AT-2006-MountainBiodiversity-Final.pdf

EPBRS (2007a). Biodiversity in the wider countryside. Recommendations of the meeting of the European Platform for Biodiversity Research Strategy held under German Presidency of the EU. EPBRS, Leipzig, Germany, 14.05.2011, Available from http://www.epbrs.org/PDF/EPBRS-DE2007-Wider%20countryside%20final.pdf

EPBRS (2007b). Biodiversity and ecosystem services: the Millennium Ecosystem Assessment framework in a European perspective. Recommendations of the meeting of the European Platform for Biodiversity Research Strategy held under German Presidency of the EU. EPBRS, Leipzig, Germany, 14.05.2011, Available from http://www.epbrs.org/PDF/EPBRS-DE2007-Mill%20Ecosystem%20final.pdf

EPBRS (2008). Water for Life: Research priorities for sustaining freshwater biodiversity. Recommendations of the meeting of the European Platform for Biodiversity Research Strategy held under Slovenian Presidency of the EU. EPBRS, Brdo, Slovenia, 14.05.2011, Available from http://www.epbrs.org/PDF/EPBRS-SI2008-Freshwater_Final_.pdf

EPBRS (2010). European Biodiversity Research Strategy 2010-2020. Version 1. EPBRS, Palma de Mallorca, Spain, 14.05.2011, Available from http://www.epbrs.org/PDF/EPBRS_StrategyBDResearch_May2010.pdf

EPBRS (2011). Recommendations of the meeting of the European Platform for Biodiversity Research Strategy held under the Hungarian Presidency of the EU concerning ecosystem services. EPBRS, Budapest, Hungary, 14.05.2011, Available from http://share.bebif.be/data/EPBRS/EPBRS-HU2011-EcosystemServices_Final.pdf

Essl, F.; Dullinger, S.; Rabitsch, W.; Hulme, P.E.; Hülber, K.; Jarošík, V.; Kleinbauer, I.; Krausmann, F.; Kühn, I.; Nentwig, W.; Vilà, M.; Genovesi, P.; Gherardi, F.; Desprez-Loustau, M.-L.; Roques, A. & Pyšek, P. (2011). Socio-economic legacy yields an invasion debt. *Proceedings of the National Academy of Sciences of the USA,* Vol.108, No. 1 (January 2011), pp.203-207, ISSN 1091-6490

Gillespie, T.W.; Foody, G.M.; Rocchini, D.; Giorgi, A.P. & Saatchi, S. (2008). Measuring and modeling biodiversity from space. *Progress in Physical Geography,* Vol.32, No.2 (April, 2008), pp. 203-221, ISSN 0309-1333

Gottfried, M.; Hantel, M.; Maurer, C.; Toechterle, R.; Pauli, H. & Grabherr, G. (2011). Coincidence of the alpine-nival ecotone with the summer snowline. *Environmental Research Letters,* Vol.6, No.1 (March 2011), 014013, ISSN 1748-9326

Gregory, R.D.; Willis S.G.; Jiguet F.; Voříšek P.; Klvaňová A.; van Strien, A.; Huntley, B.; Collingham, Y.C.; Couvet, D. & Green, R.E. (2009). An indicator of the impact of climatic change on European bird populations. *PLoS-ONE,* Vol.4, No.3, (March 2009), e4678, ISSN 1932-6203

Grill, A.; Raijmann, L.E.L.; van Ginkel, W; Gkioka, E. & Menken, S.B.J. (2007). Genetic differentiation and natural hybridization between the Sardinian endemic Maniola nurag and the European Maniola jurtina. *Journal of Evolutionary Biology,* Vol.20, No.4 (May, 2007), pp. 1255-1270, ISSN 1420-9101

Guisan, A. & Thuiller, W. (2005). Predicting species distribution: offering more than simple habitat models. *Ecology Letters* Vol.8, No.9 (June, 2005), pp.993-1009, ISSN 1461-0248

Haberl, H.; Winiwarter, V.; Andersson, K.; Ayres, R.U.; Boone, C.G.; Castillio, A.; Cunfer, G.; Fischer-Kowalski, M.; Freudenburg, W.R.; Furman, E.; Kaufmann, R.; Krausmann, F.; Langthaler, E.; Lotze-Campen, H.; Mirtl, M.; Redman, C.A.; Reenberg, A.; Wardell, A.D.; Warr, B. & Zechmeister, H. (2006). From LTER to LTSER: Conceptualizing the socio-economic dimension of long-term socio-ecological research. Ecology and Society, Vol.11, No.2, Article 13, ISSN 1708-3087

Hammond, A.; Adriaane, A.; Rodenburg, E.; Bryant, D. & Woodward, R. (1995). *Environmental Indicators: a Systematic Approach to Measuring and Reporting on Environmental Policy Performance in the Context of Sustainable Development.* World Resources Institute, ISBN 1-56973-026-1, Washington, DC.

Herrmann, A.; Schleifer, S. & Wrbka T. (2011). The Concept of Ecosystem Services Regarding Landscape Research: A Review. Living Reviews in Landscape Research Vol.5, No.1, (March 2011), pp. 1-37, ISSN 1863-7329

Hooper, D.U.; Chapin, F.S.III; Ewel, J.J.; Hector, A.; Inchausti, P.; Lavorel, S.; Lawton, J.H.; Lodge, D.M.; Loreau, M.; Naeem, S.; Schmid, B.; Setälä, H.; Symstad, A.J.; Vandermeer, J. & Wardle, D.A. (2005). Effects of biodiversity on ecosystem functioning: a consensus of current knowledge. *Ecological Monographs,* Vol.75, No.1 (February 2005), pp. 3-35, ISSN 0012-9615.

Kati, V.; Poirazidis, K.; Dufrêne, M.; Halley, J.M.; Korakis, G.; Schindler, S. & Dimopoulos, P. (2010). Toward the use of ecological heterogeneity to design reserve networks: a case study from Dadia National Park, Greece. *Biodiversity and Conservation,* Vol.19, No.6, (June 2010), pp. 1585-1597, ISSN 0960-3115

Kremen, C. (2005). Managing ecosystem services: what do we need to know about their ecology? *Ecology Letters,* Vol.8, No.5 (May, 2005), pp. 468-479, ISSN 1461-0248

Kuussaari, M.; Bommarco, R.; Heikkinen, R.K.; Helm, I.; Krauss, J.; Lindborg, R.; Öckinger, E.; Pärtel, M.; Pino, J.; Rodà, F.; Stefanescu, C.; Teder, T.; Zobel, M. & Steffan-Dewenter, I. (2009). Extinction debt: a challenge for biodiversity conservation. *Trends in Ecology and Evolution,* Vol.24, No.10 (August, 2009), pp. 564-571, ISSN 0169-5347

Luck, G.W.; Daily, G.C. & Ehrlich, P.R. (2003). Population diversity and ecosystem services. *Trends in Ecology and Evolution,* Vol.18, No.7 (July, 2003), pp. 331-336, ISSN 0169-5347

MEA (2003). *Ecosystems and Human Well-being: A Framework for Assessment,* Island Press, ISBN 1-55963-402- 2, Washington, DC

Mauerhofer, V. (2010). Missing links: how individuals can contribute to reserve policy enforcement on the example of the European Union. Biodiversity and Conservation, Vol.19, No.3, (March, 2010), pp. 601–618, ISSN 0960-3115

Mauerhofer, V. (2011). A bottom-up 'Convention-Check' to improve top-down global protected area governance. *Land Use Policy,* Vol. 28, No.4 (October, 2011), pp. 877-886, ISSN 0264-8377, 10.1016/j.landusepol.2011.03.004

Mirtl, M. (2010). Introducing the next generation of ecosystem research in Europe: LTER-Europe's multi-functional and multi-scale approach, In: *Long-term ecological research: between theory and application.* F. Müller, C. Baessler, H. Schubert & S. Klotz (Eds.) Springer, pp. 75-93, ISBN: 978-90-481-8781-2, Dordrecht, The Netherlands.

Mirtl, M.; Bahn, M.; Battin, T.; Borsdorf, A.; Englisch, M.; Gaube, V.; Grabherr, G.; Gratzer, G.; Kreiner, D.; Haberl, H.; Richter, A.; Schindler, S.; Tappeiner, U.; Winiwarter, V. & Zink, R. (2010). *LTER-Austria White Paper. "Next Generation LTER" in Austria.* LTER Austria-Austrian Long-Term Ecosystem research Network, ISBN 978-3-901347-94-8, Vienna, Austria. 14.05.2011, Available from http://www.lter-austria.at

Mirtl, M.; Götzl, M.; Malicky, M.; Rainer, H.; Schleidt, K.; Schindler, S. & Schentz, H. (2011). Österreichisches ESFRI Roadmap Projekt „LIFEWATCH". Erstes Konzept als Basis zur Zeichnung des LifeWatch MoI. Vienna, Austria, 07.07.2011, Available from http://www.lter-austria.at

Newton, A.C. (2011). Implications of Goodhart's Law for monitoring global biodiversity loss. *Conservation Letters Vol. 4, in press.* ISSN 1755-263X, doi: 10.1111/j.1755-263X.2011.00167.x

Pascher, K. & Gollmann, G. (1999). Ecological risk assessment of transgenic plant releases: an Austrian perspective. *Biodiversity and Conservation,* Vol.8, No.8, (August 1999), pp. 1139-1158, ISSN 0960-3115

Pascher, K.; Moser, D.; Dullinger, S.; Sachslehner, L.; Gros, P.; Sauberer, N.; Traxler, A.; Grabherr, G. & Frank, T. (2011). Setup, efforts and practical experiences of a monitoring program for genetically modified plants - An Austrian case study for oilseed rape and maize. *Environmental Sciences Europe*, Vol.13 (March, 2011), pp. 1-12, ISSN 2190-4715

Pauli, H.; Gottfried, M.; Reiter, K.; Klettner, C. & Grabherr, G. (2007). Signals of range expansions and contractions of vascular plants in the high Alps: observations 1994-2004 at the GLORIA* master site Schrankogel, Tyrol, Austria. *Global Change Biology*, Vol.13, No.1, (January 2007), pp. 147-156, ISSN 1365-2486

Platform for Biodiversity Research in Austria (2008). Plattform Biodiversität Forschung Austria (BDFA): Tätigkeitsbericht Mai 2008 für das Bundesministerium für Wissenschaft und Forschung (BWF). University of Vienna, Vienna, Austria, 14.05.2011, Available from http://131.130.59.133/biodiv_forschung/Texte/Bericht_Bioplattform_200805_complete.pdf

Poirazidis, K.; Schindler, S.; Kati, V.; Martinis, A.; Kalivas, D.; Kasimiadis, D.; Wrbka, T. & Papageorgiou, A.C. (2010). Conservation of biodiversity in managed forests: developing an adaptive decision support system, In: *Landscape ecology and forest management: challenges and solutions in a changing globe. C.* Li; R. Lafortezza & J. Chen (Eds.), Springer, pp. 380-399, ISBN 978-3-642-12753-3, New York.

Pyšek, P.; Jarosik, V.; Hulme, P.; Kühn, I.; Wild, J.; Arianoutsou, M.; Bacher, S.; Chiron, F.; Didziulis, V.; Essl, F.; Genovesi, P.; Gherardi, F.; Hejda, M.; Kark, S.; Lambdon, P.W.; Desprez-Loustau, A.-M.; Nentwig, W.; Pergl, J.; Poboljsaj, K.; Rabitsch, W.; Roques, A.; Roy, D.; Shirley, S.; Solarz, W.; Vilá, M. & Winter, M. (2010). Disentangling the role of environmental and human pressures on biological invasions across Europe. *Proceedings of the National Academy of Sciences*, Vol.107, No.27, (June, 2010), pp. 12157-12162, ISSN 1091-6490

Renetzeder, C.; Schindler, S.; Peterseil, J.; Prinz, M.A.; Mücher, S. & Wrbka, T. (2010). Can we measure ecological sustainability? Landscape pattern as indicator for naturalness and land use intensity at regional, national and European level. *Ecological Indicators*, Vol.10, No.1 (January, 2010), pp. 39-48, ISSN 1470-160X

Reyers, B.; Roux, D.J.; Cowling, R.M.; Ginsburg, A.E.; Nel, J.L. & Farrel, P.O. (2010). Conservation Planning as a Transdisciplinary Process. Conservation Biology, Vol.24, No.4 (August, 2010) pp. 957-965, ISSN 1523-1739

Sachs J.D.; Baillie J.E.; Sutherland W.J.; Armsworth, P.R.; Ash, N.; Beddington, J.; Blackburn, T.M.; Collen, B.; Gardiner, B.; Gaston, K.J.; Godfray, H.C.J.; Green, R.E.; Harvey, P.H.; House, B.; Knapp, S.; Kümpel, N.F.; Macdonald, D.W.; Mace, G.M.; Mallet, J.; Matthews, A.; May, R.M.; Petchey, O.; Purvis, A.; Roe, D.; Safi, K.; Turner, K.; Walpole, M.; Watson, R. & Jones, K.E. (2009). Biodiversity Conservation and the Millennium Development Goals. *Science*, Vol.325, No.5947, (September 2009), pp. 1502-1503, ISSN 0036-8075

Schindler, S.; Curado, N.; Nikolov, S.; Kret, E.; Cárcamo, B.; Poirazidis, K.; Catsadorakis, G.; Wrbka, T. & Kati, V. (2011). From research to implementation: nature conservation in the Eastern Rhodopes mountains (Greece and Bulgaria), European Green Belt. *Journal for Nature Conservation*, Vol.19, No.4 (September, 2011), pp. 193-201, ISSN 1617-1381, 10.1016/j.jnc.2011.01.001

Schindler, S.; Poirazidis, K.; Papageorgiou, A.C.; Kalivas, D.; von Wehrden, H. & Kati, V. (2010). Landscape approaches and GIS as a prerequisite for biodiversity management in a Mediterranean forest landscape, In: *Landscape modelling: geographical space, transformation and future scenarios.* J. Andel; I. Bicik; P. Dostal; Z. Lipsky & S.G. Shahneshin (Eds.), Urban and Landscape Perspectives Series, Vol. 8., Springer, pp. 174-184, ISBN 978-90-481-3051-1, Dordrecht, The Netherlands.

Schindler, S.; Poirazidis, K. & Wrbka, T. (2008). Towards a core set of landscape metrics for biodiversity assessments: a case study from Dadia National Park, Greece. *Ecological Indicators,* Vol.8, No.5, (September 2008), pp. 502-514, ISSN 1470-160X

Singh, S.J.; Haberl, H.; Chertow, M.; Mirtl, M. & Schmid, M. (Eds.) (in press). *Long term socio-ecological research: Studies in society-nature interactions across spatial and temporal scales.* Springer, ISBN 978-94-007-1176-1

Singh, S.J.; Haberl, H.; Gaube, V.; Grünbühel, C.M.; Lisievici, P.; Lutz, J.; Matthews, R.; Mirtl, M.; Vadineanu, A. & Wildenberg, M. (2010). Conceptualising Long-Term Socio-ecological Research (LTSER): Integrating the Social Dimension, In: *Long-term ecological research: between theory and application.* F. Müller, C. Baessler, H. Schubert & S. Klotz (Eds.) Springer, pp. 377-398, ISBN 978-90-481-8781-2, Dordrecht, The Netherlands.

Sutherland, W.J.; Armstrong-Brown, S.; Armsworth, P.R.; Brereton, T.; Brickland, J.; Campell, C.D.; Chamerlain, D.E.; Cooke, A.I.; Dulvy, N.K.; Dusic, N.R.; Fitton, M.; Freckleton, R.P.; Godfray, H.C.J.; Grout, N.; Harvey, H.J.; Hedley, C.; Hopkins, J.J.; Kift, N.B.; Kirby, J.; Kunin, W.E.; MacDonald, D.W.; Marker, B.; Naura, M.; Neale, A.R.; Oliver, T.; Osborn, D.; Pullin, A.S.; Shardlow, E.A.; Showler, D.A.; Smith, P.L.; Smithers, R.J.; Solandt, J.-L.; Spencer, J.; Spray, C.J.; Thomas, C.D.; Thompson, J.; Webb, S.E.; Yalden, D.W. & Watkinson, A.R. (2006). The identification of 100 ecological questions of high policy relevance in the UK. *Journal of Applied Ecology,* Vol.43, No.4, (August 2006), pp. 617-627, ISSN 1365-2664

Sutherland, W.J.; Bardsley, S.; Bennun, L.; Clout, M.; Coté, I.M.; Depledge, M.H.; Dicks, L.V.; Dobson, A.P.; Felmann, L.; Fleishman, E.; Gibbons, D.W.; Impex, A.J.; Lawton, J.H.; Lockorish, F.; Lindenmayer, D.B.; Lovejoj, T.E.; Mac Nally, R.; Madgwick, J.; Peck, L.S.; Pretty, J.; Prior, S.V.; Redford, K.H.; Scharlemann, J.P.W.; Spalding, M. & Watkinson, A.R. (2011). Horizon scanning of global conservation issues for 2011. *Trends in Ecology & Evolution,* Vol.26, No.1, (January 2011), pp. 10-16, ISSN 0169-5347.

Sutherland, W.J.; Clout, M.; Côté, I.M.; Daszak, P.; Depledge, M.H.; Fellman, L.; Fleishman, E.; Garthwaite, R.; Gibbons, D.W.; De Lurio, J.; Impey, A.J.; Lickorish, F.; Lindenmayer, D.; Madgwick, J.; Margerison, C.; Maynard, T.; Peck, L.S.; Pretty, J.; Prior, S.; Redford, K.H.; Scharlemann, J.P.W.; Spalding, M. & Watkinson, A.R. (2010). A horizon scan of global conservation issues for 2010. *Trends in Ecology & Evolution,* Vol.25, No.1, (January 2010), pp. 1-7, ISSN 0169-5347

Tasser, E.; Sternbach, E. & Tappeiner, U. (2008). Biodiversity indicators for sustainability monitoring at municipality level: an example of implementation in an alpine region. *Ecological Indicators,* Vol.8, No.3, (May 2008), pp. 204-223, ISSN 1470-160X

TEEB (November 2009). The Economics of Ecosystems and Biodiversity for National and International Policy Makers - Summary: Responding to the Value of Nature 2009. 14.05.2011, Available from
http://www.teebweb.org/LinkClick.aspx?fileticket=I4Y2nqqIiCg%3d&tabid=1019&language=en-US

Walpole, M.; Almond, R.E.A.; Besancon, C.; Butchart, S.H.M.; Campbell-Lendrum, D.; Carr, G.M.; Collen, B.; Collette, L.; Davidson, N.C.; Dulloo, E.; Fazel, A.M.; Galloway, J.N.; Gill, M.; Goverse, T.; Hockings, M.; Leaman, D.J.; Morgan, D.H.W.; Revenga, C.; Rickwood, C.J.; Schutyser, F.; Simons, S.; Stattersfield, A.J.; Tyrrell, T.D.; Vié, J.-C. & Zimsky, M. (2009). Tracking progress toward the 2010 Biodiversity Target and beyond. *Science,* Vol.325, No.5947, (September 2009), pp. 1503-1504, ISSN 0036-8075

Wrbka, T.; Schindler, S.; Pollheimer, M.; Schmitzberger, I. & Peterseil, J. (2008). Impact of the Austrian Agri-Environmental Scheme on diversity of landscape, plants and birds. *Community Ecology,* Vol9, No.2, (December 2008), pp. 217-227, ISSN 1585-8553

9

Conservation of Chinese Plant Diversity: An Overview

Jordi López-Pujol[1], Hua-Feng Wang[2] and Zhi-Yong Zhang[3]
[1]BioC-GReB, Botanic Institute of Barcelona (CSIC-ICUB), Barcelona, Catalonia
[2]Beijing Urban Ecosystem Research Station, State Key Laboratory of Urban and Regional Ecology, Research Center for Eco-Environmental Sciences, Chinese Academy of Sciences, Beijing
[3]Laboratory of Subtropical Biodiversity, Jiangxi Agricultural University, Nanchang, Jiangxi
[1]Spain
[2,3]China

1. Introduction

1.1 The significance of plant diversity in China

China is one of the richest countries in plant diversity, ranking third in the world (after Brazil and Colombia) in number of species, and one of the world's 17 'mega-diversity' countries (Mittermeier et al., 1997). The estimated number of vascular plant species may approach 33,000, with 30,000 angiosperms, 250 gymnosperms, and 2,600 pteridophytes (up to 12%, 27% and 20% of world's total, respectively). Furthermore, approximately 2,200 bryophytes can be found in China (López-Pujol et al., 2006; Table 1). There are more than 3,000 genera and *ca.* 350 families of vascular plants (Li et al., 2003; MacKinnon & Wang, 2008). Nevertheless, these figures refer to mainland China and do not include either Taiwan or Hong Kong. Taiwan alone harbors more than 4,000 vascular plants (over 3,300 angiosperms, about 30 gymnosperms, and about 600 pteridophytes; Hsieh, 2002). With an area of only about 1,100 km^2, Hong Kong still retains a very rich plant diversity, with more than 2,100 higher plants (Wu, 2002).

China encompasses enormous diversity in geographical, climatological and topographical features, in addition to a complex and ancient geological history (with most of its lands formed as early as the end of the Mesozoic era; Wang, 1985). The country spans five major climatic zones (cold-temperate, temperate, warm-temperate, subtropical, and tropical), and is home to the highest mountain range on Earth (the Himalayas) and perhaps the most rugged one (the Hengduan Mountains), vast plateaux such as the Tibetan (*Qinghai-Xizang*) Plateau, deserts (e.g. Taklamakan), deep depressions (e.g. Turpan Depression), large flat areas (e.g. Sichuan Basin and North China Plains), and some of Asia's largest rivers, including the Mekong (*Lancang*), Brahmaputra (*Yarlung Zangbo*), Yangtze (*Changjiang*), and Yellow (*Huanghe*) rivers. All of these features contribute to the enormous diversity of ecosystems, including almost all types of forests, grasslands, shrublands, deserts (which cover more than 25% of the Chinese territory), marshes, savannas, tundras, or alpine meadows (Hou, 1983; Enright & Cao, 2010). Thus, it is not surprising that 19 of the 238 WWF global priority ecoregions (Olson & Dinerstein, 2002) are located totally or partially within China.

Diversity of cultivated plants is equally rich. China is one of the eight original centers of crop plants in the world (Vavilov, 1951), with more than 200 originating and differentiating there (Gu, 1998), resulting from more than 7,000 years of agricultural activities. Notable examples include rice (*Oryza sativa*), which consists of some 50,000 cultivars and three wild relatives (*O. granulata, O. officinalis,* and *O. rufipogon*) and soybean (*Glycine max*), embracing about 20,000 cultivars (SEPA, 1998; MacKinnon & Wang, 2008). It is also estimated that as many as 2,200 ornamental species originated in China (SEPA, 1998), and about 15,000 vascular plant species native to China have been cultivated around the world (Zhao & Zhang, 2003). Other economic plants as well show significant figures in China: over 1,300 taxa of food plants have been recorded (Hu, 2005), as well as up to 1,200 fiber plants and about 2,000 species of timber plants (CSPCEC, 2008); in addition, nearly 11,000 species of medicinal plants have been in use since the Palaeolithic period (Hamilton, 2004; CSPCEC, 2008).

2. Patterns of species richness and endemism: Evolutionary issues

China possesses the richest flora of the North Temperate Zone (Axelrod et al., 1996; López-Pujol et al., 2006, 2011); however, more relevant are the rates of endemism: fifty to sixty percent of the total number of species (i.e. 15,000 to 18,000) might be endemic to China (SEPA, 1998; CSPCEC, 2008). This wealth of species diversity and endemism is attributable to a series of factors largely related to the biogeography, tectonics and geological history of the country, including: (i) a complex and extended geological history, with many tectonic events, (ii) a large proportion of the area of China within tropical and subtropical latitudes, (iii) the wide and persistent connection of China to tropical regions of Southeast Asia as well as with other regions, (iv) an unbroken connectivity between tropical, subtropical, temperate, and boreal forests, (v) a highly rugged and dissected topography (especially in southern China), and (vi) perhaps the most significant, reduced extinction rates during the late Cenozoic global cooling (e.g. Tiffney, 1985; Latham & Ricklefs, 1993; Axelrod et al., 1996; Guo et al., 1998; Guo, 1999; Qian & Ricklefs, 1999; Qian, 2001, 2002; Ying, 2001; López-Pujol et al., 2006, 2011; Qian et al., 2006; Wu et al., 2007; Li, 2008).

Taxa	Species in China (SC)	Species in the world (SW)	SC/SW (%)
Lichen	2,000	10,000	20.0
Fungi	8,000	70,000	11.4
Algae	8,979 - 12,500	40,000	22.4 - 31.2
Bryophyta	2,200	15,000	14.7
Pteridophyta	2,300 - 2,600	13,025	17.6 - 20.0
Gymnospermae	192 - 270	980	19.6 - 27.5
Angiospermae	25,000 - 30,000	258,650	9.7 - 11.6

Table 1. Richness of Chinese flora (data are only for mainland China). All the figures have been taken from López-Pujol et al. (2006); original sources can be found there

One of the main features of Chinese vascular flora is its ancient origin; the modern flora of China is still showing a strong relictual character (Thorne, 1999; Qian & Ricklefs, 1999; Qian, 2001; López-Pujol & Ren, 2010; López-Pujol et al., 2011). This is directly linked to the geologic history of China's landmass, with some of the events influencing the flora composition tracing back to the early Mesozoic. Taking into account that the origin of vascular plants is often situated in Gondwana (Graham, 1993; Steemans et al., 2009), the collision of Gondwanan terranes (including southern China, Indochina and Sibumasu) with what is now northern China in the Triassic-early Jurassic period, and the further impact of the Indian subcontinent during the late Paleocene/early Eocene (Metcalfe, 1988; Şengör, 1997) would have extensively supplied China with pteridophytes and gymnosperms. This may help to explain the disproportionate number of lineages in China belonging to the two most ancient groups of vascular plants (Qian & Ricklefs, 1999; Qian, 2001).

In addition, China could have been close to the center of origin of flowering plants, which for some authors was located somewhere in SE Asia (Thorne, 1963; Takhtajan, 1969, 1987; Smith, 1973). The continuous land connection with southern Asian tropical areas (dating back at least 200 million years; Hallam, 1994) would have enabled a straightforward transferring of the most primitive angiosperm lineages to China. More recent studies suggest that angiosperms might have originated instead in Gondwana, but still in palaeotropical latitudes (Morley, 2000; Barrett & Willis, 2001). With the finding of the early angiosperm *Archaefructus* fossils (Fig. 1) in Liaoning Province (north-east China) in addition to other ancient angiosperms in neighboring areas (e.g. Sun et al., 2011), Sun et al. (2008) are postulating, however, for an 'Eastern Asian origin of angiosperms'. In any case, there is general agreement that China constituted an important area for the early diversification of angiosperms (Qian & Ricklefs, 1999; Barrett & Willis, 2001; Qian, 2001; López-Pujol et al., 2006; Sun et al., 2008, 2011).

As explained below, the collision of first the Gondwanan terranes and the further impact of the Indian subcontinent are responsible for China having inherited elements of Gondwanan floras in addition to its own Laurasian ones. However, despite China's (and East Asia's) initial separation from western Eurasia by the Turgai Sea, a land bridge linking eastern and western Eurasia ensured their connection since *ca.* 35 million years ago (Hallam, 1994; Qian, 2001), permitting a floristic exchange with Central Asia (and subsequently with the Mediterranean basin and even North Africa; Qian & Ricklefs, 1999), thus providing China with Tethyan floristic lineages, in addition to other Laurasian and Gondwanan elements (Wu & Wu, 1996; Wu, 1998; Qian et al., 2006). Moreover, the connection with North America via the Bering Sea land bridge allowed the migration of numerous species from this area (Tiffney & Manchester, 2001; Qian et al., 2006).

The most important force shaping the contemporary floristic richness of China is, nevertheless, the climatic change that occurred during the Neogene period. During the late Tertiary and the whole Quaternary, that is, from the middle Miocene climatic optimum (*ca.* 15 Ma; Zachos et al., 2001) onwards, a progressive global cooling produced numerous plant extinctions in most of the Northern Hemisphere, especially Europe and North America (Tiffney, 1985; Sauer, 1988; Axelrod et al., 1996; Jackson & Weng, 1999), and these were centered on the relictual, thermophilic elements of the 'boreotropical flora' (Tiffney, 1985; Latham & Ricklefs, 1993). In contrast, the limited ice coverage in East Asia during the Quaternary glaciations (significantly milder than in Europe and North America, and limited only to northern areas north of 60° N; Qian & Ricklefs 1999), coupled with the numerous mountain ranges in the southern section of China (which would have provided long-term stable habitats through local buffering of the extreme climatic oscillations) and the lack of barriers for southwards migration enabled the country to serve as a refuge for many Tertiary plant lineages, ensuring their survival to the present day (Axelrod et al., 1996; Qian & Ricklefs, 2000; López-Pujol et al., 2011). The presence of numerous and extensive refugia is likely to be the cause of the overrepresentation of relict elements in the modern flora of China and the much larger overall taxonomical richness of the Asian country with respect to other territories from the Northern Hemisphere (Europe and the United States, with comparable areas, harbor a much poorer flora: 11,500 and 18,000 species, respectively; Axelrod et al., 1996).

Fig. 1. Fossil of *Archaefructus liaoningensis*, one of the earliest known genera of flowering plants, exposed in the Beijing Museum of Natural History (photo by Jordi López-Pujol)

The role of the Neogene plant refugia as places for plant survival (i.e. 'plant museums') has been long recognized (e.g. Ying & Zhang, 1984; Wang, 1989; Wu & Wu, 1996; Li, 2008), and this is reflected in the exceptional richness of several phylogenetically primitive vascular plant groups in China; e.g., pteridophytes, gymnosperms, magnolids, and ranunculids (Qian & Ricklefs, 1999; Qian, 2001). China has a significantly higher number of ancient endemic genera and families than the United States (Qian, 2001), and most of the endemic genera of spermatophytes of China (about 240) are of ancient origin (Ying & Zhang, 1984; Ying et al., 1993; Wu et al., 2007). Moreover, a more recent study (López-Pujol et al., 2011) has revealed that the relict (pre-Quaternary) component of the modern endemic seed flora at the infrageneric (species and subspecies) level is considerable (around 40%; the remaining 60% are taxa of Pleistocene origin). For the Mediterranean endemic flora, the weight of the relict taxa is much lower (*ca.* 25%; Thompson, 2005). Noteworthy is the occurrence, especially in the central and southern regions of China, of numerous palaeoendemic lineages whose distribution was much wider and today is restricted to a few, often disjunct refugia in East Asia. The fossil record shows that some of these lineages once existed in Europe or North America, but were extirpated from there during the Neogene as a consequence of climate deterioration (Latham & Ricklefs, 1993; Axelrod et al., 1996; Manchester, 1999; Manchester et al., 2009). Representative examples within the gymnosperms include the monotypic and/or oligotypic genera *Amentotaxus*, *Cathaya* (Fig. 2), *Cunninghamia*, *Ginkgo* (which is the unique representative of the monotypic family Ginkgoaceae but also of the entire order Ginkgoales; Fig. 2), *Glyptostrobus*, *Keteleeria*, *Metasequoia* (Fig. 2), *Pseudolarix*, *Pseudotaxus*, and *Taiwania*. From the angiosperms, several examples merit citation here: *Craigia*, *Cyclocarya*, *Davidia*, *Dipelta*, *Diplopanax*, *Emmenopterys*, *Eucommia* (the only representative of the Eucommiaceae), *Fortunearia*, *Pteroceltis*, *Sargentodoxa*, *Tapiscia*, *Tetracentron*, and *Trochodendron*. Some 'living fossils' (taxa which have remained superficially unchanged for millions of years) such as *Metasequoia glyptostroboides* or *Glyptostrobus pensilis* constitute the most conspicuous examples of these relict lineages, generally remnants of the boreotropical flora that once spanned most of the Northern Hemisphere (Kubitzki & Krutzsch, 1996).

Fig. 2. Some 'living fossils' of China. Left, type tree of *Metasequoia glyptostroboides* at Moudao Town (Hubei Province) (photo by Qin Leng); top right, leaves and cones of a wild individual of *Cathaya argyrophylla* (photo by Zhao-Shan Wang); bottom right, yellowish leaves of a planted individual of *Ginkgo biloba* in Beijing (photo by Jordi López-Pujol)

Most of these plant refugia generally also served as areas for plant evolution and speciation ('plant cradles') in addition to being places for their persistence during the Neogene cooling (Axelrod et al., 1996; López-Pujol et al., 2011). Interestingly, the main centres of evolution were located in the southwestern part of China, that is, the eastern fringe of the Tibetan Plateau. The Hengduan Mountains (Fig. 3) and the neighboring ranges (Daxue Mountains, Min Mountains) constitute the 'evolutionary front' of China, because they are home to numerous neoendemisms (Ying et al., 1993; Wu & Wu, 1996; Li & Li, 1997). Recent orogenic processes in western China (the major uplift of the Tibetan Plateau took place during the Pliocene and the Pleistocene, and it is still active; Li & Fang, 1999; Zhang D. et al., 2000; Zheng et al., 2000) created a vast array of new habitats across wide altitudinal ranges (up to 5,000 m in the Hengduan Mountains; Fig. 3), which stimulated plant differentiation and speciation (Chapman & Wang, 2002; Qian, 2002), including adaptative radiations (Liu & Tian, 2007). This can be the case of *Nannoglottis*, a relic genus which probably arrived at the Tibetan Plateau not later than the Oligocene and suffered a rapid re-diversification during the Pliocene-Pleistocene (Liu et al., 2002). In contrast, the relative tectonic stability in central, south-central and southeastern China seems to have favoured the survival of relict lineages (López-Pujol et al., 2011).

Since refugia are areas that offered many opportunities for persistence and speciation, these stand out as harboring high rates of endemism as well as overall species richness. Centres of species richness and centres of endemism are, thus, generally coincident (Ying, 2001; Tang et al., 2006; Xu et al., 2008) and almost exclusively located in the mountainous regions of central and southern China, at latitudes below 35° N, including the two main Chinese islands, Hainan and Taiwan (Fig. 4). One of the most significant areas for plant diversity are the Hengduan Mountains (Fig. 3), which span north-west Yunnan, western Sichuan and south-east Tibet, and which are also considered one of the main world's biodiversity hotspots (with a total flora of about 12,000 species, of which *ca.* 3,500 are endemic; Myers et al., 2000). It is widely acknowledged that its extremely varied topography supporting a wide array of vegetation zones (from subtropical low mountain evergreen rainforests in the deep valleys to alpine communities on the summits), have greatly contributed both to the appearance of many new species and the conservation (although in a lower scale) of relict elements (Chapman & Wang, 2002; Qian, 2002). Some examples of this 'paradox' are the genera *Primula* and *Rhododendron*, present in the region before the Himalayan uplift and which became highly diversified through allopatry with the creation of numerous new habitats (Chapman & Wang 2002); for example, more than one-quarter of the world's *Rhododendron* species (276 of *ca.* one thousand) and *ca.* 30% of *Primula* (143 out of 500) can be found in these mountains (Zhang D.-C. et al., 2009).

Central China Mountains is also one of the richest areas in plant diversity of China, with about 6,400 plant species and more than 1,500 endemics (Ying, 2001). This region, in contrast to the Hengduan Mountains, contains more relict than recently-evolved taxa (López-Pujol et al., 2011). One of the most interesting areas within this hotspot is the so-called 'metasequoia area', a region of about 800 km^2 in the juncture of Hubei and Chongqing, where there are still natural populations of *Metasequoia glyptostroboides* (Hu, 1980). In this very small land extension, at least 550 species of vascular plants occur; the most interesting, however, is that we can find many 'living fossils' here in addition to *M. glyptostroboides*, most of them belonging to monotypic or oligotypic genera, e.g. *Cunninghamia lanceolata, Eucommia ulmoides, Keteleeria davidiana, Pseudolarix amabilis, Taiwania cryptomerioides, Tapiscia sinensis,* and *Tetracentron sinense* (Hu, 1980).

Fig. 3. A view of the Hengduan Mountains. The highest peak in the picture corresponds to Mount Gongga (7,556 m a.s.l.). Some deep gorges (of about only 1,000 a.s.l.) can be found just a few km away (photo by Jordi López-Pujol)

3. Endangered species: Current status

A second trait of Chinese flora is its high level of endangerment. Most estimates show that 3,000-5,000 species could be threatened with extinction (Fu, 1992; Wang, 1992; Gu, 1998; Zhang, 2007; CSPCEC, 2008), i.e. up to 20% of the total flora. However, and according to more recent studies (Wang & Xie, 2004; Xie & Wang, 2007), this figure could be even higher: from the *ca.* 4,200 angiosperm taxa (i.e. just 14% of the total number of angiosperms in China) assessed in the first phase of the elaboration of the *China Species Red List* (Wang & Xie, 2004), over 3,600 (87%) were regarded as threatened following the 2001 IUCN criteria, and up to 651 were assigned to the highest endangerment category (CR, 'Critically Endangered'), that is, facing an imminent risk of extinction. These figures, nevertheless, should be taken with caution as the red list was biased toward rare and endangered species. At the beginning of the 2000s it was estimated that, since the 1950s at least 200 plant species had become extinct (Zhang P. et al., 2000); some conspicuous examples of these are *Otophora unilocularis* (not seen since 1935) and *Rhododendron kanehirai*, whose natural populations were flooded by a dam in Taiwan although it is extensively cultivated (IUCN, 2010). However, some plant taxa have also been lost from the wild during the last decade (2000-2009), such as *Betula halophila, Cystoathyrium chinense* or *Plantago fengdouensis* (Zhang & Ma, 2008; López-Pujol & Zhang, 2009). In addition to these losses, many species remain on the brink of extinction. An appreciable number of taxa are in an extreme situation of risk with population sizes often consisting of fewer than 100 individuals: examples include the gymnosperms *Cupressus chengiana* var. *jiangeensis* and *Abies beshanzuensis* var. *beshanzuensis* (from which only one and three individuals are remaining, respectively), and the angiosperms *Carpinus putoensis, Gleditsia japonica* var. *velutina*, and *Acer yangbiense* (with just one, two and four individuals remaining in the wild, respectively); these taxa are undoubtedly among the most endangered plants on the Earth (López-Pujol & Zhang, 2009).

Despite the fact that many species have been severely threatened by extinction for many decades, it was not until 1984 that the *National List of Rare and Endangered Plant Species* was issued, encompassing 388 species [8 listed as 'first grade' nationally protected (NPC-1), 159 species as 'second grade', and 221 species as 'third grade']. In 1992, the *China Plant Red Data Book* (Fu, 1992) was published, including all 388 of the endangered taxa warranting protection. Of these, 121 were listed as 'endangered', 110 as 'rare', and 157 as 'vulnerable'. Two additional volumes were originally planned (López-Pujol et al., 2006), but they are still awaiting completion. Meanwhile, based on the request of the *Regulations on Wild Plants Protection* (implemented in 1997), China promulgated the first batch of the *Catalogue of the National Protected Key Wild Plants* in 1999, which included about 300 plant taxa, distributed into two protection categories (Zhang & Ma, 2008). The second batch is currently in preparation and will mean that up to 1,900 plant taxa will be strictly protected in mainland China (CSPCEC, 2008).

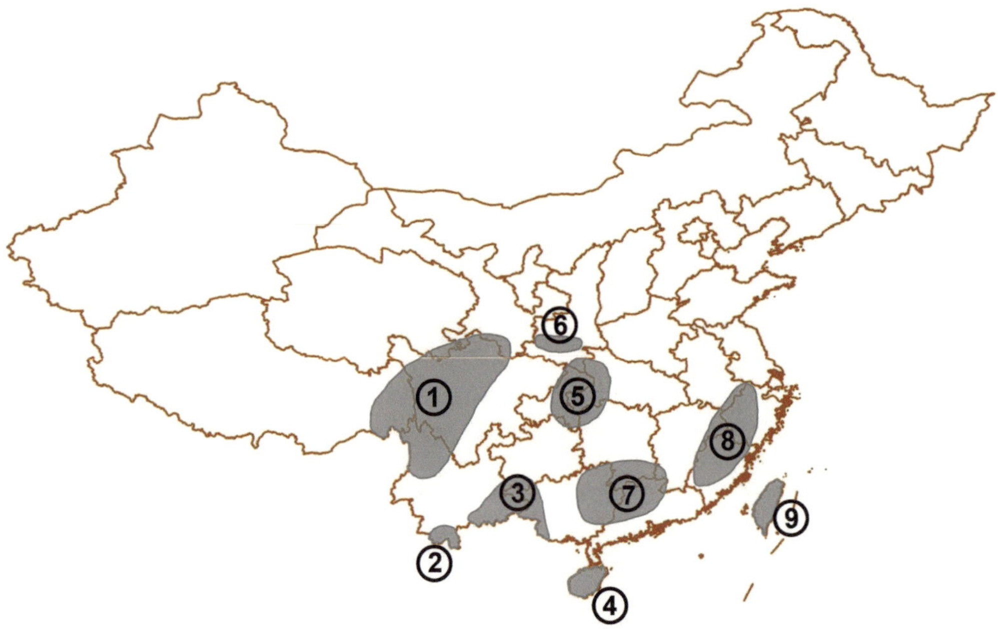

Fig. 4. Approximate location of Chinese plant diversity hotspots. The code numbers for the plant diversity hotspots correspond to those in Table 2 (see below).

In addition to these protection lists, China has also progressed in assessing plant species using the IUCN Red List Categories and Criteria. By the end of 2010 only about 740 taxa (species and subspecies) from mainland China appeared in the *IUCN Red List of Threatened Species* (http://www.iucnredlist.org/), however the publication of the first volume of *China Species Red List* (Wang & Xie, 2004) has given rise to the assessment of 4,408 plants using the 2001 IUCN criteria (IUCN, 2001). Moreover, several experts are working actively in assessing new species which are added regularly in the website of the Wildlife Conservation Society (http://www.chinabiodiversity.com/), and an updated print version of the red list is expected to be released around 2012 (CSPCEC, 2008).

Endangered plant species tend to be concentrated in the southern part of the country (e.g. Tang et al., 2006; Zhang & Ma, 2008), thus showing high congruence with both centres of species richness and centres of endemism (Xu et al., 2008). Thus, 'biodiversity hotspots' in its broadest sense (as centres of species richness, endemism and threatened species) are entirely located in the central and southern mountainous regions of China (Table 2 and Fig. 4), which are moderately populated (much less compared to the North China Plains) and where the agricultural exploitation has been relatively limited (e.g. Huang et al., 2010) due to their low suitability (too steep). However, other practices such as extensive logging and overgrazing have significantly damaged the natural ecosystems of these areas (CI, 2007; Morell, 2008).

Hotspot	Species richness	Endemism	Threatened plants
Hengduan Mountains (1)	+++	+++	++
Xishuangbanna (2)	+++	+	+++
SE Yunnan-SW Guizhou-SW Guangxi (3)	++	+++	++
Hainan Island (4)	++	+	++
Central China Mountains (5)	++	+++	++
Qinling Mountains (6)	+	++	+
Nanling Mountains (7)	++	++	++
East China Mountains (8)	++	++	++
Taiwan Island (9)	++	++	+

Table 2. Chinese plant diversity hotspots. Relative occurrences for each type of plant species (+++, high occurrence; ++, intermediate occurrence; +, low occurrence) and for each hotspot have been inferred after taking into account all the relevant related literature (e.g. Ying et al., 1993; Wang & Zhang, 1994; Ying, 2001; Li et al., 2003; Tang et al., 2006; Zhang & Ma, 2008; Li et al., 2009; López-Pujol et al., 2011). The code numbers for the different hotspots correspond to those in Fig. 4

4. Threats to Chinese plant biodiversity

The main threats to Chinese plant diversity, both directly and indirectly, can be described as below.

4.1 Habitat destruction

Destruction and/or fragmentation of natural habitats (Fig. 5) are the principal causes of species extinction throughout the world (Crooks & Sanjayan, 2006). China has experienced since the imperial times but particularly from 1950s onward, a tremendous loss of natural habitats, mainly due to the over-logging of forests, as well as from the conversion of forests, grasslands and wetlands into croplands (Liu & Diamond, 2005). Some areas have suffered a dramatic reduction in forest coverage; e.g., the rainforest on Hainan Island covered 25.5% of the total area in the early 1950s; thirty years later, this coverage had decreased to below 9%, and these were mainly replaced by rubber plantations (Francisco-Ortega et al., 2010); in Xishuangbanna, the forest cover diminished, from over 60% to less than 30% during the same period (Zhang & Cao, 1995). It is thought that up to 67 million hectares of forests were logged during the 50s, 60s and 70s (Cai, 1990). Throughout China, however, forest cover has progressively increased over the most recent decades due to large afforestation and/or reforestation campaigns (Wang et al., 2008). In 1962, forest cover measured around 9.0% of the total land, while in 1981 it had increased to 12.0%, and then to over 20% by the end of 2009 (MEP, 2010). Nevertheless, nearly all of these new plantations, which replaced logged natural forests, have been mono-specific, and have sometimes consisted of exotic species, greatly diminishing the biodiversity value of the original forestlands (Xu & Wilkes, 2004; Liu & Diamond, 2005). The progressing expansion of the desert in the northwestern provinces has been caused primarily by the massive degradation of grasslands (due to overgrazing and land reclamation for agriculture); 90% of China's total grasslands are degraded to some degree (Liu & Diamond, 2005). At the end of 20th century, China exhibited an extremely fast desertification rate (with an annual increase of almost 3,500 km^2), although this trend seems to have been mitigated during the last decade due to re-vegetation programs (Wang et al., 2007; but see Wang et al., 2010). Loss of wetlands (which may reach 50% of China's total; Yu, 2010), mainly through conversion to cropland, has also contributed to the rampant erosion processes experienced in China since the 1950s: severe soil erosion affects today about 20% of the country land area (Liu & Diamond, 2005; Li, 2010). Deforestation and soil erosion definitely contributed to the devastating floods in the Yangtze River basin in 1998 (Liu, 2010), and the increasing incidence of dust storms in northern China (He, 2009) is also one of the catastrophic consequences of land vegetation cover degradation (Fig. 6).

4.2 Environmental contamination and global climate change

Until recent times, the lack of any ecological consciousness by China's leaders and/or the poverty of the country (which made the government focus on development policies) have caused air, water and land to be extremely polluted. This has lead not only to the degradation of natural habitats and the loss of many species but has also brought huge economic losses and numerous effects on public health (e.g. Liu & Diamond, 2005; Fu et al., 2007; Zhang J. et al., 2010). For instance, severe defoliation (up to 50%) and increased mortality rates have been detected in forests of native pines due to acid rain (Larssen et al., 2006), and it is estimated that 2.4 million premature deaths are produced in China every year due to these environmental risks (Zhang J. et al., 2010). Despite the fact that several sources are claiming a substantial amelioration in pollution levels in recent years (Xu H. et al., 2009, 2010; MFA, 2010), concerns are still unquestionable (e.g. Vennemo et al., 2009). The most problematic issue continues to be the air pollution stemming from the extensive use of coal: in 2009 it was still by far the main energy resource in China, accounting for 69.6% of the total energy consumption; renewable sources of energy continue to be very limited (below 10%; MFA, 2010). While the growth in CO_2 emissions has proven to be significantly less than was previously projected, they still remain very significant (China became the largest emitter of CO_2 since 2006, accounting for *ca.* 20% world's total; World Bank, 2010a). Other air pollutants, whose control has been increased by the authorities, have shown a declining trend, such as the SO_2 emissions and soot and industrial dust (Xu H. et al., 2009, 2010); however, China is still a very large emitter for almost all the major air pollutants

(World Bank, 2010a), and acid rain is still affecting more than half of the Chinese cities (MEP, 2010). Regarding waters, despite the verifiable improvements produced by the important governmental efforts (e.g. Xu H. et al., 2009; MEP, 2009) these remain highly polluted: a large portion of lakes and major rivers are highly contaminated (more than 40% and over 75%, respectively; MEP, 2010; Zhang J. et al., 2010).

Fig. 5. *Sinomanglietia glauca* (=*Manglietia decidua*) is one of the most endangered species of Magnoliaceae in China, with only two isolated localities (separated by *ca.* 450 km), and it is protected by the *Catalogue of the National Protected Key Wild Plants.* The observed habitat fragmentation (left) may account for the high differentiation and low genetic diversity detected for the species (Zhang Z.-R. et al., 2009) (photos by Zhi-Yong Zhang)

The global climate change is also producing severe impacts on plant diversity, which have already been recognized by Chinese authorities (NDRC, 2007). In addition to producing a larger incidence of natural disasters (such as floods, forest fires, landslides, storms or droughts which have direct effects on biodiversity; Liu & Diamond, 2005; Yu, 2010), the melting of glaciers is of great concern (according to the latest estimations, the total surface area of glaciers of the Tibetan Plateau has decreased by at least 17% during the last 30 years; Qiu, 2010), since its potential threat to plant diversity and distribution are very significant (e.g. Xu J. et al., 2009).

Fig. 6. Comparison between a clear (left) and a sand storm (right) day in Beijing. Pictures correspond to the campus of the Research Center for Eco-Environmental Sciences (photos by Hua-Feng Wang)

4.3 Over-exploitation of species for human use

The over-exploitation of species of economic interest may seriously threaten their survival. For example, though protected by law, many plants used in Traditional Chinese Medicine (TCM) and other traditional medicines (e.g. Tibetan) are over-collected. Among the 426 herbal, unprepared, or stir-baked drugs listed in the 1995 edition of the *Pharmacopoeia of the People's Republic of China*, 28 are included in the *China Plant Red Data Book* because of their threatened status, and 49 additional plants listed in the *Red Book* are extensively used in TCM (Peng & Xu, 1997). A clear case of over-harvesting is that of the caterpillar fungus (*Cordyceps sinensis*), profusely used both in Chinese and Tibetan traditional medicines (Winkler, 2008); other examples of valuable medicinal plants which are dwindling in wild habitats due to over-collecting include *Aquilaria sinensis, Eucommia ulmoides,* and *Gastrodia elata*. Moreover, despite widespread cultivation in China there are some species whose wild populations remain severely depleted, both in number and size (e.g., *Ginkgo biloba, Houpoëa officinalis* [=*Magnolia officinalis*] and *Juglans regia*, the latter with perhaps less than 1000 individuals; Chapman & Wang, 2002). In addition to medicinal plants, other non-timber forest products have been subjected to over-collection as well, including edible mushrooms (e.g., *Tricholoma matsutake*) and some orchids (e.g. *Paphiopedilum*) and cycads for their horticultural values (Fig. 7). Fuel wood collection and timber procurement have also severely affected many forest species; some species have been pushed to the brink of extinction such as *Acer yangbiense, Picea neoveitchii,* and *Pinus squamata* (López-Pujol & Zhang, 2009). Until the logging ban of 1998, the growing timber harvests since the 1950s had resulted in the loss of extensive forest areas: in Sichuan, it is thought that *ca.* 40% of natural forests were cut (Morell, 2008). Nevertheless, and despite the ban, illegal commercial logging still occurs on a small scale (especially in SW China) as a source for heating and construction timber (Klok & Zhang, 2008; Morell, 2008).

Fig. 7. Orchids are usually over collected for their horticultural value. In China, over 1,200 taxa of Orchidaceae are included within *China Species Red List* (Wang & Xie, 2004). *Dendrobium devonianum* (left) and *Vanda cristata* (right) are included within them (photos by Hua-Feng Wang)

4.4 Introduction of exotic species

China has a long history of introducing alien species for their potential economic values or other supposed benefits (medicinal, ornamental, soil improvement, erosion control, landscaping, etc). The first species introductions took place more than 2,000 years ago, and some examples comprise *Carthamus tinctorius, Medicago sativa, Punica granatum,* or *Vitis vinifera,* which were brought through the Silk Road; Xie et al., 2001). However, its historical isolation has meant that large-scale introductions were relatively very few until the 1970s, despite China's particular vulnerability to invasive species as it harbors a wide range of suitable habitats and environmental conditions. Since the 1980s, when the country adopted market-oriented reforms and opened to international trade, the number of reported invasive alien species has significantly increased (Lin et al., 2007; Weber & Li, 2008). For plants, the number of invasive species has grown from 58 species listed in the 1990s (Ding & Wang, 1998) to 270 recently reported by Weber et al. (2008). This rapid increase, although it may reflect a better degree of knowledge of biological invasions (but see Lin et al., 2007), should be attributed to the unprecedented economic development of China, including an explosive growth of international trade which has enhanced the opportunities for alien species arrival (Ding et al., 2008). In addition, the domestic boom of the industrial and transportation infrastructures, coupled with an unparalleled rate of urbanization and a rampant ecological degradation, have promoted both alien species establishment and their spread within China (Lin et al., 2007; Ding et al., 2008; Weber et al., 2008).

The more than 400 invasive alien species detected in China include at least half of the 100 world's worst invasive alien species compiled by the IUCN (CSPCEC, 2008). These include plants (such as *Eichhornia crassipes, Lantana camara,* or *Mikania micrantha*), invertebrates (e.g. *Achatina fulica* and *Bemisia tabaci*) and vertebrates (*Bufo marinus, Gambusia affinis,* or *Myocastor coypus*). Some alien species have been introduced to restore vegetation without any previous assessment of their potential damage to local ecosystems, such as *Rhus typhina* (Fig. 8), massively planted throughout Beijing Municipality including the Olympic Park (Wang et al., 2011). Other invasives have been introduced unintentionally, such as the banana moth (*Opogona sacchari*), which entered inadvertently accompanying an ornamental plant (*Dracaena fragans*) (Ding et al., 2008). A preliminary estimation of the economic losses caused by invasive alien species gave a figure of *ca.* 14.5 billion USD one decade ago, a considerable figure but much smaller than that for the United States (Weber & Li, 2008) because of the much larger number of invasives in the latter. Thus, more biological invasions can be anticipated in China if the current pace of urbanization and infrastructure development is maintained.

4.5 Lack of effective environmental scope of government policies and ineffective legal protection

The Chinese authorities bear a major responsibility for the tremendous losses in plant diversity in the past and even now. Development policies as well as attitudes towards wildlife in Chinese society have historically been focused on the exploitation of natural resources since imperial times, but this gained special relevance during the second half of the twentieth century (Shapiro, 2001; Liu, 2010; Yu, 2010). For example, the 'Great Leap Forward' (1958-1961), a period in which communities were encouraged to be self-sufficient in steel, involved the cutting down of at least 10% of China's forests to fuel backyard furnaces (Liu, 2010). The development policies that followed were not much better, and during the Cultural Revolution (1966-1976) numerous forests, grasslands and wetlands were transformed into farmlands (McBeath & Leng, 2006). Following the adoption of the country's 'open-door' policies in 1978, although the government began to pass numerous laws for environmental protection and biodiversity conservation, and many protected areas have been set up (see following sections), economic development has taken priority over nature conservation (Liu & Diamond, 2005; Wang et al., 2007). Laws and regulations have shown many problems regarding their implementation and enforcement, sound monitoring and management systems are generally lacking, whereas financial resources are often insufficient (López-Pujol et al., 2006; Yu, 2010).

Fig. 8. Individuals of *Rhus typhina*, an invasive species of North American origin, grow in mountainous areas of Beijing Municipality (photos by Hua-Feng Wang)

4.6 Economic and population growth

China's economic growth over the last three decades has given rise to the fastest rate of GDP (gross domestic product) growth among the world's major economies (an average of 10% since the late 1970s; Huang & Luo, 2009), leading to strongly increased demands on their own (and their external) natural resources. China is already the world's biggest consumer of all major industrial commodities, although their per capita figures are still low compared to the developed nations (Grumbine, 2007). This means that if recent rates of economic growth are maintained, much more pressure will be put on natural resources. Economic development has involved the construction of numerous industries and power plants (the energy production increased by 105% during the period 1990-2007; World Bank, 2010a) in addition to a huge increase of the communication network (expressway length was less than 300 km in 1989 but it had reached 60,000 km in 2008, whereas more than 20,000 km of new railway were built during the same period; NBSC, 2009). All these infrastructures have contributed to the fragmentation of natural habitats; environmental impacts of some of the world's largest development projects hosted by China (e.g. Three Gorges Reservoir, South-to-North Water Transfer Project, West-East Gas Pipepine, Qinghai-Tibet Railway) cannot still be adequately quantified but would be huge (Liu & Diamond, 2005).

The unprecedented rise in the per capita income and living conditions experienced by Chinese people (the poverty incidence has declined from 84% in 1981 to 16% in 2005; World Bank 2010a) has generated a new industry, national (and international) tourism (national tourism has tripled in just 15 years; NBSC, 2009), with the subsequent construction of new hotels and holiday resorts, often located close to areas of natural and scenic interest or even inside nature reserves (López-Pujol et al., 2006; Yu, 2010). Improving accesses to these natural areas (e.g. new roads, cable cars and even airports—Jiuzhaigou-Huanglong and Shangri-La [Zhongdian] airports) has contributed to habitat degradation, since it has created an enormous influx of visitors. The number of visitors in Jiuzhaigou Nature Reserve has increased from *ca.* 32,000 in 1984 to over 2 million at present (Hendrickson, 2009); limitation of mass tourism has been called for by scholars due to its pervasive effects on Jiuzhaigou's biodiversity (e.g. Zhu et al., 2006; Morell, 2008). Another illustrative collateral effect of the emergence of a prosperous middle-class in China is the number of private cars, which has skyrocketed from only 19,000 in 1985 to *ca.* 30 million in 2008 (NBSC, 2009), with the associate severe pollution.

Population growth, which has slowed down since the implementation of the one-child policy, still remains significant. At present, the estimated total Chinese population is about 1,330 million, and is predicted to rise to 1,470 million by 2035, when it would start to decrease (Chen, 2010). Moreover, given the expected economic growth (China may become the world's largest economy by 2020 in terms of GDP; Hawksworth, 2010), the pressure on natural resources will continue. However, the number of households has grown much faster than the total population because of the increase of divorces and the reduction in the number of multigenerational families sharing the same home (Liu, 2010): the average household size has decreased from 3.5 in 1990 to 2.9 persons in 2008 in urban areas; NBSC, 2009). By 2030, 250 million new homes may be needed in China (that is, more than the total number of homes in the Western Hemisphere at the beginning of the present millennium; Liu & Diamond, 2005), which will suppose an additional threat to biodiversity since smaller households are less efficient in resource use.

5. Current conservation measures

China has a long history of nature conservation; the first rules concerning wildlife protection may well predate the Zhou dynasty (1046-256 BC; Edmonds, 1994). Historical conservation of the lands surrounding Buddhist and Taoist temples, as well as their 'sacred mountains' by monks has preserved these areas intact until the present day. Moreover, there is a clear link between biodiversity conservation and ethnic minorities in China. The conservation of Holy Places, Holy Mountains, and Holy Trees by several ethnic minorities is well known (Yang et al., 2004; Xu et al., 2005), e.g. the 'spirit mountains' in Xishuangbanna (Yunnan) have historically been protected by local communities of Dai nationality (Xu et al., 2005). However, 'modern' nature conservation began late in China, and can be divided into *in situ* and *ex situ* measures, as described below.

5.1 *In situ* conservation

The development of modern protected areas (PAs) in mainland China can be divided into four main stages: 1956-1965, the initiation; 1966-1974, stagnation and devastation; 1975-1979, restoration; and 1980-present, a period of rapid growth (Li et al., 2003). China's first nature reserve was founded in 1956 at Dinghu Mountain in Guangdong Province. Nine years later, up to 19 nature reserves had been established, encompassing an area of about 0.7% of the total Chinese land surface (Table 3). By the end of 2008, there were a total of 2,538 nature reserves (Taiwan, Hong Kong and Macao not included), covering a total area of 1,489,430 km^2 (MEP, 2009), i.e., more than 15% of the nation's surface (Table 3), exceeding the average for developing economies (12.7%; World Bank, 2010a). Most have been established in the last 25 years, comprising the principal governmental measure for protecting China's biodiversity.

Besides nature reserves (the most comprehensive category of PAs), there are other types of protected sites which should be regarded as PAs according to the IUCN definitions (MacKinnon & Xie, 2008) and also recognized by Chinese law, such as forest parks, scenic spots, geological parks, wetland parks, and agricultural reserves, among others. Adding all these types of protected spaces to nature reserves, the total number of PAs in China is well above 5,000, which cover over 18% of the country's land area (MacKinnon & Xie, 2008). According to the CSPCEC (2008), the present network of PAs is effectively protecting 65% of higher plant communities and 70% of national key wild flora under protection. In addition, a significant part (over 30%) of the area of the 14 Key Biodiversity Land Zones with global conservation significance (which account for nearly one quarter of China's landmass), are also protected by the PAs network (CSPCEC, 2008). Some Chinese PAs have received international recognition: 28 have been designated as Biosphere Reserves under UNESCO's Man and the Biosphere Program; 12 are Natural World Heritage Sites; and 36 have been designated globally significant wetlands under the Ramsar Convention.

Year	Number	Area (km^2)	Percentage of Chinese territory
1956	1	11	0.00
1965	19	6,488	0.07
1978	34	12,650	0.13
1982	119	40,819	0.42
1985	333	193,300	2.01
1987	481	237,495	2.47
1989	573	247,630	2.58
1991	708	560,660	5.84
1993	763	661,840	6.89
1995	799	719,070	7.49
1997	926	769,790	8.02
1999	1,146	845,090	8.80
2000	1,227	982,100	10.23
2001	1,551	1,298,900	13.53
2002	1,757	1,329,500	13.85
2003	1,999	1,397,500	14.56
2004	2,194	1,422,260	14.81
2005	2,349	1,439,500	14.99
2006	2,395	1,455,350	15.16
2007	2,531	1,458,800	15.19
2008	2,538	1,489,430	15.51

Table 3. Evolution of nature reserves in mainland China

In the last 25 years, the government of Taiwan Island has developed a complex network of PAs (up to 81) geared towards conserving its plant and animal biodiversity, which accounts for over 21% of the total land area (MacKinnon & Xie, 2008). About 44% of the land area in Hong Kong SAR is protected, the highest percentage in the Asia Pacific region (ESCAP, 2010). Nevertheless, these protected areas do not adequately cover some habitat types, such as freshwater wetlands and *feng shui* forests near urban areas (Yip et al., 2004). The most emblematic protected area in Hong Kong is the Mai Po Marshes & Inner Deep Bay, which was designated as a Wetland of International Importance (Ramsar Convention) in 1995.

5.2 *Ex situ* conservation

The most widely recognized *ex situ* conservation strategy is the preservation of living plants in botanical gardens (BGs) and arboreta. Although the first modern botanical gardens (that is, those designated for plant introduction and botanical research) were not established in China until the beginning of the 20th century, we can date their origin to *ca.* 2,800 BC (Medicinal Botanic Garden of Shennong), the earliest known botanical garden in the world (Xu, 1997). The first modern botanical garden, Hengchun Tropical Botanical Garden (Taiwan), was established in 1906, followed by Xiongyue Arboretum (Liaoning) in 1915, and Taipei Botanical Garden in 1921. Nevertheless, Hong Kong's Zoological and Botanic Garden precedes these, established in 1871. At present, over 160 botanical gardens have been set up in China (CSPCEC, 2008; Huang, 2010). The BGs belonging to the Chinese Academy of Sciences (which represents about 95% of the *ex situ* collections of all Chinese BGs) are cultivating *ca.* 25,000 species of vascular plants (Table 4), of which 20,000 are species found in China (Huang, 2010), i.e. over 60% of Chinese total flora. Living collections have increased considerably during the last decade due to a CAS innovation programme which has involved the designation of three core BGs (Xishuangbanna Tropical Botanical Garden, South China Botanical Garden and Wuhan Botanical Garden). The XTBG is currently the largest BG in China in number of plant collections (almost 15,000 taxa). The three core gardens, in addition to harboring large living collections, also maintain specialized collections: for example, SCBG holds the world's largest collection of Magnoliaceae (>130 species), Zingiberaceae (>120 species), and Palmae (>380 species) (Huang, 2010), and a collection of more that 2,000 medicinal plant species from South China (Wen, 2008).

Name	Location / Data of establishment	Area (km^2)	No. of taxa[a] / No. of species (living collections)	Red list species conserved
Xishuangbanna Tropical Botanical Garden (XTBG)	Menglun (Yunnan) / 1959	11.25	14,973 / 7,420	571
South China Botanical Garden (SCBG)	Guangzhou (Guangdong) / 1929	3.00	11,512 / 7,898	710
Wuhan Botanical Garden (WBG)	Wuhan (Hubei) / 1956	0.67	7,090 / 5,023	652
Fairy Lake Botanical Garden (FLBG)	Shenzhen (Guangdong)/ 1983	8.60	6,588 /4,956	441
Beijing Botanical Garden-CAS (BBG)	Beijing / 1956	0.72	5,001 / 3,463	177
Lushan Botanical Garden (LBG)	Lushan (Jiangxi) / 1934	3.00	4,934 / 4,378	229
Kunming Botanical Garden (KBG)	Kunming (Yunnan) / 1938	0.44	4,276 / 3,330	423
Guilin Botanical Garden (GBG)	Guilin (Guangxi) / 1958	0.67	4,056 / 3,843	445
Nanjing Botanical Garden Mem. Sun Yatsen (NBG)	Nanjing (Jiangsu) / 1929	1.86	3,790 / 2,701	263
Turpan Botanical Garden (TBG)	Turpan (Xinjiang) / 1976	1.50	506 / 490	26

[a] 'Taxa' include species, subspecies, and varieties

Table 4. The 10 main BGs of Chinese Academy of Sciences. Sources: BGCI (2010) and Huang (2010)

Seed banking has also greatly progressed during recent times. The China Southwest Wildlife Germplasm Genebank project (operated by the Kunming Institute of Botany) has already stored seeds of nearly 5,000 native plant species with the next major target to store 10,000 species by 2020 (Huang, 2010), thereby aiming to secure the preservation of the germplasm resources of SW China. The KIB seedbank, which is also storing seeds for the UK Millennium Seed Bank and the World Agroforestry Center (Tsao & Zhu, 2010), is also working as a DNA bank. Regarding crop species, extant *ex situ* conservation facilities of the Ministry of Agriculture (which include long-term, medium-term and duplicate cold storage facilities) are keeping almost 400,000 accessions of seeds of *ca.* 450 crop species (Huang, 2010). In addition, perennial and vegetatively propagated crops (and their wild relatives) are preserved in 32 national field germplasm nurseries, including more than 1,300 rare and endangered species (CSPCEC, 2008; MEP, 2008a). Other *ex situ* facilities include germplasm banks specific for forest species and medicinal plants (López-Pujol et al., 2006; CSPCEC, 2008).

5.3 Environmental legislation and government planning

In addition to *in situ* and *ex situ* measures, environmental legislation and government planning (i.e. policies) are also essential to ensure adequate conservation of biodiversity. China has passed numerous laws and regulations addressing biodiversity conservation since the early 1980s (López-Pujol et al., 2006; McBeath & Leng, 2006; Yu, 2008). The most relevant laws governing plant biodiversity include the *Environmental Protection Law* (issued in 1979, revised in 1989), the *Forest Law* (issued 1984, revised 1998), the *Grassland Law* (issued 1985, revised 2002), and the *Seeds Law* (2000, revised 2004). China has also issued a significant number of regulations and rules, such as the *Regulation about Protection and Administration of Wild Medicinal Material Resources* (1987), the *Regulation about Nature Reserves* (1994), the *Regulations on Wild Plants Protection* (1996), or the recent *Regulation on the Import and Export of Endangered Wild Fauna and Flora Species* (2006) and *Regulation on Scenic Spots and Historical Sites* (2006). In addition to laws and regulations, there is some governmental supervision actively supporting biodiversity conservation in China, the most relevant being the Environmental Impact Assessment (EIA) system, which was legally implemented in 1981 and amended several times thereafter (1986, 1989, 1998, and 2002). Other mechanisms include licensing systems (such as forest logging and land use licenses), economic incentives (financial subsidies, tax-deductions, and compensation fees to enhance sustainable exploitation of natural resources, and more recently, payment for ecological and environmental services), or the quarantine system (established in the early 1980s).

Regarding government planning, China started to launch several comprehensive biodiversity-related policies from the early 1990s. Within the *Eight Five-Year Plan for Economic and Social Development* (1991-1995), China took biodiversity conservation as a key national policy. In addition to starting an inventory of biodiversity at all levels (Li, 2010), the *China Biodiversity Conservation Action Plan* (NEPA, 1994) was launched in 1994 to implement the Convention on Biological Diversity (CBD) together with *China's Agenda 21*. The Ninth Five-Year Plan (1996-2000) formally included the execution of CBD: *China's Biodiversity: A Country Study* plan was launched at the end of 1997, which analyzed the country's overall biodiversity, its economic value and benefits, the cost of implementing the CBD, and long-term objectives for biodiversity conservation and sustainable use of biological resources (SEPA, 1998). Other major plans issued before the end of the 20th century encompassed the

National Program for Nature Reserves (1996-2010) and the *National Plan for Ecological Construction* (1998-2050) (MEP, 2008a).

At the turn of the century, new environmental polices were launched to cope with the need for a more comprehensive and sustainable nature management. These new policies, commonly known as the 'Six Key Forestry Programs' (SKFP), meant an investment which exceeded the total expenditure during the period 1949-1999, and were aimed to avoid some of the pitfalls of the past in nature management (Wang et al., 2008). The SKFP, launched essentially for ecosystem rehabilitation, environmental protection and afforestation, covered more than 97% of China's counties, and included: (i) the National Forestry Protection Program (NFPP); (ii) the Shelterbelt Development Program (SDP); (iii) the Grain to Green Program (also known as the Sloping Land Conversion Program and the Cropland to Forest Program) (GTGP); (vi) the Sand Control Program for areas in the vicinity of Beijing and Tianjin (SCP), (v) the Wildlife Conservation and Nature Reserves Development Program (WCNRDP); and (vi) Fast-growing and High-yielding Timber Plantations Program (FHTPP). Most of these programs were planned to expire in 2010 except the last one, which will be alive until 2015 (Wang et al., 2007, 2008). Recent national plans include the *National Program for Conservation and Use of Biological Resources* (2007) and the *China National Environmental Protection Plan* within the Eleventh Five-Year Plan (2006-2010). At the end of 2010 the *China National Biodiversity Strategy and Action Plan* (2011-2030) was approved to replace the 1994 plan. Specific to plant biodiversity, in 2007 the *China's Strategy for Plant Conservation* (CSPC) within the *Global Strategy for Plant Conservation* (*GSPC*) was launched, aiming to pursue the CBD 2010 Biodiversity Target (CSPCEC, 2008).

All these plans have also been designed to fit other international treaties and conventions with implications for plant diversity signed by China, in addition to the CBD: the CITES Convention (1981), the Ramsar Convention (1992), the United Nations Convention to Combat Desertification (UNCCD) (1996), the UN Millennium Development Goals (2000), the Kyoto Protocol (2002), and the Cartagena Protocol on Biosafety (2005), among others. In addition, China also maintains international cooperation with governments and both public and private institutions, highlighting: the 'China Council for International Cooperation on Environment and Development' (with experts from several countries; http://www.cciced.net/encciced/), the 'China-EU Biodiversity Project' (with the European Union; http://www.ecbp.cn/en/), the 'Greater Mekong Subregion Core Environment Program (with Cambodia, Laos, Burma, Thailand, and Vietnam with the support of the Asian Development Bank; http://www.gms-eoc.org/), and the 'Sino-American joint investigation on plant diversity in Hengduan Mountains' (with Harvard University; http://hengduan.huh.harvard.edu/fieldnotes).

6. Problems, prospects, and recommendations

6.1 Habitat destruction

The huge habitat destruction suffered in China, particularly since the 1950s (e.g. Shapiro, 2001), began to receive attention by the government authorities only in recent years, due at least in part, to the occurrence of natural disasters and the fall in crop productivity associated with soil erosion and land desertification (e.g. Liu & Diamond, 2005). To redress this situation, the state implemented a series of forestation and shrub or grass-planting projects (Fig. 9). Although the first plans were ratified in the late 1970s, they consisted generally of mono-culture forest plantations (often involving exotic species), lacking a comprehensive scientific basis and failing to account for local floristic features (Zhang P. et al., 2000; Morell, 2008). However, after the disastrous floods of 1998, the National Forestry Protection Program (NFPP) was introduced, aimed at protecting the forests by logging bans and afforestation activities. In addition, the five other programs within the SKFP (Six Key Forestry Programs), launched soon after, meant that on completion, 76 million hectares should be forested (Wang et al., 2007); due to this unprecedented effort, the forest cover has increased from 16.6% in 2000 to 20.4% by the end of 2009. These new afforestation initiatives were planned to avoid past mistakes; however, several problems and limitations aroused, such as the failing of local cadres to implement the programs effectively (mostly due to corruption), the lack of clarification of land ownership (although in recent years some reforms have started to be introduced), but also a decrease in forest quality (a young plantation cannot provide the same ecological services that a mature stand can provide) and a low rate of survival of populations, sometimes due to the use of inappropriate species (Wang et al., 2008; Song & Zhang, 2010). Moreover, other shortcomings such as an overemphasis in shrub and tree planting instead of grasses (as often reported in arid zones; e.g. Cao et al., 2011) have produced undesired effects including a loss of native vegetation and the exacerbation of water shortages.

Fig. 9. Shelterbelt to protect farmland from Gobi's sand encroachment (near Jiayuguan, Gansu Province) (photo by Jordi López-Pujol)

6.2 Protected areas

The *National Program for Nature Reserves* (implemented in 1996) stipulated that the number of nature reserves must reach 1,200 by 2010 and accounting for 10% of the Chinese territory (or 12% when forest parks and scenic spots are considered) (Li et al., 2003). This goal has been widely exceeded due to the impressive rate of nature reserves establishment, particularly during the last 25 years (Table 3). Unfortunately, provisions for staffing and financing in order to manage these reserves have not increased at the same rate. For example, about one-third of all nature reserves had neither staff nor budget (i.e., are protected only 'on paper'; Liu et al., 2003; Xu H. et al., 2009). Moreover, the staff is rarely professionally-trained (with higher education) (MacKinnon & Xie, 2008; Xu H. et al., 2009). Lack of financial investment is, however, a general problem for all the reserves including those state-funded (which are somewhat better funded but represent less than 12% of the total no. of reserves). Lack of budget compromises reserves' protection duties: they are poorly or never patrolled, species and ecosystems are not satisfactorily monitored nor inventoried, and some reserves do not even have signposts delineating their borders (Qiu et al., 2009; Xu H. et al., 2009; Quan et al., 2011). In the recent study of Quan et al. (2011), some worrying figures arose, such as a mere 2% of the nature reserves had enough financial support for their daily management activities, and that only *ca.* 11% had set up comprehensive monitoring systems.

To solve the funding shortage and to cover daily operation costs, many reserves are forced to be self-sufficient through resource exploitation (e.g. over-exploitation of plant resources including medicinal and edible plants, hunting, mining, land reclamation, hydropower development, tourism and recreation), a policy inconsistent with their intended purpose and which may cause severe harm (López-Pujol et al., 2006; MacKinnon & Wang, 2008; Yu, 2010). One illustrative example is the destruction of over 1,000 km^2 of natural wetlands in the Yancheng National Nature Reserve, listed both as a Ramsar site and a Biosphere Reserve (Qiu et al., 2009).

Another consequence of lack of investment is the frequent failure of compensation schemes (subsidies, compensation fees) to the local people (a problem often aggravated by the widespread corruption among local officials), who may be against the establishment of new PAs because they feel that their interests are in conflict with nature preservation. Tourism creates opportunities for local people, but this should evolve towards sustainability, and planned and managed to combine biodiversity protection while ensuring adequate economic benefits to local communities (Quan et al., 2011). Engaging local communities in conservation activities as well as in the planning and management of reserves also constitutes a useful tool for the long-term sustainability of nature reserves, since the pressures placed on reserves by local residents are largely eased (McBeath & Leng, 2006; MacKinnon & Xie, 2008). Enhancing public participation should also include the NGOs, which in other countries have demonstrated a good performance in both assisting in the PAs management as well as resolving people-park conflicts (McBeath & Leng, 2006; Qiu et al., 2009).

Nevertheless, nature reserves are afflicted with other serious problems besides insufficient budgets. Overlapping management–in some cases involving up to seven administrations–can cause confusion, inefficiency, uncertain boundaries, and multiple designations of the same reserve (López-Pujol et al., 2006; McBeath & Wang, 2009). Another common problem of Chinese PAs is that they are too small to maintain genetic diversity or to ensure species and ecosystem viability (Liu et al., 2003; Xu H. et al., 2009). This is especially true in eastern China (see Fig. 10), where nature reserves are often just occupying a very few square kilometers; for example, the 512 smallest reserves in China (which account for about 20% of their total number), accounted for *ca.* 0.13% of their total area (MacKinnon & Xie, 2008). On the contrary, very large areas can be found in western China (Fig. 10), some exceeding 10,000 km^2 (Qiangtang Nature Reserve, in Tibet, has almost 300,000 km^2). In China, the combined area of the 20 largest nature reserves accounts for nearly 60% of the total area of all reserves (MacKinnon & Xie, 2008), which shows that the design of PAs has not been entirely rational in the past.

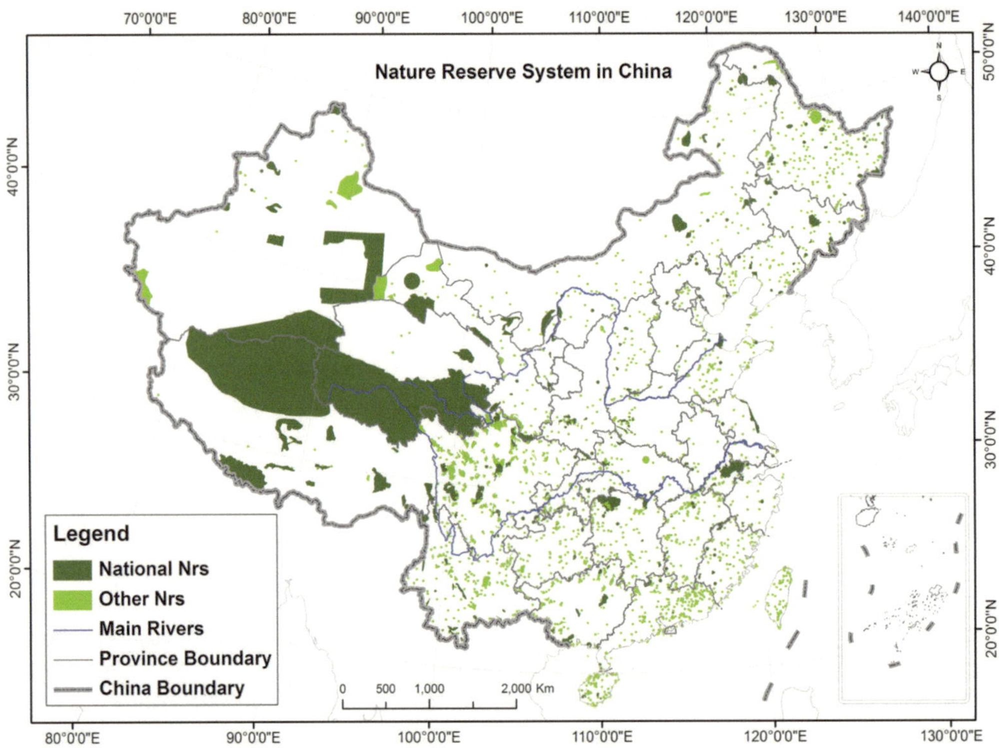

Fig. 10. Map of nature reserves in China at the end of 2008 (map elaborated by Lu Zhang, Research Center for Eco-Environmental Sciences, Beijing)

The lack of systematic planning is also evidenced when criteria of biogeographic and ecosystem representativeness are tested. Xu et al. (2008) reported a total lack of correlation between the percentage of land area occupied by nature reserves and overall species richness, endemism, or threat at provincial level; in this sense, these authors are calling for the setting up of new reserves in the provinces with less than 10% of reserves coverage (Fig. 11). Moreover, if we compare the Figs. 10 and 4 (the plant hotspots), it is clear that the areas richest in plant diversity are not protected enough. Some of the main gaps in the Chinese coverage of PAs (see Li et al., 2003; MacKinnon & Xie, 2008) are precisely those corresponding to hotspots or areas located within the hotspots (e.g. Hengduan Mountains, N Guangxi, SE Yunnan-SW Guizhou-SW Guangxi). Numerous gaps in ecosystem protection also exist, such as marine, wetland, grassland and desert vegetation (Li et al., 2003; Xu H. et al., 2009). An additional problem is their lack of connectivity through biological corridors (MacKinnon & Wang, 2008). Establishing a centralized management by a new State Agency of Nature Reserves Service at state-level, and upgrading the current *Regulation of Nature Reserves* (of 1994) to a new *Nature Reserve Law* (which is currently being drafted) seem necessary steps to achieve more comprehensive planning and management of the Chinese PAs network (Yu, 2008; McBeath & Wang, 2009).

6.3 *Ex situ* conservation measures

The increasing number of BGs during the last decades (from just 52 in 1975 to 160 at present; Fig. 12; Huang, 2010) and the launching of government programs (such as the 15-year master plan of Chinese Academy of Sciences; Huang et al., 2002) has meant a great advancement in the *ex situ* conservation of Chinese flora. Firstly, the target to conserve 21,000 native plant species by the 15-year plan has almost been totally achieved. Secondly, the GSPC (Global Strategy for Plant Conservation; CBD, 2002) target to conserve at least 60% of threatened plants has been partially achieved: virtually all the 388 species of the *National List of Rare and Endangered Plant Species* are included in the *ex situ* living collections of Chinese BGs (although some exceptions apply; López-Pujol & Zhang, 2009), but only a small fraction (less than 40%) of the 4,408 species of *China Species Red List* (Table 4; Huang, 2010). Yet another deficiency is, despite the recent progress, that BGs are not representative of the local floras across China; some regions boasting a rich plant diversity, such as western China, have too few botanical gardens (only 10% of the total; Huang, 2010), such as the Himalayas, the Qinghai-Xizang Plateau (in the Tibet Autonomous Region there are no BGs; Cram et al., 2008), and the dry-hot valleys of southwestern China, a trend also apparent in the three primary distribution centers for endemic plants in China (He, 2002; López-Pujol et al., 2006). In addition, both the number of gardens where a given plant is cultivated and the sizes of the collections are generally insufficient: of the *ca.* 25,000 plant species cultivated in the CAS BGs, about two-thirds are not duplicated (that is, present in just one BG; Huang, 2010). For the threatened species, the trend is the same: only half of the Chinese threatened species are duplicated. Furthermore, despite the claim of Xu (1997), population sizes are generally not sufficient for maintaining adequate levels of genetic diversity; for example, the only *ex situ* collection of *Picea neoveitchii* (a threatened species included within the *National List of Rare and Endangered Plant Species* and the *Catalogue of the National Protected Key Wild Plants*) consist of two individuals cultivated in the Xi'an Botanical Garden (Zhang, 2007). Other problems are related to lack of financial resources; the Three Gorges Botanic Garden of Rare Plants (which housed about 10,000 individuals belonging to 175 plant taxa) was closed in 2007 owing to a lack of funding (López-Pujol & Ren, 2009). Regarding seed storage facilities, there are still considerable gaps in conservation of Chinese native wild plants; for example, no seeds from any Tibetan plant species were stored in the Kunming seedbank until recently (Cram et al., 2008), although this is now being solved by the staff of Kunming Institute of Botany.

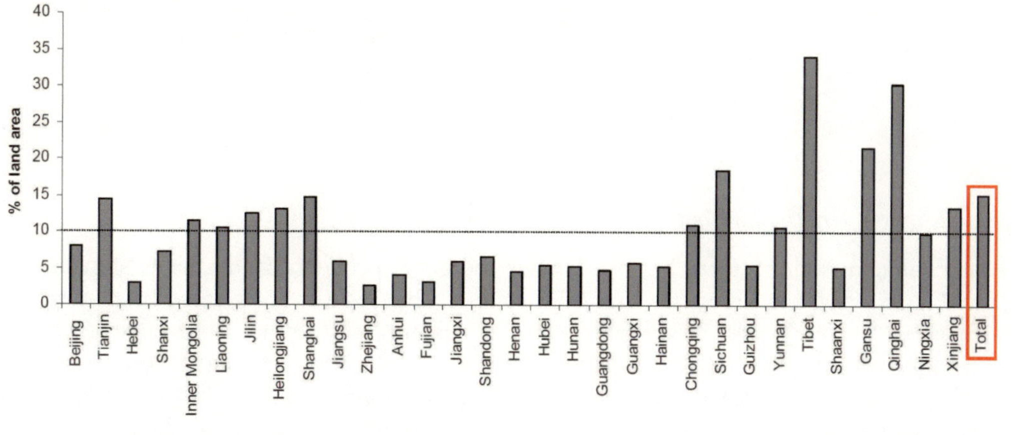

Fig. 11. Coverage of nature reserves in each province of mainland China (source: MEP, 2008b). Dotted line indicates 10% of reserves coverage.

6.4 Environmental legislation

Legislation addressing environment and biodiversity has significantly expanded in the last 25 years, and a relatively comprehensive body of laws and regulations have been enacted, some as restrictive as European laws in many aspects (Ferris, 2005). Nevertheless, two major problems remain: an insufficient and inefficient legal system and a lack of enforcement (López-Pujol et al., 2006; Johnson, 2008; Yu, 2010).

The main purpose of many biodiversity-related laws and regulations is still to manage the use of natural resources and they are poorly-focused on conservation (although this focus is increasing in recent years; McBeath & Leng, 2006) since they were mainly promulgated taking into account the natural resources' economic value and not their sustainable use (Yu, 2008). Moreover, legislation tends to focus on endangered animals and not plants (which are covered by regulations and not laws), and also do not provide explicit protection of their habitats (McBeath & Wang, 2009). In addition, specific legislation to preserve genetic resources is still very limited (Yu, 2008). There is no comprehensive law governing protected areas (although this is being drafted), and a law specifically devoted to protection of biodiversity is still pending. Another inconsistency inherent to China's legislation is the lack of a clear demarcation of responsibilities, whereas punishments only mandate damage compensation rather than ecological restoration or rehabilitation (Beyer, 2006).

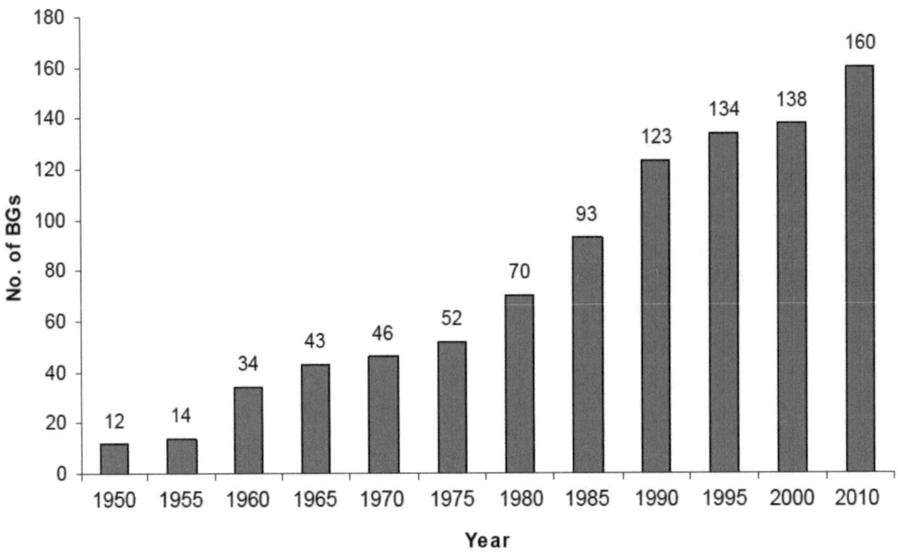

Fig. 12. Increase in the number of BGs in China. Source: Huang (2010)

Historically, law enforcement has been one of the major problems in the establishment of a sound legal system in China. As Yu (2010) states, "China is a country ruled by men rather than ruled by law", and moral precepts and customs of Confucian heritage generally outweighed formal laws (Beyer, 2006). Violations of environmental legislation are all too common and even tolerated. For example, at the beginning of the 2000s there still was a generalized lack of effective *in situ* legal protection for the nationally listed rare and endangered plant species (Xie 2003), and at present this situation is still continuing for some of them (López-Pujol & Zhang, 2009). This poor enforcement has multiple reasons apart from historical ones, including: (i) legislation is too general and largely vague; (ii) violations of environmental and nature protection laws have, with a few exceptions, no serious penalty; this means that most companies prefer paying the fines instead of following the law; (iii) a lack of capability on the part of administration for monitoring law enforcement (staff, funding and technical expertise are insufficient); (iv) lack of coordination among the different administrative levels (several agencies are sometimes responsible for the same task); (v) conflicts of interest between national-level legislation and local interests; (vi) widespread corruption among government officials; and (vii) lack of public participation (e.g. Beyer, 2006; McBeath & McBeath, 2006; Johnson, 2008; Liu & Diamond, 2008; Nagle, 2009; Yu, 2010).

Implementation of the environmental impact assessment (EIA) has experienced enormous difficulties in the past, although the rate of enforcement has been significantly increasing since the 1990s. Prior to passage of the Law on EIA in 2003, Chinese environmental legislation only focused on individual construction projects that might pollute the environment; however, the 2003 law expanded the environmental assessment to include government-proposed plans and projects (although with some exceptions) and included public participation as part of the process (Moorman & Zhang, 2007; Zhao, 2009). However, some problems remain, as EIA is still mainly applied to assess the impact of projects that might pollute the environment rather than addressing all activities affecting biodiversity (Yu, 2008). The growing conflict between the central government and local governments is a formidable obstacle to implementing the EIA, as well as the still-limited public scrutiny and the minimal violation penalties (Zhao, 2009). In order to strengthen its enforcement, the government has implemented a moratorium on EIA approvals since 2007 (You, 2008).

6.5 Scientific research

Large-scale national surveys of vegetation and flora began in the early 1960s mainly with the aid of experts from the USSR (Li, 2010). After the difficult period of the Cultural Revolution, when all academic activities were largely stopped, scientific research received a new impulse, and some major publications started to appear, such as *Vegetation of China* (ECVC, 1980) whereas other works progressed rapidly, such as *Flora Reipublicae Popularis Sinicae*, which was started in 1958 and whose 80 volumes were finally completed in 2004 (Li, 2008). Currently, the Missouri Botanical Garden (MBG) and the Chinese Academy of Sciences (CAS) are working together on the *Flora of China* project (Fig. 13), an international effort to produce a 25-volume English-language revision of the *Flora Reipublicae Popularis Sinicae.* At present, 19 volumes have already been published and they can also be browsed online (http://hua.huh.harvard.edu/china/). These same institutions (MBG and CAS) have also promoted the *Moss Flora of China* project, which aims to provide an updated, English version of the bryophyte flora of China (http://www.mobot.org/mobot/moss/china/welcome.html). In addition, an increasing number of local and provincial floras (although rarely in English) are available today (Liu et al., 2007).

Regarding conservation biology, after the first symposium on biodiversity conservation in China took place in 1990 (Wang et al., 2000), some general surveys and studies have been published since then, including the seminal *China Plant Red Data Book* (Fu, 1992), *Chinese Biodiversity – Status and Conservation Strategy* (Chen, 1993), *A Biodiversity Review of China* (MacKinnon et al., 1996), *Conserving China's Biodiversity* (2 volumes; Wang & MacKinnon, 1997; and Schei et al., 2001), *China's Biodiversity: A country Study* (SEPA, 1998), *The Plant Life*

Fig. 13. Entry for *Cathaya argyrophylla* in the online version of *Flora of China*

of China (Chapman & Wang, 2002), or the more recent *The Green Gold of China* (MacKinnon & Wang, 2008). In parallel, publication of papers in high impact factor journals has experienced a huge rise since the 1990s: China was the seventh most productive country in biodiversity research during the period 1900-2009 (and due to obvious historical reasons, almost all of this corresponds to the last two decades (Enright & Cao, 2010), whereas the Chinese Academy of Sciences is the world's most productive research institution (Liu et al., 2011). Almost all aspects of plant biodiversity are currently being explored by Chinese researchers, including the cutting-edge ones (see Enright & Cao, 2010), and this is mostly due to the great effort of the Chinese government: spending on research and development has increased to 1.5% of GDP, well above the developing economies (0.96%; World Bank, 2010a). The total research funding of the China's National Natural Science Foundation almost quadrupled during the period 2001-2008 (He, 2009).

Other biodiversity surveys include the national survey on traditional Chinese medicinal resources, conducted between 1984 and 1994, and identifying 11,146 plant species (Xu et al. 1999). The State Forestry Administration has performed five-year surveys (including censuses) of forest resources since 1973, recently completing their 7th forest survey. Moreover, the Chinese Academy of Sciences has set up a series of ecological field stations since 1988 (about 40 at present), organized into the Chinese Ecosystem Research Network (CERN) (Fu et al., 2010), whereas the Ministry of Science and Technology has set up the Chinese National Ecological Research Network (CNERN), with over 50 field stations including those of CERN (Li, 2010). It is also noteworthy the launching of *Chinese Virtual Herbarium* (http://www.cvh.org.cn/), an on-line portal which allows access to the plant specimens maintained in Chinese herbaria and to related botanical databases.

Taiwan has also made considerable efforts to study the island's biodiversity, with the publication of *Flora of Taiwan* (the 6 volumes of the second edition were compiled during the period 1993-2003, and are available online at http://tai2.ntu.edu.tw/ebook.php) and the *Red Data Book of Taiwan Region: Criteria and Measure for Rare and Threatened Plant Species* (Lai, 1991), although a revised red book is under preparation. In the Hong Kong SAR, the first surveys on flora were conducted as early as 1861, with the *Flora Hongkongensis* of G. Bentham. Most recent works on plant biodiversity are the *Check List of Hong Kong Plants 2001* (Wu, 2002) and three out of four volumes of the *Flora of Hong Kong*, produced by HK Herbarium and the South China Botanical Garden, have been already published.
Despite the significant strides made by these various administrations, universities, and research institutes, there are still significant gaps in the knowledge of plant diversity, mainly because of the lack of intensive botanical exploration in many parts of the country, specially the western mountainous areas. For example, despite some researches that were done in the floristically wealthy Hengduan Mountains since the 1970s, comprehensive inventories were not carried out until the 2000s. In addition, Chinese botanists were not able to do sizable international collaborations until the beginning of the 1980s (e.g. Dosmann & Del Tredici, 2003). The lack of information on China's threatened plant species is still worrisome despite the recent efforts; less than one-fifth of the angiosperms of China have been assessed using the 2001 IUCN criteria, and at present no modern red book (that is, following the standards of those published in western countries, which include valuable information of each studied plant species including geographic range, habitat characterization, conservation status, threats, and present and recommended conservation measures) has been published, either at regional or national level. Finally, it should be noted that, despite the fact that a sound environmental and ecological monitoring network is already in place, there is a need for a monitoring system specific to biodiversity (e.g. Xu H. et al., 2009); that is, monitoring populations and species, with special emphasis to those threatened.

6.6 Policies, financial resources, and environmental awareness

Government policies in mainland China have promoted rapid growth since the end of the 1970s, when the 'open-door' policy was adopted in order to eradicate poverty and quickly catch up with the developed economies. These policies have often implied the seeking of short-term economic benefits by an intense and inefficient use of natural resources and high levels of pollution; such unsustainable development policies have also been applied to the conservation and management of biodiversity (Liu & Diamond, 2005; Liu, 2010; Yu, 2010). However, some optimistic signs have emerged during recent years; since the 'fourth generation' of Chinese leaders took government in mainland China, environmental protection and natural resources conservation policies (which became a national strategy since the middle of the 1990s) have been emphasized (e.g. Johnson, 2008), as a part of a broader policy aimed at achieving the *xiaokang* (a moderately prosperous society) by 2020 (MFA, 2010). In 2007, the new 'scientific development' concept was proposed during the 17th Party Congress, consisting of combining the development of the economy with the protection of natural resources and the environment, with the final goal of achieving a 'harmonious society' (that is, development based on sustainability) (Fu et al., 2007; Johnson, 2008). Consequently, the Chinese economy is progressively turning from a polluting, low-efficiency economy to an environmentally-friendly, circular economy (McBeath & Wang, 2009; Liu, 2010). For example, the Eleventh Five-Year Plan (2006-2010) includes a target to reduce energy consumption per unit of GDP by about 20% (MEP, 2008a), which was likewise included in the 'China's National Climate Change Programme' launched in 2007 (NDRC, 2007). Furthermore, the ambitious goal of reducing CO_2 emissions per unit of GDP by 40-45% by 2020 from 2005 levels will be included within the Twelfth Five-Year Plan (2011-2015) (MFA, 2010). Other very positive steps include the upgrade of State Environmental Protection Administration (SEPA) to ministry status (Ministry of Environmental Protection) in 2008 (McBeath & Wang, 2009); however, it still remains a small body, with much less employees than the United States Environmental Protection Agency (Liu & Diamond, 2008; McBeath & Wang, 2009).

Despite these newly-oriented state-level policies, short-term economic and political gains still often outweigh the long-term benefits of preserving natural resources by the local governments; this is undoubtedly because the promotion of government officials at the local level still strongly relies on economic performance (Qiu et al., 2009; Liu, 2010; Yu, 2010). Therefore, the actual conflict between economic development and nature protection is the primary reason why many PAs are managed for purposes other than nature preservation and scientific research (e.g. McBeath & Leng, 2006; Qiu et al., 2009). Therefore, it is highly advisable, while being one of the major recommendations of the Task Force on Environmental Governance of the CCICED ('China Council for International Cooperation on Environment and Development'), that evaluating the performance of local government leaders should incorporate environmental performance (Xue et al., 2007).

One of the main reasons for the still dominant attitudes toward over-exploitation of natural resources and the lack of biodiversity protection is that the importance of biodiversity is not fully understood. The ideology of 'pollute first, clean later' is still seen as suitable for the development of the country by many Chinese, including officials (Liu, 2010), and this attitude is often fuelled by the poor example offered by the western countries in the past: although China is today the world's highest emitter of CO_2, the cumulative CO_2 emissions of China during the period 1850-2005 represent only about 8% of the world's total, much less than the 27,8% of the United States and also lower than those for Germany and Russia (World Bank, 2010b). In addition, traditional beliefs such as Confucian philosophy (which asserts control over nature; Shapiro, 2001; Yu, 2010) are still very present in contemporary Chinese society. A second reason is that the economic value of biodiversity (e.g. the ecosystem services) has also often been underestimated; ecological benefits are seldom contemplated because most of them cannot be reflected in the traditional marketplace (Zhang B. et al., 2010). However, some recent large catastrophes (e.g. the 1998 floods or the 2005 Songhua River spill) have shown the government and society that the toll (economic but also human) associated with lack of environmental protection is too high. The introduction of the calculation of Green GPD by the government showed that the costs of environmental pollution and ecological degradation accounted for about 3% of GDP in 2004 (the only year for which the figures have been released to the public; Liu, 2010), although some have estimated that these costs have ranged from 7% to 20% of GDP during the most recent decades (Liu & Diamond, 2005; Fu, 2008).

The lack of financial resources, which is a logical consequence of that explained in the previous paragraphs, is another of the root causes for the current biodiversity loss. The current investment of the Chinese government in environmental protection remains inadequate, although it has increased significantly since the 1990s; whereas the total environmental investment for the Ninth Five-Year Plan (1996-2000) was less than 1% of GDP, this rose to about 1.2% of GDP during the Tenth Five-Year Plan (2001-2005), and it has reached 1.5% in the Eleventh Five-Year Plan (2006-2010) (NBSC, 2009), but it is far from the average of 2.5% of GDP spent by the developed countries (McBeath & Wang, 2009). Despite the commendable effort by the Chinese government in raising the funds yearly, much more investment is expected for the country with the largest monetary reserves of the world (World Bank, 2010a) and which harbors some of the most valuable flora in the world (see section 2 of this chapter); in this sense, China should play a much more leading role in plant biodiversity conservation. In addition to increasing such funding, the government should encourage the active participation of all sectors of Chinese society, the private companies in particular (e.g., through tax-exemption), as well as non-governmental organizations (NGOs) and other active groups of civil society.

The strengthening of environmental awareness can be viewed as a sign of maturation of societies. In this sense, there is room for optimism as China has witnessed during recent years a flourishing of numerous environmental groups (including over 2,000 environmental NGOs; Yu, 2010) while the control over media and scientists has been significantly reduced. Indeed, the opinion of scientists, environmental groups, and general public are now taken very seriously by the government; for example, protests forced a moratorium on the damming of the Nujiang River in 2004 (declared by the Prime Minister himself) and also halted the construction of a chemical plant in Xiamen City in 2007 (McBeath & McBeath, 2006; MacKinnon & Wang, 2008). In addition, Chinese authorities have also implicitly acknowledged the contribution of NGOs to the protection of biodiversity; at present, many international environmental NGOs are assisting the government in the design and management of conservation activities (e.g. the WWF helped in establishing the Wolong Giant Panda Reserve). The role of the NGOs is, however, weak in China compared to the western countries, and their scrutiny by the government is still significant (e.g. Xue et al., 2007; McBeath & Wang, 2009).

Despite this progress on environmental consciousness, much more should be done. Environmental education, although expanding at all levels, is still limited (CSPCEC, 2008), and specialized courses in biodiversity conservation are badly needed (McBeath & Wang, 2009). Environmental awareness should be particularly geared towards the indigenous and local communities, as they usually have more direct contact with the natural resources in both protected and unprotected areas. More importantly, most of the plant diversity (including threatened species) is located in poorer areas of the west and the southwest; therefore, it is essential to give the highest priority to those policies oriented to conciliate poverty alleviation and biodiversity conservation (CDB, 2010).

7. Acknowledgement

The authors thank Lu Zhang for his help in elaborating Fig. 10. The authors also thank Gemma Rawsthorne for revising the English. This study has been partially subsidized by a post-doctoral grant (EX2005-0922) from the Spanish Ministry of Education and Science (MEC) for the period 02/2006-01/2008, and by a 'JAE-Doc' contract within the CSIC program 'Junta para la Ampliación de Estudios' for the period 03/2008-03/2011.

8. References

Axelrod, D.I.; Al-Shehbaz, I. & Raven, P.H. (1996). History of the modern flora of China, In: *Floristic characteristics and diversity of East Asian plants*, A. Zhang & S. Wu (Eds.), pp. 43-55, China Higher Education Press, Beijing

Barrett, P.M. & Willis, K.J. (2001). Did dinosaurs invent flowers? Dinosaur-angiosperm coevolution revisited. *Biological Reviews*, Vol. 76, No. 3, pp. 411-447

Beyer, S. (2006). Environmental law and policy in the People's Republic of China. *Chinese Journal of International Law*, Vol. 5, No. 1, pp. 185-211

BGCI (Botanic Gardens Conservation International) (2010). Garden search, In: *Botanic Gardens Conservation International*, 8.01.2011, Available from http://www.bgci.org/garden_search.php

Cai, Y. (1990). Land use and management in PR China. Problems and strategies. *Land Use Policy*, Vol. 7, No. 4, pp. 337-350

Cao, S.; Chen, L.; Shankman, D.; Wang, C.; Wang, X. & Zhang, H. (2011). Excessive reliance on afforestation in China's arid and semi-arid regions: Lessons in ecological restoration. *Earth-Science Reviews*, Vol. 104, No. 4, pp. 240-245

CBD (Convention on Biological Diversity) (2002). *Global Strategy for Plant Conservation*, Secretariat of the Convention on Biological Diversity, Montreal

CBD (Convention on Biological Diversity) (2010). *Linking biodiversity conservation and poverty alleviation: A state of knowledge review*, Secretariat of the Convention on Biological Diversity, Montreal

Chapman, G.P. & Wang, Y.-Z. (2002). *The plant life of China. Diversity and distribution*, Springer-Verlag, Berlin & Heidelberg

Chen, L. (Ed.) (1993). *Chinese biodiversity – Status and conservation strategy*, Science Press, Beijing (In Chinese)

Chen, Y. (Ed.) (2010). *Energy science & technology in China: A roadmap to 2050*, Science Press, Beijing & Heidelberg, Springer

CI (Conservation International) (2007). Biodiversity Hotspots. Mountains of Southwest China, In: *Conservation International*, 21.12.2010, Available from http://www.biodiversityhotspots.org/xp/hotspots/china/Pages/default.aspx

Cram, W.J.; Zhong, Y.; Tersing, T. & Cai, J. (2008). Tibet's seeds must be stored as climate changes. *Nature*, Vol. 452, No. 7183, pp. 28

Crooks, K.R. & Sanjayan, M. (Eds.) (2006). *Connectivity conservation*, Cambridge University Press, Cambridge

CSPCEC (China's Strategy Plant Conservation Editorial Committee) (2008). *China's Strategy for Plant Conservation*, Guangdong Press Group, Guangzhou

Ding, J.Q. & Wang, R. (1998). Invasive alien species and their impact on biodiversity in China, In: *China's biodiversity: a country study*, Compilation Group of China's Biodiversity (Eds.), pp. 58-63, China Environmental Science Press, Beijing (In Chinese)

Ding, J.; Mack, R.N.; Lu, P.; Ren, M. & Huang, H. (2008). China's booming economy is sparking and accelerating biological invasions. *BioScience*, Vol. 58, No. 4, pp. 317-324

Dosmann, N. & Del Tredici, P. (2003). Plant introduction, distribution and survival: A case study of the 1980 Sino-American botanical expedition. *BioScience*, Vol. 53, No. 6, pp. 588-597

ECVC (Editorial Committee for Vegetation of China) (1980). *Vegetation of China*, Science Press, Beijing (In Chinese)

Edmonds, R.L. (1994). *Patterns of China's lost harmony*, Routledge, London

Enright, N.J. & Cao, K.-F. (2010). Plant ecology in China. *Plant Ecology*, Vol. 209, No. 2, pp. 181-187

ESCAP (Economic and Social Comission for Asia and the Pacific) (2010). *2009 statistical yearbook for Asia and the Pacific*, Economic and Social Comission for Asia and the Pacific (United Nations), Bangkok

Ferris, R.J. (2005). China's increasing focus on Europe: trends and implications for the development of Chinese environmental law. *International Environmental Law Committee Newsletter*, Vol. 7, No. 2, pp. 1-7

Francisco-Ortega, J.; Wang, Z.-S.; Wang, F.-G.; Xing, F.-W.; Liu, H.; Xu, H.; Xu, W.-X.; Luo, Y.-B.; Song, X.-Q.; Gale, S.; Boufford, D.E.; Maunder, M. & An, S.-Q. (2010). Seed plant endemism on Hainan Island: a framework for conservation actions. *The Botanical Review*, Vol. 76, No. 3, pp. 346-376

Fu, L.K. (Ed.) (1992). *China Plant Red Data Book. Rare and endangered plants*, Vol. 1, Science Press, Beijing & New York

Fu, B. (2008). Blue skies for China. *Science*, Vol. 321, No. 5889, pp. 611

Fu, B.; Li, S.; Yu, X.; Yang, P.; Yu, G.; Feng, R. & Zhuang, X. (2010). Chinese ecosystem research network: Progress and perspectives. *Ecological Complexity*, Vol. 7, No. 2, pp. 225-233

Fu, B.-J.; Zhuang, X.-L.; Jiang, G.-B.; Shi, J.-B. & Lü, Y.-H. (2007). Environmental problems and challenges in China. *Environmental Science and Technology*, Vol. 41, No. 22, pp. 7597-7602

Graham, L.E. (1993). *Origin of land plants*, John Wiley and Sons, New York

Grumbine, R.E. (2007). China's emergence and the prospects for global sustainability. *BioScience*, Vol. 57, No. 3, pp. 249-255

Gu, J. (1998). Conservation of plant diversity in China: achievements, prospects and concerns. *Biological Conservation*, Vol. 85, No. 3, pp. 321-327

Guo, Q. (1999). Ecological comparisons between Eastern Asia and North America: historical and geographical perspectives. *Journal of Biogeography*, Vol. 26, No. 2, pp. 199-206

Guo, Q.; Ricklefs, R.E. & Cody, M.L. (1998). Vascular plant diversity in eastern Asia and North America: historical and ecological explanations. *Botanical Journal of the Linnean Society*, Vol. 128, No. 2, pp. 123-136

Hallam, A. (1994). *An outline of Phanerozoic biogeography*, Oxford University Press, Oxford

Hamilton, A.C. (2004). Medicinal plants, conservation and livelihoods. *Biodiversity and Conservation*, Vol. 13, No. 8, pp. 1477-1517

Hawksworth, J. (2010). Convergence, catch-up and overtaking: How the balance of world economic power is shifting, In: *PricewaterhouseCoopers*, 6.01.2010, Available from http://www.ukmediacentre.pwc.com/imagelibrary/downloadMedia.ashx?MediaDetailsID=1626

He, S. (2002). Fifty years of botanical gardens in China. *Acta Botanica Sinica*, Vol. 44, No. 9, pp. 1123-1133

He, F. (2009). Price of prosperity: economic development and biological conservation in China. *Journal of Applied Ecology*, Vol. 46, No. 3, pp. 511-515

Hendrickson, C.Y. (2009). *To develop or conserve? A case study on biodiversity protection and local sustainable development in Jiuzhaigou, China*. Master Thesis, Lund University Center for Sustainable Studies, Lund University, 6.01-2011, Available from http://www.lumes.lu.se/database/alumni/07.09/thesis/Hendrickson_Cary_Thesis_2009.pdf

Hou, X-Y. (1983). Vegetation of China with reference to its geographical distribution. *Annals of the Missouri Botanical Garden*, Vol. 70, No. 3, pp. 509-548

Hsieh, C.-F. (2002). Composition, endemism and phytogeographical affinities of the Taiwan flora. *Taiwania*, Vol. 47, No. 4, pp. 298-310

Hu, S.Y. (1980). The *Metasequoia* flora and its phytogeographic significance. *Journal of Arnold Arboretum*, Vol. 61, No. 1, pp. 41-94

Hu, S.-Y. (2005). *Food plants of China*, The Chinese University Press, Hong Kong

Huang, H. (2010). Ex situ plant conservation: a key role of Chinese botanic gardens in implementing China's Strategy for Plant Conservation. *BGjournal*, Vol. 7, No. 2, pp. 14-19

Huang, Y. & Luo, X. (2009). Reshaping economic geography in China, In: *Reshaping economic geography in East Asia*, Y. Huang & A.M. Bocchi (Eds.), pp. 196-217, The World Bank, Washington

Huang, H.; Han, X.; Kang, L.; Raven, P.; Jackson, P.W. & Chen, Y. (2002). Conserving native plants in China. *Science*, Vol. 297, No. 5583, pp. 935-936

Huang, K.-Y.; Zheng, Z.; François, L.; Guan, D.-S. & Cheddadi, R. (2010). Plants bioclimatic affinity groups in China: Observed vs. simulated ranges. *The Open Ecology Journal*, Vol. 3, No. 1, pp. 24-42

IUCN (International Union for Conservation of Nature) (2001). *IUCN Red List Categories and Criteria: Version 3.1.*, IUCN Species Survival Commission, Gland & Cambridge

IUCN (International Union for Conservation of Nature) (2010). *IUCN Red List of Threatened Species. Version 2010.4*, 7.01.2011, Available from http://www.iucnredlist.org

Jackson, S.T. & Weng, C. (1999). Late Quaternary extinction of a tree species in eastern North America. *Proceedings of the National Academy of Sciences of the United States of America*, Vol. 96, No. 24, pp. 13847-13852

Johnson, T.R. (2008). New opportunities, same constraints: environmental protection and China's new development path. *Politics*, Vol. 28, No. 2, pp. 93-102

Klok, C. & Zhang, T. (2008). *Biodiversity and its conservation in China. Authorities, mandates, and conventions*, Alterra, Wageningen

Kubitzki, K. & Krutzsch, W. (1996). Origins of East and South East Asian plant diversity, In: *Floristic characteristics and diversity of East Asian plants*, A. Zhang & S. Wu (Eds.), pp. 56-70, China Higher Education Press, Beijing

Lai, M.J. (1991). *The Red Data Book of Taiwan region: Criteria and measure for rare and threatened plant species, 1991 Ecological Research No. 12*, Council of Agriculture, Taipei

Larssen, T.; Lydersen, E.; Tang, D.; He, Y.; Gao, J.; Liu, H.; Duan, L.; Seip, H.M.; Vogt, R.D.; Mulder, J.; Shao, M.; Wang, Y.; Shang, H.; Zhang, X.; Solberg, S.; Aas, W.; Okland, T.; Eilertsen, O.; Angell, V.; Liu, Q.; Zhao, D.; Xiang, R.; Xiao, J. & Luo, J. (2006). Acid rain in China. *Environmental Science and Technology*, Vol. 40, No. 2, pp. 418-425

Latham, R.E. & Ricklefs, R.E. (1993). Continental comparisons of temperate-zone tree species diversity, In: *Species diversity in ecological communities: historical and geographical perspectives*, R.E. Ricklefs & D. Schluter (Eds.), pp. 294-314, University of Chicago Press, Chicago

Li, D.-Z. (2008). Floristics and plant biogeography in China. *Journal of Integrative Plant Biology*, Vol. 50, No. 7, pp. 771-777

Li, W. (2010). Progresses and perspectives of ecological research in China. *Journal of Resources and Ecology*, Vol. 1, No. 1, pp. 3-14

Li, J. & Fang, X. (1999). Uplift of the Tibetan Plateau and environmental changes. *Chinese Science Bulletin*, Vol. 44, No. 23, pp. 2117-2124

Li, X.-W. & Li, J. (1997). The Tanaka-Kaiyong line – an important floristic line for the study of the flora of East Asia. *Annals of the Missouri Botanical Garden*, Vol. 84, No. 4, pp. 888-892

Li, G.; Shen, Z.; Ying, T. & Fang, J. (2009). The spatial pattern of species richness and diversity centers of gymnosperm in China. *Biodiversity Science*, Vol. 17, No. 3, pp. 272-279 (In Chinese)

Li, D.; Song, Y. & Ouyang, Z. (2003). *Research on the national forestry nature reserve system plan*, China Land Press, Beijing

Lin, W.; Zhou, G.; Cheng, X. & Xu, R. (2007). Fast economic development accelerates biological invasions in China. *PLoS ONE*, Vol. 2, No. 11, pp. e1208

Liu, J. (2010). China's road to sustainability. *Science*, Vol. 328, No. 5974, pp. 50

Liu, J. & Diamond, J. (2005). China's environment in a globalizing world. *Nature*, Vol. 435, No. 7046, pp. 1179-1186

Liu, J. & Diamond, J. (2008). Revolutionizing China's environmental protection. *Science*, Vol. 319, No. 5859, pp. 37-38

Liu, J.-Q. & Tian, B. (2007). Origin, evolution, and systematics of Himalaya endemic genera. *Newsletter of Himalayan Botany*, Vol. 40, pp. 20-27

Liu, J.-Q.; Gao, T.-G.; Chen, Z.-D. & Lu, A.-M. (2002). Molecular phylogeny and biogeography of the Qinghai-Tibet Plateau endemic *Nannoglottis* (Asteraceae). *Molecular Phylogenetics and Evolution*, Vol. 23, No. 3, pp. 307-325

Liu, J.; Ouyang, Z.; Pimm, S.L.; Raven, P.H.; Wang, X.; Miao, H. & Han, N. (2003). Protecting China's biodiversity. *Science*, Vol. 300, No. 5623, pp. 1240-1241

Liu, Q.-R.; Yu, M. & Ma, J.-S. (2007). Review on the Chinese local floras. *Guihaia*, Vol. 27, No. 6, pp. 844-849 (In Chinese)

Liu, X.; Zhang, L. & Hong, S. (2011). Global biodiversity research during 1900-2009: a bibliometric analysis. *Biodiversity and Conservation*, Vol. 20, No. 4, pp. 807-826

López-Pujol, J. & Ren, M.-X. (2009). Biodiversity and the Three Gorges Reservoir: a troubled marriage. *Journal of Natural History*, 43, No. 43 & 44, pp. 2765-2786

López-Pujol, J. & Ren, M.-X. (2010). China: a hot spot of relict plant taxa, In: *Biodiversity Hotspots*, V. Rescigno & S. Maletta (Eds.), pp. 123-137, Nova Science Publishers, New York

López-Pujol, J. & Zhang Z.-Y. (2009). An insight into the most threatened flora of China. *Collectanea Botanica*, Vol. 28, pp. 95-110

López-Pujol, J.; Zhang, F.-M. & Ge, S. (2006). Plant biodiversity in China: richly varied, endangered and in need of conservation. *Biodiversity and Conservation*, Vol. 15, No. 12, pp. 3983-4026

López-Pujol, J.; Zhang, F.-M.; Sun, H.-Q.; Ying, T.-S. & Ge, S. (2011). Centres of plant endemism in China: places for survival or for speciation? *Journal of Biogeography*, Vol. 38, No. 7, pp. 1267-1280

Manchester, S.R. (1999). Biogeographical relationships of North American Tertiary floras. *Annals of the Missouri Botanical Garden*, Vol. 86, No. 2, pp. 472-522

Manchester, S.R.; Chen, Z.D.; Lu, A.M. & Uemura, K. (2009). Eastern Asian endemic seed plant genera and their paleogeographic history throughout the Northern Hemisphere. *Journal of Systematics and Evolution*, Vol. 47, No. 1, pp. 1-42

McBeath, J. & McBeath, J.H. (2006). Biodiversity conservation in China: policies and practice. *Journal of International Wildlife Law and Policy*, Vol. 9, No. 4, pp. 293-317

McBeath, G.A. & Leng, T.-K. (2006). *Governance of biodiversity conservation in China and Taiwan*, Edward Elgar Publishing, Cheltenham & Northampton

McBeath, J. & Wang, B. (2009). Endangered species protection in China, In: *Endangered species: New research*, A. Columbus & L. Kuznetsov (Eds.), pp. 17-44, Nova Science Publishers, New York

MacKinnon, J. & Wang, H. (2008). *The green gold of China*, China Federation of Literary and Art Circles Publishing Corporation, Beijing

MacKinnon, J. & Xie, Y. (2008). *Regional action plan for the protected areas of East Asia. 2006-2010*, IUCN, Gland & Bangkok

MacKinnon, J.; Sha, M.; Cheung, C.; Carey, G.; Xiang, Z. & Melville D. (1996). *A biodiversity review of China*, WWF-International, WWF China Programme, Hong Kong

MEP (Ministry of Environmental Protection of the People's Republic of China) (2008a). *China's fourth national report on implementation of the Convention of Biological Diversity*, Ministry of Environmental Protection of the People's Republic of China, Beijing

MEP (Ministry of Environmental Protection of the People's Republic of China) (2008b). *2007 Report on the state of the environment in China*, Ministry of Environmental Protection of the People's Republic of China, Beijing

MEP (Ministry of Environmental Protection of the People's Republic of China) (2009). *2008 Report on the state of the environment in China*, Ministry of Environmental Protection of the People's Republic of China, Beijing

MEP (Ministry of Environmental Protection of the People's Republic of China) (2010). *2009 Report on the state of the environment in China*, Ministry of Environmental Protection of the People's Republic of China, Beijing (In Chinese)

Metcalfe, I. (1988). Origin and assembly of south-east Asian continental terranes, In: *Gondwana and Tethys. Geological Society Special Publication No. 37*, M.G. Audley-Charles & A. Hallam (Eds.), pp. 101-118, Oxford University Press, Oxford

MFA (Ministry of Foreign Affairs of the People's Republic of China) (2010). *China's progress towards the Millenium Development Goals. 2010 Report*, Ministry of Foreign Affairs of the People's Republic of China, Beijing

Mittermeier, R.A.; Gil, P.R. & Mittermeier, C.G. (1997). *Megadiversity. Earth's biologically wealthiest nations*, Conservation International (CI), Washington D.C.

Moorman, J.L. & Zhang, G. (2007). Promoting and strengthening public participation in China's environmental impact assessment process: comparing China's EIA law and U.S. NEPA. *Vermont Journal of Environmental Law*, Vol. 8, No. 2, pp. 281-335

Morell, V. (2008). Letting 1000 forests bloom. *Science*, Vol. 320, No. 5882, pp. 1442-1443

Morley, R.J. (2000). *Origin and evolution of tropical rain forests*, Wiley, Chichester

Myers, N.; Mittermeier, R.A.; Mittermeier, C.G.; Da Fonseca, G.A.B. & Kent, J. (2000). Biodiversity hotspots for conservation priorities. *Nature*, Vol. 403, No. 6772, pp. 853-858

Nagle, J.C. (2009). The effectiveness of biodiversity law. *Journal of Land Use*, Vol. 24, No. 2, pp. 203-252

NBSC (National Bureau of Statistics of China) (2009). China *statistical yearbook 2009*, China Statistics Press, Beijing

NDRC (National Development and Reform Comission, People's Republic of China) (2007). *China's National Climate Change Programme*, National Development and Reform Comission, Beijing

NEPA (National Environmental Protection Agency) (1994). *China Biodiversity Conservation Action Plan*, China Environmental Science Press, Beijing

Olson, D.M. & Dinerstein, E. (2002). The Global 200: priority ecoregions for global conservation. *Annals of the Missouri Botanical Garden*, Vol. 89, No. 2, pp. 199-224

Peng, H. & Xu, Z. (1997). The threatened wild plants used for medicine as Chinese medicinal herbs, In: *Conserving China's biodiversity*, S. Wang & J. MacKinnon (Eds.), pp. 175-189, China Environmental Science Press, Beijing

Qian, H. (2001). A comparison of generic endemism of vascular plants between east Asia and North America. *International Journal of Plant Sciences*, Vol. 162, No. 1, pp. 191-199

Qian, H. (2002). A comparison of the taxonomic richness of temperate plants in east Asia and North America. *American Journal of Botany*, Vol. 89, No. 11, pp. 1818-1825

Qian, H. & Ricklefs, R.E. (1999). A comparison of the taxonomic richness of vascular plants in China and the United States. *The American Naturalist*, Vol. 154, No. 2, pp. 160-181

Qian, H. & Ricklefs, R.E. (2000). Large-scale processes and the Asian bias in species diversity of temperate plants. *Nature*, Vol. 407, No. 6801, pp. 180-182

Qian, H.; Wang, S.; He, J.S.; Zhang, J.; Wang, L.; Wang, X. & Guo, K. (2006). Phytogeographical analysis of seed plant genera in China. *Annals of Botany*, Vol. 98, No. 5, pp. 1073-1084

Qiu, L. (2010). Measuring the meltdown. *Nature*, Vol. 468, No. 7321, pp. 141-142

Qiu, W.; Wang, B.; Jones, P.J.S. & Axmacher, J.C. (2009). Challenges in developing China's marine protected area system. *Marine Policy*, Vol. 33, No. 4, pp. 599-605

Quan, J.; Ouyang, Z.; Xu, W. & Miao, H. (2011). Assessment of the effectiveness of nature reserve management in China. *Biodiversity and Conservation*, Vol. .20, No. 4, pp. 779-792.

Sauer, J.D. (1988). *Plant migration: The dynamics of geographic patterning in seed plant species*, University of California Press, Berkeley & Los Angeles

Schei, P.J.; Wang, S. & Xie, Y. (2001). *Conserving China's Biodiversity (II)*, China Environmental Science Press, Beijing

Şengör, A.M.C. (1997). Asia, In: *Encyclopedia of European and Asian regional geology*, E.M. Moores & R.W. Fairbridge (Eds.), pp. 34-51, Chapman & Hall, London

SEPA (State Environmental Protection Administration) (1998). *China's Biodiversity: a country study*, China Environmental Science Press, Beijing

Shapiro, J. (2001). *Mao's war against nature: Politics and the environment in revolutionary China*, Cambridge University Press, Cambridge

Smith, A.C. (1973). Angiosperm evolution and the relationship of the floras of Africa and America, In: *Tropical forest ecosystems in Africa and South America: a comparative review*, B.J. Meggers, E.S. Ayensu & W.D. Duckworth (Eds.), pp. 49-61, Smithsonian Institution Press, Washington D.C.

Song, C. & Zhang, Y. (2010). Forest cover in China from 1949 to 2006, In: *Reforesting landscapes: Linking pattern and process*, H. Nagendra & J. Southworth (Eds.), pp. 341-356, Springer, Dordrecht

Steemans, P.; Le Hérissé, A.; Melvin, J.; Miller, M.A.; Paris, F.; Verniers, J. & Wellman, C.H. (2009). Origin and radiation of the earliest vascular land plants. *Science*, Vol. 324, No. 5925, pp. 353.

Sun, G.; Dilcher, D.L.; Wang, H. & Chen, Z. (2011). A eudicot from the Early Cretaceous of China. *Nature*, Vol. 471, No. 7340, pp. 625-628

Sun, G.; Dilcher, D.L. & Zheng, S.L. (2008). A review of recent advances in the study of early angiosperms from northeastern China. *Paleoworld*, Vol. 17, No. 3-4, pp. 166-171

Takhtajan, A.L. (1969). *Flowering plants: Origin and dispersal*, Oliver & Boyd, Edinburgh

Takhtajan, A. (1987). Flowering plant origin and dispersal: the cradle of the angiosperms revisited. In: *Biogeographical evolution of the Malay Archipelago*, T.C. Whitmore (Ed.), pp. 26-31, Clarendon Press, Oxford

Tang, Z.; Wang, Z.; Zheng, C. & Fang, J. (2006). Biodiversity in China's mountains. *Frontiers in Ecology and the Environment*, Vol. 4, No. 7, pp. 347-352

Thompson, J.D. (2005). *Plant evolution in the Mediterranean*, Oxford University Press, New York

Thorne, R.F. (1963). Biotic distribution patterns in the tropical Pacific. In: *Pacific Basin biogeography*, J.L. Gressitt (Ed.), pp. 311-354, Bishop Museum Press, Honolulu

Thorne, R.F. (1999). Eastern Asia as a living museum for archaic angiosperms and other seed plants. *Taiwania*, Vol. 44, No. 4, pp. 413-422

Tiffney, B.H. (1985). The Eocene North Atlantic land bridge: its importance in Tertiary and modern phytogeography of the Northern Hemisphere. *Journal of the Arnold Arboretum*, Vol. 66, No. 2, pp. 243-273

Tiffney, B.H. & Manchester, S.R. (2001). The use of geological and paleontological evidence in evaluating plant phylogeographic hypotheses in the Northern Hemisphere Tertiary. *International Journal of Plant Sciences*, Vol. 162 , No. S6, pp. S13-S17

Tsao, L. & Zhu, P. (2010). 'Noah's Ark' gives rare plants seeds of survival. *China Daily (US Edition)*, No. 24.08.2010, Available from http://www.chinadaily.com.cn/usa/2010-08/24/content_11194227.htm

Vavilov, N.I. (1951). The origin, variation, immunity and breeding of cultivated plants. *Chronica Botanica*, Vol. 13, No. 1-6, pp. 1-366.

Vennemo, H., Aunan, K., Lindhjem, H. & Seip, H.M. (2009). Environmental pollution in China: status and trends. *Review of Environmental Economics and Policy*, Vol. 3, No. 2, pp. 209–230

Wang, H. (Ed.) (1985). *Atlas of the palaeogeography of China*, Cartographic Publishing House, Beijing

Wang, H. (1989). A study on the origin of spermatophytic genera endemic to China. *Acta Botanica Yunnanica*, Vol. 11, No. 1, pp. 1-16 (In Chinese)

Wang, S. (Ed.) (1992). *Biodiversity in China: status and conservation needs*, Science Press, Beijing & New York

Wang, S. & MacKinnon, J. (Eds.) (1997). *Conserving China's Biodiversity*, China Environmental Science Press

Wang, S. & Xie, Y. (Eds.) (2004). *China Species Red List. Vol. I. Red List*, Higher Education Press, Beijing

Wang, H.-S. & Zhang, Y.-L. (1994). The bio-diversity and characters of spermatophytic genera endemic to China. *Acta Botanica Yunnanica*, Vol. 16, No. 3, pp. 209-220 (in Chinese)

Wang, G.; Innes, J.L.; Lei, J.; Dai, S. & Wu, S.W. (2007). China's forestry reforms. *Science*, Vol. 318, No. 5856, pp. 1556-1557

Wang, G.; Innes, J.L.; Wu, S.W. & Dai, S. (2008). Towards a new paradigm: the development of China's forestry in the 21st century. *International Forestry Review*, Vol. 10, No. 4, pp. 619-631

Wang H.-F.; López-Pujol, J.; Meyerson, L.A.; Qiu, J.-X.; Wang X.-K. & Ouyang, Z.-Y. (2011). Biological invasions in rapidly urbanizing areas: a case study of Beijing, China. *Biodiversity and Conservation*, Vol. 20, in press

Wang, S.; Wang, X.; Xie, Y.; Jin, J.; Chen, C.; Wang, Z. & Qing, J. (2000). Developing and implementing National Biodiversity Strategy and Action Plan: lessons from China, *Proceedings of the 1st Workshop on National Biodiversity Strategies and Action Plans in Northeast and East central Asia. Experiences and Lessons*, Beijing, April 26-28, 2010, Biodiversity Planning Support Programme (UNDP-UNEP), Available from http://bpsp-neca.brim.ac.cn/calendars/workshop-1/9.html

Wang, X.M.; Zhang, Z.X.; Hasi, E. & Dong, C.B. (2010). Has the Three Norths Forest Shelterbelt Program solved the desertification and dust storm problems in arid and semiarid China? *Journal of Arid Environments*, Vol. 74, No. 1, pp. 13-22

Weber, E & Li, B. (2008). Plant invasions in China: What is to be expected in the wake of economic development? *BioScience*, Vol. 58, No. 5, pp. 437-444

Weber, E., Sun, S.G. & Li, B. (2008). Invasive alien plants in China: diversity and ecological insights. *Biological Invasions*, Vol. 10, No. 8, pp. 1411-1429

Wen, X. (2008). The importance of botanic gardens in Chinese cities. *BGjournal*, Vol. 5, No. 2, pp. 19-22

Winkler, D. (2008). *Yartsa gunbu (Cordyceps sinensis*) and the fungal commodification of Tibet's rural economy. *Economic Botany*, Vol. 62, No. 3, pp. 291-305

World Bank (2010a). *2010 world development indicators*, The World Bank, Washington D.C.

World Bank (2010b). *World Development report 2010 – Development and climate change*, The World Bank, Washington D.C.

Wu, C.-Y. (1998). Delineation and unique features of the Sino-Japanese floristic region, *The University Museum, the University of Tokyo Bulletin*, Vol. 37 (*Sino-Japanese flora: its characteristics and diversification*, D.E. Boufford & H. Ohba H, Eds.), pp. 1-35

Wu T.-L. (Ed.) (2002). *Check List of Hong Kong Plants 2001*, Agriculture, Fisheries and Conservation Department, The Government of the Hong Kong Special Administrative Region, Hong Kong

Wu, Z. & Wu, S. (1996). A proposal for a new floristic kingdom (realm), In: *Floristic characteristics and diversity of East Asian plants*, A. Zhang & S. Wu (Eds.), pp. 3-42, China Higher Education Press, Beijing

Wu, Z.; Sun, H.; Zhou, Z.; Peng, H. & Li, D. (2007). Origin and differentiation of endemism in the flora of China. *Frontiers of Biology in China*, Vol. 2, No. 2, pp. 125-143

Xie, Z. (2003). Characteristics and conservation priority of threatened plants in the Yangtze Valley. *Biodiversity and Conservation*, Vol. 12, No. 1, pp. 65-72

Xie, Y. & Wang, S. (2007). Conservation status of Chinese species: (1) Overview. *Integrative Zoology*, Vol. 2, No. 1, pp. 26-35

Xie, Y.; Li, Z.; Gregg, W.P. & Li, D. (2001). Invasive species in China - an overview. *Biodiversity and Conservation*, Vol. 10, No. 6, pp. 1317-1341

Xu, Z. (1997). The status and strategy for ex situ conservation of plant diversity in Chinese botanic gardens – Discussion of principles and methodologies of ex situ conservation for plant diversity. In: *Conserving China's Biodiversity*, S. Wang & J. MacKinnon (Eds.), pp. 79-95, China Environmental Science Press. Beijing

Xu, J. & Wilkes, A. (2004). Biodiversity impact analysis in northwest Yunnan, southwest China. *Biodiversity and Conservation*, Vol. 13, No. 5, pp. 959-983

Xu, H.; Ding, H. & Wu, J. (2010). National indicators show biodiversity progress. *Science*, Vol. 329, No. 5994, pp. 900

Xu, J.; Grumbine, R.E.; Shrestha, A.; Eriksson, M.; Yang, X.; Wang, Y. & Wilkes, A. (2009). The melting Himalayas: cascading effects of climate change on water, biodiversity, and livelihoods. *Conservation Biology*, Vol. 23, No. 3, pp. 520-530

Xu, J. ; Ma, E.T.; Tashi, D.; Fu, Y.; Lu, Z. & Melick, D. (2005). Integrating sacred knowledge for conservation : cultures and landscapes in Southwest China. *Ecology and Society*, Vol. 10, No. 2, pp. 7

Xu, H.; Tang, X.; Liu, J.; Ding, H.; Wu, J.; Zhang, M.; Yang, Q.; Cai, L.; Zhao, H. & Liu, Y. (2009). China's progress toward the significant reduction of the rate of biodiversity loss. *BioScience*, Vol. 59, No. 10, pp. 843-852

Xu, H.; Wang, S. & Xue, D. (1999). Biodiversity conservation in China: legislation, plans and measures. *Biodiversity and Conservation*, Vol. 8, No. 6, pp. 819-837

Xu, H.; Wu, J.; Liu, Y.; Ding, H.; Zhang, M.; Wu, Y.; Xi, Q. & Wang, L. (2008). Biodiversity congruence and conservation strategies: a national test. *BioScience*, Vol. 58, No. 7, pp. 632-639

Xue, L.; Simonis, U.E. & Dudek, D.J. (2007). Environmental governance for China: major recommendations of a Task Force. *Environmental Politics*, Vol. 16, No. 4, pp. 669-676

Yang, Y.; Tian, K.; Hao, J.; Pei, S. & Yang, Y. (2004). Biodiversity and biodiversity conservation in Yunnan, China. *Biodiversity and Conservation*, Vol. 13, No. 4, pp. 813-826

Ying, T.-S. (2001). Species diversity and distribution pattern of seed plants in China. *Biodiversity Science*, Vol. 9, No. 4, pp. 393-398 (In Chinese).

Ying, T.-S. & Zhang, Z.-S. (1984). Endemism in the flora of China - Studies on the endemic genera. *Acta Phytotaxonomica Sinica*, Vol. 22, No. 4, pp. 259-268 (In Chinese)

Ying, T.S.; Zhang, Y.L. & Boufford, D.E. (1993). *The endemic genera of seed plants of China*, Science Press, Beijing

Yip, J.Y.; Corlett, R.T. & Dudgeon, D. (2004). A fine-scale gap analysis of the existing protected area system in Hong Kong, China. *Biodiversity and Conservation*, Vol. 13, No. 5, pp. 943-957

You, M. (2008). Moratorium on EIA approvals: China's new environmental law enforcement tool. *Natural Resources Journal*, Vol. 48, No. 1, pp. 163-187

Yu, W. (2008). Assessment on biodiversity-related policies and legislations in China, In: *EU-China Biodiversity Programme (ECBP)*, 12.01.2011, Available from http://www.ecbp.cn/upload/pdf/02%20Legal%20Assessment%20Yu%20Wenxuan%20EN.doc

Yu, X. (2010). Biodiversity conservation in China: barriers and future actions. *International Journal of Environmental Studies*, Vol. 67, No. 2, pp. 117-126

Zachos, J.; Pagani, M.; Sloan, L.; Thomas, E. & Billups, K. (2001). Trends, rhythms, and aberrations in global climate 65 Ma to present. *Science*, Vol. 292, No. 5517, pp. 686-693

Zhang, D. (2007). *Plant germplasm resources and conservation*, Shanghai Scientific & Technical Publishers, Shanghai

Zhang, J. & Cao, M. (1995). Tropical forest vegetation of Xishuangbanna, SW China and its secondary changes, with special reference to some problems in local nature conservation. *Biological Conservation*, Vol. 73, No. 3, pp. 229-238

Zhang, Y.-B. & Ma, K.-P. (2008). Geographic distribution patterns and status assessment of threatened plants in China. *Biodiversity and Conservation*, Vol. 17, No. 7, pp. 1783-1798

Zhang, B.; Li, W. & Xie, G. (2010). Ecosystem services research in China: Progress and perspective. *Ecological Economics*, Vol. 69, No. 7, pp. 1389-1395

Zhang, D.; Liu, F. & Bing, J. (2000) Eco-environmental effects of the Qinghai-Tibet Plateau uplift during the Quaternary in China. *Environmental Geology*, Vol. 39, No. 12, pp. 1352-1358

Zhang, Z.-R.; Luo, L.-C.; Wu, D. & Zhang, Z.-Y. (2009). Two genetically distinct units of *Sinomanglietia glauca* (Magnoliaceae) detected by chloroplast PCR-SSCP. *Journal of Systematics and Evolution*, Vol. 47, No. 2, pp. 110-114

Zhang, J. ; Mauzerall, D.L. ; Zhu, T. ; Liang, S. ; Ezzati, M. & Remais, J.V. (2010). Environmental health in China: progress towards clean air and safe water. *The Lancet*, Vol. 375, No. 9720, pp. 1110-1119

Zhang, P.; Shao, G.; Zhao, G.; Le Master, D.C.; Parker, G.R.; Dunning Jr., J.B. & Li, Q. (2000). China's forest policy for the 21st century. *Science*, Vol. 288, No. 5474, pp. 2135-2136

Zhang, D.-C. ; Zhang, Y.-H. ; Boufford, D.E. & Sun, H. (2009). Elevational patterns of species richness and endemism for some important taxa in the Hengduan Mountains, southwestern China. *Biodiversity and Conservation*, Vol. 18, No. 3, pp. 699-716

Zhao, Y. (2009). Assessing the environmental impact of projects: a critique of the EIA legal regime in China. *Natural Resources Journal*, Vol. 49, No. 2, pp. 485-524

Zhao, L. & Zhang, D. (2003). Ornamental plant resources from China. *Acta Horticulturae*, Vol. 620, pp. 365-375

Zheng, H.; Powell, C.M.; An, Z.; Zhou, J. & Dong, G. (2000). Pliocene uplift of the northern Tibetan Plateau. *Geology*, Vol. 28, No. 8, pp. 715-718

Zhu, Z.; Bao, W.; Pang, X.; Yan, X. & Li, Y. (2006). Tourism effect on species composition and diversity of understory plants in *Abies fargesii* var. *faxoniana* forest in Jiuzhaigou, Sichuan. *Biodiversity Science*, Vol. 14, No. 4, pp. 284-291 (In Chinese)

10

Biodiversity Conservation in Costa Rica - An Animal and Plant Biodiversity Atlas

Bert Kohlmann
EARTH University
Costa Rica

1. Introduction

Biodiversity conservation has become one of the most urgent tasks facing humanity because of the accelerating rates of biodiversity loss (Pimm et al., 1995). An appropriate action to this end would be the establishment of global inventories, although the time required for both surveying and documenting this plethora of taxa far outreaches our present capacity. Availability of adequate data is also a limiting factor (Prendergast et al., 1999). Therefore, the writing of biogeographic atlases can be proposed as a practical tool for biodiversity conservation (Prendergast et al., 1993; Morrone, 2000) and hotspot identification (area that combines a high biodiversity with a high threat degree by humans; Myers, 1988; Kappelle, 2008). A very important task of biogeography atlases is the study of diversity and endemicity patterns in order to protect rare and endangered species. As Lomolino et al. (2006) indicate two major tasks of this process are: (1) to document the intensities and locations of hotspots for a particular taxonomic group and (2) to determine to what degree do different taxon-specific hotspots overlap. Although levels of endemism and species richness are frequently positively correlated (Balmford & Long, 1995), unfortunately, many times there is little overlap in the species richness and endemicity areas (Bibby et al., 1992; Prendergast et al., 1993; Araujo, 2002; Cox & Moore, 2005; Lomolino et al., 2006). This fact forces the analysis of distribution patterns region by region in order to understand what the approximate situation is and being able to identify biodiversity hotspots (Myers et al., 2000). As Gaston (2000) and Gaston & Spicer (2004) indicate, species distribution is not very uniform across the world and must therefore be mapped. Peaks of diversity exert widespread fascination, especially regarding the origin of high numbers. After all, conservation planning is based on spatial biodiversity distribution (Margules & Pressey, 2000).

Another aspect of the use of biodiversity atlases that has not been previously mentioned is their possible importance as a tool for following distributional changes caused by climatic effects, especially in mountainous areas, such as Costa Rica. The study and knowledge of the aforementioned situation in Costa Rica is of utmost importance. Costa Rica belongs to one (Middle America) of the 36 world hotspots, as defined by Mittermeier et al., (2004). Costa Rica is not a big country (Fig. 1). It has 51 042.8 km^2 of continental and insular land surface, representing 0.03 % of the Earth surface (Jiménez, 1995; Ministerio del Ambiente y Energía, 2000). In the ranking of world diversity, Costa Rica occupies the 20th place, approximately. As such, it is not considered a megadiverse country, since only twelve countries build up the list in this category. However, what makes Costa Rica special is its species density (number of species per unit of area) (Valerio, 1999; Obando, 2002). Using this measure, Costa Rica could probably occupy the first place in the world (Valerio, 1999, 2006; Obando, 2002, 2007). This country possesses approximately 3.6 % of the total expected world diversity, and if the total number of described species is considered, this number jumps then to 4.5 %, with more than 90 000 known species (insects: 66 946 species, plants: 11 451 species, other invertebrates: 5253 species) (Obando 2007). To give a comparative idea of species density, Costa Rica has 234.8 plant species per 1000 km^2, whereas Colombia, in second place, has only 43.8 plant species per 1000 km^2 (Obando, 2007). If we consider orchids alone, Costa Rica has 25.5 species per 1000 km^2, whereas Colombia has 2.6 species per 1000 km^2 (Valerio, 1999). Similarly, Costa Rica has 28.2 species of vertebrates (excluding fishes) per 1000 km^2, whereas Ecuador, the second most biodiverse vertebrate country per km^2 in the world, has 9.2 species per 1000 km^2, and the third most biodiverse vertebrate country, Malaysia, has only 4.4 vertebrate species per 1000 km^2 (Valerio, 2006). This enormous biodiversity in Costa Rica is now under protection by a world-class national system of protected areas, which began in the 1970's and today protects almost 27 % of the national territory (Vaughan, 1994; Vaughan et al., 1998). Interestingly, Costa Rica is also the country with most ecotourists per km^2 worldwide, 22.47 international ecotourists/km^2 in the year 2007, with the African sub-Saharan countries as the next places with most ecotourists per km^2 (Kohlmann et al., 2008).

Costa Rica is considered to have a moderate degree of endemics (Obando, 2007); approximately 1.3 % of the known species are endemics. It is estimated that around 10 % of the total plant species are endemics (1102 species), whereas the different vertebrate groups vary from a minimum of 0.7 % in birds to a maximum of 25 % for the amphibians (Obando, 2007). Using these two great groups, four great areas of endemism have been identified for continental Costa Rica: the Central Volcanic Cordillera, the Talamanca Cordillera, the Central Pacific Region and the Osa Peninsula Region (Fig. 2); a fifth area has been identified in Coco Island, in the Pacific Ocean (Elizondo et al., 1989). From the ecosystem point of view, cloud forests are the most endemic ecosystems (Obando, 2002). This study defined a biodiversity atlas indicating the areas of high species richness and endemism for Costa Rica, using freshwater fishes (Pisces), insects (Coleoptera: Scarabaeidae: Dynastinae and Coleoptera: Scarabaeidae: Scarabaeinae) and plants (Araceae, Arecaceae and Bromeliaceae).

Adequate representation of biodiversity is ideally achieved by the use of multiple taxonomic groups (Stork & Samways, 1995; Pawar et al., 2007; Larsen et al., 2011). However, due to time funding, collection and taxonomic constraints for many of the groups, especially in tropical regions, many area-prioritization studies assume some similarity levels in species geographical distributions and consequently available groups are used as surrogates for others (Garson et al., 2002; Rondini & Boitani, 2006; Pawar et al., 2007). Despite the popularity of the surrogacy approach, its efficacy remains unclear (Moore et al., 2003; Graham & Hijmans, 2006; Lamoreux et al., 2006).

A recurrent question is whether plant and vertebrate distribution patterns are reflected by those of invertebrates as well (Howard et al., 1988). Moritz et al. (2001) found high levels of congruence with data on tropical insects, snails, plants, and vertebrates only in areas with a clear history of geographical vicariance. In some other cases, like in tiger beetles, there seems to be also congruence; in other cases the relationships are not clear (Mittermeier et al., 2004).

The analysis focused on continental Costa Rica; Coco Island was not included because neither Scarabaeinae, nor Dynastinae material has been collected from that locality. This atlas represents an effort to help define those areas most in need of conservation and sustainable use in Costa Rica. The atlas will also help define those areas that have been under sampled and therefore future collecting efforts can be directed to these information voids. This study is also an expansion and continuation of previous gap analyses which used beetles and plants (Kohlmann & Wilkinson, 2007; Kohlmann et al., 2007; Kohlmann et al., 2010); the 2007 study has been considered a pioneer study in Costa Rica by Arias et al. (2008), because it represents the first attempt to use the actual distribution of all species of a specific taxonomic group.

2. Biodiversity mapping

2.1 Taxon information

Information regarding dung scarab beetle distribution (coordinates) was taken from the collections and electronic database of the National Biodiversity Institute (INBio, www.inbio.ac.cr). This institution has been collecting plants and insects in Costa Rica for the last 20 years. The dung beetles (Coleoptera: Scarabaeidae: Scarabaeinae) have been particularly well studied in relation to their systematics (Kohlmann et al., 2007, 2010). So far, 177 native taxa of Scarabaeinae have been reported in Costa Rica. In relation to the dynastine scarab beetles (Coleoptera: Scarabaeidae: Dynastinae) information was also taken from the electronic database of INBio, as well from the regional study written by Ratcliffe (2003). A total of 125 species were considered. Fish information and distributions were based on the seminal study made by Bussing (1998). Different to Bussing's holistic study that also considers marine species that penetrate into freshwater or contrariwise, like sharks, sawfishes, tarpons and eels, only strict freshwater fishes (111 species) were considered in this study in order to better reflect local biodiversity. Regarding the three plant families, Araceae, Arecaceae and Bromeliaceae, their distribution information (coordinates) was also taken from the collections and electronic database of the National Biodiversity Institute (INBio).

Additional information was incorporated from the publication "Manual de Plantas de Costa Rica. Gimnospermas y Monocotiledóneas (Agavaceae-Musaceae)", published by Hammel et al. (2003). Furthermore, extra information was obtained from the Missouri Botanical Garden electronic database (www.mobot.org). The total number of plant species considered for this study was: Araceae, 229 species; Arecaceae, 107 species; Bromeliaceae, 187 species. No introduced species, neither animals nor plants, were considered for this study. Concluding, the six chosen groups have been particularly well sampled in Costa Rica, as well as systematically studied in great taxonomic detail; their analyzed distributional areas are relatively smaller than the study area, therefore complying with Müller's (1981) three tenets for making this group particularly well-suited for the present biogeographic analysis.

2.2 Vegetation base map

One of the most popular systems used in Costa Rica and in twelve other countries (Meza, 2001) for the classification of vegetation is the Life Zone System developed by Holdridge (1967). This system divides Costa Rica into 12 Life Zones and 11 Transition Zones based on environmental factors such as humidity, rainfall and temperature (Fig. 3). This system is thus independent of floristic relationships and the same zones can then reappear in different regions of the world. According to Hall (1984), this system takes into account not only variations caused by latitude, but also by altitude, and is therefore especially useful for tropical mountainous countries (Meza, 2001).

Fig. 1. View of Costa Rica looking northwest as seen from the Space Shuttle (taken from Kohlmann et al., 2002).

According to this classification, the five most extensive vegetation types are: tropical wet forest (wf-T) (10.5 % of the total country area), premontane wet forest (wf-P) (7.2 %), lower montane wet forest (wf-LM) (5.9 %), premontane rain forest (rf-P) (5.6 %) and tropical moist forest (mf-T) (5.5 %) (Obando, 2002). There are some limitations to this system. The Holdridge life zone system can potentially vary along other environmental axes, besides total precipitation and temperature, such as edaphic conditions and this could impact species abundance and endemism. For example, bioclimatic regions such as the Pacific dry forest comprise long belts along mountain/volcanic ranges, and by assuming that these long belts share the same biodiversity category a potential risk can be generated of loosing resolution when assigning conservation priority zones.

2.3 GIS Analysis

Some of the advantages of digital mapping techniques using Geographic Information Systems (GIS), comprise that they are faster, more efficient, and more powerful and versatile than traditional analog cartography. Some of many advantages of these techniques are the storage of large amounts of spatial information, the ease for mapping many map layers, and their use in modeling and predicting species distributions. For that reason we have followed a GIS-oriented process for the elaboration of our biogeography atlas. For the GIS analysis the following processes were done using Arcview®3.1 (ESRI, 2002), ArcGIS®9.2 (ESRI 2006) and Microsoft Excel® (2002):

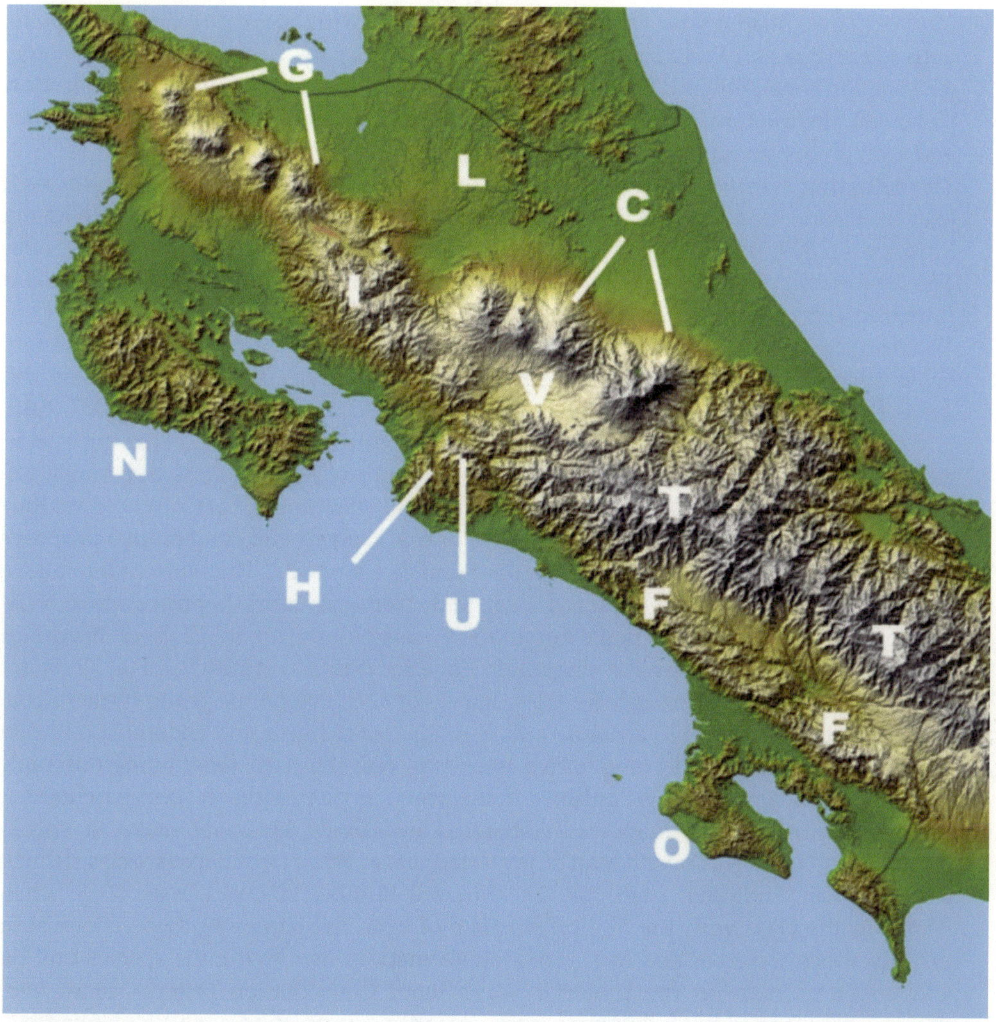

Fig. 2. Geographical areas in Costa Rica: C, Central Cordillera; F, Coastal mountain range; G, Guanacaste mountain range; H, Herradura mountain; I, Tilarán mountain range; L, northern plains; N, Nicoya peninsula (Pacific Northwest); O, Golfo Dulce/Osa Peninsula; P, Central Pacific; T, Talamanca mountain range; U, Turrubares mountain; V, Central Valley (taken from Kohlmann et al., 2002).

1. Establishment and cleansing of the data bases for each taxon in relation to taxon names, type of endemism and location of collecting sites. Information layers were generated using the collecting sites for each species.
2. Depuration of referential geographic information. The layers containing the National System of Coordinates were transformed to geographic coordinates (the same datum

was always used: Fundamental de Ocotepeque). For distributional referencing, each Holdridge life zone polygon was numbered (Fig. 3).

3. For each taxon the collecting sites were superimposed on the Holdridge life zones and out of this product the number of collections and taxa, as well as the total number of taxa and endemics and type of endemicity (endemics known to occur only in Costa Rica, endemics shared with Panama, endemics shared with Nicaragua, endemics shared with Nicaragua and Panama, total number of endemics for Costa Rica) were obtained for each life zone polygon. Each polygon was associated with the number of collections per taxon, the total number of collections and taxa. Layers for the total number of taxa and each type of endemism were produced for the groups.
4. The collecting sites were overlaid on the Holdridge life zones for the total number of taxa and endemics, as well as each of the possible endemism situation, following Morrone's (2000) suggestions regarding the formal preparation of a biogeographic atlas. The base electronic map was derived from the one presented in Atlas Costa Rica 2000 (Instituto Tecnológico de Costa Rica, 2000).
5. To create comparable maps for the different taxonomic groups of this study, the rank levels of species richness and endemism by life zone were calculated in accordance with a previous classification used for Costa Rica, as defined by Kohlmann et al. (2007, 2010). Accordingly, five levels were distinguished for both categories. Concerning species richness these limits are: up to 7 % of the maximum species richness in a single Life Zone (class one), up to 20 % (class two), up to 44 % (class three), up to 70 % (class four) and more than 70 % (class five). For the sake of this analysis and comparative purposes, only the two most numerous ranks (ranks 4 and 5) were used. Thus this system allows us to focus the analysis on the richer and therefore more representative areas. Concerning endemism the limits for the five classes are: up to 12 % of maximum number of endemic species in a single Life Zone for class 1, with 24 %, 46 %, 72 %, and over 72 % for classes 2 through 5, respectively. For each taxonomic group these relative values were converted into absolute values of species richness and endemicity. For the sake of this analysis and comparative purposes, only the two most numerous ranks were used, following the logic outlined in the previous discussion on species richness.
6. A conservation priority map was elaborated overlaying layers of maps of species numbers and endemics over a map of protected areas. These two maps (species richness and endemics) indicated each one five different number of taxa classes (1-5), where class 5 is the class with the highest number of taxa. Subsequently, two conservation priority zones were defined in a gap analysis map by overlaying the species number and endemics map on the protected areas map. Conservation priority zones were defined according to the following scheme: priority conservation zone 1 is defined by a species richness and endemicity rank of 5, conservation priority zone 2 is defined by an endemicity rank of 5 and a species richness rank of <5.

This method of priority definition using complementarity (degree to which an area contributes otherwise unrepresented species to a set of areas), picturing the combination of areas of greatest species and endemicity richness, was chosen following the suggestion made by Williams et al. (1996). They found that the areas chosen by using complementarity represented all the species many times over rather than by either choosing species or endemicity areas separately. They also found that it is also a well suited method for supplementing an existing conservation network, in their case British birds. Equally, the decision to prioritize endemicity over species richness in the definition process follows well established recommendations expressed by Mittermeier et al. (2004), because the endemics are irreplaceable.

3. Conclusion

3.1 Distribution of collection localities

The collection localities indicate that the northern part of Costa Rica, as well as the Central Pacific, are under collected; due mostly to the fact that these areas have been highly altered by agricultural activities. Other areas that also require more collecting effort are the Nicoya peninsula of Northwestern Costa Rica and the higher parts of the Talamanca Cordillera to the southeast; the lack of roads in these regions is one of the main barriers to collecting in these areas. The selection of collecting sites is often biased. Unfortunately, as already indicated, not all areas of Costa Rica have been collected with equal intensity. In order to deal with under sampled areas, as well as to know the areas with a good collecting record, and for comparative purposes, regions with a collecting effort of five or more years were arbitrarily chosen for this study (Fig. 3). The subsequent analyses will be based on these regions. Life zones areas depicted in grey in several maps (Figs. 7-8), represent zones where no collecting efforts have been undertaken, thus indicating regions where collecting should be directed in the future.

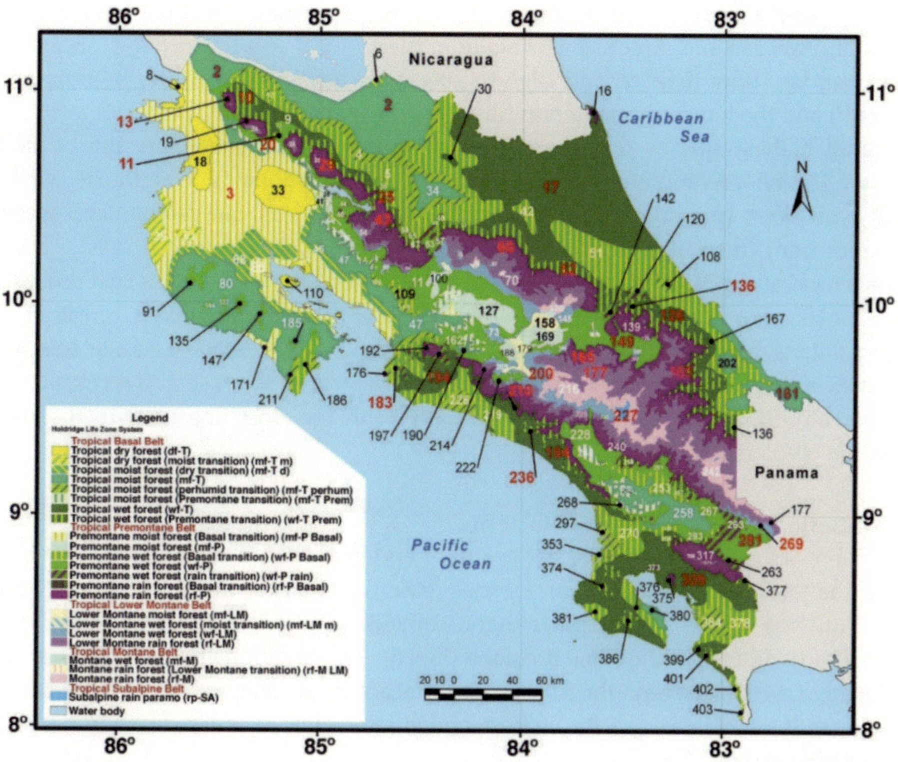

Fig. 3. Numbering the Holdridge Life Zone polygons in Costa Rica. Numbers in red represent life zones with 5 or more years of collecting, which are considered in this study as well represented (taken from Kohlmann et al., 2010).

3.2 Protection of life zone areas

Costa Rica has a total mainland area of 51 042.8 km^2. Out of these, 12 422.4 km^2 (24.3 %) are under some sort of official governmental protection. Noteworthy is that a 100 % of the total area of the montane rain forest (lower montane transition) (rf-M LM) and the subalpine rain paramo (rp-SA) are protected. Other life zones with a high percentage of area under protection are: premontane rain forest (basal transition) (rf-P Basal) (99.9 %), montane rain forest (rf-M) (89.8 %), and lower montane rain forest (rf-LM) (78.6 %). All other life zones have less than 50 % of their area under protection.

3.3 Distribution of species richness by life zone

Figure 4 indicates species richness per group per life zone in Costa Rica. Only the life zones highlighted with a star have been well sampled (*i.e.* more than five years of collecting), therefore, they can be adequately compared. Figure 4 also clearly shows where no members of the taxa under study have been found so far. It should be noted that the highest species richness areas do not coincide for all the taxa in the same place. Araceae, Arecaceae, and Bromeliaceae (Fig. 4) show the same areas of highest species richness in the premontane rain forest (rf-P) and the Scarabaeinae and Dynastinae show an area of highest species diversity coinciding in the premontane wet forest (wf-P) along the different mountain systems; whereas Pisces shows its highest species richness area in the premontane wet forest (Basal transition) (wf-P Basal). In the second highest species richness rank we have that Pisces, Dynastinae, Araceae and Arecaceae coincide in the tropical wet forest (wf-T). Scarabaeinae and Dynastinae show their second highest species richness levels in the premontane rain forest (rf-P) and the lower montane rain forest (wf-LM), respectively.

The overall highest species richness life zones in descending order are: the tropical wet forest (wf-T), the premontane rain forest (rf-P), the premontane wet forest (wf-P), the tropical wet forest (premontane transition) (wf-T Prem), and the premontane wet forest (basal transition) (mf-P Basal). The tropical wet forest (wf-T) (approx. 0 masl – 500 masl, average temperature 24 °C) is generally considered to be the most species rich ecosystem in Costa Rica (Fogden & Fogden, 1997; Valerio, 1999). The premontane rain forest (rf-P) and the premontane wet forest (wf-P) (approx. 500 masl – 1750 masl, average temperature between 17 °C and 24 °C) cover one of the largest geographical areas in the country, where the upper altitudinal limit corresponds spatially with the frost line or with the so-called "coffee line" (Valerio, 2006).

3.4 Distribution of endemicity by life zone

Figure 5 shows the overall number per group of endemic species per life zone. We basically mapped the total number of endemic species (strictly endemic plus shared with Nicaragua and/or Panama) by life zone. Interestingly, coincidences of the highest endemicity areas exist for almost all taxa in the same life zone (Fig. 5). The common areas are the premontane rain forest (rf-P) between the Araceae, Arecaceae, Bromeliaceae, Dynastinae, and Scarabaeinae. Only Pisces has its highest endemicity area in the premontane wet forest (Basal transition) (wf-P Basal).

The overall highest number of endemics by life zone in descending order are: the premontane rain forest (rf-P), the tropical wet forest (wf-T), the premontane wet forest (wf-P), the premontane wet forest (Basal transition) (wf-P Basal), and the lower montane rain forest (rf-LM). Several coincidences occur for the second rank. Pisces, Araceae, and Arecaceae coincide in the tropical wet forest (wf-T), Scarabaeinae and Dynastinae in the premontane wet forest (wf-P). The premontane rain forest (rf-P) is present on the Pacific, as well as on the Caribbean slopes, and although Valerio (2006) indicates that few endemic species are present in this forest type, the results end up supporting Obando´s (2002) conclusion that the cloud forest is the most endemics-rich ecosystem of Costa Rica.

3.5 Representativeness of the protected areas

As previously indicated above, 24.3 % of the total mainland area represents some sort of a governmentally protected area. An analysis of the totality of the species for each of the studied groups (Araceae, 229; Arecaceae, 107; Bromeliaceae, 187; Dynastinae, 125; Scarabaeinae, 177; and Pisces, 111) indicates that: 205 (89.5 %), 95 (88.8 %), 156 (83.3 %), 108 (86.4 %), 165 (93.2 %), and 99 (89.2 %) species, respectively, are present in protected areas. Likewise, an analysis for the total number of endemics for each of the six groups under study (Araceae, 116; Arecaceae, 57; Bromeliaceae, 80; Dynastinae, 68; Scarabaeinae, 68; and Pisces, 62) indicates that 97 (83.6 %), 40 (70.2 %), 64 (80 %), 55 (80.8 %), 64 (94.1 %), and 53 (85.5 %) species, respectively, are present in protected areas.

3.6 Areas of highest species richness per life zone

There are three zones with highest species richness (Fig. 6) according to the overlay of the six groups under study: the first two are the tropical wet forests (wf-T) (approximately 0 masl - 500 masl) in the northeastern corner, bordering Nicaragua (although most probably the central and southern Caribbean coast might also have high numbers that shall become evident after a more intense collection programme is applied), and the Osa Peninsula region. It would appear that the high species richness of these lowland forests tend to diminish inland, as is the case for the tropical moist forest (mf-T) in the northern Caribbean plains, and the tropical wet forest (wf-T) along the piedmont of the Caribbean versant. Both versants share naturally a very high number of common elements to the South with Panama. The third area of highest species richness is the premontane wet forest (wf-P) (approximately 500 masl - 1750 masl) along the Pacific versant of the Guanacaste, Tilarán and Central mountain ranges

This same approximate area was named the Pacific mid-elevation region by DeVries (1987, 1997) and was considered by him to be a very complex area because of its multiplicity of habitats and microhabitats. The same author considered this zone to be very species-rich and a major migrational corridor between the Atlantic and Pacific slopes, as well as a mixing zone for species of both slopes. This area has more species than the Talamanca mountain range to the South, which has a greater extension and is much older (Eocene) than the mountain ranges to the North (Eocene-Pleistocene) (Coates, 1997; Bergoeing, 1998; Valerio, 1999; Alvarado, 2000; Denyer & Kussmaul, 2000), thus contradicting all the tenets (time, species-area, and modified species-area relationship) of the island biogeography theory. The northwestern dry Pacific area of Costa Rica has been well sampled by many institutions throughout the years. However, it is evident that this area does not have a species richness level comparable with the Caribbean and South Pacific coasts or with the mid-elevation areas of the mountain ranges. Clearly, a dry climate with less precipitation can reduce the number of species (Townsend et al., 2008).

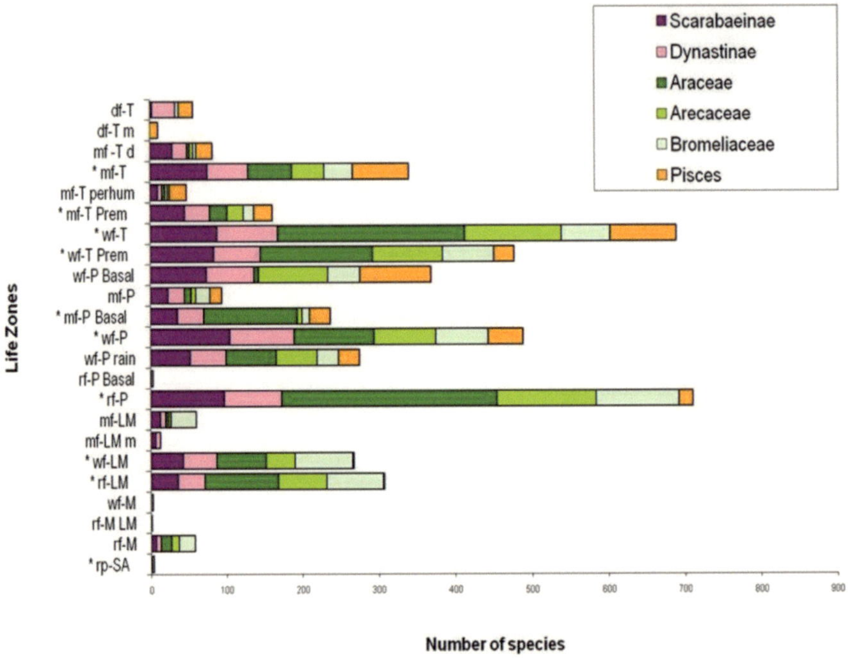

Fig. 4. Distribution of species richness by life zone. Life zones highlighted with a star have been sampled for more than five years and are therefore well-sampled.

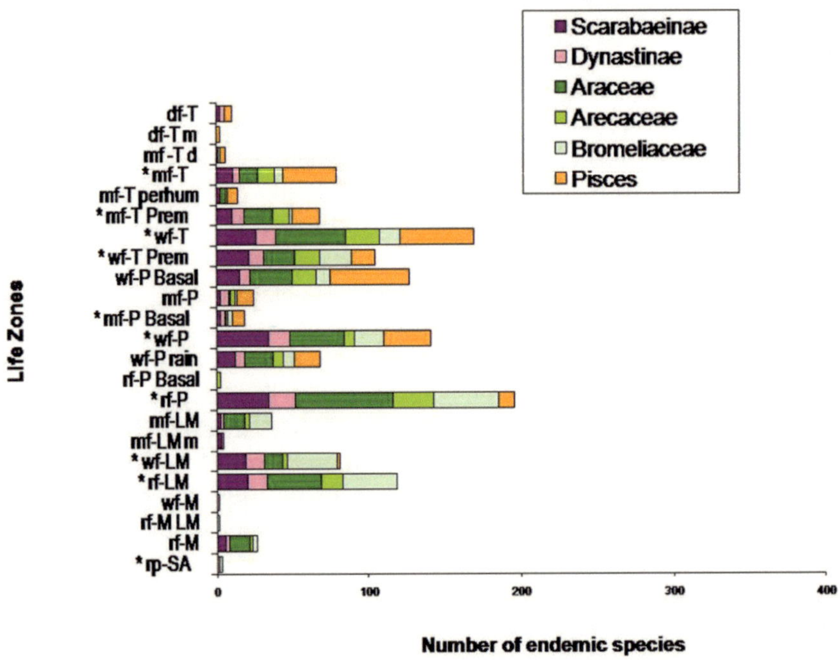

Fig. 5. Distribution of endemic species by life zone. Life zones highlighted with a star have been sampled for more than five years and are therefore well-sampled.

3.7 Areas of highest endemism per life zone

The areas of highest endemism (Fig. 7) according to the overlay analysis show a great spatial correspondence with the previous analysis, containing the same aforementioned three areas. A similar situation had been reported by Campbell (1999), who found that the majority of amphibians' species are endemic to Middle America and therefore there is a tendency of areas of high species diversity to overlap with areas of high endemism. However, for this analysis there is also a fourth area, the lower montane rain forest (rf-LM) (approx. 1000 - 2000 masl) on both slopes of the Talamanca mountain range. The northwestern Pacific with a dry tropical forest, although well sampled, is not an area of high endemism, at least for dung beetles, contrary to the high dung beetle endemism levels found in dry tropical forests along the Mexican Pacific coast (Kohlmann & Solís, 2006).

Obando (2002) reports in her study the existence of five major areas of endemism in Costa Rica. These areas are represented by Coco Island, which was not considered in this study; the Golfo Dulce region (Fig. 2, O), the Cordillera Central (Fig. 2, C), the Talamanca mountain range (Fig. 2, T), and the Central Pacific region (Fig. 2, P). This study supports previously proposed areas of endemism, with the exception of the Central Pacific region. However, three new important areas of endemism are proposed here: the premontane wet forests (wf-P) of the Tilarán and Guanacaste mountain ranges and the tropical wet forest (wf-T) of the northeastern Caribbean (Fig. 7). These last results are important because they contradict a previous study by Elizondo et al. (1989), based on vertebrates and plants, in which the authors found no reason to support the hypothesis that the Tilarán and Guanacaste mountain ranges could represent areas for the generation of endemics. DeVries (1987) had already defined the Guanacaste mountain range as a species pocket area, a place with rare and unusual species (not necessarily an area of endemism). At the same time, the Caribbean lowlands have a relatively recent origin (Pliocene-Pleistocene) according to Bergoeing (1998), yet are rich in endemics. The Tilarán and Guanacaste mountain ranges, as well as the Caribbean lowlands, were reported for the first time to be of importance in the generation of endemics by using dung beetles (Kohlmann et al., 2007).

3.8 Distribution of priority conservation areas

Two conservation categories were defined (Fig. 8). As a reminder suffice to say that a complementarity system was used (Williams et al., 1996), where areas of greatest species and endemicity richness defined the priority conservation categories. In the present case, endemicity was prioritized over species richness, following Mittermeier et al., (2004). Tropical wet forests (wf-T) on the Nicaraguan border and the Osa Peninsula, as well as premontane wet forest (wf-P) along the Central, Tilarán and Guanacaste Cordilleras were determined as priority 1 conservation areas. On the other hand, premontane rain forest (rf-P) on the Tilarán Cordillera and lower montane rain forest (rf-LM) on both slopes of the Talamanca Cordillera were determined as priority 2 conservation areas.

3.9 Number of endemics

Regarding the percentage (26 % - 45 %) of regional (Nicaragua-Costa Rica-Panama) (Table 1) endemic values, they are fairly high, as compared to the regional (28 %) endemism that Savage (2002) reported for the herpetofauna of Costa Rica, which was considered to be the group with the highest endemism for the country. These figures also compare well with the estimates that Obando (2002) established for plant endemism (12 %) in Costa Rica.

Mammals and birds on the contrary present low values of endemism of 0.8 % and 2.5 %, respectively, according to Obando (2002), being one reason for not developing a conservation analysis using only these groups, as is usually done. The use of other groups, like insects, plants and freshwater fishes, can give a much more detailed picture of areas of endemism.

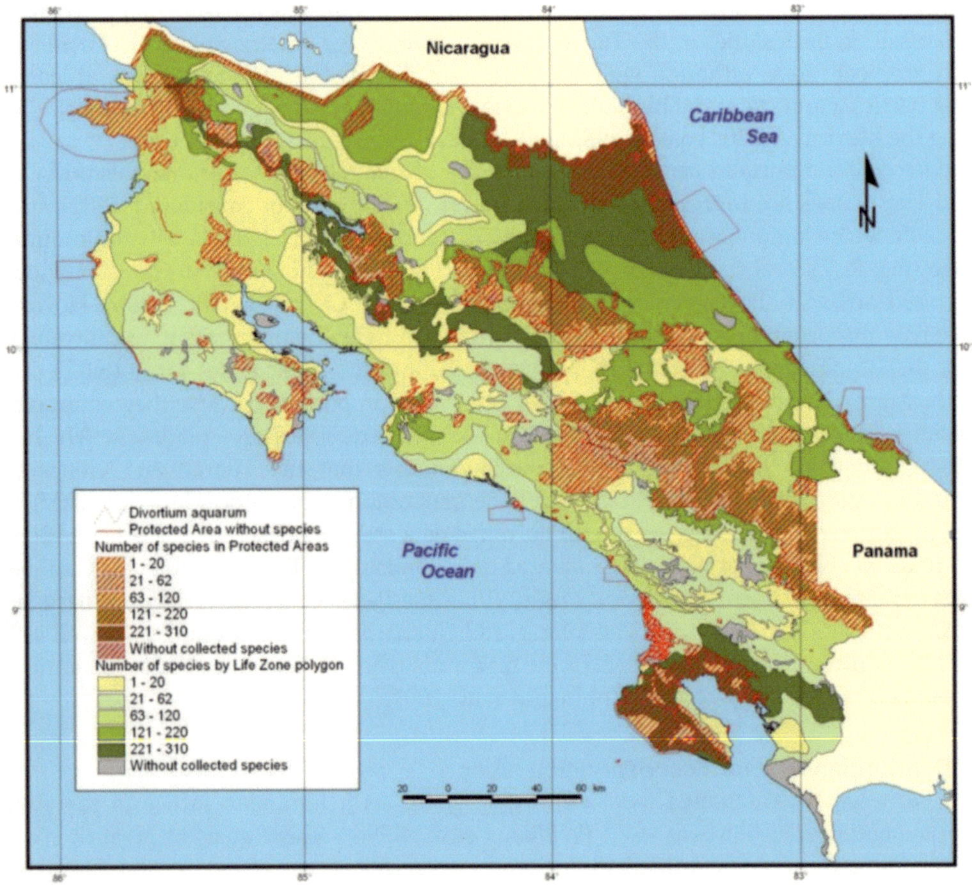

Fig. 6. Overall species richness ranks, based on the totality of the studied groups and their overlap with the established protected areas. Areas in grey represent zones where no collecting has been undertaken. Divortium aquarum = watershed divide.

4. Discussion

4.1 Representativeness of protected areas

The representativeness analysis indicates that a high number (Araceae 89 %, Arecaceae 89 %, Bromeliaceae 83 %, Dynastinae 86 %, Scarabaeinae 95 %, and Pisces 89 %) of the total species are already included by the established protected area system. A similar analysis concerning endemic species also shows the presence of high numbers (Araceae 86 %, Arecaceae 80 %, Bromeliaceae 80 %, Dynastinae 80 %, Scarabaeinae 97 %, and Pisces 85 %) in these protected areas. It is possible that the number for plants may be slightly underestimated, because the dung beetles have been more thoroughly collected. It can be argued that the representation of both, species richness and endemics, in protected areas is already high.

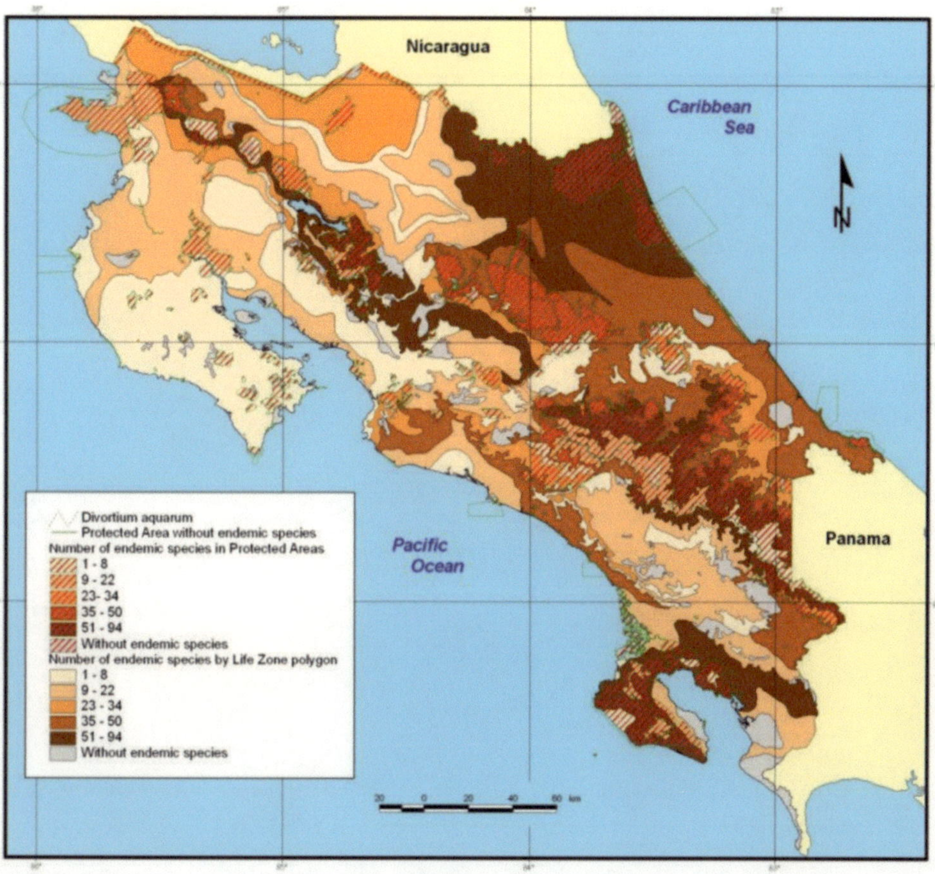

Fig. 7. Overall endemic species richness ranks, based on the totality of the studied groups and their overlap with the established protected areas. Areas in grey represent zones where no collecting has been undertaken. Divortium aquarum = watershed divide.

However, this fact does not guarantee their safeguarding or viability in the long run, because a range collapse could still occur. The endemic population or the community, to which it pertains, could still be marginal or vulnerable to natural or human-induced processes. At present we do not have the necessary information in order to establish the minimum required area to ensure species protection. It is interesting to compare the above results with a similar analysis undertaken by McLean & Meyer (2010), where they estimate that 71 % of the original biodiversity of Costa Rica is still preserved inside protected areas and only 46 % of this same biodiversity is preserved in the whole country! Their analysis was not based on actual counting, but using a model called Mean Species Abundance (MSA), that combines various pressures on biodiversity (land use, infrastructure, fragmentation, and climate change). This study by McLean & Meyer (2010) does arrive to the same conclusion presented by Kohlmann et al. (2010), that land use is the main threat impacting the loss of biodiversity, especially pastures and agriculture. This same study (McLean & Meyer, 2000) also concludes that the zones indicated in this study as priority conservation zones are also the areas showing the greatest remaining biodiversity in Costa Rica.

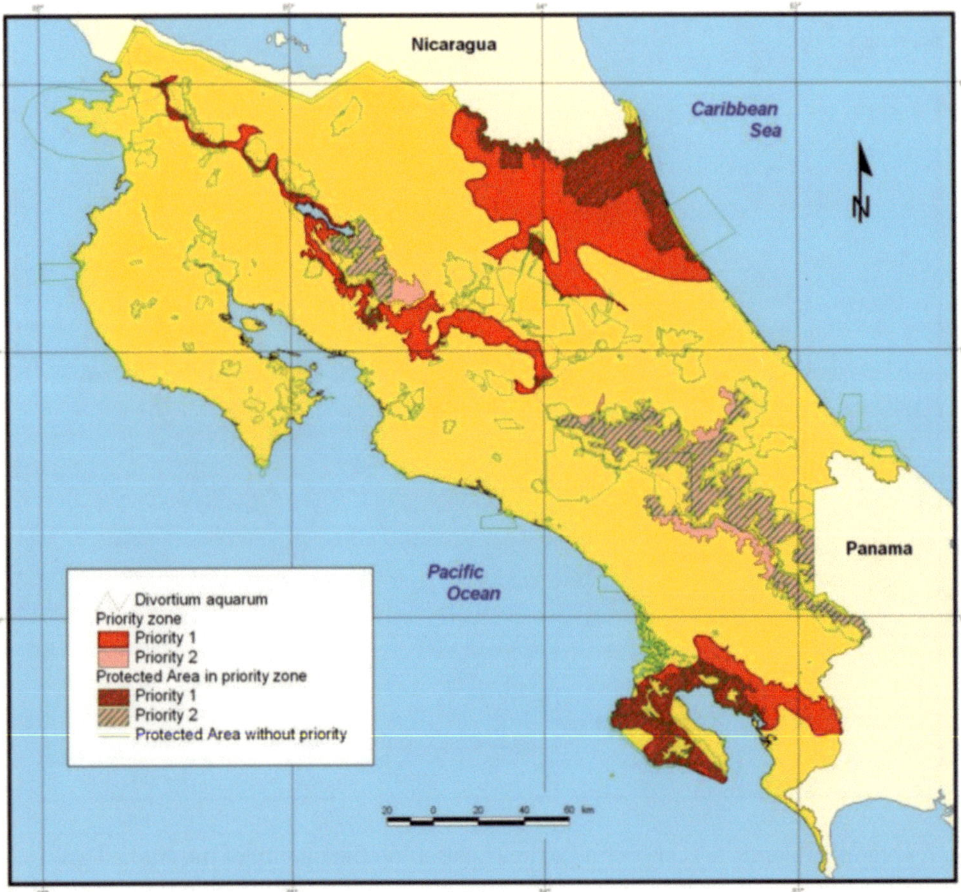

Fig. 8. Distribution of conservation priority zones based on the overlay of species and endemic species richness ranks. Divortium aquarum = watershed divide.

4.2 Distribution of priority conservation areas

Information taken from the two previous maps (Figs. 7 and 8) serves as a base for the present gap analysis creating a conservation priority map (Fig. 8). Priority zone 1 indicates the areas where the highest species richness (rank 5) and the highest endemics (rank 5) numbers coincide. Three areas are defined in this category: the tropical wet forest (wf-T) along the northeastern border with Nicaragua and in the Osa peninsula and the premontane wet forest (wf-P) along the Guanacaste, Tilarán and Central mountain ranges. Priority zone 2 indicates the areas where the highest endemicity level (rank 5) coincides with areas below the highest species richness level (rank<5). Two areas are defined in this category: the lower montane rain forest (rf-LM) on the Talamanca mountain range and the premontane rain forest (rf-P) on the Tilarán Cordillera.

4.3 Representativeness and complementarity

In the past, the majority of the species richness and endemicity studies of Costa Rica have relied basically on vertebrate distribution analysis, especially birds and big-sized vertebrates as indicator indexes of human impacts on the biodiversity, and more recently plants have been employed for this purpose (Obando, 2002; SINAC, 2007a). Insects have not been prominent in these studies.

It is shown in this paper that a different and perhaps a much more refined picture can be gained by using three plant families, two beetle subfamilies, and freshwater fishes instead. This analysis suggests the existence of three previously undetected endemicity areas (Fig. 7) that had not been registered using vertebrates. Although overlap between the different groups is nonrandom, it is not perfect, thus the need for analyzing as many taxonomic groups as possible. In this study, hotspots for species richness tended to overlap with hotspots of endemicity (Fig. 8), thus defining the different conservation priority zones generated by this study. Costa Rica is perhaps the best-collected country in Central America. Not only through the work of many foreign scientists, but lately through the incredible work done by the INBio (Obando, 2007). Still, some areas have been under collected, but the available information allows us to elucidate general patterns.

Taxonomic group	Total number of species	Endemics	
		Total by group	%
Araceae	229	116	50.6
Arecaceae	107	57	53.2
Bromeliaceae	187	80	42.7
Dynastinae	125	66	52.8
Scarabaeinae	177	68	38.4
Pisces	111	62	55.8
Total	936	449	48.9

Table 1. Total number of species and regional endemics by taxonomic group.

This analysis represents a complementary representation and contribution to the excellent proposal presented by the National System of Conservation Areas (Sistema Nacional de Áreas de Conservación (SINAC, 2007b) of Costa Rica. This analysis did follow a different conceptual and methodological approach by defining a conservation strategy oriented toward the necessity of representativeness of selected species (plant and vertebrate species listed as endemic, red list and zero extinction), ecological systems and connectivity of core areas. The SINAC (2007b) thus proposed the undertaking of the project entitled "Propuesta de Ordenamiento Territorial para la Conservación de la Biodiversidad de Costa Rica" (Proposal of Territorial Ordination for the Conservation of Biodiversity in Costa Rica). The aim of the project is to maintain representative samples of the natural richness of the country, correlating them with productive activities of national or local relevance that are conservation-compatible by basing its conservation planning strategy mostly on a phytogeographic system (Zamora, 2008), that would act as a biodiversity surrogate. In the

specific case of the terrestrial environment the aim was to identify vegetation types that are not adequately represented by the present net of conservation areas.
However, a recent study by Rodrigues & Brooks (2007) suggests that the use of environmental data (forest types, vegetation systems, ecoregions, floristic regions, species assemblages, abiotic data) as biodiversity surrogates are substantially less effective than cross-taxon surrogates ("extent to which conservation planning based on complementary representation of species surrogates effectively represents target species"; Rodrigues & Brooks, 2007: 719), where surrogacy is defined as: "extent to which conservation planning based on a particular set of biodiversity features (surrogates) effectively represents another set (targets)" (Rodrigues & Brooks, 2007: 714). Additionally, Pawar et al. (2007) did a very interesting conservation biogeography hierarchical analysis of cross-taxon distributional congruence in North-East India, using amphibians, reptiles and birds from tropical rainforest sites. They found that inherent life-history characteristics shared by certain groups contribute to observed patterns of congruence. They also found that examining biologically distinct subsets of larger groups can improve the resolution of congruence analysis, thus refining area-prioritization initiatives by revealing fine-scale discordances between otherwise concordant groups and vice versa and therefore providing a better resolution even with single-group data. This congruence can then be used as a diversity surrogate simplifying the task of area prioritization and conservation and efficient use of resources. The present paper is thus a first attempt at aiming in this direction in Costa Rica and will hopefully shed some light on the urgent need for cross-taxon analyses and area prioritization efforts.

5. Acknowledgements

We would like to thank foremost the Humboldt Foundation, who graciously provided the principal author with a Georg Foster stipend, which allowed the time and conditions necessary for a sabbatical leave in Germany, where the base of this study was laid in 1999 at the University of the Saarland. Subsequently, the principal author enjoyed two more years of funding by the Research Office of EARTH University, then under the coordination of Carlos Hernández, which helped the completion of this work. The National Institute of Biodiversity has also been most forthcoming in providing through the good offices of Randall García the base information for this study, to them our heartfelt thanks. Special thanks are due to A. Solís, who revised the Scarabaeinae list, to N. Zamora, who revised the Araceae and Arecaceae list, and to J.F. Morales for checking the Bromeliaceae list. We would also like to thank David Roderus, Ortwin Elle, Ricardo Russo, and Xinia Soto for their help in the GIS work. Last but not least, we would also like to thank NASA for the synthetic aperture radar map (SAR) (Fig. 2) of Costa Rica.

6. References

Alvarado, G. (2000). *Volcanes de Costa Rica*. Editorial EUNED, San José, Costa Rica.

Araujo, M.B. (2002). Biodiversity hotspots and zones of ecological transition. *Conservation Biology* Vol.16: 1662-1663, ISSN 0888-8892.

Arias, E. O., Chacón, G., Induni, B., Herrera, H., Acevedo, L., Corrales, J.R., Barborak, M., Coto, J., Cubero, J. & Paaby, P. (2008). Identificación de vacíos en la representatividad de ecosistemas terrestres en el Sistema Nacional de Áreas Protegidas de Costa Rica. *Recursos Naturales y Ambiente* Vol. 54: 21-27.

Balmford, A. & Long, A. (1995). A cross-country analysis of biodiversity congruence and current conservation effort in the tropics. *Conservation Biology* Vol. 9: 1539-1547, ISSN 0888-8892.

Bergoeing, J.-P. (1998). *Geomorfología de Costa Rica*. Instituto Geográfico Nacional, San José, Costa Rica.

Bibby, C.J., Collar, N.J., Crosby, M.J., Heath, M.F., Imboden, Ch., Johnson,T.H., Long, A.J., Stattersfield, A.J. & Thirgood, S.J. (1992). *Putting Biodiversity on the Map: Priority Areas for Global Conservation*. International Council for Bird Conservation, Cambridge, UK.

Bussing, W. A. (1998). *Freshwater Fishes of Costa Rica*. Editorial de la Universidad de Costa Rica, San José, Costa Rica.

Campbell, J.A. (1999). Distribution patterns of amphibians in Middle America. In: *Patterns of Distributions of Amphibians. A Global Perspective*, W. E. Duellman, (Ed.), The Johns Hopkins University Press, Baltimore, USA, pp. 111-210.

Coates, A. G. 1997. The forging of Central America. In: *Central America. A Natural and Cultural History*, A.G. Coates (Ed.), Yale University Press, New Haven, EEUU, pp. 1-37.

Cotera, J., Dengo, J.M., Lücke, O. & Orlich, D. (1998). *Propuesta de Ordenamiento Territorial. Costa Rica en el Año 2025. Comisión Terra*. San José, Costa Rica.

Cox, C.B. & Moore, P.D. (2005). *Biogeography. An Ecological and Evolutionary Approach*. Blackwell Publishing, Oxford, UK.

Denyer, P. & Kussmaul, S. (2000). *Geología de Costa Rica*. Editorial Tecnológica de Costa Rica, Cartago, Costa Rica.

DeVries, P.J. (1987). *The Butterflies of Costa Rica and their Natural History, vol. I. Papilionidae, Pieridae, Nymphalidae*. Princeton University Press, New Jersey, USA.

DeVries, P.J. (1997). *The Butterflies of Costa Rica and their Natural History, vol. II. Riodinidae*. Princeton University Press, New Jersey, USA.

Elizondo, L.H., Jiménez, Q., Alfaro, R.M. & Cháves, R. (1989). *Contribución a la Conservación de Costa Rica. 1. Áreas de Endemismo. 2. Vegetación Natural*. Fundación Neotrópica, San José, Costa Rica.

Estado de la Nación. (2007). *XIII Informe*. Programa Estado de la Nación, San José, Costa Rica.

Fogden, M. & Fogden, P. (1997). *Wildlife of the National Parks and Reserves of Costa Rica*. Editorial Heliconia, San José, Costa Rica.

Garson, J., Aggarwal, A. & Sarkar, S. (2002). Birds as surrogates for biodiversity: an analysis of a data set from southern Quebec. *Journal of Biosciences* Vol. 27: 347-360, ISSN 0939-5075.

Gaston, K. J. (2000). Global patterns in biodiversity. *Nature* Vol. 405: 220-227, ISSN 0028-0836.

Gaston, K.J. & Spicer, J.I. (2004). *Biodiversity. An Introduction*. Blackwell Publishing, Oxford, UK.

Graham, C.H. & Hijmans, R.J. (2006). A comparison for methods for mapping species ranges and species richness and climate. *Global Ecology and Biogeography* Vol. 15: 578-587, ISSN 1466-8238 .

Hall, C. (1984). *Costa Rica: Una Interpretación Geográfica con Perspectiva Histórica*. Editorial Costa Rica, San José.

Hammel, B. E., Grayum, M., Herrera, M.H.C. & Zamora, N. (Eds.). (2003). *Manual de Plantas de Costa Rica. Volumen II. Gimnospermas y Monocotiledóneas (Agavaceae-Musaceae).* Monographs in Systematic Botany from the Missouri Botanical Garden, Vol. 92., St. Louis, Missouri, USA.

Holdridge, L. (1967). *Life Zone Ecology.* Tropical Science Center, San José, Costa Rica.

Howard, P.C., Viskanic, P., Davenport, T.R.B., Kigenyi, F.W., Baltzer, M., Dickinson, C.J., Lwanga, J.S., Matthews, R.A. & Balmford, A. (1988). Complementarity and the use of indicators for reserve selection in Uganda. *Nature* Vol. 394: 472-475, ISSN 0028-0836.

Instituto Tecnológico de Costa Rica. (2000). *Atlas Costa Rica 2000.* 1 CD. Cartago, Costa Rica.

Jiménez, J. (1995). Training parataxonomists and curators to help conservation: The biodiversity inventory. In: *Integrating people and wildlife for a sustainable future, Proceedings of the First International Wildlife Congress,* J.A. Bissonette & P.R. Krausman (Eds.), The Wildlife Society, Bethesda, Maryland, EEUU, pp. 165-167.

Kappelle, M. (2008). *Diccionario de la Biodiversidad.* Editorial INBio, Santo Domingo de Heredia, Costa Rica.

Kohlmann, B. & Solís, A. (2006). El género *Canthidium* (Coleoptera: Scarabaeidae) en Norteamérica. *Giornale Italiano di Entomologia* Vol. 53: 235-295, ISSN 0392-7296.

Kohlmann, B. & Wilkinson, J. (2007). The Tárcoles Line: biogeographic effects of the Talamanca Range in lower Central America. *Giornale Italiano di Entomologia* Vol. 54: 1-30, ISSN 0392-7296.

Kohlmann, B., Wilkinson, J. & Lulla, K. (2002). *Costa Rica from Space.* EARTH University, San José, Costa Rica.

Kohlmann, B., Mitsch, M.J. & Hansen, D.O. (2008). Ecological management and sustainable development in the humid tropics of Costa Rica. In: *Ecological Management and Sustainable Development in the Humid Tropics of Costa Rica,* B. Kohlmann & M.J. Mitsch (Eds.), Ecological Engineering Vol. 34 (No. 4): 254-266, ISSN 0925-8574.

Kohlmann, B., Solís, A., Elle, O., Soto, X. & Russo, R. (2007). Biodiversity, conservation, and hotspot atlas of Costa Rica: a dung beetle perspective (Coleoptera: Scarabaeidae: Scarabaeinae). *Zootaxa* Vol. 1457: 1-34, ISSN 1175-5326.

Kohlmann, B., Roderus, D., Elle, O., Solís, A., Soto, X. & Russo, R. (2010). Biodiversity conservation in Costa Rica: a correspondence analysis between identified biodiversity hotspots (Araceae, Arecaceae, Bromeliaceae, and Scarabaeinae) and conservation priority life zones. *Revista Mexicana de Biodiversidad* Vol. 81: 511-559, ISSN 0034-7744.

Lamoreux, J.F., Morrison, J.C., Ricketts,T.H., Olson, D.M., Dinerstein, E., McKnight, M.W. & Shugart, H. (2006). Global tests of biodiversity concordance and the importance of endemism. *Nature* Vol. 440: 212-214, ISSN 0028-0836 .

Larsen, T.H., Escobar, F. & Armbrecht, I. (2011). Insects of the Tropical Andes: Diversity Patterns, Process and Global Change. In: *Climate Change and Biodiversity in the Andes,* S. Herzog, R. Martínez, P.M. Jørgensen & H. Tiessen (Eds.), IAI, 16.04.2011, Available from http://www.iai.int/files/communications/publications/scientific/Climate_Change_and_Biodiversity_in_the_Tropical_Andes/book.pdf

Lomolino, M. V., Riddle, B.R. & Brown, J.H. (2006). *Biogeography.* Sinauer Associates, Inc., Sunderland, Massachussetts, USA.

Margules, C.R. & Pressey, R.L. (2000). Systematic conservation planning. *Nature* Vol. 405: 243-254, ISSN 0028-0836.

McLean, D. & Meyer, S.T. (2010). *Current and Future Status of Biodiversity in Central America.* Strategic Biodiversity Monitoring and Evaluation Program. Technical Report. Netherlands Environmental Assessment Agency. Convention on Biological Diversity. Zamorano, Honduras.

Meza, T. A. (2001). *Geografía de Costa Rica.* Editorial Tecnológica de Costa Rica, Cartago, Costa Rica.

Ministerio del Ambiente y Energía. (2000). *Estrategia Nacional de Conservación y Uso Sostenible de la Biodiversidad.* San José, Costa Rica.

Mittermeier, R. A., Robles Gil, P., Hoffmann, M., Pilgrim, J., Brooks, T., Goettsch Mittermeier, C., Lamoreux, J. & da Fonseca, G.A.B. (2004). *Hotspots. Biodiversidad Amenazada II.* CEMEX, México, D.F.

Moore, J.L., Balmford, A., Brooks, T., Burgess, N.D., Hansen, L.A., Rahbek, C. & Williams, P.H. (2003). Performance of sub-Saharan vertebrates as indicator groups for identifying priority areas for conservation. *Conservation Biology* Vol. 17: 207-218, ISSN 0888-8892.

Moritz, C., Richardson, K.S., Ferrier, S., Monteith, G.B., Stanisic, J., Williams, S.E & Whiffin, T. (2001). Biogeographical concordance and efficiency of taxon indicators for establishing conservation priority in a tropical rainforest biota. *Proceedings of the Royal Society of London, Series B, Biological* Sciences Vol. 268: 1875-1881, ISSN 1471-2954.

Morrone, J. J. (2000). La importancia de los atlas biogeográficos para la conservación de la biodiversidad. In: *Hacia un proyecto CYTED para el inventario y estimación de la diversidad entomológica en Iberoamérica: PrIBES 2000,* F. Martín Piera, J.J. Morrone, & A. Meliá (Eds.), m3m-Monografías Tercer Milenio, vol. 1, Sociedad Entomológica Aragonesa (SEA), Zaragoza, España, pp. 69-78.

Müller, P. (1981). *Arealsysteme und Biogeographie.* Eugen Ulmer GmbH and Co., Stuttgart, Germany.

Myers, N. (1988). Threatened biotas: hotspots in tropical forests. *Environmentalist* Vol. 10: 243-256, ISSN 1573-2991 .

Myers, N., Mittermeier, R.A., Mittermeier, C.G., da Fonseca, G.A.B. & Kent, J. 2000. Biodiversity hotspots for conservation priorities. Vol. 403: 853-858, ISSN 0028-0836.

Obando, V. (2002). *Biodiversidad de Costa Rica. Estado del Conocimiento y Gestión.* Instituto Nacional de Biodiversidad, Santo Domingo de Heredia, Costa Rica.

Obando, V. (2007). *Biodiversidad de Costa Rica en Cifras.* Editorial INBio, Santo Domingo de Heredia, Costa Rica.

Pawar, S.S., Birand, A.C., Ahmed, M.F., Sengupta, S. & Raman, T.R.S. (2007). Conservation biogeography in North-east India: hierarchical analysis of cross-taxon distributional congruence. *Diversity Distribution* Vol. 13: 53-65, ISSN 1472-4642.

Pimm, S.L., Russell, G.J., Gittleman, J.L. & Brooks, J.L. (1995). The future of biodiversity. *Science* Vol. 269: 347-350, ISSN 0036-8075.

Prendergast, J.R., Quinn, R.M. & Lawton, J.H. (1999). The gaps between theory and practice in selecting nature reserves. *Conservation Biology* Vol. 13: 484-492, ISSN 0888-8892.

Prendergast, J.R., Quinn, R.M., Lawton, J.H., Eversham, B.C. & Gibbons, D.W. (1993). Rare species, the coincidence of diversity hotspots and conservation strategies. *Nature* Vol. 365: 335-337, ISSN 0028-0836.

Ratcliffe, B.C. (2003). *The Dynastinae scarab beetles of Costa Rica and Panama.* Bulletin of the University of Nebraska State Museum Vol. 16.

Rodrigues, A. S. L. & Brooks, T.M. (2007). Shortcuts for biodiversity conservation planning: The effectiveness of surrogates. *Annual Review of Ecology, Evolution, and Systematics* Vol. 38: 713-737, ISSN 0066-4170.

Rondinini, C. & Boitani, L. (2006). Differences in the umbrella effects of African amphibians and mammals based on two estimators of the area of occupancy. *Conservation Biology* Vol. 20: 170-179, ISSN 0888-8892.

Savage, J. M. (2002). *The Amphibians and Reptiles of Costa Rica. A Herpetofauna between Two Continents, between Two Seas.* The University of Chicago Press, Chicago and London.

Sistema Nacional de Áreas de Conservación SINAC del Ministerio de Ambiente y Energía (MINAE). (2007a). *Programa de monitoreo ecológico de las Áreas Protegidas y corredores biológicos de Costa Rica (PROMEC-CR) Etapa I (2007-2011): Resumen Ejecutivo/SINAC-MINAE.* Asociación Conservación de la Naturaleza, San José, Costa Rica.

Sistema Nacional de Áreas de Conservación (SINAC) Grúas II. (2007b). *Propuesta de ordenamiento territorial para la conservación de la biodiversidad de Costa Rica: Vol. 1. Análisis de vacíos en la representatividad e integridad de la biodiversidad terrestre/SINAC-MINAE.* Asociación Conservación de la Naturaleza, San José, Costa Rica.

Stork, N.E. & Samways, M.J. (1995). Inventorying and Monitoring. In: *Global Biodiversity Assessment*, V.H. Heywood and R.T. Watson (Eds.), 453-553, Cambridge University Press, Cambridge, UK.

Townsend, C.R., Begon, M. & Harper, J.L. (2008). *Essentials of Ecology.* Blackwell Publishing, Oxford, UK.

Valerio, C.E. (1999). *Costa Rica. Ambiente y Sociedad.* Instituto Nacional de Biodiversidad, Santo Domingo de Heredia, Costa Rica.

Valerio, C.E. (2006). *Costa Rica. Ambiente y Sociedad.* 2ed. Instituto Nacional de Biodiversidad, Santo Domingo de Heredia, Costa Rica.

Vaughan, C. (1994). Management of conservation units: The Costa Rican national system of conservation. In: *Principles of Conservation Biology,* G. Meffe & C.R. Carroll (Eds.), Sinauer Press, Sunderland, Massachussetts, USA, pp. 395-404.

Vaughan, C., Fallas, J. & McCoy, M. (1998). Conservation mapping in Costa Rica. In: *GIS methodologies for developing conservation strategies,* B.G. Savitsky & T.E. Jr. Lacher (Eds.), Tropical Forest Recovery and Wildlife Management in Costa Rica. Columbia University Press, New York and Chichester, West Sussex, UK, pp. 13-25.

Williams, P., Gibbons, D., Margules, C., Rebelo, A., Humphries, C. & Pressey, R. (1996). A comparison of richness hotspots, rarity hotspots, and complementary areas for conserving diversity of British birds. *Conservation Biology* Vol. 10: 155-174, ISSN 0888-8892 .

Zamora, N. (2008). Unidades fitogeográficas para la clasificación de ecosistemas terrestres en Costa Rica. *Recursos Naturales y Ambiente* Vol. 54: 14-20, ISSN 1659-1216.

11

Multiple Approach for Plant Biodiversity Conservation in Restoring Forests

Federico Vessella, Bartolomeo Schirone and Marco Cosimo Simeone
Department of Forest and Environment (D.A.F.) – University of Tuscia
Italy

1. Introduction

The current extinction crisis requires dramatic action to save the Earth' s biological diversity. In the mid-1980's the word "biodiversity" was coined to catch the essence of research into the variety and richness of life on Earth, that is, the variety of life expressed at many levels (Wilson & Peter, 1986). These levels include the genetic diversity within species as well as the array of genera, families, and still higher taxonomic levels that, taken together, comprise communities of organisms within particular habitats and physical conditions that form entire ecosystems. It is widely demonstrated that more species contribute to a greater ecosystemic stability. Moreover, individuals, populations and ecosystems are tightly linked and interact to maintain landscapes, large socio-economic systems and man's health. As a consequence, biodiversity maintenance is fundamental for the planet life, and should be carried out with "passive" conservation measures implemented with "active" procedures using the most recent progress in technique and policy. In this context, reforestation programmes have to be considered as dynamic actions devoted to the biodiversity conservation toward the recovery and/or the enlargement of such areas essential for coenosis' evolution. This concept of reforestation is relatively new and still has difficulty to be established.

Most conservation biologists recognize that although we can not save everything, we should at least ensure that all ecosystem and habitat types are represented within regional conservation strategies that have been applied at a number of geographical scales, from single watersheds to entire continents (Hummel, 1989; Eriksson et al., 1993; Caldecott et al., 1994; Krever et al., 1994; Noss & Cooperrider, 1994; BSP et al., 1995; Dinerstein et al., 1995; UNEP, 1995; Ricketts et al., 1999; Abell et al., 2000).

Forests are the single most important repositories of terrestrial biological diversity. They provide a wide range of products and services to people throughout the world. Forest trees and other woody plants help support many other organisms, and have developed complex mechanisms to maintain high levels of genetic diversity. This genetic variation, both inter- and intraspecific, serves a number of fundamentally important purposes. It allows trees and shrubs to react to changes in the environment, including those brought about by pests, diseases and climatic change. It provides the building blocks for future evolution, selection and human use in breeding for a wide range of sites and uses. And, at different levels, it supports the aesthetic, ethical and spiritual values of humans. Forest management for productive and protective purposes can and should be rendered compatible with conservation through sound planning and coordination of activities at different geographical levels. Accordingly, the conservation of these resources should be seen as an attempt to preserve groups of genotypes or populations, and their various combinations of genes (Gregorius, 1991). Therefore, the aim of forest resource management is to maintain conditions in which the genetic makeup of a species can continue to evolve in response to changes in its environment (Eriksson, 2001). At the same time, management for conservation aims at reducing the rates of genetic erosion (FAO et al., 2004).

Different conservation strategies and practices have been developed. *In situ* ('in place') conservation implies the continuing maintenance of a population within the environment where it originally evolved, and to which we assume it is adapted (Frankel, 1976); this type of conservation is most frequently applied to wild populations regenerated naturally in protected areas or managed forests, but can include artificial regeneration whenever planting or sowing is carried out, without directional selection, in the same area where the seed was collected. *In situ* conservation in general has the advantage of conserving the function of an ecosystem rather than just species. This means that *in situ* programmes for conservation of selected target species often result in valuable conservation of a number of associated animal and plant species (Thomson et al., 2001). *Ex situ* ('out of place') conservation measures are mainly concerned with sampling and maintaining as much of the genetic variation as possible that resides within and among populations of selected target species. *Ex situ* conservation requires substantial levels of human intervention, in the form either of simple seed collections, storage and field plantings or of more intensive plant breeding and improvement approaches. Unlike breeders of agricultural crops, forest tree breeders cannot rapidly produce new varieties, nor can they quickly breed for new variations among populations. Therefore, the existing genetic diversity among populations is important and fundamental to the conservation of forest genetic resources, particularly as it may relate to maintaining genetic diversity in viable populations in the long term. This also suggests that special attention must be given to conserving intraspecific genetic variation in peripheral or isolated populations, as they could possess higher levels of characteristics such as drought resistance, tolerance to various soil conditions (Stern & Roche, 1974), or features that will help to protect them from future climate change (Muller-Starck & Schubert, 2001). The important features of an *ex situ* conservation programme for any particular species are: to be an important backup measure should other *in situ* conservation means be unworkable or unavailable, to ensure that a wide range of the diversity (phenotypic and genotypic) available in a species is conserved, and to manage the regeneration of the species outside its original natural range (provenance) in a more controlled way (which is likely to further develop the population(s) for use or conservation) (Amaral & Yanchuk, 2004).

Recently, a European funded project, EUFGIS (European Information System on Forest genetic Resources) established a web-based information system to serve as a documentation platform for national forest genetic resources inventories and to support practical implementation of gene conservation and sustainable forest management in Europe (further information at http://www.eufgis.org). The main purpose was to assess pan-European minimum requirements and data standards for the dynamic gene conservation units of forest trees, i.e. selected areas which emphasizes the maintenance of evolutionary processes within tree populations to safeguard their potential for continuous adaptation.

Those tools are adopted in programmes devote to afford issues such as the conversion of forest land to other uses. Increasing pressure from human populations who aspire to higher standards of living, without balancing the sustainability of resource utilization underpinning such developments, raises concerns in this regard. It is inevitable that changes of land use will occur in the future, but such changes should be planned to help ensure that the complementary goals of conservation and development are achieved. In recent times there has been a growing awareness in this topic, and the research community is looking for moving away from the dominant focus on deforestation and resources' conservation to examine the patterns and the processes associated with reforesting landscapes (Rudel, 2005). Developing a more comprehensive understanding of the factors that can help to promote reforestation is therefore critical, if we are to increase useful policy interventions to arrest or reverse deforestation, and encourage forest regrowth. Yet, it is important to recognize that forests are embedded within larger-level ecological, socio-economic and political settings, which have the capacity to significantly influence outcomes. Thus, discussions of context (biophysical, geographic, ecological, socio-economic and institutional) are essential to the development of our understanding of this area of study (Nagendra & Southworth, 2009). This implies awareness of the availability of efficient tools to comply with traditional management strategies, as well as action plans and guide lines at large scale. Under these circumstances the Council of the European Union promoted a legislative tool in 1999 that recognized social, economic, environmental, ecological and cultural functions of forests. Both the restocking of these forests and new afforestation require a sustainable forest management in relation to the Forestry Strategy for the European Union, that include the use of reproductive material which is genetically and phenotypically suited to the site and of high quality (European Council, 1999). In this context, the definition and delimitation of Regions of Provenance have been proposed as fundamental to select reproductive material and to approve basic material with highest possible standards.

However, an improvement of genetic knowledge about forest plants is surely required to accomplish the requirements of the Directive, and should also contribute to better define what inter- and intra-specific biodiversity is. One of the latest standardized molecular approach is DNA Barcoding (Hebert et al., 2003) that identifies living organisms by joining specific sequences of DNA and electronic information retrieval. Biodiversity characterization and improvements in genetic knowledge would be two of the main benefits of the widespread application of Barcoding, in terms of speed, low cost, reliability, and improved resolution power. Besides taxonomy, a powerful research complement for molecular ecology, diversity studies and population genetics is clearly to be expected. DNA Barcoding may lead to many useful applications in forestry sciences, such as community ecology (to describe plant-animal interactions and vegetation dynamics/changes), biodiversity surveys (aimed at habitat and species protection), silviculture (to assess forest regeneration), and nursery activities and market regulation (to establish wood, secondary products and germplasm certification). Conversely, it must be emphasized that some species-rich tree genera may prove very difficult to barcode, especially those in which species circumscription is affected by complicated taxonomies, biogeographies and/or reproductive biology.

Since plant biodiversity is strictly related to natural restoration and rehabilitation of ecosystem functions, with respect to its health, integrity, and sustainability, all the tools mentioned above are linked to the reforestation techniques proposed by scientists and

experts from decades. New insights both in theoretical and in practical actions have been developed as innovative methods to foresters and ecological specialists. Among these, the Miyawaki method based on the vegetation-ecological theories, seems to be a reliable approach that include the principles of self-organized criticality and cooperation theories in forest ecosystems, also fulfilling the objectives anticipated by the Regions of Provenance.
The tools mentioned above face the same issues, and try to get practical solutions for conserving and/or restoring forests. However, they approach plant biodiversity conservation in a piecemeal fashion with projects and management plans focusing narrowly on one or a small range of techniques that coincide with the responsibilities, philosophies, and capabilities of the people working in a given setting.
In this chapter we would like to propose a multiple approach as a potentially powerful system for facing the challenge of conserving, but mainly expanding forests over long time horizons. A critical review on the methods mentioned above toward an holistic point of view will be discussed. We believe that multidisciplinary would be the way to follow out, and our effort regarded the setting up of a linkage between the mentioned strategies and practices. In the next paragraphs a deeper description of Regions of Provenance, DNA Barcoding, and reforestation using Miyawaki method will be presented, focusing on recent achieved results, improvements and proposals. Some study cases in Italy will be also shown, in order to shed some light on the criteria for detecting best actions in the Mediterranean Basin. Finally, we will attempt to logically order these tools in an improved and well-organized "toolbox".

2. Regions of Provenance in Europe

The relationship between genetic variability and adaptability for a species is particularly important if we refer to forest plants, because they are characterized by long life cycle and consequently more exposed to environmental changes. Looking at biodiversity within a single species, very important are those populations with specific adaptations that could characterize local ecotypes. When populations are geographically separated and genetic flux is interrupted, differentiation processes can lead to speciation.
In many countries, the uncontrolled use of germplasm of unknown origin favoured serious phenomena of genetic erosion and pollution, in particular after the implementation of the Regulation EEC 2080/92 which encouraged the reforestation of agricultural land. In Italy, for instance, the Rural Development Programme 2000-2006 promoted reforestation on huge surfaces, and many land owners joined the program. The lack of enough autochthonous propagation material to supply the demand, leaded the operators to use plant material from several ecologically different geographical areas; as a result , many reforestation plans failed because of diverse pedoclimatic requirements of the adopted material, and with the rising up of infestations by new parasites.
For these reasons, the use of high quality propagation material, phenotipically and genetically appropriate to the plantation area is fundamental. Such principle, previously introduced in two European Directives (EEC 404/66 and EEC 161/71) was finally integrated in the Directive EC 105/99 about the marketing of forest reproductive material. Moreover, the Directive establishes that the basic material for reforestation has to be harvested from selected stands, and underlines the importance of delimiting Regions of Provenance, defined as "the area or group of areas subject to sufficiently uniform ecological conditions in which stands or seed sources showing similar phenotypic or genetic characters are found,

taking into account altitudinal boundaries where appropriate". It is also specified that forest reproductive material (seeds, cones, fruits, parts of plants, planting stocks) is classified in the following categories: Source-identified, Selected, Qualified and Tested. About the source-identified and the selected materials, they belong to seed sources, stands, seed orchards, parents of family, clones or clonal mixtures located in a single Region of Provenance. Selected materials include also the phenotypic ones, identified at population level, and fulfilling the requirements of origin, isolation, population dimension, age and development, homogeneity, phytosanitary status, quali-quantitative production, status and morphology. Indeed, the Directive highlights the need for each Member State to define the Regions of Provenance for a correct use of reproductive material, in order to ensure forest biodiversity conservation with specific regards to the nursery activities. However, for reforestation practices, the Directive's contents does not suggest the use of basic material in accordance with the Regions of Provenance.

In the last twenty years, many European countries developed management systems based on ecoregions, adopting national measures accordingly to the Communitarian legislation. Anyway, the delimitation of the Regions of Provenance is very elaborate, since it requires the definition of the actual relationships between the ecological features of an area, the ecophysiological characteristics of each species, the peculiar propagation dynamics (pollination, dissemination, diffusion methods) of the species, and the intra-specific genetic diversity at both the individual and the population levels.

2.1 Common criteria for the definition of the Regions of Provenance

The subdivision of the distribution range for a species in spatially and genetically homogeneous regions complies the hypothesis of an intra-specific differentiation according with the environmental selection effects. This argument is valid only if populations have enough genetic variability to face, in terms of adaptability, and possibly to mild the environmental changes that may occur within a certain physical area. Some reproductive isolation derived from the genetic differentiation is a necessary prerequisite for allowing adaptability processes at a local scale. Therefore, the delimitation of the Regions of Provenance plays a key role in identifying those basic materials from which harvesting forest reproductive materials. Despite the environmental and genetic homogeneities are essential requirements to define different provenances, a weak point is detectable: the genetic composition of a population, i.e. the main indicator of adaptability derived from the evolutive processes, is commonly assessed throughout the analysis of the phenotypic performance, while a description through the use of molecular markers would be more appropriate. The adaptability at local environmental conditions, together with a genetic peculiarity, are essential features to reveal the autochthony of a population, possibly witnessed by historical documents. Recommendations by national and regional measures underline the importance of autochthonous resources for environmental restoration, starting from considerations about species' adaptability. Referring to forest populations, "autochthony" indicates the continuous occurrence of a species, in terms of genealogy, in a defined site since the last post-glacial migration. However, adaptability, as a peculiar feature of autochthonous populations, raises further considerations about the surface size where populations occur, as well as the time they passed under the same environmental conditions. For these reasons, the meaning of adaptability has been redefined several times, but always focusing on the spatial and temporal continuity in constant environment settings. Such quantitative characterization allows to consider the autochthony of a population as a phenomenon in terms of degree rather than presence/absence. This suggests the need for defining the population extension, the size and structure, and the assessment of a continuous occurrence through the time in a specific area. At the same time, it is important to specify meta-populations' structure preserving their reproductive coherence by genetic flux. According to the spatial scale of investigation, the local genealogic continuity could appear relatively low in some stands, because of punctual extinction events or other disturbances. An appropriate spatial scale should be only established after the understanding of the reproductive coherence within the species' distribution range, and the analysis of the homogeneity of the environmental characteristics. Temporal and spatial scales, and the degree of environmental heterogeneity could be indirectly observed in the genetic structure of an autochthonous population, as the consequence of evolutive processes of adaptability. This circumstance derived from the presence of heterogeneity variation within a population; such variation has to be heritable, so the availability of genetic diversity is fundamental.

As mentioned above, it is often hard to check for the main adaptability determinants, as well as to accurately measure the features of autochthony. There is a significant mass of literature about the most commonly applied methods for delimiting Regions of Provenance according to the factors mentioned above (e.g. Geburek & Konrad, 2008; Kleinschmit et al., 2004; Lindgren & Ying, 2000; Krusche & Geburek, 1991; Raymond & Lindgren, 1990). They usually refer to the division of the territory (*divisive method*), if ecological parameters are considered, instead of joining of similar populations (*agglomerative method*) according to common biological features. Three clustering approaches are generally followed:

1. clusters according to homogeneous environmental conditions;
2. clusters according to genetic markers;
3. clusters according to phenotypic response.

The first procedure consists in grouping areas that share congruent ecological conditions. The selected parameters useful to characterize these conditions are supposed to be important for maintaining and expanding the referred species. The complexity of the growth regulation phenomena make difficult the selection of such parameters; however, some artificial delimitation of the Regions of Provenance based on administrative boundaries have been adopted to facilitate the management procedures. A specific problem raises when the reproductive coherence within a Region does not match with the potential effects on the intra-regional genetic differentiation; this difficulty is typical of areas with plantations, where the individuals have significant differences in terms of geographic origin, and no genetic information resulting from the adaptation at the local conditions.

The second clustering procedure gives more emphasis on the genetic variability of every single species, intended as peculiar feature for observing intra-specific differentiation at level of regional areas, populations, individuals, etc., and needed for defining the Regions of Provenance. Unfortunately, the correspondence between adaptability and neutral molecular markers (e.g., isoenzymes, plastid and mitochondrial microsatellites) may not be sufficiently clear to mirror differences between Regions of Provenance, and the more variable nuclear markers (e.g., microsatellites, AFLPs) may prove too difficult and expensive for large scale investigations (Karp et al., 1997; Duminil et al., 2007). On the other hand, adaptive traits could be more efficiently dissected by use of QTLs analyses (Lewontin, 1984; Borevitz and Chory., 2004), but the technical requirements of the method (and the complex biology of trees) has limited the number of available information on a short number of world species.

Finally, the most recent advancements of molecular biology (identification and characterization of ecologically important Candidate Genes, transcriptomics and metagenomics) are highly promising but still hindered by prohibitive costs and difficulties (Pflieger et al., 2001; Eveno et al., 2008; Derory et al., 2010).

The last procedure is based on statistical researches, starting from test areas, about relationships between growth performance and environmental variables, such as stem growth, bud set, flowering time, and geographic coordinates, photoperiod, altitude, etc. These affinities are used to cluster populations according with specified range of values calculated by statistical functions. Resulting clusters are built minimizing the distance between the origin of propagation material and the area to under investigation.

In conclusion, all these criteria focus on reducing the risk connected with the transfer of the material in ecologically heterogeneous areas. Such risks are evaluated in terms of adaptive failure and undesired phenotypic traits (habitus, seed productivity, etc.), but the methods applied are still under discussion and development. Indeed, every useful strategy devoted to minimize the mentioned risks has to be based on a spatial delimitation consistent with the real and/or potential adaptability of a population through the time.

The main goals to define provenances concern the species' range, metapopulation and subpopulation delimitation, the estimation of adaptability, and the assessment of the adaptive effectiveness in terms of evolution. Therefore, the common methods to achieve these goals involve a spatial-genetic clustering of at least a second generation of adult individuals, the heterogeneity analysis of life mechanisms and functions, and the connexion of adaptive phenotypic variability with the genetic one, developing transplant tests with adaptive differentiation study.

In many Communitarian experiences, the definition of Regions of Provenance leaded to delimitate large areas (e.g. oak species in Germany, Scots pine in the Baltic region), because of the results obtained on experimental fields where the adaptive flexibility of several populations were tested at different environmental conditions. In any case, Regions of Provenance have to be directed to the preservation of the natural mechanisms of persistence for a species within its range.

2.2 State of the art

In the European Union, several countries actuated the delimitation of the Regions of Provenance for forest species after the adoption of the Directive EEC 161/71. The progress project reports differ State by State, but practically match the requirements suggested by the Communitarian normative. An important step concerns the mapping of species' occurrence in each country, and the characterization of homogeneous areas using ecological indexes. In many cases, the Regions of Provenance have been delimited summarizing the results obtained by the application of several clustering approaches, and taking also into account regional and/or provincial administrative boundaries; such last conformation is particularly useful for those countries which have commissioned the competence for environmental topics to the local authorities.

In the following table the Member States adopting the Regions of Provenance in accordance with the Directive 105/99 have been listed (thus, Norway has been excluded even if adopting the Directive), providing synthetic information about used parameters and number of target species (Table 1, modified from Alía et al., 2008). This summary table could be subjected of updating, as the Regions of Provenance is a dynamic process still in progress in several countries (e.g. in Italy).

Parameters	AT	BE	CZ	DE	DK	ES	FI	FR	GB	GR	HU	IT	IR	LT	NL	PL	RO	SE	SI	SK
	Countries																			
Geographical information	X	X	X	X		X	X	X	X	X	X	X	X	X		X	X	X	X	X
Altitude	X	X	X	X		X	X	X	X	X	X	X	X	X		X	X	X	X	X
Climate	X	X		X		X	X	X	X	X	X	X	X	X		X	X	X	X	X
Soil	X	X	X	X	X	X		X		X	X	X	X	X		X	X	X	X	X
Neutral markers		X			X	X		X				X			X		X			X
Field test					X	X		X	X			X	X	X	X	X				
Nursery/phytotron test						X						X		X	X					
Growth				X		X		X				X	X	X		X				
Phenology			X	X				X				X	X	X		X			X	
Resistance to disease/pests			X	X					X											
Vegetation/ phytogeography	X			X		X		X	X	X	X	X							X	
Overall adaptation				X	X	X	X	X	X			X	X	X	X	X	X			
Administrative boundaries		X					X		X	X		X		X				X	X	
N. of target species	24	37	23	50	36	56	14	53	54	10	25	78	28	9	22	10	35	17	11	7

Table 1. Overview of the criteria adopted by the countries in the European Union that have defined the Regions of Provenance and total amount of forest target species. Short abbreviations of country names are given according to the ISO 3166-1-alpha-2 code.

The differences highlighted between countries suggest some critical remarks. Against a common acceptance of the Directive 105/99, there is not a pan-European strategy for the Regions of Provenance; some difficulties could be detected such as the available information, the possibility to exchange, and the different formats of the data. Moreover, large differences regard the size and the methodologies for the delineation, the number of target species and the knowledge about their biological parameters that varies from species to species.

A standardization process devoted to set up a common approach in the delineation of the Regions of Provenance surely will take long time, as it requires both technical and political actions. However, it is possible to suggest improvements for the methods, using additional ecological variables and/or techniques to explore phenological and biological behaviours of forest species not considered yet. A study case is presented in the next paragraph and regards the delineation of Regions of Provenance in the Latium District (Italy).

2.3 New proposals and improvements for delineating Regions of Provenance: The case of Latium

The clustering procedure according to homogeneous environmental conditions could be considered as the easiest approach for defining Regions of Provenance, because of the large databases about chemo-physical parameters that each country has stored since the beginning of the last century. Generally, we have more information about the property of a territory instead of the living species that occur there. These simple assumptions could justify why the divisive method is often used. Nevertheless, additional variables could be considered for improving the delimitation of the Regions of Provenance, for instance phytoclimatic indexes as the Mitrakos Winter Cold Stress (WCS) or Summer Drought Stress (SDS), and the Emberger coefficient (Mitrakos, 1980; Emberger, 1955). These parameters demand for time-series climatic data, and refer to a data point network of weather stations widespread in an area; but, it is possible to spatially extend them by using numerical and mathematical techniques dealing with the characterization of spatial phenomena, using geostatistic analyses that rely on statistical approaches based on random function theory to model the uncertainty associated with spatial estimation and simulation. Using the geostatistic methods, as implemented in many GIS softwares, it is also possible to go beyond the interpolation problem by considering the studied phenomenon at unknown locations as a set of correlated random variables. In the case of Latium, both Mitrakos indexes and Emberger coefficient have been spatialized using Kriging interpolation from 85 data points recording precipitation and temperature for 15 years at least; topography and continentality have been also considered as supplementary variables extrapolating data from the Digital Elevation Model (DEM) of Italy with 75x75 m grid cells. The resulting outputs have been overlapped to other chemo-physical variables, i.e. mean annual temperature, minimum temperature of the coldest month, maximum temperature of the warmest month, annual precipitation, geomorphology, soil, etc. A summary layer storing all the spatialized variables has been performed and areas with homogeneous ecological features have been finally detected. Moreover, the boundaries of these areas have been buffered to better represent the gradual spatial shifting from an ecological context to another (Figure 1).

According to the main phytoclimatic parameters, as well as the vegetation maps proposed by several authors (Blasi, 1994; Tomaselli, 1973; Pavari, 1916), Latium has been divided in 3 Primary Regions of Provenance and subsequently in 17 Secondary Regions including also the geomorphology and the soil characteristics (Figs. 2, 3). This procedure basically follows the common strategies adopted by the other European countries, but increases the number of variables to be considered for a deeper ecological investigation that includes plant response to climatic conditions.

Since the evaluation of the effects of natural selection and bioclimatic responses across space is at the base of the definition of the Regions of Provenance, a better characterization of basic material should be achieved by combining already showed results with parameters related to species performance (biological responses to ecological factors) and to the altitudinal gradient (as suggested by the Directive 105/99), in order to provide homogeneous material both for afforestation and genetic preservation. In those countries where the knowledge about forest species' biological and genetic features are studied from years, or for peculiar species with great economic impact (e.g. *Populus spp.*, *Castanea sativa* Mill., *Quercus petraea* Liebl., *Picea abies* (L.) Karst., *Pinus sylvestris* L., *Fagus sylvatica* L., *Quercus suber* L., etc.) such improvements have been made; in particular, as showed in Table 1, phenology, growth performance, and neutral markers for genetic characterization are the most used parameters. Nevertheless, it is possible to consider further investigations focusing on bioindicator species' behaviours, and extending the method to all forest species. Dendroecology can contribute to these studies by improving the analysis of tree growth response to environmental gradients, thereby refining the classifications that are based on climate-vegetation interactions. This approach was previously taken on 17 beech forests in central Italy (Latium and Abruzzi) to obtain horizontal and vertical gradients in tree-climate

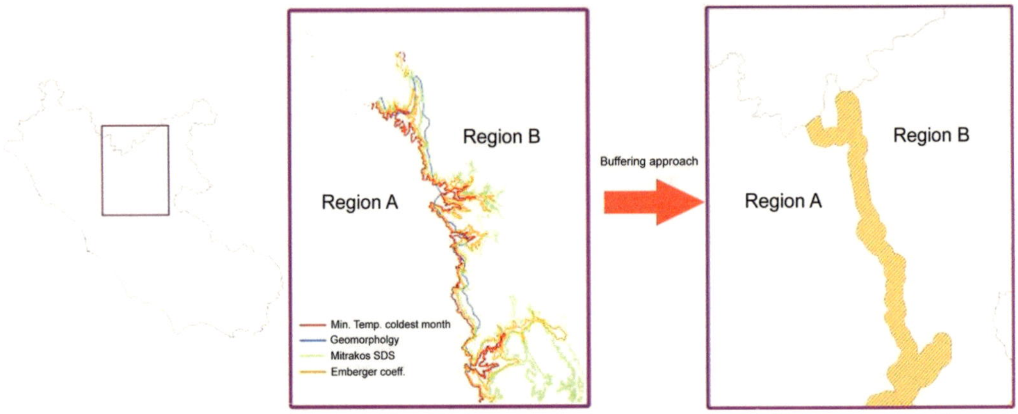

Fig. 1. Example of buffered boundary between two Regions of Provenance in Latium. The buffer zone represents a gradual transition of the ecological characteristics from Region A to B and vice-versa.

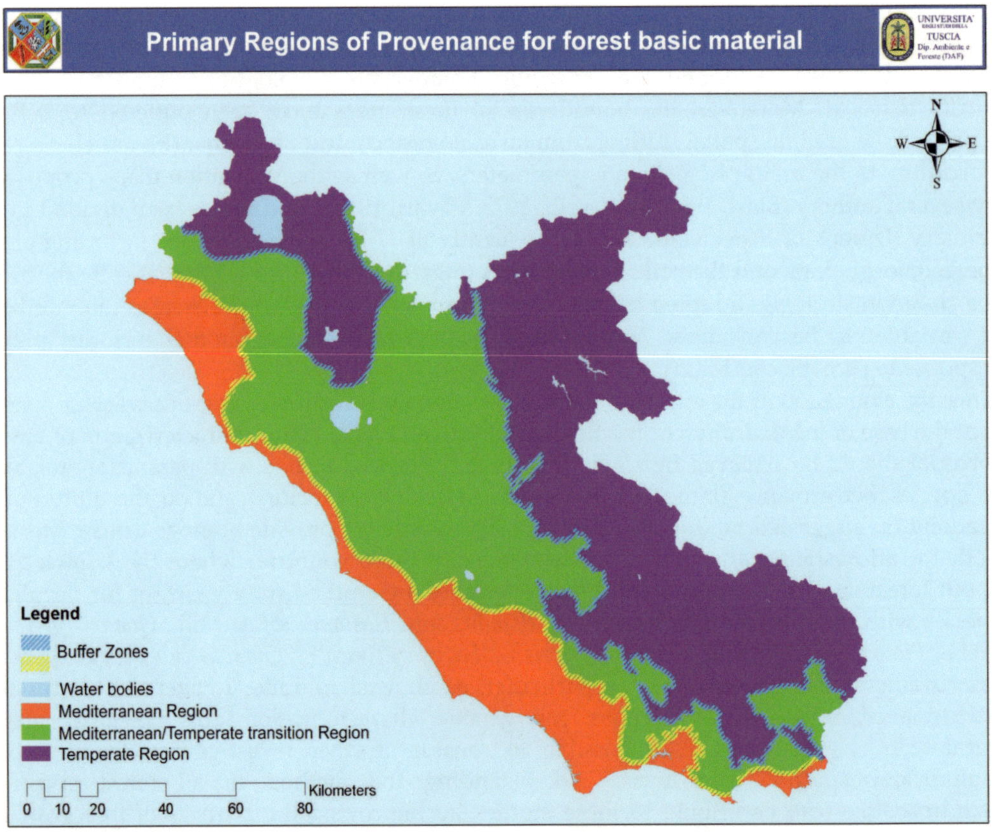

Fig. 2. Delineation of Primary Regions of Provenance for Latium using the improvements cited in the text.

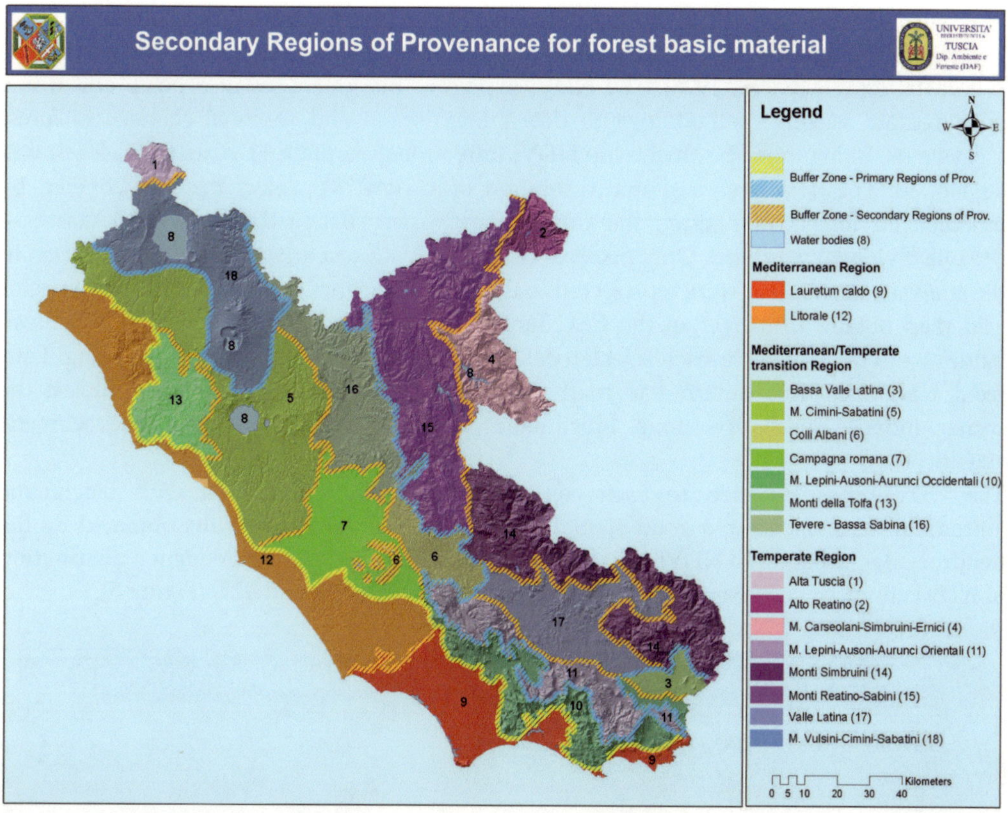

Fig. 3. Map of Secondary Regions of Provenance for Latium derived from phytoclimatic, vegetation and chemo-physical variables.

relationships, thus providing the basis to assess bioclimatic units in terms of the leading dendroclimatic signals (Piovesan et al., 2005). Evidence from tree-ring analysis demonstrated that tree growth is strictly related to elevation, generating distinct beech forest types. In agreement with previous studies (e.g. Biondi, 1992; Biondi & Visani, 1996; Dittmar et al., 2003), distinctive radial growth–climate relationships uncovered in the tree-ring network are organized along altitudinal and latitudinal gradients. Since beech could be considered a good bioindicator, i.e. its dendroecological features are significantly related to elevation, comparisons were extended to all the Latium forest surfaces, including the altitudinal belt where beech is not present in the landscape, i.e. a non-beech belt with bioclimatic traits that generally do not allow the growth of beech. The following results could be considered a starting point for the selection of basic material used in genetic and provenance studies, to accomplish the definition of the Regions of Provenance for Latium previously showed. It is a new approach that checks the agreement between the dendroclimatic classification and the phenological traits analyzed by remote sensing measurements, expressed by the normalized difference vegetation index (NDVI).

The NDVI allows decadal (10 day) monitoring of terrestrial vegetation, at regional to global scales, using the spectral reflectance measurements acquired in the red and near-infrared regions. These spectral reflectances are themselves ratios of the reflected over the incoming radiation in each spectral band. NDVI reflects the chlorophyll and carotenoid content in the leaves (Tucker & Sellers, 1986), but it is also related to leaf area index (LAI) (Fassnacht et al., 1997) and the fraction of photosynthetically active radiation absorbed by leaves (fPAR) (Veroustraete & Myneni, 1996). The NDVI expresses the greenness of a pixel, and it is a good remote sensing methodology to detect interannual and seasonal changes in forest ecosystems. Using the GIS software an NDVI time series spanning 11 years (1998–2008) was developed. The data have a spatial resolution of $1x1km^2$. To detect the area covered by broadleaved forests in Latium, the Corine Land Cover (CLC) database (3.1.1 classes—Broadleaf woods) was used. Only pixels ($1x1km^2$) with a forest area above 60% were used in the analysis. Raster data were reprojected to the same coordinate system of the subset vector grid map obtained, to overlap the CLC forest surface with satellite images. NDVI mean values were calculated for each selected cell and partitioned using k-mean clustering. Four fixed *a priori* clusters (referred to in the text as NDVI classes) were chosen to test the correspondence with the four bioclimatic zones obtained by the dendroclimatic classification.

The NDVI class assigned to each cell was graphically overlapped with bioclimatic altitudinal belts, showing a good spatial correspondence between results obtained by the dendroecological and the NDVI classification (Figure 4). The relative frequency distribution of NDVI clustered cells per altitudinal belt ranged between 61% and 92% (Figure 5).

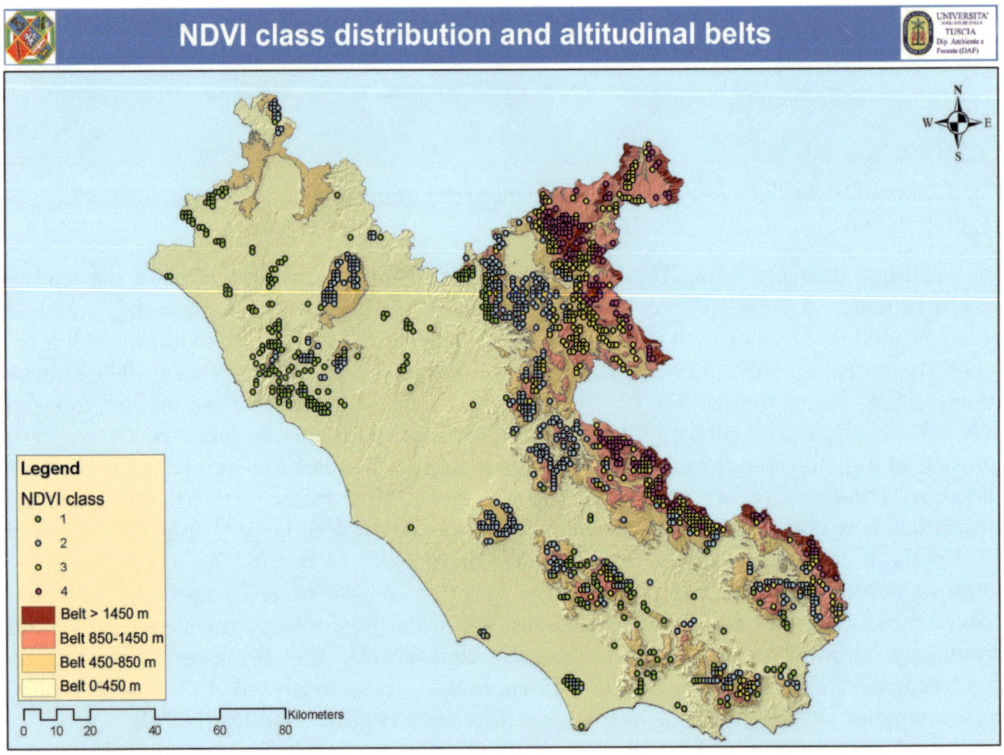

Fig. 4. Map of the spatial overlapping between normalized difference vegetation index (NDVI) classes and altitudinal belts detected using the dendroclimatic approach.

Physical parameters, such as aspect or edaphic conditions, could play a fundamental role where other non-correspondent NDVI class cells were present. This confirmed the general role of elevation as a key factor in controlling both the growth and phenological behaviour of forest stands in central Italy. The NDVI varied greatly among months and NDVI classes, stressing the difference in photosynthetic activities throughout the growing season of distinct forest bioclimatic belts; in particular the growing season length shortened according to increasing elevation (Figure 6). These results assess that it is possible to link tree-ring climate signals to phenology for each altitudinal belt by combining the two methods, adding important clues to the further comprehension and modelling of the bioclimatic organization of these forests. The two methods were mutually validated, and therefore would be useful in defining Regions of Provenance as agglomerative approach. The main benefit is in providing an automated approach at local spatial scale useful to map these regions. The coupled dendroecological application and NDVI can offer a prompt, economic and operative tool to check and manage homogeneous ecological areas, objectively identifying Regions of Provenance according to plant responses. Moreover, this approach could be combined with other biological and genetic parameters, e.g. growth performance, resistance to diseases, DNA markers, for a wider scenario of species' behaviour. At the same time, matching the full dataset of ecological, biological and genetic variables a more completed delineation of Regions of Provenance could be achieved (Alessandrini et al., 2010).

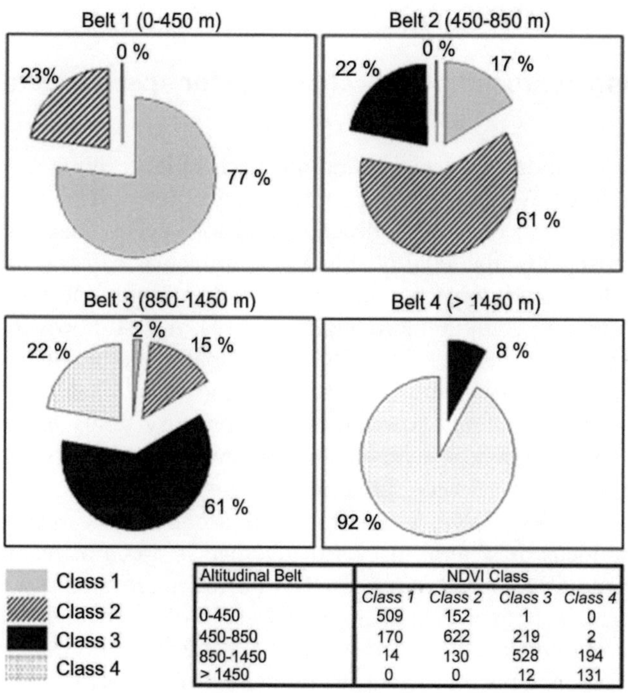

Altitudinal Belt	NDVI Class			
	Class 1	Class 2	Class 3	Class 4
0-450	509	152	1	0
450-850	170	622	219	2
850-1450	14	130	528	194
> 1450	0	0	12	131

Fig. 5. Pie charts of percentage correspondence between assigned normalized difference vegetation index (NDVI) classes and tree-ring altitudinal belts. The panel below the charts shows the number of cells per class and belt.

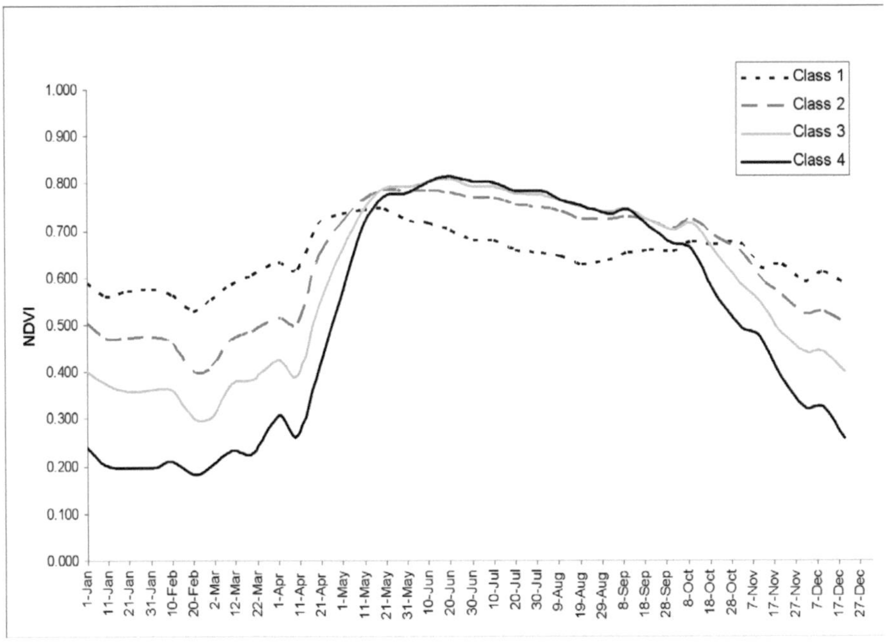

Fig. 6. Annual normalized difference vegetation index (NDVI) profile of the four classes obtained by k-means clustering.

3. DNA barcoding approach: A new challenge for species identification and conservation

DNA barcoding is a standardized molecular approach to label living organisms by joining specific sequences of DNA and electronic information retrieval (Hebert et al., 2003), and it has recently become an increasingly attractive tool for species identification in terms of accuracy, speed, cost and functionality. Ideally, a universal barcode system would be a valuable resource to provide objective and worldwide comparable results, which can be efficiently used in turn to compile biodiversity surveys in local floras (Lahaye et al., 2008; Gonzalez et al., 2009; Kress et al., 2009, 2010). Additionally, the method allows the analysis of poor, fragmented samples at any life stage (Chase et al., 2005) and it can be easily repeated even by non-taxonomist specialists. The primary goals of barcoding are thus species identification of known specimens and discovery of unnoticed species to enhance taxonomy for the benefit of science and society (Kress & Erickson, 2008). The term "DNA barcode" refers to a short DNA sequence-based identification system which may be constructed of one locus or several loci used together as a complementary unit (Kress & Erickson, 2007). Necessary prerequisites of DNA barcodes are ease of application across a broad range of *taxa*, sufficient sequence variation to distinguish between species, and absence of intra- and inter-specific diversity overlaps which would prevent rank definition. Many studies have proved the efficacy of the mitochondrial cytochrome c oxidase 1 (COI or cox1) gene sequence in barcoding animal groups such as birds (Hebert et al., 2004), fishes (Ward et al., 2005), spiders (Greenstone et al., 2005), lepidopterans (Hajibabaei et al., 2006), and amphibians (Smith et al., 2008), as well as in red algae (Robba et al., 2006) and fungi (Seifert et al., 2007). In plants, the difficulty of finding a single-locus barcode has suggested a

multilocus approach, focusing on the plastid genome as currently the most effective strategy (see Hollingsworth et al., 2009, and citations therein), although there is still much debate concerning the most suitable regions to be used. From the broad pool of loci recently considered (Kress et al., 2005; Chase et al., 2007; Newmaster et al., 2008; Ford et al., 2009), the greatest interest was aroused by seven candidate plastid loci: rpoB, rpoC1 and rbcL (three easy-to-align coding regions), a section of matK (a rapidly evolving coding region), and trnH-psbA, atpF-atpH and psbK-psbI (three rapidly evolving intergenic spacers).
Various biological contexts (e.g., sampling strategies) have been used to compare the performance of plant barcoding loci, and/or the efficacy of the method. A sound assessment of the universality of regions is usually given by the "species pairs" and "floristic" approaches. The former involves analysing pairs of related species from multiple phylogenetically divergent genera and may be defined as a "methodological" protocol; the latter involves sampling multiple species within a given geographical area and represents an example of how barcoding might be applied in practice. However, only limited insights into species-level resolution is usually provided by both approaches, as individual genera are not sampled in depth to provide estimates of intra- and interspecific genetic distances to achieve species identification. Conversely, a third method, the "*taxon*-based" approach, involves sampling multiple species within a given taxonomic group, in a global geographical context. This provides limited insights into universality and local applicability, but offers more definitive information on discrimination power at species level. To date, the species pairs (e.g. Kress et al., 2005; Kress & Erickson, 2007), and the *taxon*-based (e.g. Newmaster et al., 2008; Newmaster & Ragupathy, 2009) sampling designs have provided useful insights into the potential performance of varying combinations of barcoding loci, whereas the floristic approach (e.g. Fazekas et al., 2008; Lahaye et al., 2008; Gonzalez et al., 2009), has showed strong potential applicability in as many diverse research fields as biodiversity inventories, community assembly, food and medicine identification, ethno- and forensic botany. Based on the relative ease of amplification, sequencing, multi-alignment, and on the amount of variation displayed (sufficient to discriminate among sister species without affecting their correct assignation through intra-specific variation), the most frequently recommended marker combinations for broad future applications appear to be: rbcL + trnH-psbA (Kress & Erickson, 2007), matK + trnH-psbA (Newmaster et al., 2008; Lahaye et al., 2008), rbcL + trnH-psbA + matK, and rbcL + matK (Consortium for the barcode of Life, Plant Working Group [CBOL PWG], 2009).

3.1 DNA Barcoding of forest tree species

In forestry science, DNA barcodes is highly promising for the detection, monitoring and management of biodiversity (von Crautlein et al., 2011). In addition to resolving many taxonomic uncertainties, enhancing clear and more accurate biodiversity assessments, DNA barcoding may provide a boost to efficient management and conservation practices, mainly focusing on community ecology (to describe plant-animal interactions and vegetation dynamics/changes, to discriminate native vs. alien germplasm), biodiversity inventories (aimed at habitat and species protection), silviculture (to assess forest regeneration), and nursery activities and market regulation (to establish wood, secondary products and germplasm certification). The applications might be particularly relevant to manage correctly the over-exploited and the newly identified species, to adequately protect those having limited ranges and relatively small population sizes, as well as for mending damaged landscapes by planning and monitor congruent reforestation programmes.

Indeed, one of the future challenges for DNA barcoding in plants is to increase the number of practical studies, and validation of the method for forestry purposes is still to be demonstrated. Priority should be given to the use of markers with universal primers and uniform PCR conditions. Under these criteria, the most updated recommendation from the CBOL PWG is that rbcL+matK is adopted as the core DNA barcode for land plants (CBOL PWG, 2009), with trnH-psbA (the next best performing plastid locus) as a supplementary barcode option for difficult plant groups. However, success in angiosperms is often perceived by the majority as the most important issue. For gymnosperms (and cryptogams) the universality criterion has received little consideration up to date, and clade specific/multiple primer sets were often used to evaluate matK and other putative barcode markers (including rbcL and rpoC1). For instance, in the few currently available gymnosperm-based barcoding studies, only 24% PCR success was obtained in Cycads (Sass et al., 2007) with matK universal primers, whereas Hollingsworth et al. (2009) and Ran et al. (2010) obtained 100% PCR and sequencing success in *Araucaria* and *Picea* by use of a combined set of specific primers and under non-standard PCR conditions. More recently, a *taxon* based study on *Taxus* was attempted with new matK specific primers (Liu et al., 2011). Clearly, matK universality across both gymnosperms and angiosperms is still a matter of concern, while rbcL and trnH-psbA have repeatedly shown strong rates of sequence recovery in both clades but their use still requires some technical adjustments (see for instance Hollingsworth et al., 2009).

The efficacy of the method is still under question, too. Pooled sequence data from 445 angiosperm, 38 gymnosperm, and 67 cryptogam species indicated that overall species discrimination was successful in 72% of cases (CBOL PWG, 2009), in agreement with the upper limit of ca. 70% resolution pointed out in previous studies (Fazekas et al., 2009; Hollingsworth et al., 2009). Large-scale plant diversity inventories conducted at a local or regional context matched this limit or revealed even higher percentages, although absence/scarcity of gymnosperms in their datasets is still noticeable. Irrespective to the statistical methods used to cluster sequences into taxonomic units, and to the marker combinations used, <70% of species resolution was achieved on 254 angiosperm species from an environmental sampling in Amazonia (Gonzalez et al., 2009), ca. 71% on 92 primarily angiosperm species (including 7 conifers) from selected locations of Southern Ontario (Fazekas et al., 2008), ca. 90% on 32 angiosperm species and over 1000 orchid species from two national parks (Lahaye et al., 2008), and 93-98% on 143 and 296 angiosperm species in community studies in tropical forest dynamics plots in Puerto Rico and Panama (Kress et al., 2009, 2010). However, it has been shown that woody plant lineages have consistently lower rates of molecular evolution as compared with herbaceous plant lineages (Smith & Donoghue, 2008), suggesting that the application of DNA barcoding concepts should be more difficult for tree than for non-woody floras (Fazekas et al., 2009). Moreover, the discrimination rate of plastid barcoding loci varies greatly among different plant lineages. In tree species, no resolution was achieved in 12 *Quercus* (Piredda et al., 2011), 18 *Betula* and 26 *Salix* species (von Crautlein et al., 2011), whereas 30%, 63% and 100% were achieved in *Berberis* (16 species), *Alnus* (26 species), and *Compsoneura* (8 species), respectively (Roy et al., 2011; Ren et al., 2010; Newmaster et al., 2008). In Gymnosperms, all extant five *Taxus* species (Liu et al., 2011) were fully discriminated with a non-standard barcode (trnL-F); in 32 *Picea* species (Ran et al., 2010), the highest rate of successful discrimination was 28.57% for a three-locus barcode (trnH-psbA, matK, atpF-atpH). A slightly higher percentage was obtained by Hollingsworth et al. (2009) in *Araucaria* (32%).

Available data show that some limitations are predictable, matching the view of Fazekas et al. (2009). Limitations are mostly due to polyploidy, hybridization/introgression phenomena, shares of ancestral polymorphism, which would prevent the correct match between DNA variation at the plastid level and species identity. Such phenomena probably affect many tree species; in addition, trees are known to have markedly slower mutation, nucleotide substitution and speciation rates than other plants, seemingly owing to longer generation times and slower metabolic rates (see Petit & Hampe, 2006 for a review). At the same time, biogeographic patterns of species, lineages and area relationships can strongly affect the resolution of *taxa*. Together with this assumption, the barcoding efficiency of tree *taxa* is still to be demonstrated, and it appears to be most hardly challenged by the peculiar evolutionary history and intrinsic biology of each *taxon*, and in those areas where recent explosive radiations have taken place, or where a high number of only slightly diversified congenrics co-exist.

3.2 Barcode application in the Italian flora

A summary of explorative data on the foreseeable barcoding efficacy in the Mediterranean area, with specific regard to Italian forest flora is reported in Table 2. With the aim to provide a test for future *in situ* applications of DNA barcodes by evaluating the efficacy of species discrimination under the criteria of uniformity of methods and natural co-occurrence of the species in the main forest ecosystems, we examined whether four marker regions (trnh-psba, rbcL, rpoc1, matK) proposed by the Consortium for the Barcode Of Life matched species taxonomy in a biodiversity survey of Italian forested land.

Seventy-eight species were included in a floristic study, including 53 Angiosperm and 25 gymnosperm species (trees, shrubs and vines from the Alpine timberline to the Mediterranean sea dunes; 68 native and 10 introduced/naturalized *taxa*); in addition, taxon-based studies were performed on Quercus (15 species, 30 individuals), Acer (8 species, 15 individuals) and Pinus (10 species, 30 individuals). individuals) and Pinus (10 species, 30 individuals). We observed total universality of the rbcL+trnH-psbA marker combination across all *taxa*, and an overall 78.4% of species discrimination, with 100% in gymnosperms and 66.7% in Angiosperms,whereas matK and rpoC1 showed incomplete, or limited, applicability due to some primer specificity, Differences in the biology/evolutionary history of tree genera are represented by the contrasting results obtained in the three taxon-based studies: Quercus exhibited an exceptional 0% of species resolution, whereas Acer and Pinus reached 100% discrimination success. As a main result, the barcoding approach provided molecular tools for the identification of all *taxa* co-occurring in most of the Italian forest ecosystems, from the Alpine timberline, to montane, submontane, humid/riparian, Mediterranean evergreen forest/maquis and sea dunes, including some ubiquitous vines and shrubs, with the exception of oaks and willows. The approach was also useful for the molecular identification of all the rare endemics investigated (*Fontanesia phylliraeoides, Acer lobelii, Abies nebrodensis, Pinus heldreichii* ssp. *leucodermis*), and all native vs. allochtonous germplasm (*Aesculus hippocastanum,* Quercus *rubra,* Acer *negundo, Abies pinsapo, A. cephalonica, Pinus radiata, P. brutia, Cupressus arizonica, Pseudotsuga menziesii, Gingko biloba*). Concerning the intraspecific *taxa*, ssp. *nigra* was clearly differentiated from all other *Pinus nigra* subspecies, as well as ssp. *turbinata* within *Juniperus phoenicea*. Lastly, two vines and four shrubs were efficiently discriminated from co-occurring arboreal *taxa*. Investigated *taxa* could be efficiently barcoded in most ecosystems, with the exception of those forests where a high number of willows and oak species co-occurred.

Among the species-rich genera, those which would benefit most from molecular identification (*Quercus, Salix*) because of their complex morphology, showed little or no variation at the plastid genome. Remarkably, none of the markers used could resolve 12 Italian *Quercus* species below the sectional level (i.e., Sclerophyllodrys, Cerris and Quercus), due to large haplotype sharing between closely related species. On the other hand, intra-specific variation in Italian conifers appears to correspond to some regional patterns reflecting important prints of species survival during glaciations and post-glacial recolonization (Follieri, 2010). Specific haplotypes were found in Southern Italy (Apulia), Central Italy (Tuscany, Latium), Northern Italy (Eastern and Western Alps), and main islands (Sicily), all falling within the 52 biodiversity refugia recently indicated on a regional scale in the Mediterranean basin (Medail & Diadema, 2009). Variation in the barcoding loci also evidenced the occurrence of two distinct haplotypes of *Taxus baccata* in Italy, one shared with other European provenances and a second exclusive of South East Italy. Finally, our results confirmed the genetic diversity existing between Southern and Central Italy provenances of *Cupressus sempervirens* (Bagnoli et al., 2009), and divergence between Eastern and Western Alps provenances of *Picea abies* (Collignon & Favre, 2000), as well as between Eastern and Western Mediterranean provenances of *P. halepensis* (Korol et al., 2002), all previously detected with other molecular markers.

Major Clade	Familia	Genus	Species in Italy	Species investigated	Species identification	Notes
Angiosperms	Aceraceae	Acer	7	8*	Yes	Possible haplotype sharing between A. obtusatum and A. monspessulanum
	Oleaceae	Ligustrum	1	1	Yes	
		Olea	1	1	Yes	
		Fraxinus	3	3	Yes	
		Phyllirea	3	3	Yes	Possible haplotype sharing between P. angustifolia and P. latifolia
		Fontanesia	1	1	Yes	
	Fagaceae	Fagus	1	1	Yes	
		Castanea	1	1	Yes	
		Quercus	10-14	15*	No	No species resolution at National scale
	Salicaceae	Populus	3	2	Yes	Possible haplotype sharing between P. nigra and P. alba
		Salix	>30	2	No	No species resolution at National scale (**)
	Ulmaceae	Ulmus	3	1	Yes	
	Rosaceae	Prunus	9	1	Yes	Possible haplotype sharing
		Craetegus	2-3	1	Yes	Possible haplotype sharing
		Rosa	>20	2	Yes	Possible haplotype

Major Clade	Familia	Genus	Species in Italy	Species investigated	Species identification	Notes
						sharing
		Rubus	>20	2	Yes	Possible haplotype sharing
	Betulaceae	Corylus	1	1	Yes	
		Alnus	4	1	Yes	
	Araliaceae	Hedera	1	1	Yes	
	Sapindaceae	Aesculus	0	1*	Yes	
	Cannabaceae	Humulus	1	1	Yes	
	Moraceae	Ficus	1	1	Yes	
		Morus	0	1*	Yes	
	Tamaricaceae	Tamarix	10	1	Yes	
Gimnosperms	Pinaceae	Pinus	8	10*	Yes	Possible haplotype sharing between P. mugo and P. sylvestris
		Larix	1	1	Yes	
		Pseudotsuga	0	1*	Yes	
		Abies	2	4*	Yes	
		Picea	1	1	Yes	No species resolution at National scale (**)
	Cupressaceae	Juniperus	4	4	Yes	
		Cupressus	1	2*	Yes	
	Taxaceae	Taxus	1	1	Yes	
	Gingkoaceae	Gingko	0	1*	Yes	

Table 2. Barcoding efficacy on some of the most important tree species in Italy. Asterisk indicate non native species included (*), and results implemented with literature data (**).

We therefore conclude that, despite some failures, the DNA barcoding approach will continue to be useful in some applications, especially when applied at local contexts, with some plant groups and for some peculiar investigations. Ideally, an important technological advancement to improve the method would include the achievement of primer universality for the main plastid markers, and eventually the opportunity to cope information from both organellar DNA and the more informative nuclear genome.

Organisms identification is essential to many disciplines, and the scientific community has recently come to realize the importance of integrated approaches to organism identification (Steele & Pires, 2011). Indeed, conservation planners and government agencies would need well defined species boundaries to protect ecosystems and writing effective laws (Primack, 2008), and restoration ecologists must accurately identify native plant species suitable for rebuilding damaged ecosystems (Guerrant et al., 2004). As well, conservation biologists must be able to correctly identify plant species for fighting invasive, reseeding restoration areas with appropriate species, monitor the regeneration processes of a community after their intervention, protecting native and/or threatened ecosystems by preserving all life forms. Finally, seed harvesters and germplasm traders must ensure the end-users that the right species are produced before distribution to the public. Nevertheless, the role that DNA barcoding might play in these views still relies heavily on experimentation and tests.

Our data suggest that forest biodiversity can be efficiently barcoded at a local level, or in well characterized regions of the world which have comparatively low numbers of species; conversely, the barcoding efficiency of tree *taxa* might rather be under question in large areas where peculiar genera (e.g., *Betula, Quercus, Salix,* etc.) occur with multiple species. Future large breadth *taxon*-based studies will help clarify the efficacy of DNA barcoding to inspect the biological diversity of forest tree species. However, factors suggested to contribute toward limiting the efficacy of barcoding tree species such as longevity, complex reproductive strategies, and slow mutation and speciation rates (Petit & Hampe, 2006) may not affect the barcoding efficacy at a local context.

4. From conservation to restoration: The Miyawaki method

It is widely known that global climatic changes, together with recent rapid urbanization and industrialization, have been the main anthropogenic effects worldwide in destroying natural environments, changing land use, reducing biodiversity, and modifying ecosystems. They suggest the need for performing more environmental conservation strategies, as well as using innovative environmental recovery activities. We have seen in the Introduction as *in situ* gene conservation measures ecosystem functions and species interactions, rather than individual tree species; however their conservation may require specific management measures, which could be ensured through the establishment of genetic conservation areas. From a theoretical point of view, a network of genetic resource conservation areas should be an efficient way to conserve the genetic resources of target species, if they follow the patterns of distribution of genetic variation (Eriksson et al., 1995). Practical experience suggests that sound management of genetic resources must include conservation efforts based on two overlapping strategies: management of natural forests with due respect to their genetic resources, and the establishment of networks of smaller gene conservation areas (Thomson et al., 2001). Nevertheless, it should be remembered that *in situ* conservation is only a technical option in a broader approach to conservation of the diversity between species and within species. In several cases, conserving forest trees *in situ* may be the only method that is socially and economically possible. In other cases, a combination of protected areas, managed reserves, clone banks, research plantations and breeding programmes may be better suited to different conditions and objectives.

In the last years, the greatest challenge is to move from the conservation of existing resources, toward a rationale restoration ecology, increasing efforts to rehabilitate degraded lands. Often the preliminary objective is to re-establish tree cover for environmental purposes, especially for control of soil erosion and for watershed protection. Facing these items, scientists have developed new insights both in theoretical and in practical actions for restoration and reconstruction of natural ecosystems (Clewell & Aronson, 2008; Falk et al., 2006; Jordan et al., 1987; Perrow & Davy 2002a, b; Soulé & Wilcox, 1980; Miyawaki, 1975, 1981). Natural restoration is strictly related to increased sustainability and includes rehabilitation of ecosystem functions, enlargement of specific ecosystems, and enhancement of biodiversity restoration (Stanturf & Madsen, 2004). At the ecological level, restoration is also defined as "an intentional activity that initiates or accelerates recovery of an ecosystem with respect to its health, integrity and sustainability" (Society for Ecological Restoration International Science & Policy Working Group [SER], 2002). Degraded plant communities are generally quite difficult or sometimes impossible to restore (Van Diggelen & Marrs, 2003). More than 200 years of reforestation practice has demonstrated that forest recovery takes a very long time, frequently with unsatisfying results. Nowadays, it is possible to plant plantations of several species, but the transition from the simple plantation to a forest community able to evolve and sustain itself, according to the natural successional pattern, is still a rare event. Moreover, a number of "regreening" projects in the past have paid scant attention to the source of planting materials used and their biological requirements, and have failed because of poor species choice. Use of non-local seed sources of indigenous species can result in the contamination of genepools of nearby populations (Thomson, 2001). On the other hand, the mere superficial appearance of vegetation restoration should be avoided. It is essential to restore the natural vegetation using a combination of native species that conform to the potential trend of the habitat and to try to restore the whole specific ecosystem of a region (Miyawaki, 1992). Currently, most forest reforestation programs adopt a scheme of planting one or more early successional species; after successful establishment, they are gradually replaced by intermediate species (either naturally or by planting), until late successional species arise. This pattern tries to simulate natural processes of ecological succession, from pioneer species to climax vegetation. However, it requires several silvicutural practices and normally takes a long time; because we live in a world where industry and urbanization are developing very rapidly, improvement of an alternative reforestation technique that reduces these times could be a useful tool (Miyawaki, 1999).

One reliable forest restoration method is the "native forests by native trees", based on the vegetation-ecological theories (Miyawaki, 1993a, b, 1996, 1998b; Miyawaki & Golley, 1993; Miyawaki et al., 1993; Padilla & Pugnaire, 2006) proposed by Prof. Akira Miyawaki and applied first in Japan. Restoring native green environments, multilayer forests, and natural biocoenosis is possible, and well-developed ecosystems can be quickly established because of the simultaneous use of intermediate and late successional species in plantations. The Miyawaki method involves surveying the potential natural vegetation (sensu Tüxen, 1956) of the area to be reforested and recovering topsoil to a depth of 20-30 cm by mixing the soil and a compost from organic materials. In this way, the time of the natural process of soil evolution, established by the vegetational succession itself, is reduced. Tree species must be chosen from the forest communities of the region in order to restore multilayer natural or quasi-natural forests. For a correct choice, based on reconstructing the potential natural vegetation, several analyses (e.g., phytosociological investigation) are required. Detection of the soil profile, topography, and land utilization can improve our grasp of the potential natural vegetation. After these field surveys, all intermediate and late successional species are mixed and densely planted, with as many companion species as possible (Kelty, 2006; Miyawaki, 1998a), and soil between them is mulched. In fact, biocoenotic relationships involve autoregulations between species, favouring a dynamic equilibrium and avoiding any further silvicultural practice and need no insecticides or herbicides (with some exceptions). Indeed, in the Miyawaki method, the principles of self-organized criticality and cooperation theories have been essentially applied (Bak et al., 1988; Callaway, 1997; Camazine et al., 2003; Padilla & Pugnaire, 2006; Sachs et al., 2004). If compared to traditional methods, some known restrictions regard the requirement of specialists for botanical and ecological investigation of the sites, a higher need of manpower for planting, and higher costs of plant material due to the plant density. On the other hand, no human care is required after 1-2 years from planting, the undergrowth with late-successional species is immediately on site, and forest stands become quickly part of the natural ecosystems. Moreover, the theoretical principle at the base of the definition of Regions of Provenance might be considered almost included in the Miyawaki method, as it suggests to use seed from the nearest natural populations. Figure 7 shows a schematic overview of the comparison between classical succession theory and the one proposed by Miyawaki.

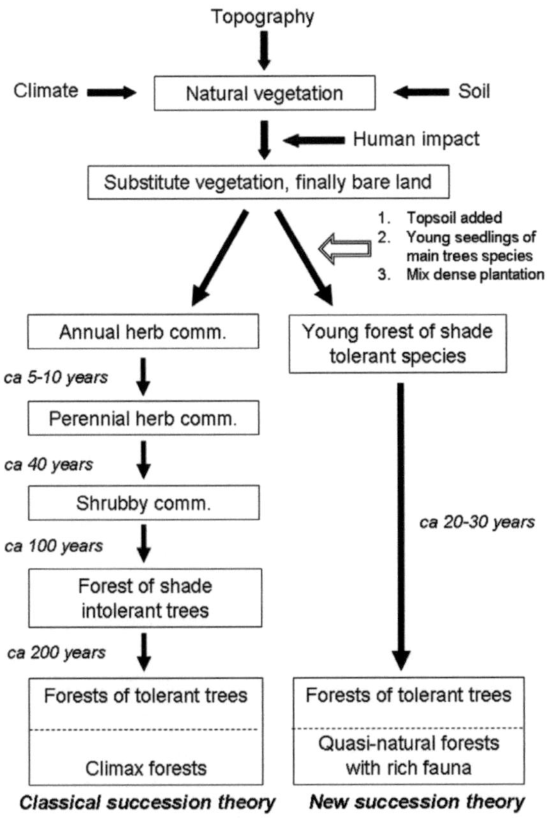

Fig. 7. Comparison between classical succession theory and the new one proposed by Miyawaki (redrawn from Miyawaki, 1999).

4.1 The adaptability of Miyawaki method to the Mediterranean environment: a case study

It has been demonstrated that multilayer quasi-natural forests can be built in 15–20 years in Japan and 40–50 years in Southeast Asia by ecological reforestation based on the system of natural forests. Results obtained by application of the Miyawaki method in about 550 locations in Japan, as well as in Malaysia, Southeast Asia, Brazil, Chile, and in some areas of China, were found to be successful, allowing quick environmental restorations of strongly degraded areas (Miyawaki, 1989, 1999). Until now, the Miyawaki method has been applied in countries characterized by cold-temperate and tropical climatic regimes, which do not experience seasonality, i.e. winter cold and summer aridity stress (cf. Mitrakos, 1980) with potential risk of desertification (increased by global change). Thus, the Mediterranean context could be considered an interesting test to assure the effectiveness of such a method in other important biomes, even with high biodiversity hotspots. Nevertheless, it could be interesting for the Mediterranean Basin, because complete environment restoration takes longer time than in tropical or cold-temperate climates. To estimate the effectiveness of

Miyawaki method in such different circumstances without altering its theoretical principles, several changes were introduced and tested in two experimental plots in Sardinia (Italy) in 1997, on target sites where traditional reforestation approaches are widely used but have mostly failed (Schirone, 1998). First, the soil condition of the planting sites was not adjusted, so no recovery of the 20-30 centimetre-deep topsoil with compost from organic materials has been done, but only a labouring of the pre-existent soil. Tillage was used to improve soil water storage over the winter and reduce water stress during the summer. Between the selected species, some autochthonous early-successional ones were planted (e.g. *Pinus pinaster* L., and shrubs) to improve plant community resilience, and no weeding after planting was done. Mulching was provided experimenting straw as in the original method, but also other types of materials (saw mill residuals, dry and green materials), and tested planting densities were assessed to 8600 and 21000 plants/hectare respectively. A particular care was dedicated to the choice of the best planting season, and watering was provided once soon after planting. Figure 8 summarizes the Miyawaki method as implemented in the mentioned experiments.

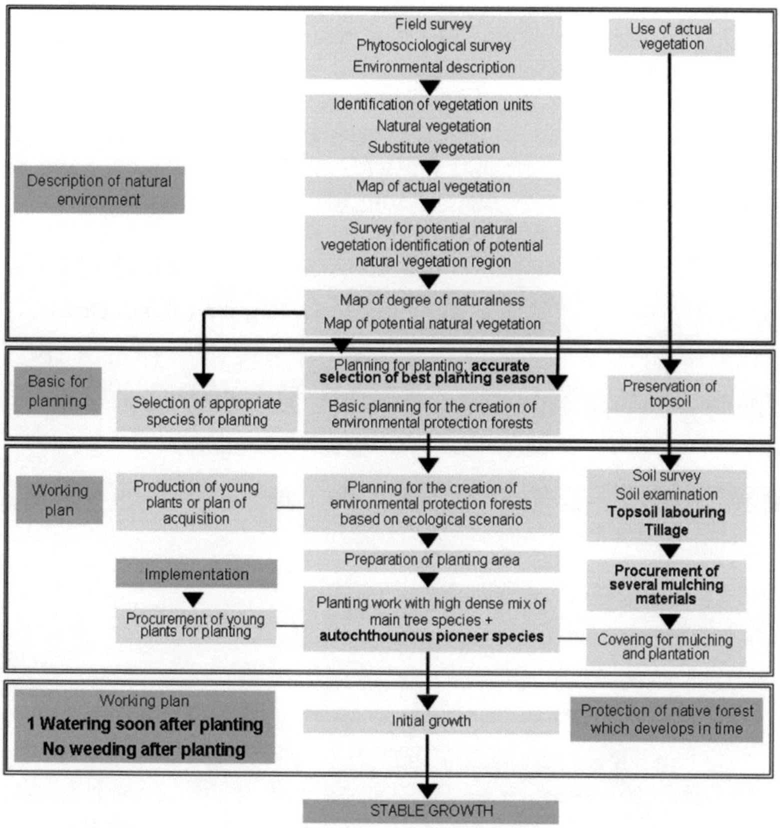

Fig. 8. Schematic overview of the Miyawaki method modified for Mediterranean environment. Dark grey text boxes describe main processes; bold texts refer to the changes to the original method.

To estimate the efficiency of this adapted Miyawaki method to Mediterranean, three surveys were performed in 1998, 1999, and 2009 in both experimental plots. Moreover, comparisons were done with two nearby coeval sites where traditional reforestation techniques were applied to better understand the differences in plants growth, forest composition, and vegetation cover in percentage (Schirone et al., 2011). The results after 12 years from the planting showed a more rapid development of trees on the Miyawaki plots, in particular early-successional species, as well as a stable assessment of species' occurrence with high level of biodiversity (Table 3). The benefits over previous methods are remarkable and comparable with those obtained by Miyawaki in Asia and South America. At the same time, the changes made to better fit the method to the Mediterranean environment seem to be particularly useful. For instance, adding some autochthonous early successional species to the intermediate- and late-successional ones the system resilience was improved; this solution was already tested by Miyawaki in Brazil, even if no benefits were recorded (Miyawaki & Abe, 2004). Looking for an optimal high plant density, it was assessed that cooperative processes (e.g. mutual shading) prevail over competitive ones (Callaway, 1997). In fact, low plant density has been traditionally retained as appropriate in arid and semiarid environments in order to avoid competition for water resources between plants (Caramalli, 1973; Bernetti, 1995), but a higher one reduces, for instance, the impact of acorn predators, thus encouraging oak regeneration, i.e., the main late-successional forest species in Mediterranean environments (Gómez et al., 2003); high plant density can also favour root anastomosis processes, that seem to influence coenosis' stability and reforestation success (Kramer & Kozlowski, 1979). In addition, excellent plant stock remains fundamental for planting success in harsh environments (Palacios et al., 2009). Finally, these results could offer a chance to introduce a new method into the Mediterranean context that is able to reduce the time for a complete environmental restoration.

	Species survival in Miyawaki sites			**Height ± (Stand. Dev.)**			
Species	n_i	n_f	n_f/n_i (%)	MS-1	MS-2	TRS-1	TRS-2
Acer monspessulanum L.	51	2	3.92%	40 ± (14.1)	0	-	-
Arbutus unedo L.	61	41	67.21%	32.7 ± (4.1)	0	500 ± (35.8)	110 ± (20.6)
Castanea sativa Mill.	42	1	2.38%	10	0	-	-
Cedrus atlantica Endl.	-	-	-	-	-	-	162 ± (54.6)
Celtis australis L.	59	3	5.08%	26.7 ± (28.9)	-	-	-
Erica arborea L.	-	-	-	-	-	115 ± (12.7)	130 ± (18.6)
Fraxinus ornus L.	17	1	5.88%	250	-	-	-
Ilex aquifolium L.	237	23	9.70%	45.2 ± (30.6)	0	-	-
Juniperus oxicedrus L.	45	30	66.67%	-	36.2 ± (18.5)	-	-
Laurus nobilis L.	41	3	7.32%	30 ± (17.3)	0	-	-
Ligustrum vulgare L.	139	33	23.74%	32.8 ± (52.6)	30 ± (8.16)	-	-
Malus domestica Borkh.	40	7	17.50%	100 ± (45.5)	0	-	-

Myrtus communis L.	114	5	4.39%	10	10 ± (1.4)	-	-
Phyllirea angustifolia L.	1	1	100.00%	70	0	-	-
Phyllirea latifolia L.	203	0	0.00%	-	0	-	-
Pinus pinaster L.	428	288	67.29%	433.2 ± (143.6)	325.5 ± (38.6)	376.4 ± (73)	425.7 ± (25.1)
Pyrus communis L.	41	20	48.78%	71 ± (65.1)	60 ± (61.2)	-	-
Quercus ilex L.	694	255	36.74%	34.2 ± (32.1)	40.8 ± (36.2)	69.4 ± (23.2)	146.2 ± (38.1)
Quercus pubescens Willd.	361	124	34.35%	23.6 ± (27.5)	10 ± (5.3)	-	-
Quercus suber L.	632	103	16.30%	174.3 ± (49.6)	77.5 ± (51.9)	-	-
Rosmarinus officinalis L.	46	15	32.61%	89.3 ± (33.9)	0	-	80 ± (14.9)
Salvia officinalis L.	9	0	0.00%	0	0	-	-
Sorbus torminalis (L.) Crantz	42	12	28.57%	35 ± (50)	40 ± (12.9)	-	-
Spartium junceum L.	74	29	39.19%	110.7 ± (62.2)	0	-	-
Taxus baccata L.	377	9	2.39%	33.3 ± (38)	0	-	-
Thymus vulgaris L.	24	0	0.00%	-	0	-	-
Viburnum tinus L.	84	3	3.57%	10	0	-	-

Table 3. Total number of individuals in the Miyawaki sites, at the beginning of the experiment (n_i, 1997), after 12 years (n_f, 2009), percentage of species' survival (n_f/n_i), and comparison of plant height (cm) between Miyawaki sites (MS-1, MS-2) and the traditional reforested ones (TRS-1, TRS-2) in 2009. Dashes indicate species not planted, and *zero values* refer to planted species that did not survive in 2009. Successional position of each species is indicated by the row color: white (early successional), light grey (middle-successional), dark grey (late-successional).

5. Conclusion

The conservation of biodiversity has become a major concern for resource managers and conservationists worldwide, and it is one of the foundation principles of ecologically sustainable forestry (Carey & Curtis, 1996; Hunter, 1999). Many efforts were dedicated to set aside networks of reserves and protected areas advocated by scientists, governments, etc. to preserve the extraordinary biodiversity that characterizes forest ecosystems, perpetuating their integrity, their evolutionary patterns and yet providing social and environmental benefit . At the same time, a strategic value has been assigned also to biodiversity in terms of genetic resources, through the conservation of plant populations in their natural habitats (*in situ*) to better evolve and adapt to physical environmental trends and to changes in the web of interactions with other life forms. Generally, the simplest way forward in economic and political terms is for countries to locate genetic resources in existing protected areas, as this likely to provide benefits to local people communities. However, despite the critical role of conservation sites, a large debate arose about the combination of protection, management, and restoration of forests and woodland landscapes as pivotal starting points of sustainable development in many of the world's ecoregions (e.g. Pierce et al., 2003; Norton, 2003; Aldrich et al., 2004; Loucks et al., 2004). At pan-European level, several legislative tools emphasized the need of facing habitat fragmentation, biodiversity loss, genetic pollution, and invasive species use, throughout the definition of certified basic material and ecologically homogeneous areas.

Some strategies have been included in the Directive 105/99, with the definition of Regions of Provenance and the requirements for an appropriate marketing of forest reproductive material. Unfortunatently, there was an heterogeneous achievement of the Directive by the European countries in time, as well as in adopting common methods. Mainly according to the available data, the chosen parameters for detecting the Regions of Provenance differed case by case. However, it is also evident that both agglomerative and divisive approaches could be improved by adding further variables and/or methods. Nowadays, the need for models implemented with biological parameters is suggested by a changing climate, in which bioclimatic shifts could characterize vegetation arranged along altitudinal gradients or at ecotonal boundaries (e.g. Peñuelas & Boada, 2003; Steltzer & Post, 2009). Data analysis at different temporal scales could allow to understand the effects of climate trends on species success and survival, and thus to choose the most appropriate genetic material for reforestation actions. In this view, genetic approaches must certainly be refined and made uniform through countries in order to speed up detection of diversity and comparability of results (Aguinagalde et al., 2005). At the same time, given the rapid pace of environmental degradation in many biologically species-rich parts of the world, a clear organism identification is essential for restoration experts to define species' distribution range, native plants for restoring damaged ecosystems or afforesting new ones, invasive species to fight. Moreover, it is important to check the phases of the regeneration processes of a community after an intervention, and protect native and/or threatened ecosystems. These items could be achieved by using a standardized molecular approach as DNA Barcoding, once its actual efficacy is demonstrated with preliminary study cases.

Recently, the need to understand the development and the spatial dynamics of pattern in ecological phenomena leaded to the concepts of landscape ecology, i.e. broad scale investigations strictly linked to the vegetation occurring at local scale. The Committee of Ministers of the Council of Europe adopted the European Landscape Convention on 2000, aiming to promote European landscape protection, management and planning and to organise European cooperation (European Council, 2000). The Convention is the first international treaty exclusively devoted to all aspects of European landscape, but the importance of reforestation and genetic fundamentals of landscape is not well considered yet (Granke et al., 2008).

Since the main goal is to guarantee not only simple conservation measures, but also the expansion of forest surfaces throughout reforestation actions, we need methods able to provide forest quality and reduce the time for a complete environmental restoration. This is particularly true in those areas where the environment has been modified and exploited by humans over the course of thousands of years, as in the Mediterranean Basin. In particular, forests have experienced many processes that have led to degradation and consequent soil loss as reported since the fourth century B.C. by Plato in Critias. The Miyawaki method could take up the challenge, but its effectiveness will be increased if it is joined with other tools, like well defined Regions of Provenance, *in situ* and/or *ex situ* networks of reserves for providing the most suitable genetic resources, and DNA Barcoding to assess and monitor the trend of the intervention.

It is undoubted that we have to move toward a holistic approach, in order to improve the present methods with as many criteria as possible, and define a unique project design. For these reasons, a toolbox based on this multidisciplinary concept is presented as ideal guideline attending the gained experiences in the Mediterranean Basin (Figure 9).

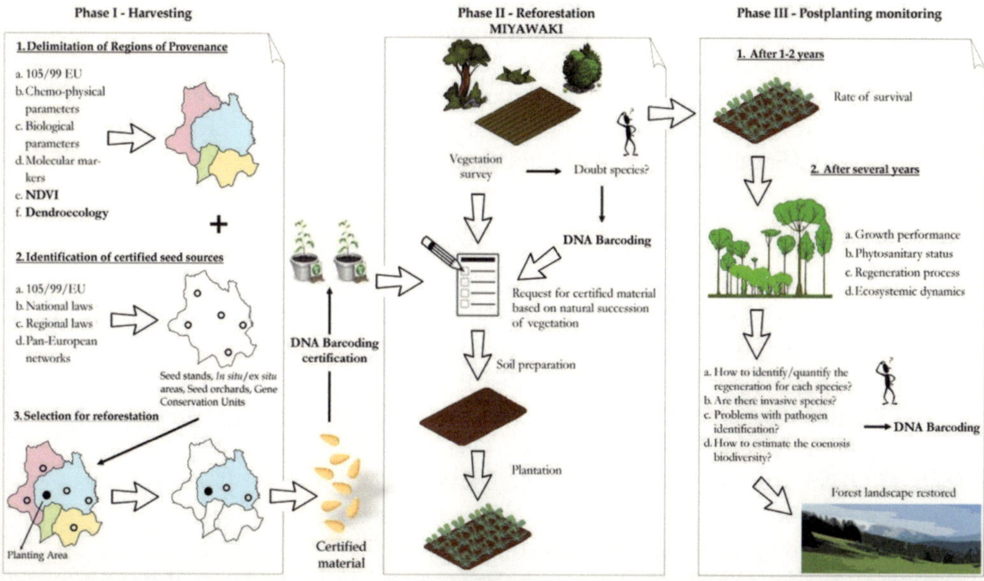

Fig. 9. Theoretical example of reforestation process implemented with the tools (bold texts) discussed in the chapter.

Computer-based methods existed since 1980 to assist tree species and provide information about uses, distribution, environment, and silviculture; nowadays The Forestry Compendium developed by CAB International (CABI) is probably the most impressive tool that has been developed (CABI, 2010). However, further developments of this tool should include information on selection systems about requirements of particular genotypes, including provenances, hybrids, clones, and genetically modified material choice. The toolbox we propose is composed of the mentioned actions and methods, including latest informatics supports, and it has been developed to be applied in reforestation activities, starting from the delimitation of the Regions of Provenance with the detection of adequate seed sources, the correct identification of plant species, the environmental and vegetation surveys, the selection of certified basic materials, up to the reforestation technique and the checks after planting. Basically, each mentioned step poses a specific question, and the toolbox would provide the answer or the best tool to achieve it. However, this proposal is a preliminary tentative to create a logic framework of actions that will require a validation measure also throughout a socio-economic analysis to estimate the costs of each step. For instance, it would be useful to understand the costs for data capture and development of further biological indexes retrieved from satellite images, the expenses for extracting and analysing DNA, including molecular markers and Barcoding of plants, the costs of manpower and plantlets for the reforestation practices.

6. Acknowledgements

We are indebted to Regional Forest Directorate of Sardinia for conceding the logistic support in providing the Miyawaki test areas. Special thanks to Dr. Carmine Sau, Dr. Antonello Salis, and Dr. Francesco Mazzocchi for their valuable help and commitment on the field work performed in Pattada Municipality. The study about NDVI remote sensing analysis was partially supported by the project PRIN 2007AZFFAK and developed in collaboration with Prof. Gianluca Piovesan and Dr. Alfredo Alessadrini. Special thanks to Dr. Despina Paitaridou of the Greek Ministry of the Environment for providing information about the Regions of Provenance in Greece. For DNA Barcoding bioinformatic study in Italy, we are grateful to Dr. Roberta Piredda, Dr. Laura Armenise, and Dr. Alessio Papini. Our warmest thanks to Prof. Rosanna Bellarosa, Prof. Luca Santi, Dr. Avra Schirone, Mr. Armando Parlante, Mr. Luigi Sandoletti, and Ms. Giulia Sandoletti.

7. References

Abell, R., Olson, D.M., Dinerstein, E., et al. (2000). *Freshwater ecoregions of North America: A conservation assessment.* Island Press, ISBN: 978-1559637343, Washington, DC, USA.

Aguinagalde, I., Hampe, A., Mohanty, A., Martin, J.P., Duminil, J. & Petit, R.J. (2005). Effects of life-history traits and species distribution on genetic structure at maternally inherited markers in European trees and shrubs. *Journal of Biogeography*, Vol. 32, No. 2, (February 2005), pp. 329-339, ISSN 1365-2699.

Aldrich, M., Belokurov, A., Bowling J., et al. (2004). *Integrating Forest Protection, Management and Restoration at a Landscape Scale.* WWF, Retrieved from http://wwf.panda.org/what_we_do/footprint/forestry

Alessandrini, A., Vessella, F., Di Filippo, A., Salis, A., Santi, L., Schirone, B. & Piovesan, G. (2010). Combined dendroecological and normalized difference vegetation index analysis to detect regions of provenance in forest species. *Scandinavian Journal of Forest Research*, Vol. 25, No. 1, (June 2010), pp. 121-125, ISSN 0282-7581.

Alía, R., Barba, D., Climent, J. & Iglesias, S. (2008). Delineation of breeding zones and provenance regions in Europe. *Abstract book of the workshop "Adaptive traits needed for breeding zones delineation"*, p. 18. TREEBREDEX, Activity 3, Madrid, Spain, 25-26 February 2008, Retrieved from http://treebreedex.eu/IMG/pdf/abstract-2.pdf

Amaral, W. & Yanchuk, A. (2004). Integrated approaches for *ex situ* conservation and use of forest genetic diversity, In: *Forest genetic resources conservation and management. Vol. 3: In plantations and genebanks (ex situ)*, FAO, DFSC, IPGRI (Eds.), pp. 1-3, International Plant Genetic Resources Institute, ISBN: 92-9043-649-2, Rome, Italy.

Bagnoli, F., Vendramin, G.G., Buonamici, A., et. al. (2009). Is *Cupressus sempervirens* native in Italy? An answer from genetic and palaeobotanical data, *Molecular Ecology*, Vol.18, No. 10, (May 2009), pp. 2276–2286, ISSN 1365-294X.

Bak, P., Tang, C., Wiesenfeld, K. (1988). Self-organized criticality. *Physical Review A*, Vol. 38, No. 1, (July 1988), pp. 364–374, ISSN 1094-1622.

Bernetti, G. (1995). *Selvicoltura speciale.* UTET, ISBN 8802048673, Torino, Italy.

Biondi, F. & Visani, S. (1996). Recent developments in the analysis of an Italian tree-ring network with emphasis on European beech (*Fagus sylvatica* L.), *Proceedings of the*

International Conference on Tree rings, environment and humanity. Special Volume. Radiocarbon, pp. 713-725, ISBN: 0-9638314-2-9, Tucson, Arizona, 17-21 May 1994.

Biondi, F. (1992). Development of a tree-ring network for the Italian Peninsula. *Tree-ring Bulletin*, Vol. 52, pp. 15-29, ISSN 1536-1098.

Blasi, C. (1994). Il fitoclima del Lazio. *Fitosociologia*, Vol. 27, No. 1, pp. 151-175, ISSN 1125-9078.

Borevitz, J.O. & Chory, J. (2004). Genomic tools for QTL analysis and gene discovery. *Current Opinion in Plant Biology*, Vol. 7, No. 2, (April 2004), pp. 132–136, ISSN 1369-5266.

BSP/CI/TNC/WRI/WWF. (1995). *A regional analysis of geographic priorities for biodiversity conservation in LAC. A report for USAID*. Biodiversity Support Program, ISBN: 1887531173, Washington, DC, USA.

CAB International (2010). Forestry Compendium. Wallingford, UK: CAB International. Available as CD-ROM or online at: www.cabicompendium.org/fc

Caldecott, J.O., Jenkins, M.D., Johnson, T. & Groombridge, B. (1996). Priorities for conserving global species richness and endemism. *Biodiversity and Conservation*, Vol. 5, No. 6, (June 1996), pp. 699-727, ISSN 0960-3115.

Callaway, R.M. (1997). Positive interactions in plant communities and the individualistic-continuum concept. *Oecologia*, Vol. 112, No. 2, (October 1997), pp. 143–149, ISSN 0029-8549.

Camazine, S., Deneubourg, J.L., Franks, N.R., Sneyd, J., Theraulaz, G. & Bonabeau, E. (2003). *Self-organization in biological systems*. Princeton University Press, ISBN 0691012113, Princeton, USA.

Caramalli, C. (1973). Influenza della preparazione del terreno nei rimboschimenti. *L'Italia Forestale e Montana*, Vol. 28, No. 1, pp. 74–83, ISSN 0021-2776.

Carey, A.B., and R.O. Curtis. 1996. Conservation of biodiversity: a useful paradigm for forest ecosystem management. *Wildlife Society Bulletin*, Vol. 24, No. 4, (December 1996), pp. 610-620, ISSN 0091-7648.

Chase, M.W., Cowan, R.S., Hollingsworth, P.M. et al. (2007). A proposal for a standardised protocol to barcode all land plants. *Taxon*, Vol. 56, No. 2, (May 2007), pp. 295–299, ISSN 0040-0262.

Chase, M.W., Salamin, N., Wilkinson, M., et al. (2005). Land plants and DNA barcodes: short-term and long-term goals. *Philosophical Transactions of the Royal Society*, Series B, Vol. 360, No 1462 (October 2005), pp. 1889–1895, ISSN 0800-4622.

Chaw, S.M., Walters, T.W., Chang, C.C., Hu, S.H. & Chen, S.H. (2005). A phylogeny of cycads (Cycadales) inferred from chloroplast matK gene, trnK intron, and nuclear rDNA ITS region. *Molecular Phylogenetics and Evolution*, Vol. 37, No. 1, (October 2005), pp. 214–234, ISSN 1095-9513.

Clewell, A.F. & Aronson, J. (2007). *Ecological restoration: principles, values, and structure of an emerging profession*. Island Press, ISBN 978-1597261692, Washington DC, USA.

Collignon, A.M. & Favre, J.M. (2000). Contribution to the postglacial history at the western margin of *Picea abies*' natural area using RAPD markers. *Annals of Botany*, Vol. 85, No. 6, (January 2000), pp. 713-722, ISSN 1095-8290.

Consortium for Barcode of Life Plant Working Group (CBOL PWG). (2009). A DNA barcode for land plants, *Proceedings of the National Academy of Sciences of the USA*, Vol. 106, No. 31, (May 2009),pp. 12794-12797, ISSN 1091-6490.

Derory, J., Scotti-Saintagne, C., Bertocchi, E., et al. (2010). Contrasting relationships between the diversity of candidate genes and variation of bud burst in natural and segregating populations of European oaks. *Heredity*, Vol. 105, No. 4, (October 2009), pp. 401-411, ISSN 0018-067X.

Dinerstein, E., Olson, D.M., Graham, D.J., et al. (1995). *A conservation assessment of the terrestrial ecoregions of Latin America and the Caribbean*. The World Bank, ISBN 978-0821332955, Washington DC, USA.

Dittmar, C., Zech, W. & Elling, W. (2003). Growth variations of common beech (*Fagus sylvatica* L.) under different climatic and environmental conditions in Europe - A dendroecological study. *Forest Ecology and Management*, Vol. 173, No. 1-3, (February 2003), pp. 63-78, ISSN 0378-1127.

Duminil, J., Fineschi, S., Hampe, A., et al. (2007). Can population genetic structure be predicted from life history traits? *American Naturalist*, Vol. 169, No. 5, (May 2007), pp. 662-672, ISSN 0003-0147.

Emberger, L. (1955). Une classification biogeographique des climates. *Recueil des Travaux du Laboratoire de Botanique, Géologie et Zoologie de la Faculté des Sciences de l'Université de Montpellier, série Botanique*, Vol. 7, pp. 3-43, ISSN 11528346.

Eriksson, G. (2001). Conservation of noble hardwoods in Europe. *Canadian Journal of Forest Research*, Vol. 31, No. 4, (April 2001), pp. 577–587, ISSN 0045-5067.

Eriksson, G., Namkoong, G. & Roberds, J. (1995). Dynamic conservation of forest tree gene resources. *Forest Genetic Resources*, FAO Annual publication, Vol. 23, pp. 2-8, ISSN 259-2894.

Eriksson, G., Namkoong, G. & Roberds, L. H. (1993). Dynamic gene conservation for uncertain futures. *Forest Ecology and Management*, Vol. 62, No. 1-4, (December 1993), pp. 15-37, ISSN 0378-1127.

European Council (1971). Council Directive 71/161/EEC of 17 April 1971 on external quality standards for forest reproductive material marketed within the Community. *Official J. Eur. Commun.*, L 87, pp. 14.

European Council (1999). Council Directive 1999/105/EC of 22 December 1999 on the marketing of forest reproductive material. *Official J. Eur. Commun.*, L 11, pp. 17-40.

European Council (2000). European Landscape Convention of 20 October 2000.

European Economic Community (1966). Council Directive 404/66 of 14 June 1966 on the marketing of forest reproductive material. *Official J. Eur. Commun.*, L 125, pp. 2326-2332.

European Economic Community (1992). Council Regulation (EEC) n. 2080/92 on instituting a community scheme for forestry measures in agriculture. *Official J. Eur. Commun.*, L 215, pp. 96-99.

Eveno, E., Collada, C., Guevara, M.A., et al. (2008). Contrasting Patterns of Selection at *Pinus pinaster* Ait. Drought Stress Candidate Genes as Revealed by Genetic Differentiation Analyses. *Molecular Biology and Evolution*, Vol. 25, No. 2, (December 2007), pp. 417–437, ISSN 1537-1719.

Falk, D.A., Palmer, M.A. & Zedier, I.B. (Eds.). (2006). *Foundations of restoration ecology*. Island Press, ISBN 978-1597260176, Washington DC, USA.

FAO, FLD, IPGRI. (Eds.). (2004). *Forest genetic resources conservation and management. Vol. 1: Overview, concepts and some systematic approaches*. International Plant Genetic Resources Institute, ISBN 92-9043-648-4, Rome, Italy.

Fassnacht, S., Gower, S. T., MacKenzie, M. D., Nordheim, E. V. & Lillesand, T. M. (1997). Estimating the leaf area index of north central Wisconsin forests using the Landsat Thematic Mapper. *Remote Sensing of Environment*, Vol. 61, No. 2, (August 1997), pp. 229-245, ISSN 0034-4257.

Fazekas, A.J., Burgess, K.S., Kesanakurti, P.R., et al. (2008). Multiple multilocus DNA barcodes from the plastid genome discriminate plant species equally well. *PLoS ONE*, Vol. 3, No. 7, (July 2008), e2802, ISSN 1932-6203.

Fazekas, A.J., Kesanakurti, P.R., Burgess, K.S., et al. (2009). Are plant species inherently harder to discriminate than animal species using DNA barcoding markers? *Molecular Ecolology Resources*, Vol. 9, No. Suppl. 1, (May 2009), pp. 130-139, ISSN 1755-0998.

Follieri, M. (2010). Conifer extinctions in Quaternary Italian records. *Quaternary International*, Vol. 225, No. 1, (September 2010), pp. 37–43, ISSN 1040-6182.

Ford, C.S., Ayres, K.L., Toomey, N., et al. (2009). Selection of candidate coding DNA barcoding regions for use on land plants. *Botanical Journal of the Linnean Society*, Vol. 159, no. 1, (January 2009), pp. 1-11, ISSN 1095-8339.

Frankel, O.H. (1976). Natural variation and its conservation, In: *Genetic diversity in plants*, A. Muhammed, R. Aksel & R.C. von Borstel (Eds.), pp. 21–44, Plenum Press, ISBN 978-0306365089, New York, USA.

Geburek, T. & Konrad, H. (2008). Why the conservation of forest genetic resources has not worked. *Conservation Biology*, Vol. 22, No. 2, (April 2008), pp. 267-274, ISSN 1523-1739.

Gómez, J.M., García, D. & Zamora, R. (2003). Impact of vertebrate acorn and seedling predators on a Mediterranean *Quercus pyrenaica* forest. *Forest Ecology and Management*, Vol. 180, No. 1-3, (July 2003), pp. 125–134, ISSN 0378-1127.

Gonzalez, M.A., Baraloto, C., Engel, J., et al. (2009). Identification of Amazonian trees with DNA barcode. *PLoS ONE*, Vol. 4, No. 10, (October 2009), e7483, ISSN 1932-6203.

Granke, O., Kenter, B., Kriebitzsch, W.U., Köhl, M., Köhler, R. & Olschofsky, K. (2009). Biodiversity assessment in forests – from genetic to landscape diversity. *iForest*, Vol. 2, No. 1-3, (March 2008), ISSN 1971-7458.

Greenstone, M.H., Rowley, D.L., Heimbach, U., Lundgren, J.G., Pfannenstiel, R.S. & Rehner, S.A. (2005). Barcoding generalist predators by polymerase chain reaction: carabids and spiders. *Molecular Ecology*, Vol. 14, No. 10, (September 2005), pp. 2347–3266, ISSN 1365-294X.

Gregorius, H-R. (1991). Gene conservation and the preservation of adaptability. In: *Species conservation: a population-biological approach*, Seitz, A. & Loeschcke, V. (Eds.), pp. 31-47, Birkhäuser Verlag, ISBN 3-7643-2493-7, Berlin, Germany.

Guerrant , E.O., Havens, K. & Maunder, M. (Eds.). (2004). *Ex situ plant conservation: Supporting species survival in the wild*. ISBN 1-55963-875-3. Island Press, Covelo, USA.

Hajibabaei, M., Janzen, D.H., Burns, J.M., Hallwachs, W. & Hebert, P.D.N. (2006). DNA barcodes distinguish species of tropical Lepidoptera. *Proceedings of the National Academy of Sciences of the USA*, Vol. 103, No. 4, (January 2006), pp. 968–971, ISSN 1091-6490.

Hebert, P.D.N., Ratnasingham, S. & deWaard, J.R. (2003). Barcoding animal life: cytochrome c oxidase subunit 1 divergences among closely related species. *Proceedings of the Royal Society of London*, series B, Vol. 270, No. Suppl. 1, (August 2003), pp. S96–S99, ISSN 0962-8452.

Hebert, P.D.N., Stoeckle, M.Y., Zemlak, T.S. & Francis, C.M. (2004). Identification of birds through DNA barcodes. *PLoS Biology*, Vol. 2, No. 10, (October 2004), pp. e312, ISSN 1545-7885.

Hollingsworth, M.L., Clark, A.A., Forrest, L.L., et al. (2009). Selecting barcoding loci for plants: evaluation of seven candidate loci with species-level sampling in three divergent groups of land plants. *Molecular Ecology Resources*, Vol. 9, No. 2, (March 2009), pp. 439-457, ISSN 1755-0998.

Hummel, M. (Ed.). (1989). *Endangered spaces: the future for Canada's wilderness*. Key Porter Books, ISBN 978-1550131017 Ontario, Canada.

Hunter, M.L. (Ed.). (1999). *Maintaining biodiversity in forest ecosystems*. Cambridge University Press, ISBN 0521-63104-1, Cambridge, UK.

Jordan, W.R. III, Gilpin, M.E. & Aber, J.D. (1987). *Restoration ecology. A synthetic approach to ecological research*. Cambridge University Press, ISBN 0-521-33728-3, Cambridge.

Karp, A., Edwards, K.J., Bruford, M., et al. (1997). Molecular technologies for biodiversity evaluation: opportunities and challenges. *Nature Biotechnology*, Vol. 15, No. 7, (July 1997), pp. 625-628, ISSN 1546-1696.

Kelty, M.J. (2006). The role of species mixtures in plantation forestry. *Forest Ecology and Management*, Vol. 233, No. 2-3, (September 2006), pp. 195–204, ISSN 0378-1127.

Kleinschmit, J.R.G., Kownatzki, D. & Gregorius H-R. (2004). Adaptational characteristics of autochthonous populations - consequences for provenance delineation. *Forest Ecology and Mangement*, Vol. 197, No 1-3, (August 2004), pp. 213-224, ISSN 0378-1127.

Korol, L., Shklar, G. & Schiller, G. (2002). Diversity among circum-Mediterranea populations of Aleppo pine and differentiation from Brutia pine in their isoenzymes: additional results. *Silvae Genetica*, Vol. 51, No. 1, (June 2001), pp. 35-41, ISSN 0037-5349.

Kramer, P.J. & Kozlowski, T.T. (1979). *Physiology of Woody plants*. Academic Press, ISBN 978-0124250505, New York, USA.

Kress, W.J. & Erickson, D.L. (2007). A two-locus global DNA barcode for land plants: the coding *rbcL* gene complements the non-coding *trnH-psbA* spacer region. *PLoS One*, Vol. 2, No. 6, (June 2007), pp. e508, ISSN 1932-6203.

Kress, W.J. & Erickson, D.L. (2008). DNA barcodes: Genes, genomics, and bioinformatics. *Proceedings of the National Academy of Science of the USA*, Vol. 105, No. 8, (February 2008), pp. 2761-2762, ISSN 1091-6490.

Kress, W.J., Erickson, D.L., Jones, F.A., et al. (2009). Plant DNA barcodes and a community phylogeny of a tropical forest dynamics plot in Panama. *Proceedings of the National Academy of Science of the USA*, Vol. 106, No. 44, (November 2009), pp. 18621–18626, ISSN 1091-6490.

Kress, W.J., Erickson, D.L., Swenson, N.G., et al. (2010). Advances in the use of DNA Barcodes to build a community phylogeny for tropical trees in a Puerto Rican forest dynamics plot. *PLoS ONE*, Vol. 5, No. 11, (November 2010), pp. e15409, ISSN 1932-6203.

Kress, W.J., Wurdack, K.J., Zimmer, E.A., Weight, L.A. & Janzen, D.H. (2005). Use of DNA barcodes to identify flowering plants. *Proceedings of the National Academy of Sciences of the USA*, Vol.102, No. 23, (June 2005), pp. 8369–8374, ISSN 1091-6490.

Krever, V., Dinerstein, E., Olson, D.M. & Williams, L. (1994). *Conserving Russia' s biological diversity: an analytical framework and initial investment portfolio.* WWF, Washington, DC, USA.

Krusche, D. & Geburek, T. (1991). Conservation of forest gene resources as related to sample size. *Forest Ecology and Management*, Vol. 40, No. 1-2, (May 1991), pp. 145-150, ISSN 0378-1127.

Lahaye, R., van der Bank, M., Bogarin, D. et al. (2008). DNA barcoding the floras of biodiversity hotspots. *Proceedings of the National Academy of Sciences of the USA*, Vol. 105, No. 8, (February 2008), pp. 2923–2928, ISSN 1091-6490.

Lewontin, R.C. (1984). Detecting population differences in quantitative traits as opposed to gene frequencies. *American Naturalist*, Vol. 123, No. 1, (January 1984), pp. 115-124, ISSN 1537-5323.

Lindgren, D. & Ying, C.C. (2000). A model integrating seed source adaptation and seed use. *New Forests*, Vol. 20, No. 1, (July 2000), pp. 87-104, ISSN 1573-5095.

Liu, J., Moller, M., Gao, L.M., Zhang, D.Q. & Zhuki, D.E. (2011). DNA barcoding for the discrimination of Eurasian yews (*Taxus* L., Taxaceae) and the discovery of cryptic species. *Molecular Ecology Resources*, Vol. 11, No. 1, (January 2011), pp. 89-100, ISSN 1755-0998.

Loucks, C., Springer, J., Palminteri, S., Morrison, J. & Strand, H. (2003). From the Vision to the Ground: A Guide to implementing ecoregion conservation in priority areas. World Wild Fund US Conservation Science Program, Retrieved from http://www.worldwildlife.org/science/ecoregions/WWFBinaryitem6507.pdf

Medail, F. & Diadema K. (2009). Glacial refugia influence plant diversity patterns in the Mediterranean basin. *Journal of Biogeography*, Vol. 36, No. 7, (July 2009), pp. 1333-1345, ISSN 1365-2699.

Mitrakos, K. (1980). A theory for mediterranean plant Life. *Acta Oecologica*, Vol. 1, No. 15, pp. 245-252, ISSN 0243-7651.

Miyawaki, A. & Abe, S. (2004). Public awareness generation for the reforestation in Amazon tropical lowland region. *Tropical Ecology*, Vol. 45, No. 1, (Summer 2004), pp. 59–65, ISSN 0564-3295.

Miyawaki, A. & Golley, F.B. (1993). Forest reconstruction as ecological engineering. *Ecological Engineering*, Vol. 2, No. 4, (December 1993), pp. 333–345, ISSN 0925-8574.

Miyawaki, A. (1975). Entwicklung der Umweltschtz-Pflanzungen und-Ansaaten in Japan, In: *Sukuzessionsforschung Berichte Über Internationales Symposium der Internationalen Vereinigung für Vegetationskunde*, Tüxen, R. (Ed.), pp. 237-254, J. Cramer, Vaduz, Liechtenstein.

Miyawaki, A. (1981). Energy policy and green environment on the base of ecology, In: *Beyond the energy crisis opportunity and challenge*, Fazzolare, R.A. & Smith, C.B. (Eds.), pp. 581-587, Oxford University Press, ISBN 978-0080275895, Oxford, UK.

Miyawaki, A. (1989). Restoration of evergreen broadleaved forest ('1aurel forest') in Japan, In: *The world community in post industrial society. The human encounter with nature: destruction and reconstruction, Vol 5*, Christian Academy (Ed.), pp. 130-147, Wooseok, Seoul, Korea.

Miyawaki, A. (1992). Restoration of evergreen broadleaved forests in the Pacific region, In: *Ecosystem rehabilitation, ecosystem analysis and synthesis, vol* 2, Wali, M.K. (Ed.), pp. 233-245, SPB Academic Publishing, ISBN 9051030711, The Hague, Netherlands.

Miyawaki, A. (1993a). Restoration of native forest from Japan to Malaysia, In: *Restoration of tropical forest ecosystems*, Lieth, H. & Lohmann, M. (Eds.), pp. 5–24, Kluwer Academic, ISBN 978-0-7923-1945-0, Dordrecht, USA.

Miyawaki, A. (Ed.). (1993b). Global perspective of green environments restoration of native forests from Japan to Malaysia and South America, based on an ecological scenario. In: *Geoscience and Remote Sensing Symposium, 1993. IGARSS '93. Better Understanding of Earth Environment*, pp. 6–8, Kogakuin University, ISBN 0-7803-1240-6, Tokyo, Japan.

Miyawaki, A. (1996). Restoration of biodiversity in urban and peri-urban environments with native forest, In: *Biodiversity, science and development*, Di Castri, F. & Younès, T. (Eds.), pp. 558-565, CAB International, ISBN 978-0851989730, Wallingford, USA.

Miyawaki, A. (1998a). Restoration of urban green environments based on the theories of vegetation ecology. *Ecological Engineering*, Vol. 11, No. 1, (October 1998), pp. 157–165, ISNN 0925-8574.

Miyawaki, A. (1998b). Vegetation ecological study for restoration of forest ecosystems, In: *A vegetation ecological study for the restoration and rehabilitation of green environment based on the creation of environmental protection forests in Japanese Archipelago*, Fujiwara, K. (Ed.), pp. 267-298, Institute of Vegetation Science, Institute of Environmental Science and Technology, Yokohama National University, Japan.

Miyawaki, A. (1999). Creative ecology: restoration of native forests by native trees. *Plant Biotechnology*, Vol. 16, No. 1, (April 1999), pp. 15–25, ISSN 1342-4580.

Miyawaki, A., Fujiwara, K., Ozawa, M. (1993). Native forest by native trees: restoration of indigenous forest ecosystem: (reconstruction of environmental protection forest by Prof. Miyawaki's method). *Bulletin of Institute of Environmental Science and Technology*, Yokohama National University, Vol. 19, No. 1, (April 1993), pp. 73–107, ISSN 0286-584X.

Muller-Starck, G. & Schubert, R. (Eds.). (2001). *Genetic Responses of Forest Systems to Changing Environmental Conditions.* Kluwer Academic Publishers, ISBN 1-4020-0236-X, Dordrecht, The Netherlands.

Nagendra, H. & Southworth, J. (Eds.). (2009). *Reforesting Landscapes: Linking Pattern and Process.* Springer, ISBN 978-1-4020-9655-6, Dordrecht, Netherlands.

Newmaster, S.G. & Ragupathy, S. (2009). Testing plant barcoding in a sister species complex of pantropical *Acacia* (Mimosoideae, Fabaceae). *Molecular Ecology Resources*, Vol. 9, No. S1, (May 2009), pp. 172-180, ISSN 1755-0998.

Newmaster, S.G., Fazekas, A.J., Steeves, R.A.D. & Janovec, J. (2008). Testing candidate plant barcode regions in the Myristicaceae. *Molecular Ecology Resources*, Vol. 8, No. 3, (May 2008), pp. 480–490, ISSN 1755-0998.

Norton, B.G. (2003). *Searching for Sustainability: Interdisciplinary Essays in the Philosophy of Conservation Biology*. Cambridge University Press, ISBN 0-521-80990-8, Cambridge, UK.

Noss, R.F. & Cooperrider, A.Y. (1994). *Saving nature' s legacy: Protecting and restoring biodiversity*. Defenders of Wildlife and Island Press, ISBN 978-1559632485, Washington, DC, USA.

Padilla, F.M. & Pugnaire, F.I. (2006). The role of nurse plants in the restoration of degraded environments. *Frontiers in Ecology and the Environment*, Vol. 4, No. 4, (May 2006), pp. 196-202, ISSN 1540-9295.

Palacios, G., Navarro-Cerrillo, R.M., del Campo, A. & Toral, M. (2009). Site preparation, stock quality and planting date effect on early establishment of Holm oak (*Quercus ilex* L.) seedlings. *Ecological Engineering*, Vol. 35, No. 1, (January 2009), pp. 38–46, ISSN 0925-8574.

Pavari, A. (1916). *Studio preliminare sulla coltura di specie forestali esotiche in Italia*. Annali del Regio Istituto Superiore Forestale Nazionale, Tipografia Ricci, Florence, Italy.

Peñuelas, J. & Boada, M. (2003). A global change-induced biome shift in the Montseny Mountains (NE Spain). *Global Change Biology*, Vol. 9, No. 2, (February 2003), pp. 131-140, ISSN 1365-2486.

Perrow, M.R., Davy, A.J. (Eds.). (2002a). *Handbook of ecological restoration. 1: Principles of restoration*. Cambridge University Press, ISBN 0521791286, Cambridge, UK.

Perrow, M.R., Davy, A.J. (Eds.). (2002b). *Handbook of ecological restoration. 2: Restoration in practice*. Cambridge University Press, ISBN 0521791294, Cambridge, UK.

Petit, R.J. & Hampe, A. (2006). Some evolutionary consequences of being a tree. *Annual Revue of Ecology Evolution and Systematics*, Vol. 37, (December 2006), pp. 187-214, ISSN 1543-592X.

Pflieger, S., Lefebvre, V. & Causse, M. (2001). The candidate gene approach in plant genetics: a review. *Molecular Breeding*, Vol. 7, No. 4, (December 2001), pp. 275–291, ISSN 1380-3743.

Pierce, S.M. (2003). *The STEP Handbook and Mapbook: integrating the natural environment into land-use decisions at the municipal level: towards sustainable development*. Terrestrial Ecology Research Unit Report No. 47, University of Port Elizabeth, South Africa. Available from: http://cpu.uwc.ac.za

Piovesan, G., Biondi, F., Bernabei, M., Di Filippo, A. & Schirone, B. (2005). Spatial and altitudinal bioclimatic zones of the Italian peninsula identified from a beech (*Fagus sylvatica* L.) tree-ring network. *Acta Oecologica*, Vol. 27, No. 3, (May-June 2005), pp. 197-210, ISSN 1146-609X.

Piredda, R., Simeone, M.C., Attimonelli, M., Bellarosa, R. & Schirone, B. (2011). Prospects of barcoding the Italian wild dendroflora: oaks reveal severe limitations to tracking species identity. *Molecular Ecology Reources*, Vol. 11, No. 1, (January 2011), pp. 72-83, ISSN 1755-0998.

Primack, R.B. (2008). *A primer of conservation biology*, Sinauer, ISBN 978-0-87893-692-2, Sunderland, Massachusetts, USA.

Ran, J.H., Wang, P.P., Zhao, H.J. & Wang, X.Q. (2010). A Test of Seven Candidate Barcode Regions from the Plastome in *Picea* (Pinaceae). *Journal of Integrative Plant Biology*, Vol. 52, No. 12, (December 2010), pp. 1109-1126, ISSN 1744-7909.

Raymond, C.A. & Lindgren, D. (1990). Genetic Flexibility - a Model for Determining the Range of Suitable Environments for a Seed Source. *Silva Genetica*, Vol. 39, No. 3-4, (June 1990), pp. 112-120, ISSN 0037-5349.

Ren, B.Q., Xiang, X.G. & Chen, Z.D. (2010). Species identification of *Alnus* (Betulaceae) using nrDNA and cpDNA genetic markers. *Molecular Ecology Resources*, Vol. 10, No. 4, (July 2010), pp. 594-605, ISSN 1755-0998.

Ricketts, T.H., Dinerstein, E., Olson, D.M., et al. (1999). *Terrestrial ecoregions of North America: A conservation assessment*. Island Press, ISBN 978-1559637220, Washington DC, USA.

Rieseberg, L.H. (1997). Hybrid origins of plant species. *Annual Rewiews of Ecology and Evolution*, Vol. 28, (December 1997), pp. 359–389, ISSN 1543-592X.

Robba, L., Russell, S.J., Barker, G.L. & Brodie, J. (2006). Assessing the use of the mitochondrial *cox1* marker for use in DNA barcoding of red algae (Rhodophyta). *American Journal of Botany*, Vol. 93, No. 8, (August 2008), pp. 1101–1108, ISSN 1537-2197.

Roy, S., Tyagi, A., Shukla, V., et al. (2010). Universal Plant DNA Barcode loci may not work in complex groups: a case study with Indian Berberis species. *PLoS ONE*, Vol. 5, No. 10, (October 2010), pp. e13674, ISSN 1932-6203.

Rudel, T. (2005). *Tropical forests: regional paths of destruction and regeneration in the late twentieth century*. Columbia University Press, ISBN 978-0231131957, New York, USA.

Sachs, J.L., Mueller, U.G., Wilcox, T.P. & Bull, J.J. (2004). The evolution of cooperation. *The Quarterly Review of Biology*, Vol. 79, No. 2, (June 2004), pp. 135–160, ISSN 0033-5770.

Sang, T., Crawford, D.J. & Stuessy, T.F. (1997). Chloroplast DNA phylogeny, reticulate evolution, and biogeography of Paeonia (Paeoniaceae). *American Journal of Botany*, Vol. 84, No. 9, (September 1997), pp. 1120–1136, ISSN 1537-2197.

Sass, C., Little, D.P., Stevenson, D.M. & Specht, C.D. (2007). DNA barcoding in the cycadales: testing the potential of proposed barcoding markers for species identification of Cycads. *PLoS ONE*, Vol. 2, No. 11, (November 2007), pp. e1154, ISSN 1932-6203.

Schirone, B. (1998). Teoria e pratica dei rimboschimenti: alcune prospettive possibili, In: *Nuove frontiere nella gestione forestale*, pp. 101-115, Ciancio, O. (Ed.), ISBN 8-8875-5300-9, Accademia Italiana di Scienze Forestali, Florence, Italy.

Seifert, K.A., Samson, R.A., deWaard, J.R., et al. (2007). Prospects for fungus identification using *CO1* DNA barcodes, with *Penicillium* as a test case. *Proceedings of the National Academy of Sciences of the USA*, Vol. 104, No. 10, (March 2007), pp. 3901–3906, ISSN 1091-6490.

Shaw, J., Lickey, E.B., Beck, J.T., et al. (2007). The tortoise and the hare II: relative utility of 21 noncoding chloroplast DNA sequences for phylogenetic analysis. *American Journal of Botany*, Vol. 92, No. 1, (January 2007), pp. 142-166, ISSN 1537-2197.

Smith, M.A., Poyarkov, N.A.J. & Hebert, P.D.N. (2008). *CO1* DNA barcoding amphibians: take the chance, meet the challenge. *Molecular Ecology Resources*, Vol. 8, No. 2, (March 2008), pp. 235–246, ISSN 1755-0998.

Smith, S.A. & Donoghue, M.J. (2008). Rates of molecular evolution are linked to life history in flowering plants. *Science*, Vol. 322, No. 5898, (October 2008), pp. 86–89, ISSN 0036-8075.

Society for Ecological Restoration International Science & Policy Working Group (SER). (2004). *The SER International Primer on Ecological Restoration*. Available from: www.ser.org & Tucson: Society for Ecological Restoration International.

Soulé, M.E. & Wilcox, B.A. (1980). Thresholds for survival: mantaining fitness and evolutionary potential, In: *Conservation Biology: an evolutionary-ecological perspective*, Soulé, M.E. & Wilcox, B.A. (Eds), pp. 151–170, Sinauer Associates Inc, ISBN 978-0878938001, Sunderland, UK.

Stanturf, J. A. & Madsen, P. (Eds.). (2004). *Restoration of boreal and temperate forests (integrative studies in water management and land development)*, CRC Press, ISBN 1-56670-635-1, Boca Raton, Florida, USA.

Steele P.R. & Pires, J.C. (2011). Biodiversity assessment: state-of-the-art and techniques in phylogenomics and species identification. *American Journal of Botany*, Vol. 98, No. 2, (February 2011), pp. 415–425, ISSN 1537-2197.

Steltzer, H. & Post, E. (2009). Seasons and life cycles. *Science*, Vol. 324, No. 5929, (May 2009), pp. 886-887, ISSN 0036-8075.

Stern, K. & Roche, L. (1974). *Genetics of Forest Ecosystems*. Springer-Verlag, ISBN 978-0387060958, New York, USA.

Thomson, L., Graudal, L. & Kjær, E. (2001). Conservation of genetic resources in their natural environment, In: *Forest genetic resources conservation and management. Vol. 2: In managed natural forests and protected areas (in situ)*, FAO, DFSC, IPGRI (Eds.), pp. 1-3, International Plant Genetic Resources Institute, ISBN 92-9043-472-4, Rome, Italy.

Thomson, L., Graudal, L. & Kjær, E. (2001). Management of natural forests for conservation of forest genetic resources, In: *Forest genetic resources conservation and management. Vol. 2: In managed natural forests and protected areas (in situ)*, FAO, DFSC, IPGRI (Eds.), pp. 13-39, International Plant Genetic Resources Institute, ISBN 92-9043-472-4, Rome, Italy.

Tomaselli, R. (1973). *La Vegetazione Forestale d'Italia*. Corpo Forestale dello Stato, Collana Verde 33, Rome, Italy.

Tucker, C. J. & Sellers, P. J. (1986). Satellite remote sensing of primary production. *International Journal of Remote Sensing*, Vol. 7, No. 11, (November 1986), pp. 1395-1416, ISSN 0143-1161.

Tüxen, R. (1956). Die heutige potentielle natiirliche Vegetation als Gegenstand der Vegetationskartierung. Angew. *Pflanzensoziologie*, Vol. 13, pp. 5–42 Stolzenau/Weser.

UNEP. (1995). *Global biodiversity assessment*. Cambridge University Press, ISBN 0-521-56481-6 Cambridge, UK.

Van Diggelen, R. & Marrs, R.H. (2003). Restoring plant communities: introduction. *Applied Vegetation Science*, Vol. 6, No. 2, (February 2003), pp. 106–110, ISSN 1654-109X.

Veroustraete, J.P. & Myneni, R.B. (1996). Estimating net ecosystem exchange of carbon using the normalized difference vegetation index and an ecosystem model. *Remote Sensing of Environment*, Vol. 58, No. 1, (October 1996), pp. 115-130, ISSN 0034-4257

Schirone, B., Salis, A. & Vessella, F. (2011). Effectiveness of the Miyawaki method in Mediterranean forest restoration programs. *Landscape and Ecological Engineering*, Vol. 7, No. 1, (January 2011), pp. 81-92, ISSN 1860-188X.

von Cräutlein, M., Korpelainen, H., Maria Pietiläinen, H. & Rikkinen, J. (2011). DNA barcoding: a tool for improved *taxon* identification and detection of species diversity. *Biodiversity and conservation*, Vol. 20, No. 2, (February 2011), pp. 373-389, ISSN 1572-9710.

Ward, R.D., Zemlak, T.S., Innes, B.H., Last, P.R. & Hebert, P.D.N. (2005). DNA barcoding Australia's fish species. *Philosophical Transactions of the Royal Society, Series B*, Vol. 360, No. 1462, (October 2005), pp. 1847–1857, ISSN 1471-2970.

Wilson, E.O. & Peter, F.M. (1986). *Biodiversity (Papers from the 1st National Forum on Biodiversity, September 1986, Washington, D.C.)*, National Academy of Sciences (October 6, 1996), ISBN 978-0309037396, Washington DC, USA.

12

Environmental Assessment and Aquatic Biodiversity Conservation of Amazonian Savannas, Marajó Island, Brazil

Luciano Fogaça de A. Montag, Tiago Magalhães da S. Freitas, Ana Cristina Mendes-Oliveira and Ronaldo B. Barthem
Universidade Federal do Pará/Museu Paraense Emílio Goeldi
Brazil

1. Introduction

Marajó Island is located in the mouth of the Amazon River, the World's largest river, at its interface with the Atlantic Ocean (Goulding et al., 2003). Marajó is the World's largest riverine island (Cruz, 1987), with a total area of approximately fifty thousand square kilometers (Fig. 1). In addition to the Amazon, the island is also influenced by the Tocantins River and other small rivers of the south (Barthem & Schwassmann, 1994; UNEP, 2004).

The island is surrounded by sandy-clayey beaches, and encompasses a hydrographic system (rivers, channels, and creeks) which both drains and floods the *terra firme* forest and *várzeas* (swamps) (Marques-Aguiar et al., 2002). In addition to these forest ecosystems, the island's lakes, lagoons, beaches, rivers, and mangroves contribute to its diversity of habitats and organisms (Marques-Aguiar et al., 2002; Montag et al., 2009; Almeida et al., 2009). This diversity of habitats influences the richness of the region's fauna, including its fish, and plays an important role in habitat preferences and use (Lowe-McConnell, 1999; Carrier et al., 2004). Most of these environments are heavily influenced by the inundation cycle of the estuarine region, which Junk (1997) characterizes as a polymodal cycle of low amplitude, but highly predictable flood pulses, influenced primarily by the local tides.

The climate of Marajó Island can be characterized by two categories of the Köppen classification system - Af (humid tropical, with mean precipitation in the driest month at least 60 mm) and Am, tropical monsoon, with excessive rainfall between February and May. During this period, two-thirds of the island are completely flooded (Cardoso & Pereira, 2002; Lima et al., 2005; Fig. 1). Annual precipitation on Marajó Island ranges between 2500 mm and 4000 mm, with a mean temperature of around 27°C, and relative humidity of 81% to 94%. Rainfall is distributed in two distinct periods, with a marked rainy season between January and June, and a dry season, between September and November.

Despite this rigorous hydrological regime, the vegetation covering the part of the island that is flooded annually is referred to as a savanna (or flooded grassland; Fig. 2), which is known for the diversity and abundance of its fishery resources (Barthem & Fabré, 2003; Montag et al., 2009; Schaan, 2010). While the term savanna is more associated with specific floristic characteristics, comparable with those of the savannas of central Brazil (Eiten, 1972; Ratter et

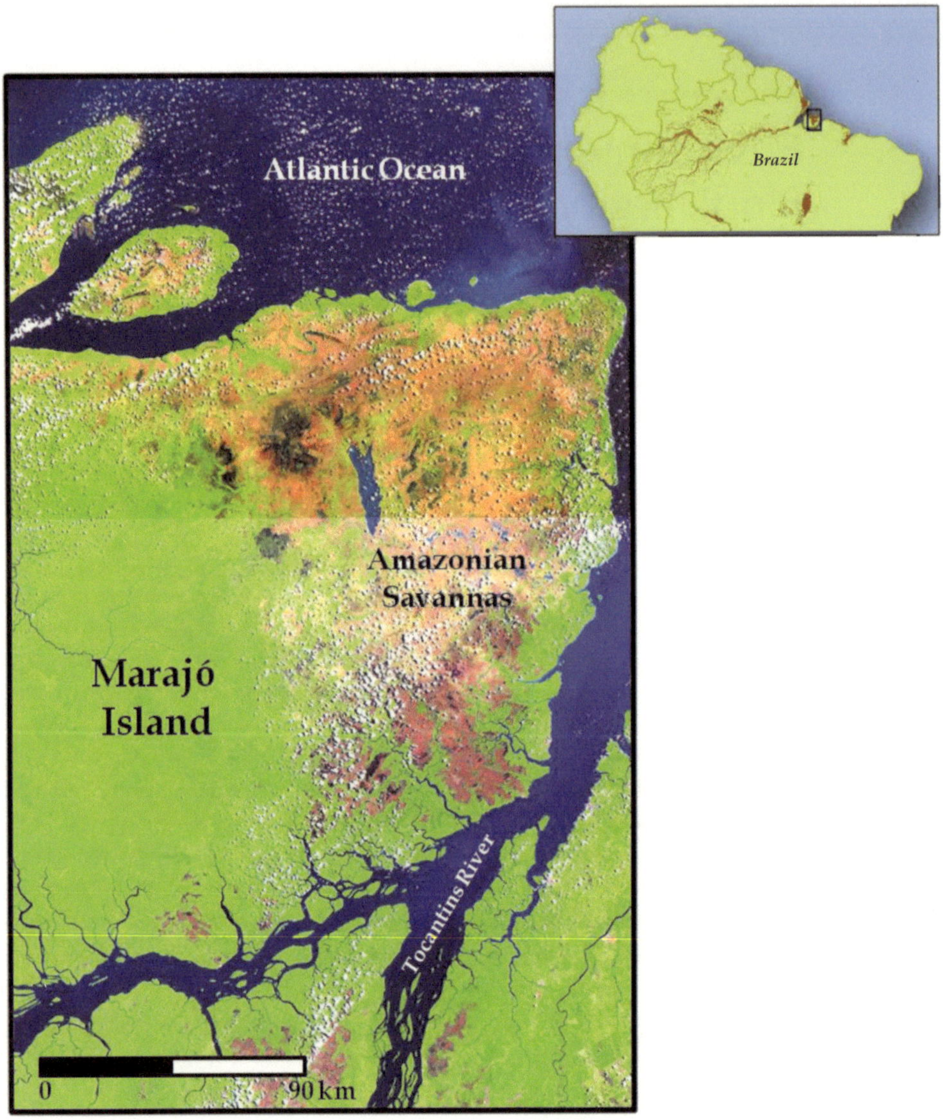

Fig. 1. LandSat Image (2001) of Marajó Island in the Eastern Amazon. The eastern of the island shows the seasonally-flooded savannas (shaded pink).

al., 1997; Sanaiotti et al., 1997; Ribeiro & Walter, 1998, Barbosa et al., 2007), Harris (1980) and Schaan (2010) recognize the flooded grasslands of Marajó Island as a typical example of seasonally-flooded savanna, more typical of the Amazonian *várzeas*, given the duration of the flooding period. However, there are also important differences, especially in the dynamics of inundation. While the Marajó grasslands are inundated by rainwater, the floodplains of the lower Amazon basin are encroached by rising river levels (Sombroek, 1966). The influence of oceanic tides is an additional factor contributing to the unique ecology of the Marajó savannas (Moran, 1995), demanding specific adaptive responses from the living organisms that inhabit this environment.

Fig. 2. Aerial view of the Marajó landscape, with the savannas or grassland in lighter green (left fig.); a typical scenario of the savannas in the Marajó Island during the dry season (right fig.). Photo by Miguel von Behr.

Humans have occupied Marajó Island for at least the past five to seven thousand years (Simões, 1981; Roosevelt et al., 1991). These first inhabitants probably subsisted on game, fish, and small-scale agriculture (Schaan, 2010), although, as in the present day, this activity would have been limited by the low fertility of the local soils and their inadequate drainage. However, these soils are appropriate for a number of native palm species, such as the buriti (*Mauritia flexuosa* L.), also known locally as miriti, and the açaí (*Euterpe oleracea* Mart.). These two plants provide the raw material on which much of the local economy is based. The buriti supplies a fiber used for the production of artifacts and household items, while the fruit of the açaí is the principal source of nutrients for much of the island's riverside population (Schaan, 2010). A third plant species, the wild nutmeg, *Virola surinamensis* (Rol.) Warb. provides the raw material for the local plywood industry (Homma et al., 1998), which is the principal source of income for many of the island's municipalities (Lima et al., 2005).

Ranching, in particular of water buffalo, is another important source of income for the local inhabitants (Lima et al., 2005; Fig. 3). According to data from the Brazilian Institute for Geography and Statistics, IBGE (Barbosa, 2005), the Marajó stocks represent approximately one quarter of all the buffalo raised in Brazil. However, traditional ranching on the island has declined in recent years due to competition from cattle ranchers from southern Pará, who produce a more marketable product (Schaan, 2010).

These economic activities, which are developed without conservation management guidelines, have contributed to the progressive degradation of the island's ecosystems. The shipping of *açaí* in motorized vessels also contributes to the contamination of the local rivers with oil and plastics, while the production of beef and pork leads to the silting up of bodies of water (Schaan, 2010). A number of authors have associated these anthropogenic impacts with the decline in the numbers of fish and species, and the abundance of fishery stocks (Ackermann, 1963; Smith, 2001, 2002). These stocks have also suffered direct impacts from overfishing, the use of gillnets of illegal mesh size, and the discarding of catches of inadequate size or low market value (Hilbert, 1952; Barthem & Goulding, 2007).

Fig. 3. The ranching of water buffaloes is an important source of income for the local inhabitants. Photo by Miguel von Behr.

Despite the unique features and ecological importance of Marajó Island in the local context, it has not been included in the environmental and hydrological management programs that have been developed for the Amazon region (Lima et al., 2005). One of the reasons for this is probably the lack of information on the region's fauna. Given this preoccupying scenario, this chapter focuses on the ecological evaluation of the fish diversity of the flooded savannas of Marajó Island in Brazil. The historical and biological aspects of the local ichthyofauna are analyzed, and the threats to this fauna are examined. The challenges for the implementation of effective conservation and management of one of the Amazon region's most important natural resources - its fish - are also discussed.

2. What do we know?

2.1 The fishes

Despite having one of the richest ichthyofaunas found anywhere on the planet (Reis et al., 2003), our scientific knowledge of the diversity of the fish of Marajó Island is restricted to just two studies, Boulenger (1897) and Montag et al. (2009), more than one hundred years apart. In the more recent study, Montag et al. (2009) revised the data available from fieldwork conducted between 2003 and 2005 in the municipalities of Chaves, Muaná, and Ponta de Pedras. Additional data were obtained from the ichthyological collections of the Goeldi Museum (MPEG) and the Museum of Zoology of São Paulo University (MZUSP) in Brazil, as well as international institutions, such as the Swedish Natural History Museum (NMR) and the Florida State Museum (UF, USA). Most of the specimens available in these collections were obtained during the 1980s and 1990s in the municipalities of Cachoeira do Arari, Muaná, Ponta de Pedras, Salvaterra, Santa Cruz do Arari, and Soure (Montag et al., 2009). This study was part of the PROBIO Marajó project entitled "Ecological Evaluation and Selection of Priority Areas for the Conservation of the Amazonian Savannas of the Marajó Archipelago in the Brazilian State of Pará", which was financed by the Program for Biodiversity Research (PPBio) and the Brazilian Environment Ministry (MMA), through the Goeldi Museum in Belém.

A total of 254 fish species were recorded from the flooded savannas of Marajó Island, including the first documentation of eight families for this region - Cetopsidae, Ctenoluciidae, Paralichthyidae, Poeciliidae, Polycentridae, Rivulidae, and Trichomycteridae (Montag et al., 2009; see Appendix 1). However, the bootstrap estimate of the region's total species richness was approximately 310 (Fig. 4), which represents an additional 50 species that may potentially be catalogued on Marajó Island.

Of these 254 species, 44 (17% of the total) are commercially important as food species, while 46 (18%) are targeted by the ornamental fish market. In fact, the marshlands of Marajó Island represent one of the most important fishing grounds of the lower Amazon basin (Barthem & Fabré, 2003).

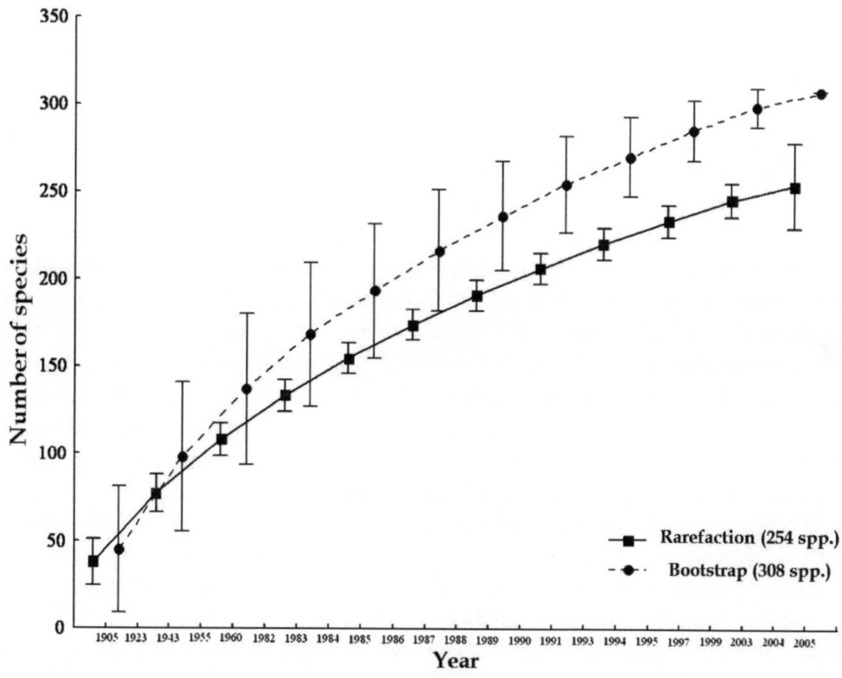

Fig. 4. Species accumulations curve for the number of fish species recorded in the Marajó savannas using the rarefaction technique and the species richness estimated by Bootstrap analysis for the data available for the period between 1905 and 2005. Source: Montag et al. (2009).

The principal species targeted by commercial fisheries include the tamuatá or brown hoplo (*Hoplosternum littorale*), thraira (*Hoplias malabaricus*), cangati or driftwood catfish (*Trachelyopterus galeatus*) and aracu or headstanders (family Anostomidae). The island's most popular ornamental fish include lebiasinids, caracids of the genera *Hemigrammus* and *Hyphessobrycon*, dwarf cichlids of the genera *Apistogramma, Crenicichla* and *Geophagus*, cascudos or plecos of the Loricariidae family, and stingrays of the genus *Potamotrygon*.

Until now, few endemic species have been recorded from the flooded savannas of Marajó Island, although this may be at least partly related to the deficiency of conclusive systematic and taxonomic studies for many species groups. One case in point here is swamp eel, *Synbranchus lampreia* Favorito, Zanata and Assumpção (2005), which is known only from the Goiapi River on Marajó Island. Given the gaps in our knowledge of the region's ichthyofauna and the unreliability of the taxonomic arrangements of most groups, Montag et al. (2009) proposed that the island should be considered among the regions with the highest priority for both biological inventories and the implementation of protected areas, based on well-defined empirical criteria.

2.2 Environmental patterns

The savannas of Marajó Island are seasonally flooded by rain, forming a complex system of temporary or permanent lakes and swamps. The aquatic biodiversity of this region is adapted to the extreme conditions of drought and flood, together with high temperatures associated and low levels of dissolved oxygen (Almeida et al., 2009). The most common fish species found in these lakes and swamps are well adapted to the extremes of oxygen depletion (Kramer et al., 1978), such as the temporary development of a dermal lip protuberance for surface respiration (Winemiller, 1989) or permanent accessory respiratory organs, like the swimbladder of the pirarucu (*Arapaima gigas*) (Greenwood & Liem, 1984), and the vascularization of the posterior intestine in the tamuatá or brown hoplo (*Hoplosternum littorale*) (Persaud et al., 2006) or the oral mucosa in the poraquê or electric eel *Electrophorus electricus* as cited by Johansen et al. (1968).

The reproduction of these fish species is synchronized with the flood pulse. The electric eel (*Electrophorus electricus*) spawns during the late dry season, between September and December (Assunção & Schwassmann, 1995), while *Hoplosternun littorale* spawns at the beginning of the rainy season, i.e. January and February (Oliveira, 2000). This species builds nests in the flooded vegetation and presents parental care. For many fishes the environmental characteristics is an important factor for species distribution, and this is also applied for potamotrigonids stingrays. However, according to Almeida et al. (2009) the distribution of the stingray *Potamotrygon motoro* is related to body size, with the largest specimens being found in the center of the island, and the smaller individuals in peripheral areas.

Fishery activities are also synchronized with the flood pulse (Welcomme, 1979). The body of water expands during the rainy season, dispersing the fish populations over the floodplain, reducing fishery productivity. By contrast, during the dry season, the fish are restricted to isolated pools or shallow channels, facilitating their capture. Overall, 90% of the island's commercial catch is based on four species, with *Hoplosternum littorale* accounting for approximately two-thirds of total fishery production (Barthem & Goulding, 2007). Fishery activities are based on cast- and gillnetting. Gillnets are employed either in a fixed manner, or as a seine, which is drawn towards a beach or a second gillnet, which acts as a barrier. Hooks and harpoons are also used by artisanal fishermen (Albuquerque & Barthem, 2008).

2.3 Pattern and process

The flood pulse is the main ecological factor that determines the productivity and biological processes of the aquatic environments of the Amazon Basin (Junk et al., 1989; Junk & Wantzen, 2004). The flood pulse of the central portion of Marajó Island is determined basically by local precipitation levels, while water levels in marginal areas are determined primarily by tidal cycles.

Data on precipitation and water levels are available from the website of the Brazilian National Waters Agency (Agência Nacional de Águas, or ANA). Rainfall is more intense during the first half of the year, between January and June, reaching a peak normally in March. The precipitation is not distributed homogeneously, and tends to be higher in the eastern and western portions of the island, and lower in its center (Lima et al., 2005). The highest precipitation - 200 mm in a single day - was recorded in the east of the island, the part closest to the Atlantic Ocean (Soure). The lowest precipitation rates coincide with the area of savanna, with the lowest values being recorded in September, when the monthly maximum is less than 100 mm (Cachoeira do Arari) (Fig. 5). The center of the island is flat and shallow, and the rainfall floods the savannas between March and May, forming permanent or temporary lakes and swamps (Fig. 6). In subsequent months, the local rivers drain off this water to coastal areas.

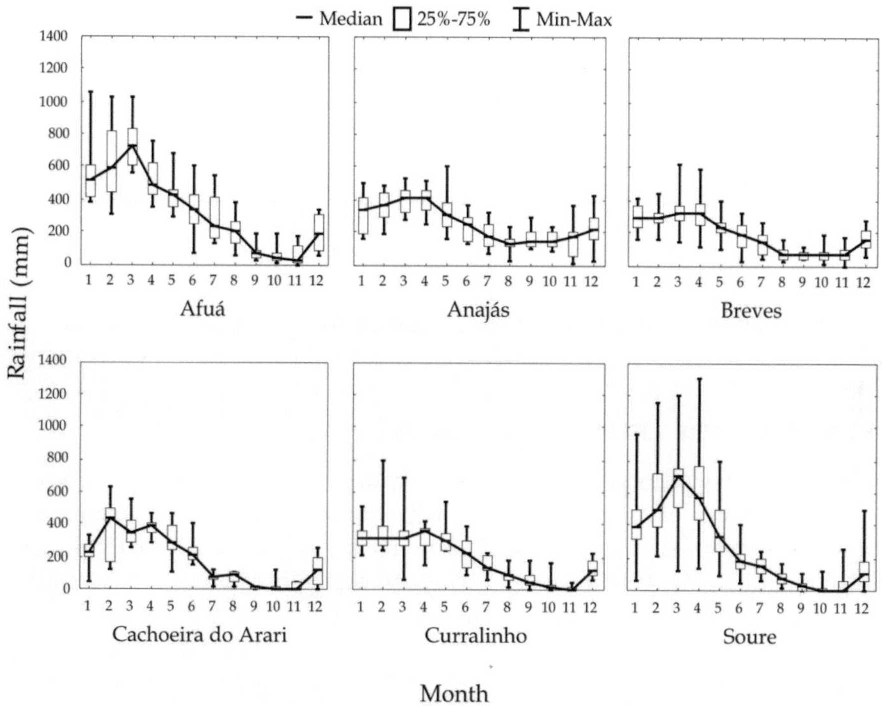

Fig. 5. Median monthly precipitation at different localities on Marajo Island (Source: Brazilian National Waters Agency - ANA).

The tidal surge moves upstream into the Amazon from east to west, when it is deformed, delayed and amortized. The high tide increases the level of the river and dams or inverts its flow, which returns to normal during the ebb tide (Kosuth et al., 1999). Rivers can be classified in three categories, according to the predominant hydrological force, i.e. river flow, tide or a combination of the two effects (Volker, 1966). The rivers of Marajó Island may be classified in all three categories. The rivers in the center of the island are affected only by local rainfall, while those on the coast are dominated by the tides, and those in intermediate areas are affected by a combination of these processes. The proximity of the ocean results in an increased intrusion of brackish water during the second half of the year, raising salinity levels near the coast and in the small rivers in the east of the island (Barthem & Schwassmann, 1994).

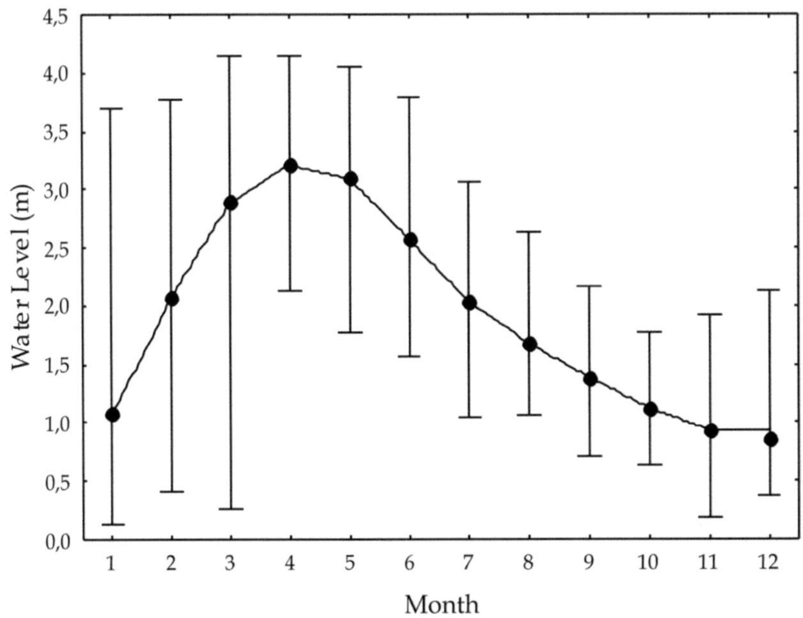

Fig. 6. Monthly variation in the water of the Cachoeira do Arari River, in the center of Marajó Island.

3. What are the threats to local biodiversity?

Throughout its recent history, the Amazon basin has been subjected to a series of rapid processes of environmental degradation, which have had negative consequences for the local biota. On a broad scale, the principal economic activities that have contributed to this process have traditionally been cattle ranching and agriculture. However, logging and the harvesting of forest products, such as fruits, oils, latex, game, and fish, have also contributed to the depletion of natural resources. On Marajó Island, these processes have been typical of those observed in the rest of the Amazon basin. First occupied by Europeans in the 17th century, Marajó Island Marajó Island was involved in the principal economic cycles of the Amazon basin. During the colonial period, the island's principal economic activities were large-scale (by the standards of the period) cattle ranching, and the production of sugar, tobacco, and cotton in plantations established primarily in coastal locations (Teixeira, 1953; Miranda Neto, 1968). During this period, the trampling of the grasslands by the cattle already contributed to a decline in local fishery productivity (Brasil, 1990). Following a period of decadence, and the rubber boom at the turn of the twentieth century, ranching began to grow on Marajó Island in the 1920s, although it is currently in decline once again. Currently, ranching on the island is based on its herd of water buffalo, which arrived at the end of the 19th century, when animals of the Carabao breed, originally from the Phillipines (see Fig. 3). However, the weight of these animals results in hoofprints of up to 10 cm in depth (Santos, 2006) which, together with their habit of excavating holes in which they cool themselves during the dry season, provoke significant alterations in the structure of the superficial soil layers and the composition of the vegetation, which favor erosive processes (Dias, 1999; Fig. 7). In addition, the compaction of the soil creates an environment unfavorable for plant growth, reducing the productivity and longevity of forage species (Imhoff et al., 2000).

Fig. 7. Water buffaloes cooling themselves during the dry season, which provokes significant alterations in the structure of the superficial soil layers and changes in the composition of the vegetation. Photo by Miguel von Behr.

Buffalo ranching also contributes with the nitrogen enrichment of water bodies through urine and feces, resulting in high turbidity and reduced oxygen levels. These alterations of hydrological resources may also affect the abundance and quality of fishery resources, as well as increasing the potential for the transmission of water-borne diseases. This process also affects the local biodiversity through the destruction of the vegetation, in which many species seek refuge when breeding (Bernardi, 2005).

Most of the economic activities developed on Marajó Island provoke some form of impact on the local biota. Activities that result in deforestation, such as ranching, agriculture, and unregulated logging, result in the loss of riparian vegetation and the silting up of river beds (Goulding et al., 2003). The extraction of fruit-producing lumber species and the commercial exploitation of the palm-heart and fruit of the açaí result in a reduction in the abundance of fruit for many frugivorous animals, including fish (Barthem, 2001). While commercial fisheries are an extremely profitable activity in eastern Amazonia, they have been exerting increasing pressure on local fish stocks. Species such as the pirarucu *Arapaima gigas* and the tamuatá *Hoplosternum littorale* are being overfished, but are marketed prominently in the Ver-o-Peso market in Belém, the capital of the Brazilian state of Pará (Fig. 8). Commercial fishing by so-called "ice boats" (equipped with icebox), in particular in *várzea* lakes and other sheltered areas of the estuary (bays, channels, and creeks), threatens the reproduction of commercial species, and thus their stocks over the long term (Goulding et al., 2003; UNEP, 2004). The negative consequences of this process may eventually throughout the local ecosystem in a cascade effect.

Fig. 8. The tamuatá *Hoplosternum littorale* are being marketed prominently in the Ver-o-Peso market in Belém, the capital of the Brazilian state of Pará. Read up on the broad "Large tamuatá from Marajó Island".

In addition to these impacts on the physical-chemical characteristics of the water and the vegetation, the compaction of the soil threatens the survival of the fish species that inhabit these seasonally-flooded habitats, such as the South American lungfish (*Lepidosiren paradoxa*), which may asphyxiate during the estivation period. During the dry season, the lungfish bury themselves in the humid soil of the savannas, covering themselves in mucous, which is produced in the tegument in large quantities to protect the animal from dehydration. They are also able to reduce their metabolism and body temperature in a process of estivation. There is an opening in the upper part of the mucous "cocoon", which allows the fish to breathe atmospheric air. The lungfish may remain in this state for many months, until the following rainy season, when more favorable conditions arise (Mesquita-Saad et al., 2002; Kramer et al., 1978).

In contrast with the fluctuations of the ranching industry, the harvesting of natural plant and animal resources has always been the mainstay of the economy of Marajó Island. The commercial exploitation of the fruit and palm-heart of the açaí, and fish stocks, constitute the most important economic activities for the local communities of the Amazon estuary (Brasil, 1990).

The native vegetation of Marajó Island has been modified by a combination of ranching, fires, and direct anthropogenic impacts (see Prance & Shaller 1982). Barthem (2001) considered the Marajó savannas to be the most threatened environment of the whole of the Amazon basin in terms of the conservation status of its ichthyofauna. In addition to these other anthropogenic pressures, this environment is being further threatened by the possible introduction of a new variety of rice developed specifically for the conditions on Marajó Island by EMBRAPA, the agricultural research agency of the Brazilian government.

One of the most dramatic anthropogenic impacts on the aquatic biota of the island's savannas was probably caused by the digging of the Tartaruga canal at the end of the 19th century. This canal crossed the savanna in a north-south direction for approximately 25 km, and was originally conceived as a means of reducing the time necessary to navigate between Macapá and Belém, the two principal cities on the Amazon estuary (Teixeira, 1953). The canal provoked the relatively rapid draining of the water accumulated in the savannas at the end of the rainy season (Ackermann, 1963), accentuating the effects of the dry season on the aquatic biota and many terrestrial animals.

In addition to all these threats to the local biodiversity, a large number of other projects are currently being planned or are under development in the Marajó region. One of these is the construction of a new canal, with the same objective of the Tartaruga canal, which is to reduce navigation time between Macapá and Belém. In addition to the intrinsic physical effects of the increase in the river-borne traffic, the degradation of aquatic habitats and the reduction or elimination of feeding resources will affect the plant and animal communities directly (Bucher et al., 1993).

The region has also been regional initiatives for the development of economic activities, including tourism, industrial fisheries, industrialized pineapple plantations, palm-heart processing plants, and cattle ranching. The Pará state government, through its Science, Technology, and Environment ministry (SECTAM), has implemented technical and scientific cooperation agreements, and other institutional covenants, with the aim of elaborating environmental management strategies for the sustainable development of all the island's municipalities which have instated official policies for the protection of the environment. Up to now, agreements have been signed with ten municipalities - Afuá, Anajás, Breves, Cachoeira do Arari, Chaves, Curralinho, Muaná, Ponta de Pedras, Salvaterra, and Soure.

Eleven of the 16 municipalities of Marajó Island have already obtained the seal of tourism quality from the Brazilian Tourism Agency (EMBRATUR), and are among the "flagships" of the tourism development program of the Pará State Tourism Company (PARATUR, 2001). The Marajó region is also included as a main hub in the Amazon Basin Ecotourism Development Program (PROECOTUR), a federal program administered by the Environment Ministry (MMA) through its secretary for Amazonian affairs, with financial support from the Inter-American Development Bank (BID). This program aims to promote the expansion of ecotourism in the Brazilian Amazon region, as one of the principal strategies for the sustainable development of the nine states had make up the region.

4. What do we still need to know?

The first step towards the effective mitigation of any of the environmental impacts affecting the biodiversity of the flooded grasslands of Marajó Island is the consolidation of our knowledge of the ecological processes that underpin this unique system. However, few data are available in the literature on the status and integrity of local habitats and biodiversity, even in the case of the ichthyofauna, which is of direct interest to local human populations.

The 254 fish species already recorded on Marajó Island represent only 12% of the total diversity of the Amazon basin. Most of the records collated by Montag et al. (2009) were derived from sites located in the eastern portion of the island, and represent a relatively reduced sampling effort, considering its total area. Furthermore, the western portion of the island, with its many channels and tidal creeks, appears to be a more propitious environment for a high diversity of fish species. This part of the island is probably inhabited by a number of endemic species that are typical of the brackish estuarine and coastal systems that extend along the northern coast of South America, between the mouth of the Orinoco River in Venezuela and São Marcos Bay, in the Brazilian state of Maranhão.

The Amazon estuary is a unique, but still poorly-studied environment inhabited by both freshwater and marine fish species, many of which fundamental to both the regional and the national fishery industries (Barthem, 2001). The estuary and the adjacent vegetation constitute an extremely important nursery for the local ichthyofauna, providing both refuges and rich feeding resources for the juveniles of an ample variety of species. Many of these species also move through the flooded forest, where they eat fruits and seeds, and thus play an important role in the dispersal of the seeds of many *várzea* plant species (Gottsberger, 1978; Goulding, 1980; Kubitzki & Ziburski, 1994; Correa et al., 2007; Pollux, 2011).

As mentioned above, insufficient information is available on the composition of the region's ichthyofauna and the biology of the species that make up this fauna. At the present time, it is essential to conduct more definitive surveys, and to evaluate the status of populations and their long-term viability in both time and space, especially in the case of migratory species and/or those with an ample geographic range. In addition, understanding the basic functioning of the Amazonian ecosystems affected by seasonal flooding will be fundamentally important for the development of effective conservation strategies for the ichthyofauna of Marajó Island (Roberts, 1993).

5. Challenges: Applying what we already know

The destruction and degradation of the rivers and creeks of the flooded savannas of Marajó Island, combined with the overfishing throughout the estuarine region, are the principal threats to the integrity of the local ichthyofauna. To be effective, any conservation strategy must take into consideration the complex life cycles of different species, and their specific habitat requirements. Given this, population-level studies will be essential for the understanding of the ecological niche of each species, so that more effective conservation strategies can be developed. Data on the habitat use of these species will also be essential for the development of effective *in situ* conservation strategies for target species.

In a wider context, the strategies applied to the conservation of species of special interest – such as the pirarucu or freshwater stingrays – will undoubtedly also favor most of the other species that share the same habitat or have similar niches. Unfortunately, far too few data are currently available for any such process to be systematic and dependable, although the identification of the key species and habitats may nevertheless provide an adequate baseline for the development of effective public policies for the conservation of the aquatic biodiversity of Marajó Island.

One of the greatest challenges in the Amazon today is to integrate the exploitation of natural resources by the local human communities with the conservation of the biota. To achieve this, the economic, social, and ecological benefits of managed ecosystems must be understood by the local residents if there is to be any possibility of the sustainable use of resources. In the specific case of the savannas of Marajó Island, the understanding by residents and managers of the benefits of the conservation of these habitats for the maintenance or even the improvement of fishery stocks, will be essential for the success of any conservation efforts.

An important step towards the integration of local populations in any conservation scheme is the understanding of conservation initiatives, however minor or incipient, developed by the local populations themselves. However, conservation strategies should be diversified on both a local and regional scales due to the complexity of anthropogenic impacts and their differential effects on different trophic levels of ecosystems. Integrated conservation initiatives involving governments, non-governmental organizations, and in particular, the active participation of local populations, should be given the highest priority.

In addition to the integrated management of natural resources, the creation of protected areas has been an important practical conservation strategy in Brazil. In fact, the whole of Marajó Island is located within a single state conservation unit, the Marajós Archipelago Environmental Protection Area, or Marajó EPA. Considered to be the largest state EPA in Brazil (MMA, 2007), this unit was created by article 13, paragraph 2 of the Pará state constitution, decreed on October 5th, 1989. However, this category of protected area is highly flexible in terms of the exploitation of natural resources and definitely does not guarantee the maintenance of the ecological functions of the local ecosystems over the long term.

The Marajó EPA was created as part of the ecological-economic zoning of the state of Pará, with the objectives of conserving the region's biodiversity, development and improvement of the quality of life of the island's population, the preservation of endangered species and representative areas of the local ecosystems, and the establishment of scientific, environmental education, and ecotourism projects. However, up to now, no management plan or administration program has been developed for this protected area.

An alternative proposal for the more effective conservation of the fauna and flora of Marajó Island would be the creation of a biosphere reserve. Biosphere reserves are areas of terrestrial and/or marine-coastal ecosystems recognized by the Man and the Biosphere Program (MaB), which was created in the 1970s by UNESCO, the United Nations Educational, Scientific and Cultural Organization. This program supports the development of a balanced relationship between people and nature, and determines that areas of special environmental and human value be recognized (designated) as biosphere reserves by the international coordinating council of the MaB program, at the request of the interested state. Following this recognition, the area of the reserve comes under the exclusive sovereignty of the state within which it is located.

6. Conclusions

The considerable lacunas in our current knowledge of the ichthyofauna of the region of Marajó Island indicate the need for more thorough, complementary surveys of the flooded grasslands (savannas) of the eastern portion of the island, studied by Montag et al. (2009), as well as new inventories of the western portion of the island. The anthropogenic impacts observed on local fish stocks, together with the dependence of local communities on the

fishery resources of the Amazon estuary combine to make this region a priority for scientific studies of the aquatic biota and associated environments.
There is still a long way to go before the conservation of the aquatic biodiversity of Marajó Island can be assured and the maintenance of fishery stocks guaranteed over the long term. In both cases, management strategies must be well diversified. As part of this process, it will be the responsibility of the region's researchers to undertake scientific studies of local diversity (taxonomic revision with the identification of species and cataloguing of specimens), the relationships between species and the physical environment, spatial distribution of species, and the identification of possible endemics and key species for conservation. These studies will provide a theoretical and empirical database for the systematic development of public policies that integrate authorities, administrators, and local residents in the quest for the effective conservation and management of the natural resources of the aquatic environments of Marajó Island over the long term.

Appendix 1

Fish species of the savannas of Marajó Island, Pará, Brazil. Modified from Montag et al. (2009).

Fish species from Marajó Island
BELONIFORMES: Belonidae
Potamorrhaphis guianensis (Jardine, 1843)
Potamorrhaphis sp.
CHARACIFORMES: Acestrorhynchidae
Acestrorhynchus altus Menezes, 1969
Acestrorhynchus cf. *altus* Menezes, 1969
Acestrorhynchus falcatus (Bloch 1794)
Acestrorhynchus falcirostris (Cuvier, 1819)
Acestrorhynchus microlepis (Schomburgk, 1841)
Acestrorhynchus sp.
Anostomidae
Leporinus affinis Günther, 1864
Leporinus fasciatus (Bloch, 1794)
Leporinus friderici (Bloch, 1794)
Leporinus sp.
Rhytiodus microlepis Kner, 1858
Schizodon fasciatus Spix and Agassiz, 1829
Schizodon sp.
Schizodon vittatus (Valenciennes, 1850)
Characidae
Astyanax bimaculatus (Linnaeus, 1758)
Bryconops alburnoides Kner, 1858
Bryconops caudomaculatus (Günther, 1864)
Bryconops giacopinii (Fernández-Yepez , 1950)
Bryconops sp.
Charax pauciradiatus (Günther, 1864)
Cheirodon sp.
Hemigrammus bellottii (Steindachner, 1882)

Hemigrammus levis Durbin, 1908
Hemigrammus ocellifer (Steindachner, 1882)
Hemigrammus rhodostomus Ahl, 1924
Hemigrammus sp.
Hemigrammus unilineatus (Gill, 1858)
Hyphessobrycon heterorhabdus (Ulrey, 1894)
Hyphessobrycon sp.
Metynnis luna Cope, 1878
Metynnis sp.
Moenkhausia collettii (Steindachner, 1882)
Moenkhausia lepidura (Kner, 1858)
Moenkhausia pyrophthalma Costa, 1994
Moenkhausia sp.
Paragoniates alburnus Steindachner, 1876
Phenacogaster sp.
Poptella sp.
Pristella maxillaris (Ulrey, 1894)
Pristobrycon aureus (Spix and Agassiz, 1829)
Pristobrycon calmoni (Steindachner, 1908)
Pygocentrus nattereri Kner, 1858
Roeboides myersii Gill, 1870
Roeboides sp.
Serrasalmus elongatus Kner, 1858
Serrasalmus rhombeus (Linnaeus, 1766)
Serrasalmus sp.
Triportheus aff. *angulatus* (Spix and Agassiz, 1829)
Triportheus albus Cope, 1872
Triportheus angulatus (Spix and Agassiz, 1829)
Triportheus elongatus (Günther, 1864)
Triportheus sp.

Crenuchidae

Crenuchus spilurus Günther, 1863
Microcharacidium sp.

Ctenoluciidae

Boulengerella lucius (Cuvier, 1816)

Curimatidae

Curimata inornata Vari, 1989
Curimata sp.1
Curimata sp.2
Curimatopsis sp.

Erythrinidae

Erythrinus erythrinus (Bloch and Schneider, 1801)
Hoplerythrinus unitaeniatus (Spix and Agassiz)
Hoplias malabaricus (Bloch, 1794)
Hoplias sp.

Gasteropelecidae

Carnegiella sp.

Carnegiella strigata (Günther, 1864)
Gasteropelecus sternicla (Linnaeus, 1758)
Thoracocharax stellatus (Kner, 1858)

Hemiodontidae
Hemiodus unimaculatus (Bloch, 1794)

Lebiasinidae
Copella arnoldi (Regan, 1912)
Copella nattereri (Steindachner, 1876)
Nannostomus eques Steindachner, 1876
Nannostomus nittidus Wetziman, 1978
Pyrrhulina filamentosa Valenciennes, 1847

CLUPEIFORMES: Engraulididae
Anchovia sp.
Anchovia surinamensis (Bleeker, 1866)
Lycengraulis batesii (Günther, 1868)
Lycengraulis sp.
Pterengraulis atherinoides (Linnaeus, 1766)

Pristigasteridae
Pellona castelnaeana (Valenciennes, 1847)
Pellona flavipinnis (Valenciennes, 1836)
Pellona sp.

CYPRINODONTIFORMES: Anablepidae
Anableps microlepis Muller and Troschel, 1844

Poeciliidae
Pamphorichthys sp.
Poecilia sp.1
Poecilia sp.2
Poecilia sp.3
Poecilia sp.4
Poecilia sp.5

Rivulidae
Rivulus sp.

GYMNOTIFORMES: Apteronotidae
Adontosternarchus sp.
Apteronotus albifrons (Linnaeus, 1766)
Apteronotus sp.
Sternarchella cf. *terminalis* (Eigenmann and Allen, 1942)
Sternarchella sp.
Sternarchogiton nattereri (Steindachner, 1868)
Sternarchogiton porcinum Eigenmann and Allen, 1942
Sternarchogiton sp.
Sternarchorhamphus muelleri (Steindachner, 1881)
Sternarchorhamphus sp.

Gymnotidae
Electrophorus electricus (Linnaeus, 1766)
Gymnotus carapo Linnaeus, 1758
Gymnotus sp.

Hypopomidae
Brachyhypopomus brevirostris (Steindachner, 1868)
Brachyhypopomus pinnicaudatus (Hopkins, 1991)
Brachyhypopomus sp.
Hypopomus sp.
Hypopygus lepturus Hoedeman, 1962
Hypopygus sp.
Microsternarchus bilineatus Fernández-Yépez, 1968
Steatogenys elegans (Steindachner, 1880)
Steatogenys sp.
Rhamphichthyidae
Rhamphichthys marmoratus Castelnau, 1855
Rhamphichthys rostratus (Linnaeus, 1766)
Rhamphichthys sp.
Sternopygidae
Distocyclus conirostris (Eigenmann and Allen)
Eigenmannia humboldtii (Steindachner, 1878)
Eigenmannia sp.1
Eigenmannia sp.2
Rhabdolichops sp.
Rhabdolichops troscheli (Kaup, 1856)
Sternopygus macrurus (Bloch and Schneider, 1801)
Sternopygus sp.
LEPIDOSIRENIFORMES: Lepidosirenidae
Lepidosiren paradoxa Fitzinger, 1837
OSTEOGLOSSIFORMES: Arapaimidae
Arapaima gigas (Schinz, 1822)
Osteoglossidae
Osteoglossum bicirrhosum (Cuvier, 1829)
PERCIFORMES: Centropomidae
Centropomus sp.
Cichlidae
Acaronia nassa (Heckel, 1840)
Aequidens pallidus (Heckel, 1840)
Aequidens sp.
Apistogramma luelingi Kullander, 1976
Apistogramma sp.
Astronotus ocellatus (Agassiz, 1831)
Chaetobranchopsis orbicularis (Steindachner, 1875)
Chaetobranchopsis sp.
Chaetobranchus sp.
Cichla melaniae Kullander and Ferreira, 2006
Cichla monoculus Spix and Agassiz, 1831
Cichla nigromaculata Jardine, 1843
Cichla ocellaris Bloch and Schneider, 1801
Cichla sp.1
Cichla sp.2

Cichla temensis Humboldt, 1821
Cichlasoma sp.
Crenicichla cincta Regan, 1905
Crenicichla macrophthalma Heckel, 1840
Crenicichla notophthalmus Regan, 1913
Crenicichla regani Ploeg, 1991
Crenicichla reticulata Heckel, 1840
Crenicichla sp.1
Crenicichla sp.2
Crenicichla sp.3
Crenicichla strigata Günther, 1862
Geophagus camopiensis Pellegrin, 1903
Geophagus sp.
Geophagus surinamensis (Bloch, 1791)
Satanoperca jurupari (Heckel, 1840)

Gobiidae
Awaous flavus (Valenciennes, 1837)
Bathygobius sp.
Eleotris sp.
Evorthodus sp.
Gobioides sp.
Gobionellus sp.

Mugilidae
Mugil curema Valenciennes, 1836

Polycentridae
Monocirrhus polyacanthus Heckel, 1840
Polycentrus schomburgkii Müller and Troschel, 1848

Sciaenidae
Cynoscion microlepidotus (Cuvier, 1830)
Cynoscion sp.
Macrodon ancylodon (Bloch and Schneider, 1801)
Pachypops fourcroi (Lacepède, 1802)
Pachypops sp.
Pachypops trifilis (Müller and Troschel, 1849)
Plagioscion auratus (Castelnau, 1855)
Plagioscion sp.
Plagioscion squamosissimus (Heckel, 1840)
Plagioscion surinamensis (Bleeker, 1973)

PLEURONECTIFORMES: Achiridae
Achirus achirus (Linnaeus, 1758)
Achirus sp.

Paralichthyidae
Paralichthys brasiliensis (Ranzani, 1842)
Paralichthys sp.

RAJIFORMES: Potamotrygonidae
Potamotrygon motoro (Müller and Henle, 1841)
Potamotrygon orbignyi (Castelnau, 1855)

Potamotrygon scobina Garman, 1913
Potamotrygon sp.

SILURIFORMES: Aspredinidae
Aspredinichthys filamentosus (Valenciennes, 1840)
Aspredo aspredo (Linnaeus, 1758)
Bunocephalus aleuropsis Cope, 1870

Auchenipteridae
Ageneiosus inermis (Linnaeus, 1766)
Ageneiosus sp.1
Ageneiosus sp.2
Ageneiosus ucayalensis Castelnau, 1855
Asterophysus sp.
Auchenipterichthys longimanus (Günther, 1864)
Auchenipterus nuchalis (Spix and Agassiz, 1829)
Centromochlus heckelii (De Filippi, 1853)
Pseudauchenipterus nodosus (Bloch, 1794)
Tatia intermedia (Steindachner, 1877)
Trachelyopterus galeatus (Linnaeus, 1766)
Trachelyopterus sp.

Callichthyidae
Callichthys callichthys (Linnaeus, 1758)
Hoplosternum littorale (Hancock, 1828)
Megalechis personata (Ranzini, 1841)
Megalechis thoracata (Valenciennes, 1840)

Cetopsidae
Cetopsis sp.
Hemicetopsis sp.

Doradidae
Acanthodoras sp.
Anadoras sp.
Doras eigenmanni (Boulenger, 1895)
Doras sp.
Hassar sp.
Lithodoras dorsalis (Valenciennes, 1840)

Heptapteridae
Gladioglanis machadoi Ferraris and Mago-Leccia, 1989
Pimelodella altipinnis (Steindachner, 1864)
Pimelodella cristata (Müller and Troschel, 1848)
Pimelodella gracilis (Valenciennes, 1835)
Pimelodella sp.
Rhamdia quelen (Quoy and Gaimard, 1824)
Rhamdia sp.

Insertae sedis
Phreatobius cisternarum Goeldi, 1905

Loricariidae
Ancistrus sp.1
Ancistrus sp.2

Chaetostoma sp.
Farlowella aff. *knerii* (Steindachner, 1882)
Farlowella amazona (Gunther, 1864)
Farlowella sp.
Hemiancistrus sp.
Hemiodontichthys acipenserinus (Kner, 1853)
Hemiodontichthys sp.
Hypoptopoma sp.
Hypostomus sp.
Lasiancistrus sp.
Limatulichthys sp.
Liposarcus pardalis (Castelnau, 1855)
Loricaria catphracta Linnaeus, 1758
Panaque sp.
Pseudoloricaria sp.
Pterygoplichthys sp.
Reganella sp.

Pimelodidae
Brachyplatystoma vaillantii (Valenciennes, 1840)
Hypophthalmus marginatus Valenciennes, 1840
Pimelodus blochii Valenciennes, 1840
Pimelodus ornatus Kner, 1858
Pimelodus sp.1
Pimelodus sp.2
Pinirampus pirinampu (Spix and Agassiz, 1829)
Zungaro zungaro (Humboldt, 1821)

Pseudopimelodidae
Batrochoglanis raninus (Valenciennes, 1840)

Trichomycteridae
Henonemus taxistigmus (Fowler, 1814)

SYNBRANCHIFORMES: Synbranchidae
Synbranchus lampreia Favorito, Zanata and Assumpção, 2005
Synbranchus marmoratus Bloch, 1795
Synbranchus sp.

TETRAODONTIFORMES: Tetraodontidae
Colomesus asellus (Müller and Troschel, 1849)
Colomesus psittacus (Bloch and Schneider, 1801)

7. Acknowledgments

We are grateful to the PROBIO program (Project for the Conservation and Sustainable Use of Biological Diversity) of the Brazilian Environment Ministry, which is subsidized by the Global Environment Facility (GEF), which financed the PROBIO Marajó project. We are also grateful to Mr. Miguel von Behr for providing the pictures that make up this work and Dr. Stephen Ferrari for his help in correcting text, particularly the english. Finally, we dedicate this paper to the memory of the coordinator of the PROBIO Marajó project, Samuel Almeida.

8. References

Ackermann, F.L. (1963). O lago Arari da ilha de Marajó e seus problemas. *Revista Brasileira de Geografia*, Vol.25, pp. 273-276, ISSN 0034-723X

Albuquerque, A.A. & Barthem, R.B. (2008). A pesca do tamoatá *Hoplosternum littorale* (Hancock, 1828) (Siluriformes: Callichthyidae) na ilha de Marajó. *Boletim do Museu Paraense Emílio Goeldi*, série Ciências Humanas, Vol.3, No.3, pp. 359-372, ISSN 1981-8122

Almeida, M.P.; Barthem, R.B.; Viana, A.S.; Charvet-Almeida, P. (2009). Factors affecting the distribution and abundance of freshwater stingrays (Chondrichthyes: Potamotrygonidae) at Marajó Island, mouth of the Amazon River. *Pan-American Journal of Aquatic Sciences*, Vol.4, No.1, pp. 1-11, ISSN 1809-9009

Assunção, M.I.S. & Schwassmann, H. (1995). Reproduction and larval development of *Electrophorus electricus* on Marajó Island (Pará, Brazil). *Ichthyological Exploration of Freshwaters*, Vol.6, No.2, pp. 175-184, ISSN 0936-9902

Barbosa, N.G.S. (2005). Bubalinocultura no Estado do Pará. *Revista Brasileira de Reprodução Animal*, Vol.29, No.1, pp. 34-38, ISSN 0102-0803

Barbosa, R.I.; Campos, C.; Pinto, F.; Fearnside, P.M. (2007). The "Lavrados" of Roraima: Biodiversity and Conservation of Brazil's Amazonian Savannas. *Functional Ecosystems & Communities*, Vol.1, pp. 29-41, ISSN 1749-0502

Barthem, R.B. & Fabré, N.N. (2003). Biologia e diversidade dos recursos pesqueiros da Amazônia, In: *A pesca e os recursos pesqueiros na Amazônia brasileira*, M.L., Rufino (Ed.), 11-55, Pro-Várzea, Manaus, Amazonas, Brazil, ISBN 85-7401-124-X

Barthem, R.B. & Golding, M. (2007). *An Unexpected Ecosystem: The Amazon revealed by the fisheries*, Lima Amazon Conservation Association (ACA) and Missouri Botanical Garden Press, Lima, Peru, ISBN 978-9972-2912-4-1.

Barthem, R.B. & Schwassmann, H.O. (1994). Amazon river influence on the seasonal displacement of the salt wedge in the Tocantins river estuary, Brazil, 1983-1985. *Boletim do Museu Paraense Emílio Goeldi*, série Zoologia, Vol.10, pp. 119-130, ISSN 0077-2232

Barthem, R.B. (2001). Componente Biota Aquática, In: *Biodiversidade na Amazônia Brasileira*, J.P.R. Capobianco, A. Veríssimo, A. Moreira, D. Sawyer, I. Santos and L.P. Pinto (Eds.), 60-78, Câmara do Livro, São Paulo, Brazil, ISBN 85-85994-13-4

Bernardi, C.C. (2005). *Conflitos sócio-ambientais decorrentes da bubalinocultura em territórios pesqueiros artesanais: o caso Olinda Nova do Maranhão*, Mastering Dissertation, Universidade Católica de Brasília, Brasília, Brazil

Boulenger, G.A. (1897). On a collection of fishes from the island of Marajó, Brazil. *Annals and Magazine of Natural History*, Vol.20, No.117, pp. 294-299

Brasil, República Federativa do Brasil. (1990). *Censos econômicos 1985 - Censo agropecuário*, Fundação Instituto Brasileiro de Geografia e Estatística (IBGE), Rio de Janeiro, No.1

Bucher, E.; Bonetto, A.; Boyle, T.; Canevari, P.; Castro, G.; Huszar, P.; Stone, T. (1993). *Hidrovia: Un examen ambiental inicial de la via fluvial Paraguay-Parana*, Humedales para Las Americas, Manomet, Mass, USA and Buenos Aires, Argentina, ISBN ISBN 91-865510-01-0

Cardoso, E.C. & Pereira, W.L.A. (2002). *Mineral deficiency of buffaloes from Marajó Island, North of Brazil: current situation and perspectives*, I Buffalo Symposium of Americas, Belém, Brazil, 2002

Carrier, J.C.; Pratt, H.L.; Castro, J.I. (2004). Reproductive biology of elasmobranchs, In: *Biology of sharks and their relatives*, J.C. Carrier, J.A. Musick and M.R. Heithaus M.R. (Eds.), 269-286, CRC Press, Texas, USA, ISBN-10: 084931514X

Correa, S.B.; Winemiller, K.O.; López-Fernández, H.; Galetti, M. (2007). Evolutionary perspectives on seed consumption and dispersal by fishes. *BioScience*, Vol.57, pp. 748–756, ISSN 0916-8451

Cruz, M.E.M. (1987). *Marajó: Essa imensidão de ilha*, Parma, São Paulo, Brazil

Dias, M.C.O. (Ed). (1999). *Manual de impactos ambientais: orientações básicas sobre aspectos ambientais de atividades produtivas*, Banco do Nordeste, Fortaleza, Ceará, Brazil

Eiten, G. (1972). The cerrado vegetation of Brazil. *Botanical Review*, Vol.38, pp. 201-341, ISSN 0006-8101

Favorito, S.E.; Zanata, A.M.; Assumpção, M.I. (2005). A new *Synbranchus* (Teleostei: Synbranchiformes: Synbranchidae) from ilha de Marajó, Pará, Brazil, with notes on its reproductive biology and larval development. *Neotropical Ichthyology*, Vol.3, No.3, pp. 319-328, ISSN 1679-6225

Gottsberger, G. (1978). Seed dispersal by fish in the inundated regions of Humaita, Amazonia. *Biotropica*, Vol.10, pp. 170–183, ISSN 0006-3606

Goulding, M. (1980). *The Fishes and the Forest: Explorations in Amazonian Natural History*, University of California Press, Berkeley, California, USA, ISBN 0-520-04131-3

Goulding, M.; Barthem, R.B.; Ferreira, E. (2003). *The Smithsonian atlas of the Amazon*, Smithsonian Books, Washington D.C., USA, ISBN 1-588-34135-6

Greenwood, P.H. & Liem, K.F. (1984) Aspiratory respiration in *Arapaima gigas* (Teleostei, Osteoglossomorpha): A reappraisal. *Journal of Zoology*, Vol.203, pp. 411-425, ISSN 0952-8369

Harris, D.R. (1980) Tropical savanna environments: Definition, distribution, diversity, and development, In: *Human Ecology in Savanna Environments*, D.R. Harris (Ed.), 2-27, Academic Press, London, UK, ISBN 0-12-326550-9

Hilbert, P.P. (1952) Contribuição à Arqueologia da Ilha de Marajó: Os Tesos Marajoaras do Alto Camutins e a atual situação da Ilha do Pacoval, no Arari. *Instituto de Antropologia e Etnologia do Pará*, Vol.5, pp. 5-32

Homma, A.K.O.; Conto, A.J.; Ferreira, C.A.P.; Carvalho, R.A.; Walker, R.T. (1998). A Dinâmica da Extração Madeireira no Estado do Pará, In: *Amazônia: Meio Ambiente e Desenvolvimento Agrícola*, A.K.O. Homma (Ed.), 161-186, EMBRAPA/SPI, Brasília, Brazil, ISBN 8-57-383026-3

Imhoff, S.; Silva, A.P.; Tormena, C.A. (2000). Aplicações da curva de resistência no controle de qualidade física de um solo de pastagem. *Pesquisa Agropecuária Brasileira*, Vol.35, pp. 1493-1500, ISSN 1678-3921

Johansen, K.; Lenfant, C.; Schmidt-Nielsen, K. (1968). Gas exchange and control of breathing in the electric eel, *Electrophorus electricus*. *Zeitschrift für vergleichende Physiologie*, Vol.61, pp. 137-163

Junk, W.J. & Wantzen, K.M. (2004). The Flood Pulse Concept: New Aspects, Approaches, and Applications - An Update. *Proceedings of the Second International Simposium on the Management of Large Rivers for Fisheries*. FAO Regional Office for Asia and the Pacific, Bangkok, Thailand, 2002, ISBN 978-0-12-088449-0

Junk, W.J. (Ed.). (1997). *The Central Amazon Floodplain: Ecology of a Pulsing System*, Springer-Verlag, New York, USA, ISBN 978-3540592761

Junk, W.J.; Bayley, P.B.; Sparks, R.E. (1989). The flood pulse concept in river-floodplain systems. In: Proceedings of International Large River Symposium, D.P. Dodge (Ed.), *Canadian Special Publication of Fisheries and Aquatic Sciences*, Vol.106, pp. 110-127

Kosuth, P.; Callède, J.; Laraque, A.; Filizola, N.; Guyot, J.L.; Seyler, P.; Fritsh, J.M. (1999). Influence de la marée océanique sur le cours aval de l'Amazone. *Hydrological and Geochemical Processess in Large Scale River Basins,* Manaus, Amazonas, Brazil, 1999

Kramer, D.L.; Lindsey, C.C.; Moodie, G.E.E.; Stevens, E.D. (1978). The fishes and the aquatic environment of the central Amazon basin, with particular reference to respiratory patterns. *Canadian Journal of Zoology*, Vol.56, pp. 719-729, ISSN 0008-4301

Kubitzki, K. & Ziburski, A. (1994). Seed dispersal in flood plain forests of Amazonia. *Biotropica*, Vol.26, pp. 30–43, ISSN 0006-3606

Lima, A.M.M.; Oliveira, L.L.; Fontinhas, R.L.; Lima, R.J.S. (2005). Ilha do Marajó: revisão histórica, hidroclimatologia, bacias hidrográficas e proposta de gestão. *Holos Environment*, Vol.5, No.1, pp. 65-80, ISSN 1519-8634

Lowe-McConnell, R.H. (1999). *Estudos ecológicos de comunidades de peixes tropicais*, EDUSP, São Paulo, Brazil, ISBN 85-314-0493-2

Marques-Aguiar, S.A.; Melo, C.C.S.; Aguiar, G.F.S.; Queiróz, G.A.L. (2002). Levantamento preliminar da mastofauna da região de Anajás-Muaná, Ilha de Marajó, Pará, Brasil. *Revista Brasileira de Zoologia*, Vol.19, No.3, pp. 841-854, ISSN 0101-8175

Mesquita-Saad, L.S.B.; Leitão, M.A.B.; Paula-Silva, M.N.; Chippari-Gomes, A.R.; Almeida-Val, V.M.F. (2002). Especialização e supressão metabólicas durante períodos de estivação do peixe pulmonado sul-americano. *Lepidosiren paradoxa. Brazilian Journal of Biology*, Vol.62, No.3, pp. 495-501, ISSN 1678-4375

Miranda Neto, M.J. (1968). *A Foz do Rio-Mar, Subsídios Para o Desenvolvimento de Marajó.* Distribuidora Record, Rio de Janeiro, Brazil

MMA, Ministério do Meio Ambiente. (2007). *Unidades de Conservação do Brasil*, Secretaria de Biodiversidade e Florestas/MMA, Brasília, Brazil

Montag, L.F.A., Albuquerque, A.A., Freitas, T.M.S.; Barthem, R.B. (2009). The Ichthyofauna of Savannas from Marajó Island, State of Pará, Brazil. *Biota Neotropica*, Vol.9, No.3, ISSN 1676-0611

Moran, E.F. (1995). Rich and poor ecosystems of Amazonia: an approach to management, In: *The Fragile Tropics of Latin American: Sustainable Management of Changing Environments*, T. Nishizawa and J.I. Uitto (Eds.), United Nations University Press, New York, USA, ISBN 92-808-0877-X

Oliveira, J.C.S. (2000). *Ecologia e reprodução do tamuatá,* Hoplosternum littorale *Hancok, 1828 (Osteichthyes: Siluriformes: Callichthyidae), da área de proteção ambiental do rio Curiaú, Macapá-AP*. Doctor Thesis, Universidade Federal do Rio Grande do Norte, Rio Grande do Norte, Brazil

Persaud, D.I.; Ramnarine, I.W.; Agard, J.B.R. (2006). Ontogeny of the alimentary canal and respiratory physiology of larval *Hoplosternum littorale* (Hancock, 1828): an intestinal air-breathing teleost. *Environmental Biology of Fishes*. Vol.76, pp. 37-45, ISSN 0378-1909

Pollux, B.J.A. (2011). The experimental study of seed dispersal by fish (ichthyochory). *Freshwater Biology*, Vol.56, pp. 197–212, ISSN 0046-5070

Prance, G.T. & Schaller, G.B. (1982). Preliminary study of some vegetation types of the Pantanal, Mato Grosso, Brazil. *Brittonia*, Vol.34, pp. 228-251, ISSN 0007-196X

Ratter, J.A.; Ribeiro, J.F.; Bridgewater, S. (1997). The Brazilian cerrado vegetation and threats to its biodiversity. *Annals of Botany*, Vol.80, pp. 223-230, ISSN 0305-7364

Reis, R.E.; Kullander, S.O.; Ferraris Jr., C.J. (2003). *Check list of freshwater fishes of South and Central America*. EDIPUCRS, Porto Alegre, Brazil, ISBN 9788574303611

Ribeiro, J.F. & Walter, B.M.T. (1998). Fitofisionomias do bioma Cerrado, In: *Cerrado: ambiente e flora*, S.M. Sano and S.P. Almeida (Eds.), 89-166, EMBRAPA/CPAC, Planaltina, Goias, Brazil

Roberts, T.R. 1993. Artisanal fisheries and fish ecology below the great waterfalls in the Mekong River in southern Laos. *Natural History Bulletin of the Siam Society*, Vol.41, pp. 31-62

Roosevelt, A.C.; Housley, R.A.; Imazio da Silveira, M.; Maranca, S.; Johnson, R. (1991). Eighth millennium pottery from a prehistoric shell midden in the Brazilian Amazon. *Science*, Vol.254, pp. 1557-1696, ISSN 0036-8075

Sanaiotti, T.M.; Bridgewater, S.; Ratter, J.A. (1997). A floristic study of the savanna vegetation of the State of Amapá, Brazil, and suggestions for its conservation. *Boletim do Museu Paraense Emílio Goeldi*, série Botânica, Vol.13, No.1, ISSN 0077-2216

Santos, V.F. (2006). *Ambientes costeiros amazônicos: avaliação de modificações por sensoriamento remoto*. Doctor Thesis, Universidade Federal Fluminense, Rio de Janeiro, Brazil

Schaan, D. (2010). Long-Term Human Induced Impacts on Marajó Island Landscapes, Amazon Estuary. *Diversity*, Vol.2, pp. 182-206.

Simões, M.F. (1981). Coletores-Pescadores Ceramistas do Litoral do Salgado (Pará). *Boletim do Museu Paraense Emílio Goeldi*, série Antropologia, Vol.7, pp. 1-31, ISSN 0522-7291

Smith, N. (2001) Land use dynamics in the amazon estuary and implications for natural resource management. *Amazoniana*, Vol.16, pp. 517-537

Smith, N. (2002). *Amazon Sweet Sea: Land, Life, and Water at the River's Mouth*, University of Texas Press, Austin, Texas, USA, ISBN 0292777701

Sombroek, W.G. (1966). *Amazon Soils: A Reconnaissance of the Soils of the Brazilian Amazon Region*, Centre for Agricultural Publication and Documentation, Wageningen, Netherlands

Teixeira, J.F. (1953). *O Arquipélago de Marajó*, Instituto Brasileiro de Geografia e Estatística, Rio de Janeiro, Brazil

UNEP, United Nations Environment Programme. (2004). *Amazon Basin: GIWA Regional assessment 40b*, University of Kalmar, Kalmar, Sweden

Volker, A. (1966). Surface hydrology of deltaic areas, *Proceedings of the Dacca Symposium - Scientific problems of the humid tropical zone deltas and their implications*, 153-159, Paris, France

Welcomme, R.L. (1979). *The Fisheries Ecology of Floodplain Rivers*, Longman Press, London, UK, ISBN 978-0582463103

Winemiller, K.O. (1989). Development of dermal lip protuberances for aquatic surface respiration in South American characid fishes. *Copeia*, Vol.2, pp. 382-390, ISSN 0045-8511

13

Biodiversity Conservation Planning in Rural Landscapes in Japan: Integration of Ecological and Visual Perspectives

Yoji Natori[1], Janet Silbernagel[2] and Michael S. Adams[2]
[1]*Conservation International*
[2]*University of Wisconsin-Madison*
[1]*Japan*
[2]*USA*

1. Introduction

Conservationists worldwide have long been interested in rural landscapes (McNeely & Keeton, 1995; McNeely, 1995; Miller & Hobbs, 2002; Washitani, 2001; Yokohari et al., 2005), which can be characterized as semi-natural areas that are neither pristinely natural nor urban, maintained by appropriate level of human interventions. A great portion of the world's biodiversity is found in these landscapes (Pimentel et al., 1992). Such landscapes have gained an international attention as Satoyama, and an international partnership has been established in 2010 to promote sustainable use of human-influenced natural environment through the Satoyama Initiative (Convention on Biological Diversity Decision X/32). Japanese archipelago is one of the 34 Biodiversity Hotspots of the world (Mittermeier et al., 2004), and its biodiversity owe much to the quality of the human-influenced natural environment.

Rural landscapes are an important conservation challenge in Japan because they are being lost rapidly. The challenge is that the traditional conservation strategy of "setting aside" will not work because humans play important roles in maintaining biodiversity on such landscapes (e.g., Farina, 1995; McNeely, 1995; Melnick, 1983; Nakagoshi, 1995; Natori et al., 2005; Washitani, 2001). Conservation in rural environments faces difficulties also because the public tends to associate nature conservation with pristine, untouched nature (Miller & Hobbs, 2002). The conservation of rural environments would require approaches different from the traditional conservation strategies employed for pristine natural areas.

Many have suggested that the consideration of the sociocultural dimension is crucial to the success of the conservation of rural environments in particular, and the conservation of biodiversity in general (e.g., Miller & Hobbs, 2002; Phillips, 1995; Pimentel et al., 1992; Saunders, 1990; Yokohari et al., 1994). Naveh (2000; 2001) has explicitly included humans in his theoretical development of a holistic approach to landscape studies. Born and Sonzogni (1995) and Margerum and Born (1995) have articulated a more pragmatic means to deal with environmental problems in the framework of integrated environmental management. Trauger (1999) calls for a shift from a traditional discipline-based approach to a problem-based approach. Accordingly, transdisciplinary research is being advocated in recent years for rural landscape research (Tress et al., 2005). This chapter focuses on the problem of rural landscape conservation. One way of considering people and nature simultaneously is to incorporate the visual perspectives into ecological conservation planning (Makhzoumi & Pungetti, 1999; Natori et al., 2005). Here we define "landscape" as the collective surface features of the study area in which the spatial (biosphere and geosphere), mental (noosphere), and temporal dimensions interact under influences of natural and cultural forces (the interaction in the sense as in Tress and Tress, 2001).

We hypothesize that: *Agricultural intensification and marginalization have had effects on local biodiversity via altering habitat amount and quality for species that occur on rural landscapes. The same changes have been perceived differently by different groups of people.* In the research conducted in the Arai-Keinan Region, Niigata, Japan, we asked the following set of questions in our attempt to incorporate human perceptions and values in ecological conservation planning for rural landscapes.

1. How has the rural landscape changed in terms of land use and land cover since the 1940s?
2. Considering the habitat associations of species dependent on the rural landscapes, what are the consequences of the observed landscape changes on the area's biodiversity?
3. Do people with different relationships with rural landscapes differ in terms of their landscape preferences and perceptions? How do the people perceive the landscape changes? What characteristics of landscape account for people's landscape preferences?

2. Underlying concepts of the study

We briefly discuss the key concepts of this Arai-Keinan study to illustrate the considerations and assumptions that have gone into our approach.

2.1 Biodiversity

Biodiversity encompasses multi-dimensional, interdisciplinary concepts. There is a multitude of definitions for what biodiversity is (e.g., Kaennel, 1998). Biodiversity conservation requires clear operationalization of the abstract concept of biodiversity, which has to be considered within the frame of the purpose and spatio-temporal scale of the conservation mission.

Species- or taxa-focused approaches are popular in conservation because they are measurable (Duelli & Obrist, 2003) and have high accountability backed by a long implementation history (Clark, 1999). However, these approaches have been criticized because 1) there are too many species on any given landscapes (Franklin, 1993); 2) it is cost-ineffective in that the amount of money spent for saving a few near-extinct species may be better spent for saving many less threatened species for a larger return in biodiversity conservation (Possingham et al., 2002); 3) the habitat manipulation specifically designed for a single species can have, although unintended, negative consequences to other species in the same community (Caughley & Sinclair, 1994; De Leo & Levin, 1997); 4) many species can go extinct unnoticed while conservation attention is given to saving one species, because saving that species does little to improve factors that are affecting other species (Caro et al., 2004; Prendergast et al., 1993); and 5) the time lags between habitat deterioration (or loss) and species loss (Brooks et al., 1999; Fry, 1998; Löfvenhaft et al., 2004) may confound the problem. The species-based approach might not be a good conservation tool, especially if saving particular species is considered as an end in itself (Meffe & Carroll, 1994).

In this study, our focus is on the landscape/habitat level characteristics of biodiversity on the basis for the coarse-filter approach (Haufler, 1999; Hunter, 1991; Noss & Cooperrider, 1994). The coarse-filter approach assumes that, if habitat types are available, the species associated with them, including those about which we know little, will persist on the landscape. This approach is effective and cost-efficient, especially when information and resources are limited (Noss & Scott, 1997). Using carefully selected species as surrogates of ecological integrity is a more effective conservation tool than considering specific species themselves as conservation targets (Johnson & Hill, 2002; Noss, 1990, 1991). We will use certain species to understand the ecological implication of land-use and land-cover changes, but quantifying species richness/diversity is outside the scope of this study.

Considering the preceding discussion, this chapter considers biodiversity in the context of nature conservation, and operationally defines that to conserve biodiversity is to secure habitats for groups of species traditionally associated with the area of interest.

2.2 Rural landscapes in biodiversity conservation

Rural landscapes are important for the conservation of biodiversity for two reasons: 1) as a buffer to the protected areas; and 2) as habitat that supports biodiversity by themselves. These should hold true globally. Most discussions of biodiversity conservation planning have focused on "natural areas," and essentially treated rural environments, including agricultural areas, as inhospitable surroundings (e.g., Lambeck, 1997; Meffe & Carroll, 1994; Peck, 1998; Trauger, 1999). Recent landscape ecological studies, however, consider such environments as matrix, and argue that their quality as habitat is important for maintenance of biodiversity (e.g., Andrén, 1994; Fahrig, 2001; Franklin, 1993). Fahrig (2001) reports that a promising and practically feasible conservation strategy is to improve the quality of the matrix (i.e., the survival probability in non-habitat areas).

The buffer zone concept, as in UNESCO's Man and Biosphere Programme (MAB), also considers rural areas as one of the layers of protection around protected areas (Lynagh & Urich, 2002; Wells & Brandon, 1993). The notion of rural areas as buffers to designated reserves carries a connotation that the reserves are primary and rural areas are secondary, but MAB's Madrid Action Plan for Biosphere Reserves (2008-2013) now recognizes clearly that buffer zones also have their own stand-alone ecological and cultural values. It is not too much of stretch to argue that the similar applies to "buffer areas" inside and outside of any protected areas. It has been recognized that protected areas established to date alone cannot adequately secure habitat for the growing number of endangered species, and cannot encompass all important ecosystem types (e.g., McNeely, 1995; Miller, 1996; Scott et al., 2001). Thus, much biodiversity depends on areas outside of protected areas (e.g., Peck, 1998; Pimentel et al., 1992). A recent gap analysis revealed that this holds true for Japan (Natori et al., in review).

Not only do rural landscapes supplement the lack of protected areas, but also they provide habitats for groups of species that would not be as abundant otherwise. An example is wetland environment that rice paddies provide to frogs, in place of natural wetlands rice paddies replaced (most likely they provide more than what naturally existed). People have been the primary disturbance agents in rural landscapes and have been maintaining systems at intermediate levels in ecological succession (Washitani, 2001). The focus of this chapter is based on the importance of rural landscapes as habitat that inherently supports biodiversity by themselves.

In Japan, rice cultivation has a history of over two thousand years (Hasegawa & Tabuchi, 1995). In many parts of Japan, it is the primary form of rural land use. Agricultural land and secondary forests, which have received human influence, covered 42% of the land in Japan according to a survey conducted during 1993-1998 (Nature Conservation Bureau & Asia Air Survey, 1999). The long history of human interaction with their natural environments has created various types of ecosystems that are different from the ecosystems from which they were originally derived (Nature Conservation Bureau, 2002). For example, many frog species have adapted to traditional rice cultivation practices that are phenologically synchronized (Washitani, 2001). Rural landscapes contribute to Japan's biodiversity not only because they occupy a large proportion of the territory, but also because they have come to serve as suitable habitats for many species (Kato, 2001). The traditional agricultural systems in Japan maintained their ecological integrity (in the sense defined by Regier, 1993); i.e., they were resilient systems, rather than the state of being unaltered. Due to the adoption of new technology, the traditional, close relations among humans, agricultural land, and the coppice have grown weaker. The 1960s, the period in which Japan experienced rapid economic growth, is often cited as a turning point for human use of rural environments (e.g., Ishii et al., 1993; Kamada et al., 1991; Tsunekawa, 2003; Washitani, 2001).

In recent years, much attention has been drawn to the multiple environmental functions of rice cultivation; the conservation of biodiversity being one of them (Takeuchi et al., 2003; Washitani, 2001). There is a concern that these functions might be lost as a higher production efficiency is sought (agricultural intensification) or as less efficient lands are abandoned (agricultural marginalization). Physical land modifications to allow mechanized agriculture (Hasegawa & Tabuchi, 1995) have had negative consequences on diversity of plants (Okubo & Maenaka, 1995), frogs (Fujioka & Lane, 1997), birds (Lane & Fujioka, 1998), freshwater fish (Katano et al., 2003), and other groups (e.g., see Ezaki & Tanaka, 1998; Kato, 2001). On the other hand, the management of coppices and grasslands, which had been practiced to obtain fuel or green manure, ceased, affecting many species of plants (Washitani, 2001) and butterflies (Higuma, 1998; Ishii et al., 1993). The traditional agricultural systems are arguably among the most threatened ecosystems in Japan today.

2.3 Visual perspectives

Rural landscapes are under varying degrees of human influence. Thus, how people behave in a landscape has much bearing on rural conservation. The habitat/landscape-based approach to biodiversity conservation helps integration of the ecological dimensions with visual dimensions. The people-landscape interaction model (Tress & Tress, 2001) conceptualizes the structure of interactions between people and rural landscapes. Landscapes influence people's perceptions, which, in turn, influence people's action toward the landscapes. We expect that visual aspects of landscapes play a very strong role in formation of perceptions, which eventually manifest as management, alteration, or conservation of landscapes. For this reason, the visual aspects of the landscape can be tightly related to the state of biodiversity, especially on rural landscapes.

In rural landscape management, those who work the land and those with naturalist/ecologist-orientations may not have the same inclinations toward the landscape and yet may have to work together. Differences in their way of looking at rural landscapes should be of great interest for planners. The degrees of landscape stewardship may be more important for farmers than for non-farmers (Nassauer, 1988). This aspect of landscape's visual quality has not been addressed in previous studies in Japan (Inose et al., 2002; Kanno et al., 1998; Shinji, 1981; Suzuki & Hori, 1989; Tanokura et al., 1999), while the dimensions of scenic quality and naturalness have been given attention. Thus, it is possible that important factors in landscape perceptions for farmers have not been studied.

3. Study site

This chapter focuses on the Arai-Keinan region located in the southwestern part of the prefecture of Niigata, on the Japan Sea side of Honshu Island, Japan (Fig. 1). This region has an area of 555.58 km^2 and elevation ranges from 10 m to 2,462 m above sea level. The region has a population of approximately 50,000 people, who mostly reside below 800 m above the sea level. A plain area is surrounded by mountains, the western part of which is designated as national park. In the east, the landscape exhibits varied topography with terrace-type rice paddies interspersed among woodlands. The central plain is gently sloped to the north, where the majority of residential and industrial activities occur. Rice cultivation agriculture is a major land use in the region, even in the mountainous areas. The climate of the region is characterized by high levels of precipitation in winter, most of which falls as snow.

Woodlands at low elevations are dominated by *Quercus serrata*, characteristic of coppice vegetation in Japan. Other major species include *Magnolia obovata*, *Prunus* spp., and *Acer* spp. At higher elevations, *Q. mongolica* var. *grosseserrata*, *Betula ermanii*, and *Fagus crenata* become dominant. Most of the plantations are those of *Cryptomeria japonica*.

Fig. 1. The location of the study sites in Japan and representative landscapes

4. Methods and results

4.1 Land use transitions

We documented changes in rural landscapes from time series aerial photographs between 1947 and 1999 for four landscapes: Itakura (3 km × 3 km; average elevation: 37 m; average slope: 0.8°), Sarukuyoji (3 km × 3 km; 277 m; 10.5°), Suibara (4 km × 4 km; 423 m; 17.5°), and Takatoko (6 km × 5 km; 250 m; 8.7°). Orthophotos, produced using OrthoMapper (Image Processing Software Inc., Madison, Wisconsin, USA), were interpreted in a 50-m grid into ten categories of land-use and land-cover (LULC, Table 1). Hereinafter, this is referred to as the "LULC study".

Large proportions of the four landscapes underwent conversions during 1947-1999 (Fig. 2). By 1999, irregular rice paddies on flat areas mostly had been converted to regular paddies. LULC conversions on slope areas occurred in smaller spatial units; rice paddy abandonment in small spatial units caused the marginalized landscape to become more heterogeneous, in terms of landscape patterns defined by the LULC types. Cutting pressure on coppices for fuel use was found to have been strong in 1947, indicated by the extensive areas of grassland or shrub land. By 1999, these areas had undergone plant succession in the absence of cutting pressure, and woodlands had become a dominant cover type. Using the resultant maps as base data, we interpreted landscape transitions between two different years, and identified the cells that underwent agricultural intensification, reclamation, agricultural marginalization, and ecological succession.

LULC Types	Descriptions
1. Irregular rice paddies	Irregularly shaped paddies; rice paddies not as 2.
2. Regular rice paddies	Rice paddies that are rectangular, that have boundaries meeting at right angle, or at least whose longer sides are parallel if surrounded by other land types.
3. Dry farms	Agricultural land producing crops other than rice.
4. Developed area	Areas not considered appropriate for habitat, including residential, roads, railroads, and other areas covered with concrete or asphalt.
5. Woodland	Collection of tall, broadleaved trees.
6. Cedar plantation	Collection of *Cryptomeria japonica* (and other conifers), including plantation.
7. Shrub land	Areas with trees of low statures (taller than grasslands, but shorter than woodlands): shrubs and re-sprouting broadleaved trees, including post-logging woodlands.
8. Grassland	Vegetated areas without tree crowns or shrubs.
9. Open water	Rivers and ponds, including non-vegetated floodplains.
10. Bare ground	Areas of bare soil, with little vegetation; including fallow rice paddies, aftermath of landslides.

Table 1. Land-use and land-cover (LULC) classification scheme

To establish theoretical grounds for future habitat outlook, the associations between the LULC changes and steepness of slopes were tested statistically using Jacobs' electivity index (Jacobs, 1974; Pastor & Broschart, 1990). The tests indicated statistically significant associations between agricultural marginalization and the slopes steeper than 6 degrees. Thus, although irregular rice paddies remained on sloped lands, they are expected be reduced further.

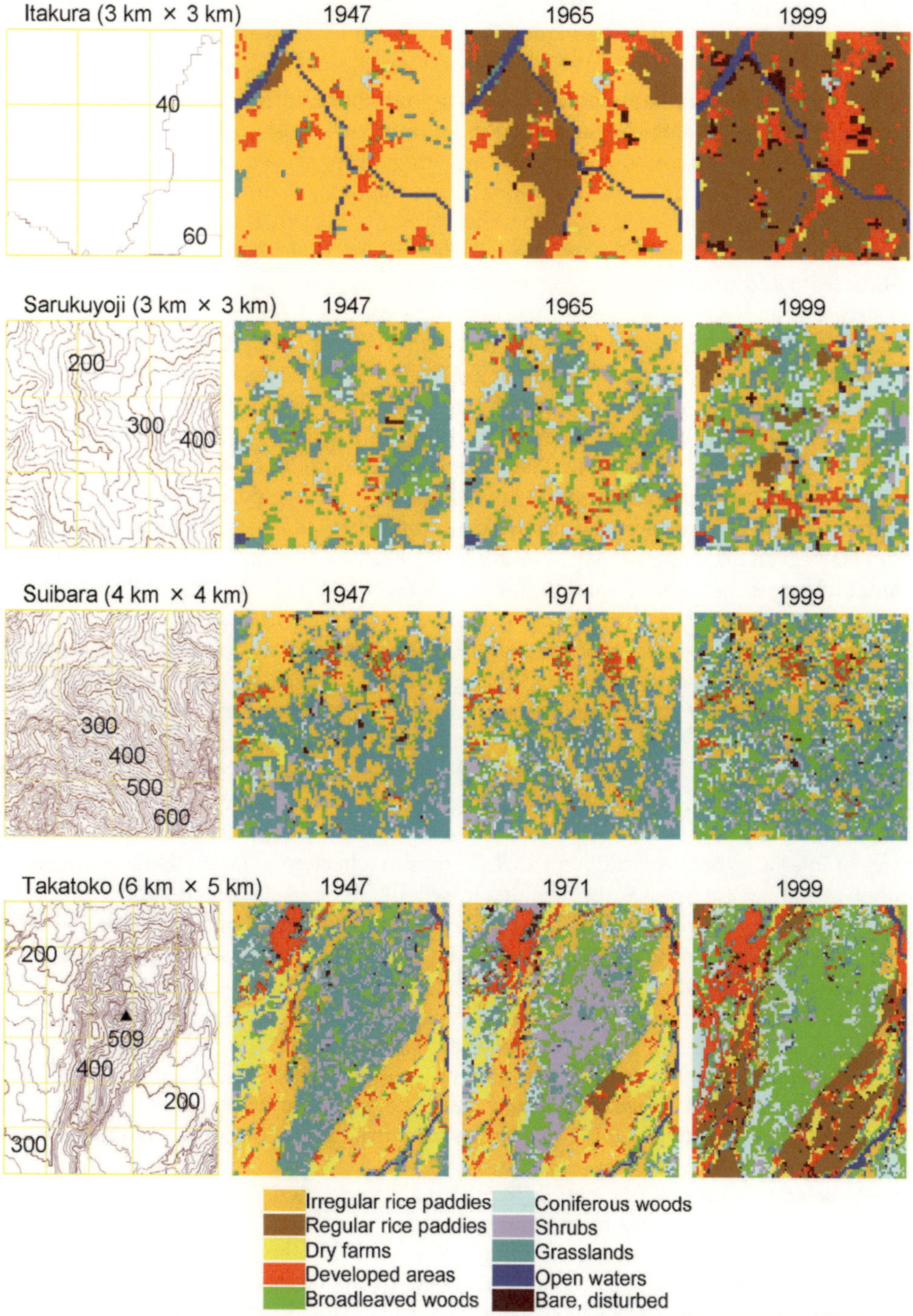

Fig. 2. Land-use and land-cover changes between 1947 and 1999 in four selected landscapes

We are then interested in the ecological and visual consequences of these changes. We addressed the hypotheses by collection of three inter-related studies: first on finer-scale ecological consequences, second on a broader scale consequences (details in Natori & Porter, 2007) and third on the visual aspects (details in Natori & Chenoweth, 2008).

4.2 Fine-scale ecological consequences: Frog habitat conditions

We investigated the ecological consequences of observed LULC changes at a fine scale, in terms of habitat quality for native frog species. Frogs were chosen as the indicator since the frog diversity characterizes the traditional rural landscapes in Japan (Hasegawa, 1998; Ministry of the Environment, 2002). From the literature, the presence of water in spring and soil ditches around the paddies were identified as the necessary conditions for suitable habitats for paddies-dependent frog species, such as *Rana japonica, R. ornativentris, R. nigromaculata,* and *R. rugosa*. These features were recorded on 96 sites (59 irregular rice paddies and 37 regular rice paddies) by repeated visits in the spring of 2005. The results confirmed that irregular rice paddies were generally wet in the spring *and* had at least some soil ditches around them (41 sites out of 59 or 70%); thus, suitable as frog habitat. Regular rice paddies tended to be dry in the spring and were irrigated by concrete ditches (25 sites out of 37 or 68%); thus, unsuitable. With this confirmation, landscape quality information (i.e., suitable or unsuitable as habitat) was associated with visual assessment of landscapes (i.e., irregular or regular rice paddies). Agricultural statistics (e.g., Statistics Department, 2005) and vegetation mapping (e.g., Nature Conservation Bureau, 1999) miss capturing the material changes in habitat quality (or habitat conversion) that takes place in rural landscapes, because they do not distinguish different types of rice paddies.

We analyzed further the effects of the LULC changes on habitat conditions for native frogs by considering the pattern of the LULC changes. The landscape patterns in the LULC maps presented above were quantified by two landscape metrics: proportion of landscape (PLAND) by LULC type and contrast-weighted edge density (CWED) between irregular rice paddies and other LULC types (McGarigal et al., 2002). The PLAND quantitatively estimated the amount of habitat for species whose life cycles complete within rice paddies and irrigation ditches; i.e., *R. nigromaculata* and *R. rugosa*. The CWED, measured in the meters of interfaces between different LULC types in a hectare, quantitatively estimated the amount of habitat for species that resides in woodlands and grasslands and come to rice paddies in springs to spawn; i.e., *R. japonica* and *R. ornativentris*. Considering the difference in habitat suitability, irregular rice paddy interface with woodland was weighted by 1.0, with grassland by 0.5, and with others by 0.0. The relative increase/decline in the CWED through time indicated the increase/decline in suitable habitat for these species.

Given the drastic decline in the amount of irregular rice paddies (Fig. 3), *Rana nigromaculata* and *R. rugosa* are expected to have lost a significant portion of their habitat. On the other hand, *R. japonica* and *R. ornativentris* may have gained more habitat during intermediate years, due to the pattern of LULC conversions from irregular rice paddies, but their habitats too are on a declining trend now.

In sum, habitat conditions for native frogs have deteriorated during 1947-1999 and will likely continue to deteriorate at the fine scale. For rice paddies to support biodiversity, efforts beyond simply continuing farming are needed. Finding and coordinating a social system that can provide key features of irregular rice paddies (i.e., standing water in the spring, and non-concrete ditches or streams) should be considered as an alternative strategy for rural conservation.

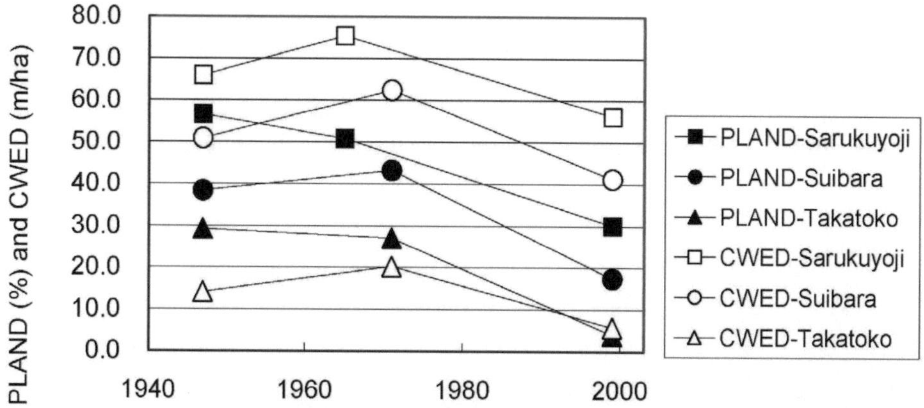

Fig. 3. Changes in proportions of landscape occupied by irregular rice paddies (PLAND) and contrast-weighted edge density between irregular rice paddies and deciduous woodlands, cedar plantation, shrubs, or grasslands (CWED)

4.3 Broad-scale ecological consequences: The Japanese serow

The Japanese serow (*Capricornis crispus*) is a medium-sized, territorial bovid (an average adult weighing 36 kg) and its habitat assessment needs to consider much larger area. Natori and Porter (2007) addressed the ecological consequences of land use changes from a broader scale to define the context in which the study area was situated. To understand the effects of the LULC changes on habitat conditions for large mammals, it assessed habitat suitability by simulating the energy budget of the Japanese serow using the energetics model (Porter et al. 1994; 2000; 2002). The energetics model determines the metabolic energy costs required to maintain homeostasis from the directly measurable properties of the environment (such as air temperature, wind speed, amount of shade, vegetation, etc.) and those of the animal (such as size, body temperature, fur density, etc.). The animal's ambient environment, modeled from weather data, was modified by vegetation. The LULC maps from Arai-Keinan Region provided time series vegetation data and the national level surveys (Nature Conservation Bureau, 2005) provided the vegetation information of 1990s for the surrounding area. Any particular locations were considered uninhabitable where the serow should not be present (urban or agricultural) or unsuitable if the animal was not able to maintain homeostasis either because of overheating because it was unable to dissipate heat efficiently enough in the summer or because the serow was unable to obtain sufficient food to sustain the level of metabolism needed in the winter.

The results of the energetics model simulation using the LULC changes described above (Fig. 2) indicated that the changes were favorable for the serow to inhabit a larger area within the region (Fig. 4). Figure 4 only shows the results for the summer because winter conditions did not limit the habitat suitability in two landscapes in Arai-Keinan Region. The forest cover, which provides shade in summer that prevents the serow from being overheated in the summer and provide thermal cover and wind moderation in the winter, is an important determinant of the suitability of landscapes as serow habitat. In winter, snow reduces the amount of forage available to the serow, but forest covers are expected to provide more forage in snow than other, more open vegetations. Thus, the increase in forested areas in the Arai-Keinan region observed during 1947-1999 is expected to have increased the amount of suitable habitat for the serow. Natori and Porter (2007) also showed, based on the simulation for the larger area, that the Arai-Keinan region, which may have been isolated when the forest cover was limited, is now a part of a contiguous patch of suitable habitat. This makes this region even a better habitat for the serow. Thus, unlike the case of the frogs, the landscape changes in the Arai-Keinan Region have been in favor of the serow.

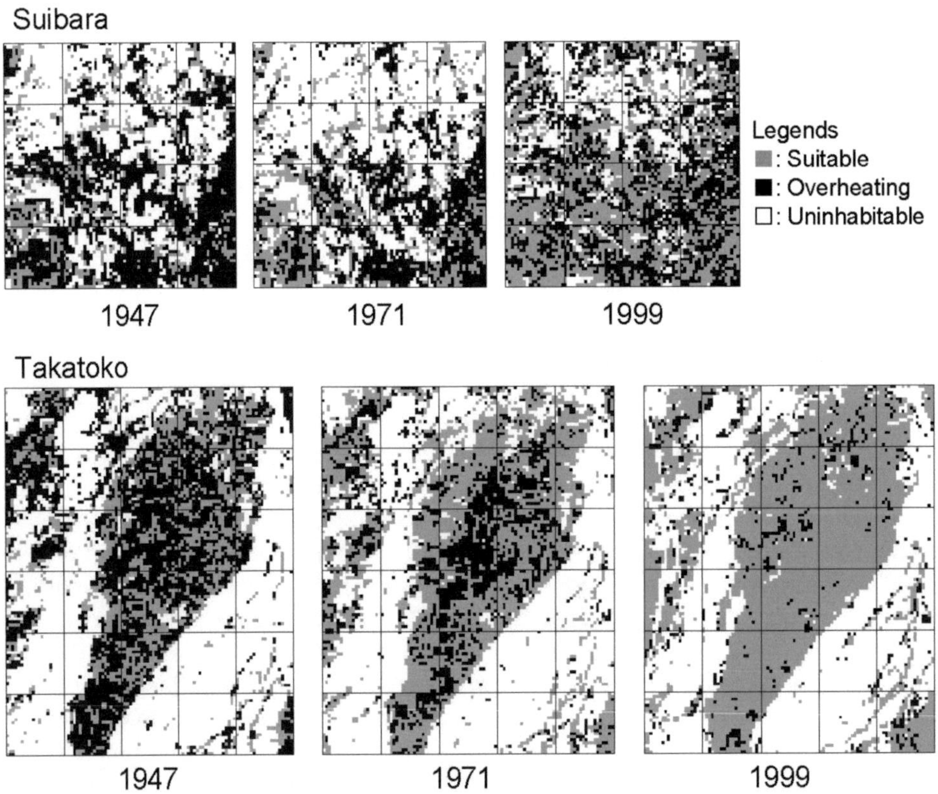

Fig. 4. Expansion of suitable habitat area in summer for the Japanese serow over the period from 1947 to 1999.

4.4 Landscape aesthetics

Natori and Chenoweth (2008) investigated the visual aspect of rural landscapes, and compared landscape preferences and perceptions among people having very different relationships to rice-paddy and woodland landscapes. The study had particular interest in revealing how local farmers (n = 41) and naturalists (non-farmers with conservation interests; n = 44) differ in their perceptions toward different states of rice-paddy and woodland landscapes. Photograph-based semantic differential (SD) surveys were conducted. Rice-paddy landscapes were represented by 2 × 3 factorial design (Fig. 5), and woodland landscapes, by 2 × 2 factorial design (Fig. 6). The SD variables quantified the participants' perceptions of naturalness, openness, stewardship, peacefulness, biodiversity and preference on seven-point scales. Observer difference was dummy-coded, and stepwise linear regressions were performed to test if farmers and naturalists differ in landscape preference and perception.

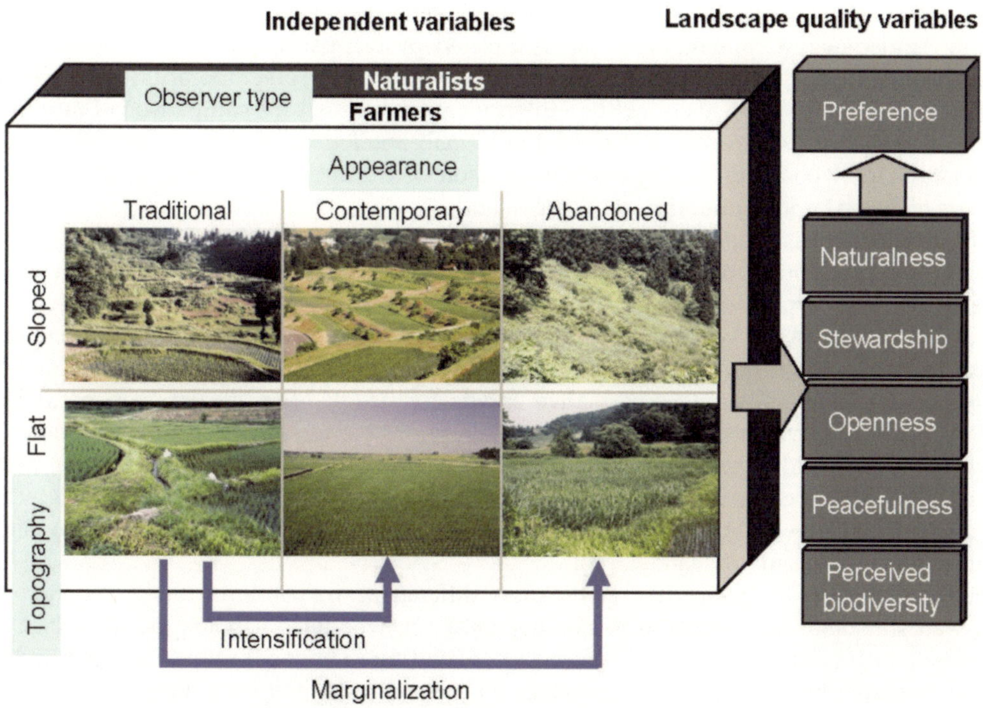

Fig. 5. Experimental design for perception survey on rice-paddy landscape

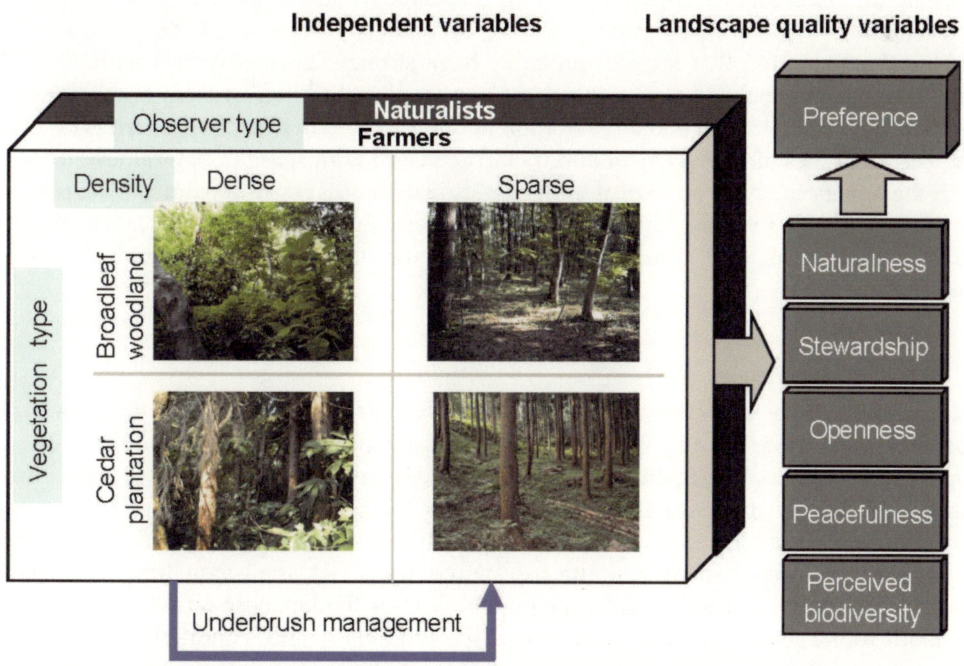

Fig. 6. Experimental design for perception survey on woodland landscape

It also investigated people's perceptions and preferences of landscape changes typical of rural landscapes in contemporary Japan inferred by comparing ratings to two different landscape types (Figs. 5 and 6). The comparison of ratings to contemporary rice paddies to those of traditional implied agricultural intensification, whereas the comparison of ratings to abandoned rice paddies to those of traditional implied agricultural marginalization. Similarly for woodlands landscapes, comparison of the ratings to sparse underbrush to those of dense implied underbrush management.

The results showed that farmers and naturalists differed in the way they look at rural landscapes. Perceptions of stewardship appeared more important for farmers, and perceptions of 'naturalness' appeared more important for naturalists with statistical significance in case of rice-paddy landscapes. With regard to changes in rice paddies, farmers disliked agricultural marginalization more strongly than naturalists did (with statistical significance), which may be attributed to farmers' stronger normative criteria for rice paddies. Farmers and naturalists disagreed about how they wanted rice paddies on sloped topography to be. Farmers preferred contemporary paddies, which the frogs study suggests to provide lesser quality habitat, while naturalists preferred traditional paddies, which provide more biodiversity benefits. This difference could be a point of conflict in biodiversity conservation as they represent two major stakeholder groups in rural landscapes.

Unlike the case of rice-paddy landscapes, differential influence of the perception of stewardship and naturalness was not apparent for woodland landscapes. This lack of difference could be because people viewed woodlands more from a third person's perspective resulting from having the lesser interaction with woodlands than with rice paddies. As for changes in woodlands, the clearing of underbrush has positive effects on both set of participants' preferences with regard to woodland landscapes. Managing underbrush in strategic locations could have a far-reaching conservation benefit by raising people's overall appreciation of woodlands, by providing an "orderly frame" (Nassauer, 1995).

The results suggested that farmers probably have stronger normative criteria for how the rural landscapes should look, and their emphasis is placed on stewardship or management. This study did not present enough variation in management to identify properties of "right" degrees and types of management and their interaction with sense of naturalness. Together with the advancement of ecological understanding on biodiversity of rural landscapes, such investigation will provide further insights for the conservation of landscapes in which natural ecosystems and human activities can sustainably coexist in today's societal conditions.

5. Discussion

The series of studies demonstrate that the LULC changes are not merely the changes in how people use the land, but that they also have consequences in ecological communities and people's perceptions. Agricultural intensification reduces habitat amount for native frogs, but it significantly increases farmers' preferences for rice paddies on sloped landscapes. Agricultural marginalization, which is significantly associated with rice paddies on sloped landscapes, reduces habitat amount for native frogs. At the same time, agricultural marginalization increases the landscape's suitability for the Japanese serow habitat because resultant forest covers provide the serow with favorable habitat conditions and the more heterogeneous landscape patterns provide more options for the serow to choose favorable environmental conditions. Agricultural marginalization, however, causes a decline in people's preference for the landscape. Increased woodland area in the rural landscape favors the Japanese serow's occurrence. Both farmers and naturalists preferred deciduous woodlands to cedar plantations, and both were in favor of underbrush management.

The two ecological studies clearly indicated that strong ecological consequences accompanied the LULC changes observed during 1947-1999, and provided support for the first hypothesis: *Agricultural intensification and marginalization have had effects on local biodiversity via altering habitat amount and quality for species that occur on rural landscapes.* The strength of support for the second hypothesis (*The same changes have been perceived differently by different groups of people*) depended on the context (e.g., rice paddies vs. woodlands).

We believe the following three mutually complementary issues to be important for biodiversity conservation in rural landscapes. This study was designed to provide concrete examples and further insights for these issues. They constitute key ingredients for identifying important areas for conservation, even in the absence of issues/projects that threaten the conservation values of landscapes. If this identification is performed, a more proactive approach, the "preemptive conservation" (Natori et al., 2005), becomes possible.

5.1 Multiple perspectives to view landscapes

If ecological considerations lacked linkages to people's landscape preferences, resultant conservation projects might not gain public support. Similarly, if only human perspectives were considered, ecological values of the rural landscape might suffer, such as in rural area development projects in which only farmers' priorities are considered.

Agricultural marginalization is more likely on sloped landscapes (the LULC study), which leads to reduction in habitat amount suitable for frogs such as *Rana nigromaculata* and *R. rugosa* (the frogs study). Depending on how agricultural marginalization occurs (especially with respect to landscape configuration), its effects on the habitat amount for *R. japonica* and *R. ornativentris* vary (the frogs study). The aesthetics study suggested that agricultural marginalization would be disliked by both farmers and naturalists, but for different reasons, which suggests that different approaches to resolve the problem might be needed. Although agricultural marginalization is viewed as unfavorable based on both ecological perspectives focused on frogs and people's landscape preferences, the serow modeling study suggested that the same change might benefit the serows by providing shade or thermal cover, especially if accompanied by plant succession to canopy forests. One may not be able to find a perfect land-use solution that satisfies all issues regarding rural landscapes, but including as many perspectives as feasible likely leads to improved efficacy of conservation measures (Born & Sonzogni, 1995; Margerum & Born, 1995).

We provide two examples to further illustrate the importance of multiple perspectives. First, current policies, such as the "direct payment system in hilly and mountainous areas," assume that multiple functions of rural landscapes will be secured if rice cultivation continues on sloped landscapes. Given that rice cultivation is the key process that supported the biodiversity in rural landscapes, and abandoning the cultivation is a threat to biodiversity conservation in Japan, maintaining rice cultivation appears to be consistent with the interest of biodiversity conservation. However, emphasis on only the continuation of agricultural production can undermine the goal of protecting multiple functions of the landscapes, because new agricultural practices may not be able to provide the functions that traditional agriculture provided (Yokohari, 2000). For rice cultivation to be economically viable, Zhou (2001) advocates that the enlargement of farm sizes, with mechanical work replacing human labor, would be the future direction for (economically) sustainable agriculture. To allow mechanization, small, irregular paddies must be converted to large, regular paddies. This conversion will result in decline of rice paddies' quality as habitat for many species that depend on traditional rice paddy landscapes. The aesthetics study considered such alteration of rice paddies from the perspective of people's landscape preferences, and revealed that it could lead to a conflict between farmers and naturalists, resulting from their different views about rice paddies on slopes.

Second, current methods of agricultural statistical record-keeping and vegetation mapping may not capture the substantial alterations of rural landscapes because they do not make a distinction between types of rice paddies. Among the environmental functions most recognized for rice paddies are flood control and groundwater recharge (Hasegawa & Tabuchi, 1995; Tabuchi & Ogawa, 1995; The Japan Environmental Council, 2005; Yokohari, 2000). For these functions, the type of rice paddy does not make a difference, though the type of rice paddy does make a difference for the conservation of biodiversity (the LULC and frogs studies). The inclusion of ecological aspects into rural development plans is critical to the effective conservation of biodiversity in rural landscapes. New agricultural policy since 1999, under the Basic Law on Food, Agriculture, and Rural Areas, can accommodate such considerations, but its focus is still on protection of farmers, rather than the environment (Yokohari, 2000). Concrete and convincing arguments from an ecological perspective are necessary for genuine inclusion of ecological considerations to happen on the ground.

Many functions that rural landscapes provide have been brought about by the process of rice cultivation, but this process is changing. We can no longer focus only on this process (i.e., continuing rice cultivation) and expect that other functions recognized for rural landscapes will be automatically provided. Thus, multiple perspectives on the same landscape must be considered.

5.2 Landscape-focused approaches and maps facilitate integration

In this chapter we conceptualized landscape as a space in which mental and ecological/physical dimensions interact under influences of natural and cultural factors (the people-landscape interaction model; Tress & Tress, 2001). Our operational definition of the conservation of biodiversity has been to secure habitats, rather than individual species. This framework helped integrate the ecological and visual aspects of rural landscapes in biodiversity conservation. The integration of ecological and visual aspects of landscapes brings us closer to reconciling nature and culture (i.e., people). Actions with spatial components can be linked to other functions if they are considered in the spatial expanse of landscapes. For example, as we demonstrated in the LULC study, one can see that patterns of association can emerge from consideration of the LULC changes and landscape features, such as steepness of slope.

Mapping is inseparably related to landscape-focused approaches; it is a tool for landscape visualization and landscape change analysis. Maps allow spatially-specific changes and properties to be visualized and understood better. Comparisons between both locations and times are also possible, and landscape metrics can facilitate quantitative comparisons (the LULC study). Maps linked the analyses of land use and land cover in the LULC study to ecological analysis in the frogs and serow modeling studies. Maps provided specificity to descriptions of changes. This is a key feature when multiple functions of rural landscapes are considered, because they are linked by landscapes. The visual features of maps are also effective in conveying information to the public (Nassauer & Corry, 2004; Rookwood, 1995), which may constitute one venue to foster cooperation and to close communication gaps that exist among ecologists and planners, policy-makers, and the public (August et al., 2002; Holdgate, 1984).

5.3 Temporal considerations are important

Knowing past changes provides information for the future. Even though we might not be able to revert to the past, we can recreate conditions that approximate the past. Multiple functions of rural landscapes (e.g., support for biodiversity, provision of scenery, etc.) can be maintained only if we know about the past. The trends observed historically in relation to physical features of the landscapes such as slopes, may allow future predictions. A caveat, which was also provided by temporal considerations, is that landscape changes have not always been unidirectional. Thus, although science cannot state what the reference for conservation should be, the dynamic nature of landscapes necessitates understanding the range of possible conditions conducive to target species or goals.

For example, our landscape assessment using aerial photograph interpretation indicated that the coppicing pressure on woodlands was strong in 1947. Some questions still require further research. Had the coppicing pressure always been at this level, or did it escalate during the World War II? Was such a land use sustainable in terms of maintenance of the ecological communities on rural landscapes as described in the literature? A more meaningful question for practical conservationists might be: what level of coppice management would it take to sustain the biodiversity of traditional rural landscapes within the modern landscapes? Answering these questions is important for the conservation of rural landscapes and requires an understanding of historic landscape conditions and dynamics. Such research could be considered in the framework of adaptive management (Holling et al., 1998; Walters & Holling, 1990).

Temporal depth is important in understanding the trajectory of changes in landscapes. In this LULC study, we demonstrated that landscape imagery, such as aerial photographs, can effectively be used to track changes in land use and land cover since 1947 (see also Ihse, 1995; Kurita & Yokohari, 2001; Yokohari & Kurita, 2003). It will be difficult to discriminate types of rice paddies in satellite imagery. Aerial photographs and other landscape imagery can effectively provide temporal depth of information on land use and land cover. For these reasons, aerial photographs will continue to be a valuable tool for rural conservation planning, even in this age of the remarkable development of satellite remote sensing technologies (e.g., Lillesand & Kiefer, 2000).

5.4 Conservation planning

These three issues just discussed can be placed in the context of conservation planning and implementation as in Fig. 7. In rural conservation planning, the direct stakeholders and the public in general need to be included in the process. Sharing a clear common goal among stakeholders is key for collaboration despite differences (Norton, 1991). The "spatial narrative" (Silbernagel, 2005), a framework that synthesize multiple perspectives, such as objective geographic space and subjective experiential place, will be particularly effective. Temporal consideration can inform what the conservation goals can be. Information is the key for effective conservation measures, but there is a limit to what can be known with the resource available. The concept of adaptive management (Walters & Holling, 1990), which considers policies as hypotheses and their implementation as experiments, would enhance the approach for biodiversity conservation. Conservation planning can start with coarse-filter approach and using surrogates, and specific measures can be refined as more information is collected through the course of the conservation activities.

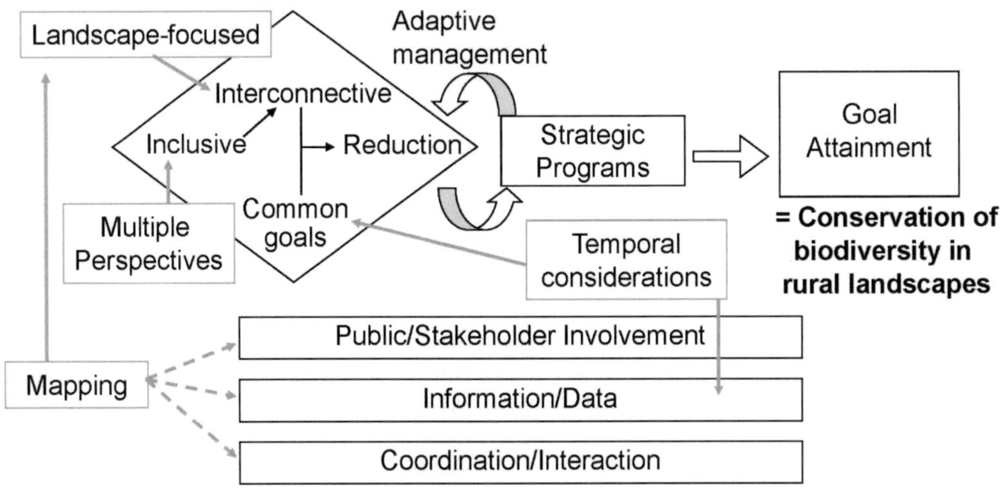

Fig. 7. A model for holistic approach to landscape management

6. Conclusion

Landscape changes observed in aerial photographs during 1947-1999 were shown to have had substantial ecological impact in the rural landscape of the study area. However, the assessment of whether the changes were positive or negative differed greatly depending on what part of biodiversity was used for the assessment; in our example, the native frogs vs. the Japanese serow. Furthermore, the human dimension, which should always be considered in rural conservation, was demonstrated to be also diverse, and a single preferred conservation direction may not automatically exist. These findings suggest that an effective conservation planning will require a transdisciplinary approach (Tress et al., 2005), in which interdisciplinary approach is combined with participation, to bring a long-lasting success to rural conservation.

7. Acknowledgements

We are grateful for the support we received from Drs. Richard Chenoweth, Warren P. Porter, Brian Ohm, and Brian Stone at the University of Wisconsin-Madison. During the period of research, YN was supported by a Japan-US Fulbright scholarship, teaching assistantship from the Botany and Zoology Departments, and a travel grant from the Nelson Institute for Environmental Studies at the University of Wisconsin-Madison.

8. References

Andrén, H. (1994). Effects of habitat fragmentation on birds and mammals in landscapes with different proportions of suitable habitat: a review. *Oikos*, Vol.71, pp. 355-366, ISSN 0030-1299

August, P., Iverson, L. and Nugranad, J. (2002). Human conversion of terrestrial habitats. In: *Applying Landscape Ecology in Biological Conservation,* Gutzwiller, K. J. (Ed.), pp. 198-224, Springer, ISBN 0-387-95322-1, New York

Born, S. M. and Sonzogni, W. C. (1995). Integrated environmental management: strengthening the conceptualization. *Environmental Management,* Vol.19, pp. 167-181, ISSN 0364-152X

Brooks, T. M., Pimm, S. L. and Oyugi, J. O. (1999). Time lag between deforestation and bird extinction in tropical forest fragments. *Conservation Biology,* Vol.13, pp. 1140-1150, ISSN 0888-8892

Caro, T., Engilis, A., Fitzherbert, E. and Gardner, T. (2004). Preliminary assessment of the flagship species concept at a small scale. *Animal Conservation,* Vol.7, pp. 63-70, ISSN: 1367-9430

Caughley, G. and Sinclair, A. R. E. (1994). *Wildlife Ecology and Management.* Blackwell Scientific Publications, ISBN 0-86542-144-7, Boston

Clark, J. R. (1999). The ecosystem approach from a practical point of view. *Conservation Biology,* Vol.13, pp. 679-681, ISSN 0888-8892

De Leo, G. and Levin, S. (1997). The multifaceted aspects of ecosystem integrity. *Conservation Ecology,* Vol. 1, pp. 3. Retrieved from http://www.consecol.org/vol1/iss1/art3

Duelli, P. and Obrist, M. K. (2003). Biodiversity indicators: the choice of values and measures. *Agriculture, Ecosystems and Environment,* Vol.98, pp. 87-98, ISSN: 0167-8809

Ezaki, Y. and Tanaka, T. (Eds.) (1998). *Mizube Kankyo no Hozen--Seibutsu Gunshu no Shiten kara.* Asakura-Shoten, ISBN 4-254-10154-6, Tokyo

Fahrig, L. (2001). How much habitat is enough? *Biological Conservation,* Vol.100, pp. 65-74, ISSN 0006-3207

Farina, A. (1995). Cultural landscapes and fauna. In: *Cultural Landscapes of Universal Value,* Droste, B. v., Plachter, H. and Rossler, M. (Eds.), pp. 60-77, Gustav Fischer, ISBN 3-334-61022-5, Jena

Franklin, J. F. (1993). Preserving biodiversity: species, ecosystems, or landscapes? *Ecological Applications,* Vol.3, pp. 202-205, ISSN 1051-0761

Fry, G. L. A. (1998). Changes in landscape structure and its impact on biodiversity and landscape values; a Norwegian perspective. In: *Key Concepts in Landscape Ecology: Proceedings of the 1998 European Congress of the International Association for Landscape Ecology, Held at the Myerscough College, 3rd-5th September 1998* Dover, J. W. and Bunce, R. G. H. (Eds.), pp. 81-92, IALE (UK), ISBN 0-9524263-5-8, Aberdeen, Scotland

Fujioka, M. and Lane, S. J. (1997). The impact of changing irrigation practices in rice fields on frog populations of the Kanto Plain, central Japan. *Ecological Research,* Vol.12, pp. 101-108, ISSN 0912-3814

Hasegawa, M. (1998). Suiden kosaku ni izon suru kaeru-rui gunshu. In: *Mizube Kankyo no Hozen (Conservation of Biological Communities in Rivers, Ponds and Paddy Fields),* Ezaki, Y. and Tanaka, T. (Eds.), pp. 53-66, Asakura-Sheten, ISBN 4-254-10154-6, Tokyo

Hasegawa, S. and Tabuchi, T. (1995). Well facilitated paddy fields in Japan. In: *Paddy Fields in the World,* Tabuchi, T. and Hasegawa, S. (Eds.), pp. 103-124, Japanese Society of Irrigation, Drainage and Reclamation Engineering, ISBN 4-88980-077-8, Tokyo

Haufler, J. B. (1999). Strategies for conserving terrestrial biological diversity. In: *Practical Approaches to the Conservation of Biological Diversity,* Baydack, R. K., Campa, H., III and Haufler, J. B. (Eds.), pp. 17-30, Island Press, ISBN 1-55963-544-4, Washington, D.C.

Higuma, S. (1998). Myoko-sankai no nantomen chiiki ni sumu cho-rui. *Transactions of Essa Entomological Society,* Vol.79, pp. 1-69, ISSN 0289-4289

Holdgate, M. W. (1984). Information needs for the decision-making process. In: *Ecology in Practice Part II: The Social Response,* Di Castri, F., Baker, F. W. G. and Hadley., M. (Eds.), pp. 249-263, Tycooly International, Dublin

Holling, C. S., Berkes, F. and Folke, C. (1998). Science, sustainability and resource management. In: *Linking Social and Ecological Systems: Management Practices and Social Mechanisms for Building Resilience,* Berkes, F., Folke, C. and Colding, J. (Eds.), pp. 342-363, Cambridge University Press, ISBN 0-52178-562-6, Cambridge

Hunter, M. L. (1991). Coping with ignorance: the coarse-filter strategy for maintaining biodiversity. In: *Balancing on the Brink of Extinction: The Endangered Species Act and Lessons for the Future,* Kohm, K. A. (Ed.), pp. 266-281, Island Press, ISBN 1-55963-006-X, Washington, D.C.

Ihse, M. (1995). Swedish agricultural landscapes--patterns and changes during the last 50 years, studied by aerial photos. *Landscape and Urban Planning,* Vol.31, pp. 21-37, ISSN 0169-2046

Inose, R., Kurita, K., Kuroyanagi, N. and Miyakawa, H. (2002). The categorization of grassland landscape features in the Mt. Aso Region of on Kyushu Island and characteristics of their public image. *Journal of the Japanese Institute of Landscape Architecture,* Vol.65, pp. 621-626, ISSN 1340-8984

Ishii, M., Ueda, K. and Shigematsu, T. (1993). *Satoyama no shizen wo mamoru.* Tsukiji Shokan, ISBN 4-8067-2346-0, Tokyo.

Jacobs, J. (1974). Quantitative measurement of food selection: a modification of the forage ratio and Ivlev's electivity index. *Oecologia,* Vol.14, pp. 413-417, ISSN 0029-8549

Johnson, B. and Hill, K. (2002). *Ecology and design : frameworks for learning.* Island Press, ISBN 1-55963-813-3, Washington, D.C.

Kaennel, M. (1998). Biodiversity: a diversity in definitions. In: *Assessment of Biodiversity for Improved Forest Planning: Proceedings of the Conference on Assessment of Biodiversity for Improved Planning, 7-11 October 1996, held in Monte Verità, Switzerland,* Bachmann, P., Köhl, M. and Päivinen, R. (Eds.), pp. 71-81, Kluwer Academic Publishers, ISBN 9780792348726, Dordrecht

Kamada, M., Nakagoshi, N. and Nehira, K. (1991). Pine forest ecology and landscape management: a comparative study in Japan and Korea. In: *Coniferous Forest Ecology from an International Perspective,* Nakagoshi, N. and Golley, F. B. (Eds.), pp. 43-62, SPB Academic Publishing, ISBN 9-05103-065-7, Hague, The Netherlands

Kanno, T., Fukuyama, M., Oku, T., Sasaki, H. and Nagamachi, M. (1998). A study of images and amenity of grassland landscape, using SD method. *Grassland Science,* Vol.44, pp. 127-137, ISSN 1744-6961

Katano, O., Hosoya, K., Iguchi, K. and Yamaguchi, M. (2003). Species diversity and abundance of freshwater fishes in irrigation ditches around rice fields. *Environmental Biology of Fishes,* Vol.66, pp. 107-121, ISSN 0378-1909

Kato, M. (2001). 'SATOYAMA' and biodiversity conservation: 'SATOYAMA' as important insect habitats. *Global Environmental Research,* Vol.5, pp. 135-149, ISSN 1343-8808

Kurita, H. and Yokohari, M. (2001). Ecological potentials of golf courses on hills for the conservation of satoyama landscapes. *Journal of the Japanese Institute of Landscape Architecture,* Vol.64, pp. 589-594, ISSN 1340-8984

Lambeck, R. J. (1997). Focal species: a multi-species umbrella for nature conservation. *Conservation Biology*, Vol.11, pp. 849-856, ISSN 0888-8892

Lane, S. J. and Fujioka, M. (1998). The impact of changes in irrigation practices on the distribution of foraging egrets and herons (Ardeidae) in the rice fields of central Japan. *Biological Conservation*, Vol.83, pp. 221-230, ISSN 0006-3207

Lillesand, T. M. and Kiefer, R. W. (2000). *Remote Sensing and Image Interpretation.* John Wiley & Sons, ISBN 0-471-25515-7, New York

Lynagh, F. M. and Urich, P. B. (2002). A critical review of buffer zone theory and practice: a Philippine case study. *Society and Natural Resources*, Vol.15, pp. 129-145, ISSN 0894-1920

Löfvenhaft, K., Runborg, S. and Sjögren-Gulve, P. (2004). Biotope patterns and amphibian distribution as assessment tools in urban landscape planning. *Landscape and Urban Planning*, Vol.68, pp. 403-427, ISSN 0169-2046

Makhzoumi, J. and Pungetti, G. (1999). *Ecological Landscape Design & Planning: The Mediterranean Context.* E & FN Spon, ISBN 0-419-23250-8, London

Margerum, R. D. and Born, S. M. (1995). Integrated environmental management: moving from theory to practice. *Journal of Environmental Planning and Management*, Vol.38, pp. 371-391, ISSN 0964-0568

McGarigal, K., Cushman, S. A., Neel, M. C. and Ene, E. (2002). *FRAGSTATS: Spatial Pattern Analysis Program for Categorical Maps.* University of Massachusetts, Amherst. Available from: http: www.umass.edu/landeco/research/fragstats/fragstats.html

McNeely, J. and Keeton, W. (1995). The interaction between biological and cultural diversity: Components of a Global Strategy. In: *Cultural Landscapes of Universal Value*, Droste, B. v., Plachter, H. and Rossler, M. (Eds.), pp. 25-37, Gustav Fischer, ISBN 3-334-61022-5, Jena

McNeely, J. A. (1995). How traditional agro-ecosystems can contribute to conserving biodiversity. In: Halladay, P. and Gilmour, D. A. (Eds.), *Conserving Biodiversity Outside Protected Areas: the Role of Traditional Agro-ecosystems.* IUCN, Gland, pp. 228.

Meffe, G. K. and Carroll, C. R. (1994). *Principles of Conservation Biology.* Sinauer Associates, Sunderland.

Melnick, R. Z. (1983). Protecting rural cultural landscapes: finding value in the countryside. *Landscape Journal*, Vol.2, pp. 85-97.

Miller, J. R. and Hobbs, R. J. (2002). Conservation where people live and work. *Conservation Biology*, Vol.16, pp. 330-337, ISSN 0888-8892.

Miller, K. R. (1996). Conserving biodiversity in managed landscapes. In: Szaro, R. C. and Johnston, D. W. (Eds.), *Biodiversity in Managed Landscapes.* Oxford University Press, New York, pp. 425-441.

Ministry of the Environment (2002). *Shin-Seibutsutayosei Kokka Senryaku. Gyos*ei, Tokyo.

Mittermeier, R. A., Gil, P. R., Hoffman, M., Pilgrim, J., Brooks, T., Mittermeier, C. G., Lamoreux, J. and Fonseca, G. A. B. d. (2004). *Hotspots Revisited: Earth's Biologically Richest and Most Endangered Terrestrial Ecoregions.* Cemex, Mexico City, Mexico.

Nakagoshi, N. (1995). Changing cultural landscapes in Japan. In: *Cultural Landscapes of Universal Value*, Droste, B. v., Plachter, H. and Rossler, M. (Eds.), pp. 128-138, Gustav Fischer, ISBN 3-334-61022-5, Jena

Nassauer, J. I. (1988). Landscape care: perceptions of local people in landscape ecology and sustainable development. In: *Landscape/land use planning.* American Society of Landscape Architects (Ed.), pp. 27-41, ASLA, Washington, D.C.

Nassauer, J. I. (1995). Messy ecosystems, orderly frames. *Landscape Journal*, Vol.14, pp. 161-170, ISSN 0277-2426

Nassauer, J. I. and Corry, R. C. (2004). Using normative scenarios in landscape ecology. *Landscape Ecology*, Vol.19, pp. 343-356, ISSN 0921-2973

Natori, Y. and Chenoweth, R. (2008). Differences in landscape perceptions and preferences between farmers and naturalists. *Journal of Environmental Psychology*, Vol.28, pp. 250-267, ISSN 0272-4944

Natori, Y., Fukui, W. and Hikasa, M. (2005). Empowering nature conservation in Japanese rural areas: a planning strategy integrating visual and biological landscape perspectives. *Landscape and Urban Planning*, Vol.70, pp. 315-324, ISSN 0169-2046.

Natori, Y., Kohri, M., Hayama, S. and De Silva, N. (in review). Key Biodiversity Areas identification in Japan Hotspot. Journal of Threatened Taxa, ISSN 0974-7907.

Natori, Y. and Porter, W. P. (2007). Japanese Serow (*Capricornis crispus*) Energetics Landscape Modeling Predicts Distribution on Honshu, Japan. *Ecological Applications*, Vol.17, pp. 1441-1459, ISSN 1051-0761

Nature Conservation Bureau (1999). *The Dataset for GIS on the Natural Environment, Japan (Ver. 2) CD-ROM.* Nature Conservation Bureau, Environment Agency, Tokyo

Nature Conservation Bureau (2002). *Living with Nature: the National Biodiversity Strategy of Japan.* Nature Conservation Bureau, Ministry of the Environment, Tokyo

Nature Conservation Bureau (2005). *The Dataset for GIS on the Natural Environment, Japan (world geodetic system) CD-ROM.* Biodiversity Center, Ministry of the Environment, Yamanashi

Nature Conservation Bureau and Asia Air Survey (1999). *Shokusei Chosa Houkokusho (5th National Survey on the Natural Environment report).* Nature Conservation Bureau, Environment Agency, Tokyo

Naveh, Z. (2000). What is holistic landscape ecology? A conceptual introduction. *Landscape and Urban Planning*, Vol.50, pp. 7-26, ISSN 0169-2046

Naveh, Z. (2001). Ten major premises for a holistic conception of multifunctional landscapes. Lands*cape and Urban Planning*, Vol.57, pp. 269-284, ISSN 0169-2046

Norton, B. G. (1991). *Toward Unity among Environmentalists.* Oxford University Press, New York

Noss, R. (1990). Indicators for monitoring biodiversity: a hierarchical approach. *Conservation Biology*, Vol.4, pp. 355-364, ISSN 0888-8892

Noss, R. (1991). From endangered species to biodiversity. In: *Balancing on the Brink of Extinction: The Endangered Species Act and Lessons for the Future*, Kohm, K. A. (Ed.), pp. 227-246, Island Press, ISBN 1-55963-006-X, Washington, D.C.

Noss, R. F. and Cooperrider, A. Y. (1994). *Saving Nature's Legacy: Protecting and Restoring Biodiversity.* Island Press, ISBN 1-55963-248-8, Washington, D.C.

Noss, R. F. and Scott, J. M. (1997). Ecosystem protection and restoration: the core of ecosystem management. In: *Ecosystem Management: Applications for Sustainable Forest and Wildlife Resources*, Boyce, M. S. and Haney, A. W. (Eds.), pp. 239-264, Yale University Press, ISBN 0-30007-858-7, New Haven

Okubo, K. and Maenaka, H. (1995). Effect of consolidation on paddy-side grassland communities and ecological niche in agro-ecosystem. *Journal of the Japanese Institute of Landscape Architecture*, Vol.58, pp. 109-112, ISSN 1340-8984

Pastor, J. and Broschart, M. (1990). The spatial pattern of a northern conifer-hardwood landscape. *Landscape Ecology*, Vol.4, pp. 55-68, ISSN 0921-2973

Peck, S. (1998). *Planning for Biodiversity: Issues and Examples.* Island Press, Washington, D. C.

Phillips, A. (1995). Cultural landscapes: an IUCN perspective. In: *Cultural Landscapes of Universal Value* Droste, B. v., Plachter, H. and Rossler, M. (Eds.), pp. 380-392, Gustav Fischer, ISBN 3-334-61022-5, Jena

Pimentel, D., Stachow, U., Takacs, D. A., Brubaker, H. W., Dumas, A. R., Meaney, J. J., O'Neil, J. A. S., Onsi, D. E. and Corzilius, D. B. (1992). Conserving biological diversity in agricultural/forestry systems: most biological diversity exists in human-managed ecosystems. *BioScience*, Vol.42, pp. 354-362, ISSN 0006-3568

Porter, W. P., Budaraju, S., Stewart, W. E. and Ramankutty, N. (2000). Calculating climate effects on birds and mammals: impacts on biodiversity, conservation, population parameters, and global community structure. *American Zoologist*, Vol.40, pp. 597-630, ISSN 0003-1569

Porter, W. P., Munger, J. C., Stewart, W. E., Budaraju, S. and Jaeger, J. (1994). Endotherm energetics: From a scalable individual-based model to ecological applications. *Australian Journal of Zoology*, Vol.42, pp. 125-162, ISSN: 0004-959X

Porter, W. P., Sabo, J. L., Tracy, C. R., Reichman, O. J. and Ramankutty, N. (2002). Physiology on a landscape scale: plant-animal interactions. *Integrative and Comparative Biology*, Vol.42, pp. 431-453, ISSN 1540-7063

Possingham, H. P., Andelman, S. J., Burgman, M. A., Medellin, R. A., Master, L. L. and Keith, D. A. (2002). Limits to the use of threatened species lists. *Trends in ecology & evolution*, Vol.17, pp. 503-507, ISSN 0169-5347

Prendergast, J. R., Quinn, R. M., Lawton, J. H., Eversham, B. C. and Gibbons, D. W. (1993). Rare species: the coincidence of diversity hotspots and conservation strategies. *Nature*, Vol.365, pp. 335-337, ISSN: 0028-0836

Regier, H. A. (1993). The notion of natural and cultural integrity. In: *Ecological Integrity and the Management of Ecosystems*, Woodley, S. and Kay, J. (Eds.), pp. 3-18, St.Lucie Press, ISBN 0-96340-301-X, Delray Beach, FL

Rookwood, P. (1995). Landscape planning for biodiversity. *Landscape and Urban Planning*, Vol.31, pp. 379-385, ISSN 0169-2046

Saunders, D. A. (1990). The landscape approach to conservation: community involvement, the only practical solution. *Australian Zoologist*, Vol.26, pp. 49-53, ISSN 0067-2238

Scott, M., Davis, F. W., McGhie, R. G., Wright, R. G., Groves, C. and Estes, J. (2001). Nature reserves: do they capture the full range of America's biological diversity? *Ecological Applications*, Vol.11, pp. 999-1007, ISSN 1051-0761

Shinji, I. (1981). Shizenkeikanchi ni okeru shikisai kankyo kanri no kihonnmondai. In: *Shizenkeikanchi ni okeru Shikisaikijun ni kansuru Kenkyu Hokokusho*, Nagashima, M., Shinji, I., Aso, M., Kodama, A. and Nishikawa, I. (Eds.), pp. 16-54, Asesu, Tokyo

Silbernagel, J. (2005). Bio-regional patterns and spatial narratives for integrative landscape research and design. In: *From Landscape Research to Landscape Planning: Aspects of Integration, Education, and Application*, Tress, B., Tress, G., Fry, G. and Opdam, P. (Eds.), pp. 107-118, Springer, ISBN 1-40203-979-4, Dordrecht

Statistics Department (2005). *The 79th Statistical Yearbook of Ministry of Agriculture, Forestry and Fisheries.* Statistics Department, Ministry of the Agriculture, Forestry and Fisheries, Tokyo

Suzuki, S. and Hori, S. (1989). A study on the estimation of naturalness and the liking in forest landscapes. *Journal of the Japanese Institute of Landscape Architecture*, Vol.52, pp. 211-216, ISSN 1340-8984

Tabuchi, T. and Ogawa, Y. (1995). Environmental characteristics of paddy fields. In: *Paddy Fields in the World*, Tabuchi, T. and Hasegawa, S. (Eds.), pp. 327-340, Japanese Society of Irrigation, Drainage and Reclamation Engineering, ISBN 4-88980-077-8, Tokyo

Takeuchi, K., Brown, R. D., Washitani, I., Tsunekawa, A. and Yokohari, M. (Eds.) (2003). *Satoyama: The Traditional Rural Landscape of Japan*. Springer, ISBN 4-43100-007-0, Tokyo.

Tanokura, N., Yokohari, M., Yamamoto, S. and Kato, Y. (1999). Local residents' impressions and perceptions on paddy field landscapes: a case study in Hiki region, Saitama Prefecture, Japan. *Journal of the Japanese Institute of Landscape Architecture*, Vol.62, pp. 727-732, ISSN 1340-8984

The Japan Environmental Council (Ed.) (2005). *The State of the Environment in Asia 2005/2006*. Springer, ISBN 9784431250289, Tokyo.

Trauger, D. L. (1999). Can we manage for biological diversity in the absence of scientific certainty? In: Practi*cal Approaches to the Conservation of Biological Diversity*, Baydack, R. K., Campa, H., III and Haufler, J. B. (Eds.), pp. 195-202, Island Press, ISBN 1-55963-544-4, Washington, D.C.

Tress, B. and Tress, G. (2001). Capitalising on multiplicity: a transdisciplinary systems approach to landscape research. *Landscape and Urban Planning*, Vol.57, pp. 143-157, ISSN 0169-2046

Tress, B., Tress, G. and Fry, G. (2005). Integrative studies on rural landscapes: policy expectations and research practice. *Landscape and Urban Planning*, Vol.70, pp. 177-191, ISSN 0169-2046

Tsunekawa, A. (2003). Transition of satoyama landscapes in Japan. In: *Satoyama: The Traditional Rural Landscape of Japan*, Takeuchi, K., Brown, R. D., Washitani, I., Tsunekawa, A. and Yokohari, M. (Eds.), pp. 41-51, Springer, ISBN 4-43100-007-0, Tokyo

Walters, C. J. and Holling, C. S. (1990). Large-scale management experiments and learning by doing. *Ecology*, Vol. 71, pp. 2060-2068, ISSN 0012-9658

Washitani, I. (2001). Traditional sustainable ecosystem 'SATOYAMA' and biodiversity crisis in Japan: conservation ecological perspective. *Global Environmental Research*, Vol.5, pp. 119-133, ISSN 1343-8808

Wells, M. P. and Brandon, K. E. (1993). The principles and practice of buffer zones and local participation in biodiversity conservation. *Ambio*, Vol.22, pp. 157-162, ISSN 0044-7447

Yokohari, M. (2000). Agrienvironmental policies in North America, Europe and Japan. *Journal of the Japanese Institute of Landscape Architecture*, Vol.63, pp. 182-185, ISSN 1340-8984

Yokohari, M., Brown, R. D. and Takeuchi, K. (1994). A framework for the conservation of rural ecological landscapes in the urban fringe area in Japan. *Landscape and Urban Planning*, Vol.29, pp. 103-116, ISSN 0169-2046

Yokohari, M. and Kurita, H. (2003). Mechanisms of satoyama landscape transformation. In:, *Satoyama: The Rural Landscapes of Japan*, Takeuchi, K., Brown, R. D., Washitani, I., Tsunekawa, A. and Yokohari, M. (Eds.), pp. 71-79, Springer-Verlag, ISBN 4-43100-007-0, Tokyo

Yokohari, M., Morimoto, Y. and Nakagoshi, N. (2005). Editorial: Ecological dynamics of urban and rural landscapes in East Asia. *Landscape and Urban Planning*, Vol.70, pp. 193-194, ISSN 0169-2046

Zhou, J.-M. (2001). *Sustainable Development in Asia, America and Europe with Global Applications: A New Approach to Land Ownership*. Edward Elgar, ISBN 1-85898-965-5, Cheltenham, UK.

14

Water for Biodiversity Conservation and Livelihoods: Protecting Northernmost Tropical Deciduous Forest Relicts in Mexico

Rolando E. Diaz-Caravantes and Erick Sánchez-Flores
Universidad Autónoma de Ciudad Juárez
México

1. Introduction

In the last two decades there has been a growing interest at a global scale to bring water resources management more squarely within biodiversity conservation policy. This interest has given rise to a diverse array of global water institutions, such as the World Commission of Dams (2000), that have emphasized the urgency of balancing human demands for water with the needs of the environment. These concerns, moreover, have begun to deeply affect the domestic policies of many countries, making scholarly evaluation of policy outcomes of extremely critical importance. However, analyses of what, exactly, might constitute effective conservation strategies in providing water for the environment, especially in developing countries, are still quite limited.

This study, based on a research in two local communities within the northernmost deciduous forest natural protected area in Sonora Mexico, "Sierra de Alamos Río Cuchujaqui" (SARC), expands greatly upon the existing work by addressing the challenges faced by developing countries seeking to protect their natural areas to maintain biodiversity and healthy watersheds, while sustaining economic livelihoods of local communities that are necessary for conservation goals in such a sensitive environment. Thus, this study aims to answer the following specific questions: What have been the role and strategies of the Mexican National Commission for Natural Protected Area (CONANP) to supply water for environmental conservation?; How do the practices of local livelihoods relate to the use of water for environmental conservation?; And how, and to what extent, have the conservation goals benefited the livelihoods of local communities?

Field and remote sensing research demonstrates that land cover in local communities is being degraded due to overgrazing and agricultural practices by local villagers. Economic development projects proposed by CONANP, such as a gila monster farm, ecotourism, and a nursery garden have not been successful. Despite their questionable economic success, these projects have been used to persuade people to introduce conservation programs using key natural resources, such as water. This is the case of the *retenidas*, which are check dams located in the streams that descend from the summit of the mountains and have the double purpose of preventing soil erosion and providing water for the surrounding environment.

Based on the theoretical concepts of Common Pool Resources and Institutional Framework (Ostrom, 2005), we examine how villagers have also found some advantages of the *retenidas* for their own, expecting the aquifer will recharge thanks to this infrastructure. In sum, the *retenidas* have been the means through which CONANP is able to secure water for environmental purposes, with a sufficient level of acceptability from the villagers. Thus, in these projects, CONANP have found a clever way to achieve conservation goals in a highly sensitive environment, providing water for both flora and fauna biodiversity in a deciduous forest relict.

2. Background

In the last decades, there has been a growing interest on a global scale about conservation and the sustainable use of the natural resources crystallized in the 1972 United Nations Conference on the Human Environment in Stockholm, the Brundtland Commission Report in 1987 and the 1992 United Nations Conference on the Environment and Development in Rio de Janeiro. This phenomenon is analyzed by Zimmerer (2006) and his collaborators in the volume: Globalization and New Geographies of Conservation, in which they offer an overview of recent geographical research in which globalization processes and environmental conservation, intersect.

According to Zimmerer, there are two premises regarding globalization processes and environmental conservation intersection: the first premise establishes that on one hand, global expansion of designated conservation areas is presented as a showcase of international and national environmentalism, and on the other hand, the designation of these areas generated disputes over the control the natural resources and the relationships with livelihood activities of the local people. The second premise of the new globalization processes of environmental conservation is the so-called "third wave," which has pushed to expand incorporation of sustainability into a global system of protected areas and has forced conservation to interface increasingly with agriculture and other types of livelihoods and resource use (Zimmerer, 2006). This new approach involves biodiversity conservation with people, instead of without them.

Regarding water resources, scholars and practitioners have emphasized the urgency of balancing human demands with the needs of the environment (Postel & Ricther, 2003; World Commission of Dams [WCD], 2000). The WCD, based in eight detailed case studies of large dams and in a survey of 125 large dams, concludes that regarding environment "[o]n balance, the ecosystem impacts are more negative than positive and they have led, in many cases, to significant and irreversible loss of species and ecosystems". In this regard, one of the seven strategic priorities, called sustaining rivers and livelihoods states that "[r]eleasing tailor-made environmental flows can help maintain downstream ecosystems and the communities that depend on them" (WCD, 2000).

Postel and Richter (2003) establish that society's relationship with freshwater ecosystems will need to change in several overarching ways. One of these ways is that "[t]he globalized trend toward the commodification and privatization of water requires a strong reaffirmation that water is a public trust to be preserved for the common good of this generation and those to come" (Postel and Richter, 2003). This suggestion corresponds with their proposal that governments must act swiftly to enact policies calling for freshwater reserves because the State, like the entity in charge of the public goods, must be the responsible of creating water reserves. Undoubtedly this proposal contradicts the vision of water as an economic good, confirmed in the World Bank's principled pragmatism, which move on direction of market and privatization of the resources. These approaches exemplified the complexity of designing water policies. Everybody agrees the importance of safeguarding water for environmental conservation, but to determine what are the adequate mechanisms for this purpose, may result highly contested. In the same way, regarding the focus of this work, the indispensable question is, what is the significance of all these proposals for developing countries, and moreover what does it mean for local people and livelihoods depending on water? Thus, the central question suggested by this work is to what degree and how the inclusion of the use of water for environmental conservation influences existing livelihoods. A fundamental framework to examine this question is provided by the theory of Common Pool Resources.

A Common Pool Resource (CPR) refers "a valued natural or human-made resource or facility that is available to more than one person and subject to degradation as a result of overuse" (Dietz et al., 2002). Forests, irrigation systems, grazing lands, fisheries, groundwater basins are typical CPRs. CPRs share two characteristics: (1) it is costly to develop institutions to exclude potential beneficiaries from them, and (2) the resource units that are appropriated by one individual are not available for others (Ostrom, 1990). According to Ostrom, the first characteristic leads to the potential problem of free riding. The second characteristic leads to the potential problems of congestion and over-harvesting (2006).

Overuse of resources is a central topic in the "tragedy of the commons" presented by Hardin (1968). This approach predicts that when individual and group interests are in opposition, narrow self-interest will win over the common-interest. In this regard, overuse of the resource will be the final and inexorable outcome, and he states "as the human population has increased, the common has had to be abandoned in one aspect after another" (Hardin, 1968). Hardin suggests that an alternative to avoid the tragedy is "mutual coercion mutually agreed upon" maybe created by a central authority or by privatization of the resources, which will result in a more sustainable manner of using the resources.

However, as Dietz et al. argue (2002), the "tragedy of the commons" could be avoided by mechanisms that act in the interest of the collective good rather than narrow self-interest. Therefore, central authority and privatization are not the unavoidable alternatives to sustainable resources. Ostrom analyzes some long-enduring self-governance Common Pool Resources. She states eight design principles that allow these CPRs continue to exist throughout time (1990).

1. Clearly defined boundaries
2. Congruence between appropriation and provision rules and local conditions.
3. Collective-Choice arrangements
4. Monitoring
5. Graduated sanctions
6. Conflict-resolutions mechanisms
7. Minimal recognition of rights to organize
8. Nested enterprises

These principles, although not a recipe, are the basic characteristics of the study cases. Despite the free rider is a potential problem against the long-enduring self-governance, is not practical to exclude a user or not possible to force him to contribute to the costs of developing and maintaining the resource. Therefore, the free-riders take advantage of this situation for their own benefit. According to Dietz et al. (2002), "one 'solves' the free-rider problem when rules to regulate individual actions are adopted and accepted so that social benefits and social costs are taken into account". Rules, according to Ostrom (1990), are closely related to the term institution; she defines institution as the shared concepts used by humans in repetitive situations organized by rules, norms and strategies.

A fundamental approach to the institutional issue is the Institutional Analysis and Development (IAD) framework. A key concept within IAD is action situation, which refers to an analytic concept used to study the immediate structure affecting a process of interest, particularly explaining regularities in human actions and results. Seven clusters of variables are used to describe the structure of action situation (Ostrom, 2005):

1. The set of participants,
2. The specific positions to be filled by participants,
3. The set of allowable actions and their linkage to outcomes,
4. The potential outcomes that are linked to individual sequences of actions,
5. The level of control each participant has over choice,
6. The information available to participants about the structure of the action situation, and
7. The cost and benefits assigned to actions and outcomes.

As CPRs scholars establish, a basic way to address the free rider problem is through the use of institutions and therefore the rules that influence the behavior of actors in diminishing the incentives to overuse, congest and destroy the resources. Consequently, rules are the perfect tool to reach an agreement about common behavior among users.

According to Ostrom (1990), rules are shared understanding among those involved that refer to enforced prescriptions about what actions are required, prohibited or permitted. Ostrom (1990) states that a first step toward identifying the working rules can be made by overtly examining how they affect each of the variables of an action situation. In this regard, she identifies seven clusters of rules that directly affect the components of their own action situations (Ostrom, 2005):

1. Boundary rules affect the characteristics of the participants.
2. Position rules differentially affect the capabilities ad responsibilities of those in positions.
3. Authority rules affect the actions that participants in positions may, must, or must not do.
4. Scope rules affect the outcomes that are allowed, mandated and forbidden.
5. Aggregation rules affect how individual actions are transformed into final outcomes.
6. Information rules affect the kind of information present or absent in a situation.
7. Payoff rules affect assigned costs and benefits to actions and outcomes.

The IAD framework and rules configurations provide an important tool to measure how the "new" uses of water, such as environmental and biodiversity conservation, can match with preexisting uses. By using this framework, we analyze what the rules for use of natural resources of local people are and how those rules match with the new environmental conservation rules, particularly related to water resources developed by the CONANP in the study natural protected area.

3. The study case

For this study, we selected two small farm communities or *ejidos* within the SARC natural protected area, located in the municipality of Alamos, along the western foothills of the Sierra Madre Occidental in the southeastern portion of the Mexican state of Sonora (Figure 1). The SARC natural protected area, with 92,889.69 hectares, was issued in 1996 funded by the Global Environment Facility (GEF) (World Bank, 1997); it has been designed a Man and Biosphere (MaB) area and is currently administrated by the National Commission for Natural Protected Areas (CONANP) in Mexico (Díaz-Caravantes & Scott, 2011).

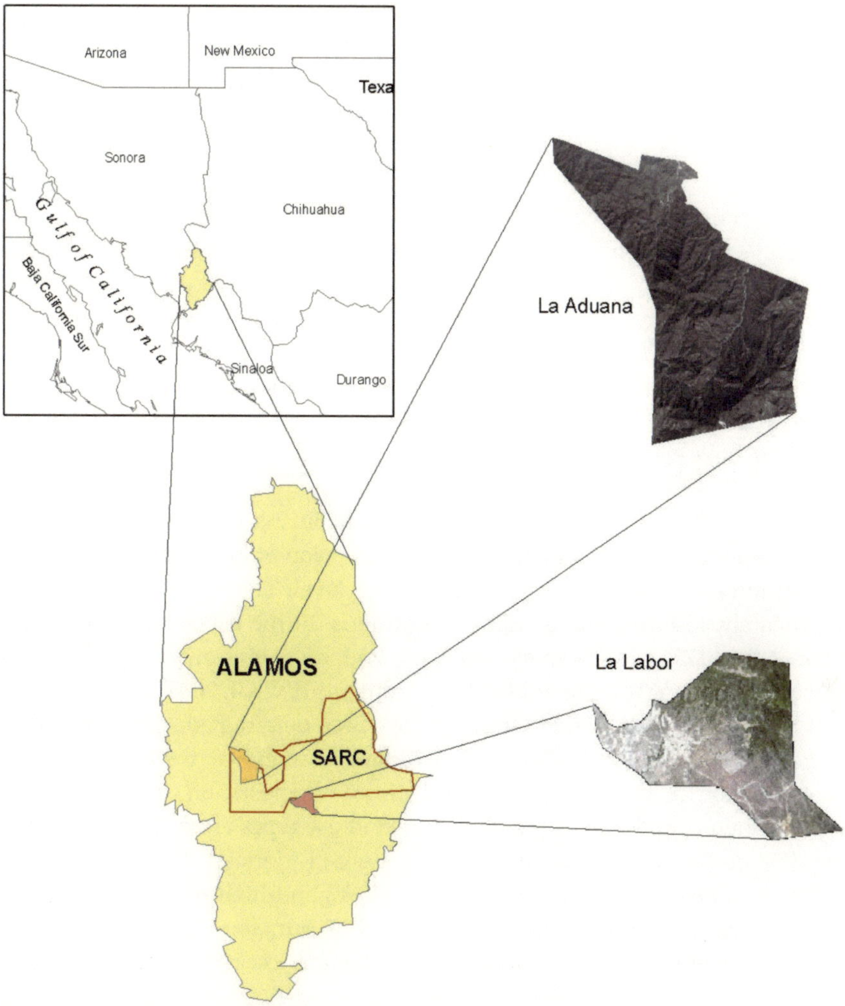

Fig. 1. Location of the ejidos La Aduana and La Labor in Alamos, Sonora, Mexico

The natural value of this area is recognized given its strategic location on the transitional zone between the Neoartic and Neotropical biogeographic regions, and its extensive geomorphic and topographic variations. This creates considerable climatic and soil conditions variations to produce a complex mosaic of vegetation (Búrquez & Yrízar 2006). The rich mix of plant communities in the area include: pine and oak forest, oak woodland, foothills thornscrub (Van Devender et al., 2005), and the northernmost relicts of tropical deciduous forest in México. The existence of a rich and varied flora in the area has been reported, with the occurrence of about 1,100 species of vascular plants divided into 148 families. Protecting this rich biodiversity of flora and fauna was the main reason for the

SARC creation and has thereafter been the principal mission of CONANP (World Bank, 2004).

Robichaux and Yetman (2000) provide a comprehensive plant list for the Río Cuchujaqui, illustrating the extraordinary biodiversity of this ecosystems mosaic. Among the most common plant species in the area are elephant tree (*Bursera microphylla*), peachwood (*Haematoxylon brasiletto*), silk cotton tree (*Ceiba acuminata*), fernleaf acacia (*Acacia millefolia*), honey mesquite (*Prosopis glandulosa*), feather tree (*Lysiloma watsonii*), Mexican cypres (*Taxodium mucronatum*), Mexican cedar (*Cedrela odorata)* and higuerillas (*Ficus sp.*). Reported animal species include: gila monster (Heloderma suspectum), mountain lion (Puma concolor), jaguar (*Panthera onca*), ocelot (*Leopardus pardalis*), jaguarundi (*Herpailurus yagouaroundi*) and military macaw (*Ara militaris*). (CONANP, 2011).

Ejidos are small-scale communal lands that were created as part of Mexico's massive land reform in the postrevolutionary decades of the 1930s and 1940s in order to redistribute productive land (via expropriation from large landowners) to landless peasants. An ejido is composed with at least 20 members called *ejidatarios*. In the past, the individual ejidatario could not use their parcel as collateral for credit, nor could they legally rent or sell their parcels. The reforms to Article 27 of the Mexican Constitution in 1992 changed these provisions, and now ejido parcels can be legally rented or sold to other ejidatarios or to private buyers.

Ejido "La Labor de Santa Lucia" (La Labor), located in the western portion of the SARC natural protected area, occupies an approximate extension of 27.6 square kilometers with an elevation ranging from 260 to 572 meters above sea level. The main settlement in La Labor has approximately 134 inhabitants. Ejido La Aduana, in the eastern portion of the SARC, ranges from 210 to 1756 meters above sea level, with an extension of 45 square kilometers. Here the main settlement is occupied by 236 inhabitants (INEGI, 2005).

Conditions of occupation and land in the area have been determined mostly by the practice of small-scale seasonal agricultural and extensive cattle breeding, with the U.S. beef market as the principal destination of the production. Agriculture, deforestation, induced grassland and secondary vegetation are, for example, the main land use types around human settlements in La Labor (Figure 2). According to remotely sensed derived land use / land cover classifications obtained from aerial photography (1998) and high resolution satellite imagery (2007), deforestation and land degradation, induced by human activities have increased in a decade, threatening biodiversity conditions in La Labor (Sánchez-Flores et al., 2009).

4. Methods

Ostrom states that in training research identifying and measuring institutions it is important to stress the concept of rule-in-use rather than focusing on rules-in-form (1990). In other words, institutions are the rules that people develop to specify the dos and don'ts related to a particular situation, even when those rules may not exist in written documents. In this regard, the field researcher cannot be a survey worker asking a random sample of respondents about their rules. These statements have important methodological implications. The IAD framework privileges the empirical field research. Thus, examining rule configurations involves basically the collection of qualitative data, consistent with standard data collection procedures used in ethnographic practice.

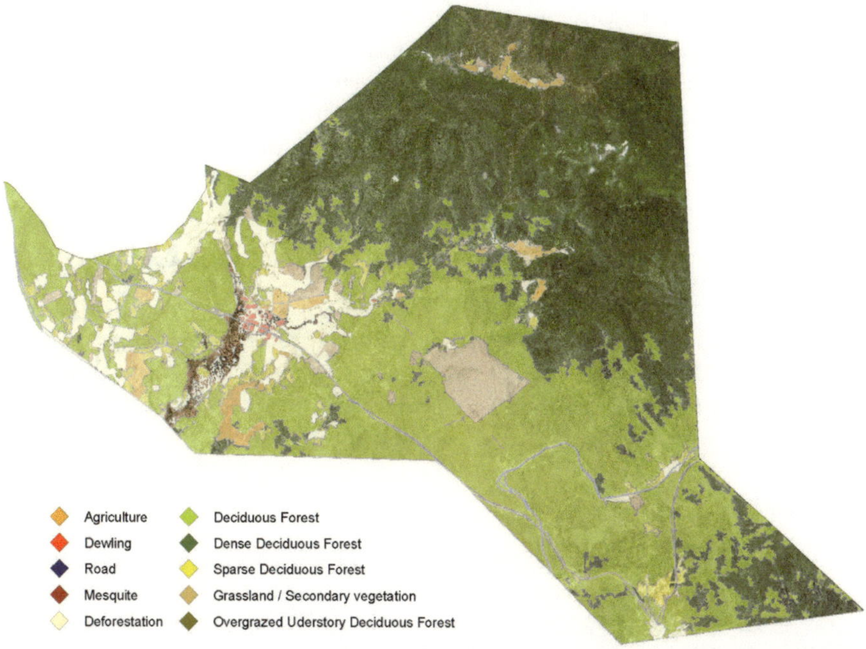

Fig. 2. Land use / land cover conditions in ejido La labor (Sánchez Flores et al., 2009).

Snowballing is one of the ways to approach potential informants. The method consists of using one contact to recruit another contact that in turn can put the researcher in touch with someone else (Flowerdew & Martin, 2005). Following the snowballing method, during the summer of 2008 and 2009 around 40 interviews were conducted with local villagers through a process of referral. In addition, we conducted around 20 interviews with key governmental agents of CONANP and environmental practitioners working in the study area.

5. Livelihoods of local communities

There are six principal productive activities in these study cases: cattle breeders in both communities, farming in the two communities, forestry exploitation in La Labor and craftsman's activities, tourism and mining in La Aduana.

An extensive cattle breeding is practiced throughout the entire SARC natural protected area. The principal destination of the calves is the U.S. beef market. According to the livestock official census, in la Labor the cattle population is 490 on average and in la Aduana the cattle population is 237 on average. Inside the SARC area farming is present. There are mainly seasonal agricultural activities practiced on a small scale. Farmland has occupied 81 hectares in La Labor and 26 hectares in La Aduana on average in the last ten years. The difference is due to the fact that in La Labor there is more flat land than in La Aduana.

In the case of La Labor there is an exploitation of the "vara blanca" (*Leucaena involucrate)* (Felger et al., 2001), which is highly demanded to make stakes that support growing plants, like tomatoes and grapes, in intensive agricultural areas of northern Mexico. In the case of La Aduana, according to the villagers and field observations, the forestry products are insignificant and the local people only use them for domestic purposes.

In La Aduana there is religious tourism due to the regional devotion of the Virgen de Balvanera, who is the patron saint of the community. The principal date of the celebration is on November 20 although, according to the villagers, there are visitors almost every weekend. Additionally, there is a famous restaurant owned by North Americans called "La Casa de la Aduana," which attracts some visitors to the community, mainly North Americans. There are two groups of craftswomen linked to the tourism in La Aduana. One was founded in 1979 and currently has 15 members. The other was formed in 2005 and has 6 members.

In La Aduana, mining activity has played a central role in the history of the community. Although now it is only practiced on a small scale (almost insignificant), at the town's foundation in the XVI century, it was the principal activity. The reopening of a mine called La Quintera in the 1950s explains why many people moved from Chihuahua and other regions along the country to La Aduana. These people helped to found the Ejido La Aduana in the 1970s.

6. The role of the CONANP

The SARC natural protected area was issued in July 1996 under the legislation and policy of the National Commission for Natural Protected Area (CONANP). As the World Bank document, "Mexico Protected Areas Program: Proposed Restructuring Project," indicates, projects of Protected Areas were financed by a combination of Global Environment Facility (GEF) with other institutions (World Bank, 1997).

According to the CONANP officers, the official version of the management program has not been issued because it is in revision in Mexico City. However, in the draft version of the management plan written on 1994 by the current director of the SARC area, there are three principal management specific programs: Conservation, Public Use and Recreation, and Community Development. On the Conservation section there are three subprograms Management and Protection of the Natural Resources, Inspection and Surveillance, and Research and Environmental Management (CONANP, 1994).

On the Public Use and Recreation program are established the objectives of promoting the nature tourism, guaranteeing the economic benefits of this activity to the Reserve and the local communities, and to make aware the visitors and local people about the role of the SARC area in preserving biodiversity and the rational use of natural resources.

On the Community Development program are stated the objectives of increasing the participation of the local communities in the protection and management of the natural resources, collaborating in the enhancement of the economic conditions of the local people and owner of the natural resources and developing, showing and spreading sustainable practices in farm, livestock, forest and tourist activities. A key benefit for the local communities designated within a Natural Protected Area was the employment generated by the projects. In the construction of the infrastructure and in the reforestation plans the CONANP has hired local workers.

The next stage is examining the CONANP projects in the local communities. The interest of this exercise, as we established above, is to analyze the interaction of CONANP with local people and resources.

CONANP has implemented five main projects in the communities, divided in two categories: economic development and environmental. The economic development projects are: the gila monster's farm in La Labor, the ecotourism in both communities, and the nursery in La Aduana. The environmental projects in both communities are the reforestation plans and the *retenidas* or stone check dams for runoff control.

The Gila monster farm was established in 2001. According to the villagers, the CONANP tried to establish this farm with the purpose of helping the community and increasing the population of this species. One villager said that supposedly they would sell 25% of the total population of the gila monsters for $1,500 dollars each, and the rest would be set free. The villagers received special training to manage these species and CONANP built the facilities for the farm. The villagers said that they were very excited and hopeful with the project. However, this project never came through because CONANP did not get the permit for the project, and now the facilities are abandoned. Additionally, it is not clear what the potential market for the gila monsters was. According to the villagers CONANP said that they would sell these species to the North Americans, however this market was never proven to exist. The CONANP tried to establish a recreation area near to the gila monster farm. However, the project has never worked and nobody has visited the area.

In La Aduana the CONANP began to build a lodge for tourists four years ago. The CONANP also planned a hiking trail by the area and an area to practice rappelling. Some people are very hopeful and excited with these projects in La Aduana. However, as one villager said, the CONANP has delayed almost four years in finishing the lodge. Four years ago the CONANP proposed a project to establish a nursery of plants in La Aduana. However, the project has not been finished and no plants have been sold at this point.

The CONANP also tried to plant native species, 15 hectares in La labor and 50 hectares in La Aduana. These species include palo mauto (*Acacia millefolia*), honey mesquite (*Prosopis glandulosa*) and palo colorado (*Caesalpinia platyloba*). However, none of these projects worked. According to some villagers, the problem is that the CONANP did not plant in the adequate season and this happened because the CONANP received the grant and it must be spend it as soon as possible because if not the "money would go to another place".

Regarding water resources, since 2000, the CONANP has implemented the constructions of around 800 *retenidas* throughout the natural protected area. These are stone check dams located in the streams that descend from the summit of the mountains. According to the CONANP agents, the *retenidas* have the double purpose of preventing and recovering soil erosion and providing water to its surrounding environment.

According to the CONANP, the ancestors of the current people had implemented this kind of work in the region, but the villagers "lost this knowledge". In an interview, the head of this project stated that the CONANP emphasizes this to encourage the villagers to accept the project.

There are two principal "benefits" relating to the *retenidas*. The first benefit of this project is the employment generated for the villagers. A second benefit, according to the CONANP, which has been assumed by the villagers, is that the *retenidas* will recharge the aquifer. However, neither CONANP nor the villagers have a clear idea of to what degree these works will contribute to this purpose. In an interview, the chief of this project stated that the CONANP told the people that the *retenidas* would contribute to the aquifer in order to encourage the villagers to accept the project. He stated, "you cannot say to the people that this project is really for the benefit of the flora and fauna, because they may not accept it".

In summary, of the five projects implemented by CONANP, the only one that that has been relatively successful is the *retenidas*.

Fig. 3. Retenidas for runoff control in La Labor

7. Using the rule configurations

The next stage is to analyze how the rule configurations for use of water resources of local people correspond or conflict with the CONANP policy. In other words, we analyze what are the rule configurations of local people regarding water resources and how the use of water for environmental purposes was included into these rule configurations. However, prior to these analyses, it is necessary to describe the formal organization of the study cases.

La Aduana y La Labor de Santa Lucia have a set of rules regarding the formal organization: La Aduana as an ejido system enacted in the 1970s, and La Labor de Santa Lucia as a producers association enacted in 1992. These types of organizations provide to these groups some significant rule configurations. In both groups, there is a document, which specifies who the members of the group are; therefore the boundary rule regarding access to the resources is defined. Additionally, it is defined that each month the members of the group should meet to agree on all the issues regarding the organization of the resources.

Moreover, this document defines what the positions rules within the group are: president, secretary, and treasurer that have the mission of solving all the problems of the group regarding the natural resources and sometimes beyond. However, it does not explain what is the position rules regarding each economic activity (i.e. grazing land and farmland) and what the authority is and scope rules of these activities. For that reason, in the following sections we analyze the principal rules of each activity related to the use of water for the environment.

Involving water resources, the most considerable productive activities are cattle breeding and farming. Here these uses and how they correspond or conflict with the use of water for environmental and biodiversity conservation are analyzed.

7.1 Cattle breeding

There are two aspects that define a CPR: it is difficult to develop institutions to exclude potential beneficiaries from them, and the resource units that are appropriate by one individual are not available for others. The cattle breeding activity fulfills these criteria. In both communities, there is a common land used to graze, in which is difficult to exclude the members of the group, and the resource used by the cows of a member cannot be used for others.

As we established before, the membership to a formal organization defines the boundary rules in the study cases. However, the members of these groups are not the only ones that can posse cows that graze in the land. It is usual that a relative, for example a son or a grandson, could be the owner of some cows. Yet, this relative must have a direct blood relation with a member of the organization in order to be accepted by the other members. Another boundary rule involving cattle breeding is the so-called *partido* system. Particularly in La Labor, there is a system in which a member of the association can include, as a part of his herd, some cows from an outsider. The deal with the outsider is that of every three calves, one is for the member of the organization and two for the outsider. The *partido* is a strategy for increasing the herd, or getting some economic benefits by selling the calves to the U.S beef market.

There are two principal sources of water for cattle breeding activities: the *represos* and the *norias*. The *represos* are like dikes made of soil (Figure 4a) that are used to provide water for the cattle. The *norias* (Figure 4b) are shallow wells used for water cattle provision. In most cases these sources are complementary. During our fieldwork, four *represos were visited*, and all of them were near a *noria*. For example, the *noria* and the *represo* shown in Figure 4 are located 300 meters apart. This strategy has the purpose of recharging the *noria*, even when the *represo* is dried, according to the villagers, the *noria* continues with water because of the recharging effects of the *represos*.

Fig. 4a

Fig. 4b

Fig. 4. The principal sources for water cattle provision. a. Represos and b. Norias

In the last years, due to the shortage of precipitation in the region, the number of *represos* has been incremented as it can be observed in Figure 5. This trend shows the importance of the *represos* for water cattle provision and also shows that the *represos* is a common practice in the municipality, and even more in the study cases.

Following the boundary rule of the access to the grazing land, in La Labor all the cattle related to a member of the group have the right to use the norias of the common land.

As mentioned above, some of the villagers have the idea that the *retenidas* will help to recharge the aquifer, and consequently the *norias*, similar to what *represos* do. We argue that this is an important reason why the villagers have opened their boundary rule to the CONANP. The villagers have the previous practice of building *represos* for water cattle provision, and for that reason the CONANP has had success in spreading the idea that the *retenidas* will help to recharge the aquifer, although the high uncertainty of this presumption.

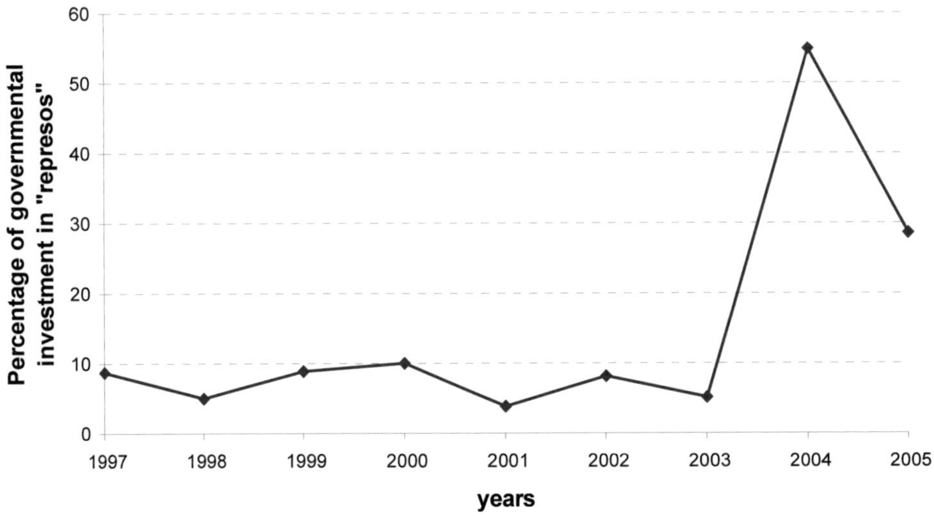

Fig. 5. Trend of the amount of *represos* in the municipality of Alamos

Of course, the jobs that the CONANP has generated and the promise of development projects to encourage the people to accept the *retenidas* are important, but we argue that this similarity between the *represos* and the *retenidas* has been the means through the CONANP was able to be incorporated as a participant of the boundary rules of La Labor and La Aduana.
However, there is a certain degree of tension between the CONANP and the villagers because of the function of the *retenidas* and the *represos*. In both communities, the producers have asked to the CONANP to build *retenidas* for cattle water provision, like the *represos* do, but the CONANP has not attended that request. A CONANP employee, who says that the people have asked *retenidas* in the lower streams, confirmed this. He states: "but this is an economic purpose and our mission is not the economic, but rather the conservation of the natural resources".

7.2 Farming

As established above, two aspects define a CPR: it is difficult of excluding users and the resource used is not available to others. The farming activity is not depending of an irrigation system; rather a seasonal agricultural activities depending of the rainfall is practiced. In this way, the study cases are not CPRs relating to water resources. Yet, regarding access to the land, the farmland can be considered a CPR because physically it is difficult to exclude the members of using farmland, and the land used by a member can be used for others.
The boundary rule regarding farming activity is defined by the membership to a formal organization similar to the cattle activity. Thus, only the ejido members can access to land

for farming purposes, and in this case it is not possible be used by the outsiders, unlike to the case of the cattle breeding throughout the *partido* system.

In the last ten years in la labor have been used for farming 81 hectares and La Aduana 26 hectares on average. Prior the project of the *retenidas* some villagers with the support of the federal agencies have built some *trincheras* located in the lower streams (similar to the *retenidas* but with a different technique) to "irrigate" the land. However the *trincheras* did not work well because the speed of the water is extremely rapid and the water moved the *trincheras*, and in some cases, they were broken. Perhaps, since the villagers observed that the technique of the *retenidas* is better than the *trincheras* because they are more resistant to the speed of the water, some villagers have asked the CONANP to build *retenidas* near the farmland. Yet the CONANP has refused this petition with the argument that the agency is focused on the conservation of the natural resources, rather than economic purposes.

Thus, similar to the cattle water provision there was a previous use of check dams, although with different techniques and different purposes. This previous use and the expectation that the *retenida* could be built for farm purposes make possible the opening of the boundary rule to include the CONANP as another participant in the appropriation of water.

Then, although the *retenidas* have been relatively successful, there is a potential problem regarding the expectation of the villagers. As well as the expectation that the *retenidas* helped to open the boundary rule, there is the risk that the perception of the people changes if they do not receive a direct benefit in the long term.

8. Conclusions

In analyzing relationship between biodiversity conservation and economic sustaining livelihoods in the SARC, we found that the institutional role of CONANP has been quite decisive to achieve a delicate balance. It is evident that the economic development projects, such as the gila monster farm, ecotourism, and the nursery, promoted by the CONANP, have not been successful. However, the *retenidas* as an environmental project have had relative success because the CONANP has achieved the implementation of this project with a sufficient level of acceptability from the villagers. A first approach suggests that although the economic development projects are empty promises in both ejidos, the projects, and also the temporal jobs generated by them, have been pragmatically used to persuade the people to introduce the conservation programs that benefit biodiversity conditions in the SARC natural protected area.

Using the rule configurations proposed by the IAD framework, we conclude that the previous use of *represos* for water cattle provision and *trincheras* for farming are fundamental explanation of why the CONANP could open the boundary rule regarding water resources. The use of *represos* (and *norias*) was a common practice among the villagers that was incremented in the context of water scarcity in the last years. In addition, the use of *trincheras* was a common practice as a way of irrigating the farmland.

However, we argue that although the *retenidas* have been relatively successful, there is a potential tension between the CONANP and the villagers due to the degree of expectation on the *retenidas*. The first expectation is that the *retenidas* will recharge the aquifer and consequently be useful for water cattle provision. Yet, the available information suggests that there is a high uncertainty about to what degree this would happen. The second expectation is regarding the potential uses for farmland. In summary, the same degree of expectation that facilitated the acceptance of the *retenidas*, and opened the boundary rule, could be a significant problem if the perception of the people changes regarding the benefits obtained of these projects. Regardless of the potential benefits for economic activity, we suggest that retenidas might produce beneficial effects to maintain de natural conditions that sustain biodiversity in the SARC natural protected area.

9. References

Búrquez, A. & Martínez Yrízar, A. (2006). *Conservación, transformación del paisaje y biodiversidad en el Noroeste de México*, in Oyama, K. & A. Castillo, Conservación y Restauración de Recursos Naturales en México. Siglo XXI Editores: México, D.F.

CONANP (2011). Ficha descriptiva del area natural protegida, In: *CONANP*, October 20, 2010, Available from: http://simec.conanp.gob.mx/Info_completa_ext.php?id_direccion=23

CONANP (1994). *Draft of Management Plan*, Unpublished Manuscript, Sonora, Mexico.

Díaz-Caravantes, R. E., & Scott, C. A. (2010). Water management and biodiversity conservation interface in Mexico: A geographical analysis. *Applied Geography*, 30(3), pp. 343-354.

Dietz, T.; Dolsak, N.; Ostrom, E. & Stern, P. (2002). The Dramma of the Commons, In: *Dramma of the commons*, Ostrom E. (ed.), National University Press, Washington DC, U.S.

Felger, R. S.; Johnson, M.B. & Wilson, M.F. (2001). *The Trees of Sonora, Mexico*, Oxford University Press, 400 pp.

Flowerdew, R. & Martin, D. (2005). *Methods in Human Geography, A guide for students doing a research project*, Pearson Education Limited, England.

Hardin, G. (1968 and reprinted in 2005) The Tragedy of the Commons, In: *Economics of the Environment, Selected Readings*, Stavins, R.N. (ed.) Fifth edition, W.W. Norton and Company Inc., U.S.

INEGI (2005) *Conteo de población y vivienda 2005*, October 9, 2009, Available in: http://www.inegi.org.mx

Ostrom, E. (2005). *Understanding Institutional Diversity*, Princeton University Press.

Ostrom, E. (1990). *Governing the Commons: The Evolution of Institutions for Collective Action (Political Economy of Institutions and Decisions)*, Cambridge University Press.

Postel, S. & Richter, B. (2003). *Rivers for Life: Managing Water for People and Nature*, Island Press, Washington, DC.

Robichaux, R. H. & Yetman, D. A. (eds.), (2000). *The tropical deciduous forest of Alamos. Biodiversity of a threatened ecosystem in Mexico*. The University of Arizona Press: Tucson.

Sánchez Flores, E.; Díaz-Caravantes, R.; Granados, A. & Chávez, J. (2009). *GIS improved object based classification for land use/cover change detection in a human altered deciduous forest environment*, ASPRS 2009 Annual Conference Proceedings, Baltimore, MD.

Van Devender T. R.; Reeder, J. R.; Reeder, C. G. & Reina A. L., (2005). Distribution and diversity of grasses in the Yécora region of the Sierra Madre Occidental of Eastern Sonora, Mexico in *Biodiversity, ecosystems and conservation in Northern Mexico*, Cartron, J-L; Ceballos, E. G. & Felger, R. S. (eds.), Oxford University Press: New York.

World Bank (1997). *Mexico, Protected areas Program: Proposed Restructuring Project, Global Environment Division*, Washington DC, U.S.

World Bank (2004) *Proposed Supplemental GEF Grant*, Washington DC, U.S.

World Commission on Dams [WCD] (2000). Overview, In: *Dams and Development: A New Framework for Decision-Making*. February 28, 2009, Available in: http://www.dams.org.

Zimmerer, K. (2006). *Globalization and New Geographies of Conservation*, University of Chicago Press.

Part 5

Uses of Biodiversity

15

Bioprospecting: Creating a Value for Biodiversity

Alan L. Harvey[1] and Nigel Gericke[2]
[1]*University of Strathclyde*
[2]*Niche Botanicals*
[1]*United Kingdom*
[2]*South Africa*

1. Introduction

Bioprospecting is the exploration of biological material for commercially valuable genetic and biochemical properties (Reid et al., 1993). This chapter will focus on the search for activities that could form the basis of new pharmaceuticals. Historically, most of the active ingredients in medicines have been natural products (Sneader, 1996), and natural products continue to form a productive source of new drugs (Newman and Cragg, 2007; Butler, 2008). Given that most drug discovery activity takes place in companies in the developed world and that most biodiversity is found in countries of the southern hemisphere, there needs to be a means whereby access to biodiversity is possible under terms and conditions that are mutually acceptable. After hundreds of years of unregulated collection of samples for many different purposes, the United Nations produced a framework for preserving the world's biodiversity while encouraging the sustainable use of biodiversity. This Convention on Biological Diversity has been widely accepted, and it is discussed in the following section. The chapter will continue with descriptions of various attempts to calculate an economic value for biodiversity, followed by an outline of current bioprospecting practices.

2. United Nations Convention on Biological Diversity

The United Nations Convention on Biological Diversity (CBD) (www.biodiv.org) was one of the major outcomes of the Earth Summit in Rio de Janeiro in June 1992. The CBD has three main goals:

- the conservation of biodiversity
- the sustainable use of the components of biodiversity
- the sharing of benefits arising from the commercial and other utilization of genetic resources in a fair and equitable way

Signatories to the CBD recognise that countries have sovereign rights over their genetic and biological resources (i.e., biodiversity) within their boundaries, and agree to the conditions in the CBD for the preservation and sustainable use of biodiversity.

In relation to accessing natural products for drug discovery, the CBD has a number of Articles (see Appendix) that set the tone for future interactions between companies and research organisations with countries with desired biodiversity. Biodiverse-rich countries that have ratified the CBD have to facilitate access to their biological resources (Article 15.2).

Such access must be in accordance with appropriate legislation (Article 15.1), and be on mutually agreed terms (Article 15.4) involving prior informed consent (Article 15.5). The source country is expected to be involved in collaborative research and development projects relating to its biodiversity (Article 15.6) and the source country should benefit from technology transfer (Article 16.2), from the results of research (Article 15.7) and from sharing of commercial benefits resulting from use of its biodiversity (Article 15.7). Article 8(j) also commits signatories to preserving the traditional knowledge of indigenous and local communities and to promoting their involvement in developing wider applications of their knowledge; however, there is little guidance on how this might be achieved.

Since 1992, 192 countries and the European Union have signed or ratified the CBD, the notable exception being the USA. However, issues relating to access to biological resources have not been fully resolved. Only about 25 countries have introduced new regulations to facilitate access, and the vast majority of countries still have to formulate the appropriate laws.

To assist the implementation of bioprospecting under the CBD, the Conference of the Parties (the official CBD body) adopted the Bonn Guidelines on 'Access to Genetic Resources and Fair and Equitable Sharing of the Benefits Arising out of their Utilization' (Secretariat, 2002). These are not legally binding, but they are intended to help all parties follow best practices in setting up bioprospecting agreements. Various professional bodies have responded to the Bonn Guidelines with their own recommendations. For example, the International Federation of Pharmaceutical Manufacturers and Associations has published its views on 'industry best practices' and the enabling steps that governments need to take with regard to regulating bioprospecting (IFPMA, 2007). The Biotechnology Industry Organisation of the USA has produced detailed guidelines for its members about engaging in bioprospecting (BIO-1, n.d.). These cover the general conduct of bioprospecting, sharing of financial benefits and of results of research, intellectual property rights, and conservation and sustainable use of biodiversity. The organisation has also published a model Material Transfer Agreement for use in bioprospecting operations (BIO-2, n.d.). There is a very useful resource published by the International Institute for Sustainable Development as an 'access and benefit-sharing management tool' and an accompanying handbook (IISD, 2007). This provides a step-by-step guide to obtaining prior informed consent, reaching mutually agreed terms, agreeing benefit-sharing arrangements, and dealing with issues relating to traditional knowledge and conservation.

The Bonn Guidelines are likely to be superseded by the Nagoya Protocol on Access to Genetic Resources and the Fair and Equitable Sharing of Benefits Arising from their Utilization to the Convention on Biological Diversity (Secretariat, 2011). This is a new treaty under the CBD that was adopted in Nagoya in October 2010. It will be operational once 50 countries ratify it (40 have done so by the end of July 2011). The aim of the Protocol is to provide greater legal certainty about all aspects of bioprospecting. In particular, it is intended to establish more predictable conditions for access to biodiversity and to ensure appropriate benefit-sharing. The Protocol deals more explicitly than previous documents with the use of traditional knowledge associated with genetic resources: contracting parties have to ensure that local communities have provided prior informed consent and that there is fair and equitable benefit-sharing with the relevant communities. The Nagoya Protocol also recognises that genetic resources are rarely confined to a single country and that traditional knowledge related to use of genetic resources is often shared by different communities. The Protocol demands involvement and cooperation of the relevant parties.

A major weakness in the implementation of the CBD with respect to bioprospecting has been the slow development of national systems for governing access to biodiversity. The

Nagoya Protocol is explicit about the responsibilities of signatories to create 'national focal points' and 'competent national authorities' to make available information on how to access genetic resources and traditional knowledge and to be responsible for granting access to biodiversity. In Africa, a diverse range of policies and laws relevant to access and benefit-sharing is in place in some countries, but these are most developed in South Africa through the National Environmental Management: Biodiversity Act (10 of 2004) ('the Biodiversity Act') and the regulations passed under this Act in 2008. The Biodiversity Act requires bioprospectors to obtain a permit from the Government for bioprospecting involving indigenous biological resources, and for the export of these resources. Prior informed consent is required with landowners and indigenous communities before a permit is issued. Benefit-sharing agreements must be entered into with indigenous communities who use the resource traditionally, or who have knowledge of its properties (Wynberg et al., 2009).

3. Value of bioresources

A key issue in bioprospecting is benefit-sharing. However, much of the debate on this topic assumes that there are benefits to share. The historical successes in drug discovery based on natural products would suggest that there should be a continued appetite for accessing natural products for use in drug discovery programmes. Despite this, the pharmaceutical industry, in general, has reduced its use of natural products (Harvey, 2008), and there are few current examples of large-scale programmes designed to access a wide variety of natural products collected from their native environment. Moreover, advances in techniques for manipulating microbes to produce novels chemicals make the use of locally sourced bacterial samples more attractive to industry (Kingston, 2011). Never-the-less, it is reasonable to ask what an appropriate price might be for biodiversity, its conservation and its availability for bioprospecting. This is an area with much debate but little consensus (see, for example, Castree, 2003).

Future royalty streams from blockbuster drugs (drugs that generate $1 billion in sales per year) directly or indirectly derived from bioresources were anticipated to contribute significantly to the conservation of biodiversity in source countries, and to the development of indigenous knowledge holder communities. However the age of blockbuster drugs seems to largely be over, and the pharmaceutical sector, as it is currently structured, is unable to deliver enough new products to market to generate revenues sufficient to sustain its own growth. Nearly all major drug developers are critically examining current R&D practices and, in some cases, considering a radical overhaul of their R&D models (Kaitlin, 2010). Only 13 natural product-derived drugs were approved in the United States between 2005 and 2007 (Harvey, 2008), and FDA approvals of new drugs reached a 24-year low in 2007 (Li & Vederas, 2009). With the contraction and consolidation seen in the pharmaceutical industry, and with only 3 in 10 new products generating revenues equal to or greater than average pharmaceutical industry R&D costs (Kaitlin, 2010), it would seem that royalty streams from natural product-derived pharmaceuticals are too uncertain to be included in the valuation of biodiversity.

A pragmatic approach to valuation is to focus on the monetary value paid by end users. In the early 1990s, Merck & Co. entered an agreement with the National Biodiversity Institute of Costa Rica, INBio, that provided a defined amount of cash (reported as $1 million) for a certain number of samples; although this deal is frequently mentioned in articles on the pros and cons of bioprospecting, key elements of the financial terms are still confidential, notably the payment per sample. More recently, GlaxoSmith Kline had an arrangement with the biotechnology company Extracta in Brazil: at its reported value of $3.2 million for a collection of 30,000 samples (Dias & da Costa, 2007), this does not seem like a high price - about $100 per sample. The Swiss company Novartis also had an agreement with a Brazilian organisation, Bioamazonia, reported to be $4 million for 10,000 samples of micro-organisms. It should be noted that political concerns in Brazil about inappropriate commercialisation of its biodiversity caused both interactions to be terminated, highlighting the need for clear national focal points and competent authorities, as called for in the Nagoya Protocol.

In a development from academic collaborations on natural products and drug discovery, the Strathclyde Institute for Drug Research, UK acted as a broker for natural product samples provided by its collaborators and offered samples under licence to companies for bioassay. Because each species could be used several times, the cumulative return on a single extract could be quite impressive, reaching $500-1,500 per gram. This can be compared with commodity prices obtained for a known herbal medicine that has proved active in double-blind clinical trials: raw material costs for *Hypericum perforatum* (St John's wort) are $8-10 per kg. However, there is currently little demand from companies for samples of natural product extracts, and the brokerage activity has stopped.

Another approach to valuation of biodiversity has argued that a value can be put on the potential contribution of an area of land to drug discovery (Simpson et al., 1996; Rausser & Small, 2000). Depending on the assumptions used, this varied from $21 per hectare (Simpson et al., 1996) to $9,177 per hectare (Rausser & Small, 2000). The latter valuation was thought to be sufficient to provide an economic reward to sustain biodiversity conservation. A later paper (Costello & Ward, 2003) sought to explain the very different valuations reached by the previous authors. They concluded that Rausser and Small's focus on species in biodiversity hotspots was largely responsible.

Another approach to valuation of biodiversity used the pharmaceutical industry figures for costs and rewards of drug discovery and development to formulate a discounted cash flow model that gave a Net Present Value for an extract in a screening programme (Artuso, 1997). This was $487 per extract. A similar approach was used to explore the impact of varying the balance between upfront payments for sample supply and long-term royalties from sales of commercialised products (Lesser & Krattinger, 2007). In theory, more value is attached to a deal with an emphasis on royalty payments, but the authors note that larger up-front fees may be more of an incentive for conservation of biodiversity. The same approach was used to explore the potential added value of traditional knowledge and of pre-screened extracts. On average, use of samples suggested by traditional knowledge was expected to double the value of the deal. However, it is unlikely that this approach, which is based on averaging the theoretical outcome of testing large collections of samples, is appropriate for use of samples from traditional medicinal uses. These are more likely to be commercialised (if at all) on an individual basis. In addition to the possibility of providing a source for new chemical entities or derivatives, traditional medicinal plants may be commercialised as standardised plant extracts developed into polymolecular botanical medicines, dietary supplements or functional foods. The value of a registered botanical medicine may be considerable. In December 2005, the German Federal Institute for Drugs and Medical Devices (BfArM, Bonn) approved a new licence for the use of a proprietary extract of the root of the South African plant *Pelargonium sidoides*, (EPs® 7630), known as Umckaloabo, as a drug (Conrad et al., 2007). This registered liquid herbal medicine has been reported to have an annual turnover in Germany alone in 2006 of €80,000,000 (Brendler & van Wyk, 2008).

In-country fractionation and screening could also, potentially, raise the value of a collection of natural product extracts on the basis of reducing the risk of failure to the commercial development partner. However, there has to be the necessary infrastructure and expertise to allow the screening, and the preliminary tests have to be relevant to the commercial partner (Lesser & Krattinger, 2007).

The discussion above has only considered direct monetary benefits. However, both the Bonn Guidelines and the Nagoya Protocol list possible non-monetary benefits in addition to the monetary ones (Table 1). The non-monetary benefits need to be looked at seriously because technology transfer and improvements in capacity may contribute to sustainable development in biodiversity-rich countries, although that assumption has been questioned (Castree, 2003).

1.	Monetary benefits may include, but not be limited to:
a.	Access fees/fee per sample collected or otherwise acquired;
b.	Up-front payments;
c.	Milestone payments;
d.	Payment of royalties;
e.	Licence fees in case of commercialization;
f.	Special fees to be paid to trust funds supporting conservation and sustainable use of biodiversity;
g.	Salaries and preferential terms where mutually agreed;
h.	Research funding;
i.	Joint ventures;
j.	Joint ownership of relevant intellectual property rights.
2.	Non-monetary benefits may include, but not be limited to:
a.	Sharing of research and development results;
b.	Collaboration, cooperation and contribution in scientific research and development programmes, particularly biotechnological research activities, where possible in the Party providing genetic resources;
c.	Participation in product development;
d.	Collaboration, cooperation and contribution in education and training;
e.	Admittance to ex situ facilities of genetic resources and to databases;
f.	Transfer to the provider of the genetic resources of knowledge and technology under fair and most favourable terms, including on concessional and preferential terms where agreed, in particular, knowledge and technology that make use of genetic resources, including biotechnology, or that are relevant to the conservation and sustainable utilization of biological diversity;
g.	Strengthening capacities for technology transfer;
h.	Institutional capacity-building;
i.	Human and material resources to strengthen the capacities for the administration and enforcement of access regulations;
j.	Training related to genetic resources with the full participation of countries providing genetic resources, and where possible, in such countries;
k.	Access to scientific information relevant to conservation and sustainable use of biological diversity, including biological inventories and taxonomic studies;
l.	Contributions to the local economy;
m.	Research directed towards priority needs, such as health and food security, taking into account domestic uses of genetic resources in the Party providing genetic resources;
n.	Institutional and professional relationships that can arise from an access and benefit-sharing agreement and subsequent collaborative activities;
o.	Food and livelihood security benefits;
p.	Social recognition;
q.	Joint ownership of relevant intellectual property rights.

Table 1. Monetary and non-monetary benefits to be considered in bioprospecting agreements (from the Nagoya Protocol)

Under the CBD, there is a clear need to reach agreement with the source of biodiversity on appropriate sharing of benefits arising from any commercialisation. There is also a commitment in the CBD to recognise and protect indigenous knowledge about uses of biodiversity. When it comes to benefit-sharing from commercial developments from such traditional knowledge, agreements can be hard to reach (see Boyd, 1996; Mays & Mazan, 1996). In part, this can be because of a cultural clash, e.g. where traditional knowledge is regarded as communal and not capable of being owned in a Western sense (see Cotton, 1997; Prathapan & Rajan, 2011). There can certainly be arguments over inventorship when the natural products in question are in widespread use, and there are frequent disputes about what constitutes appropriate benefits. In some cultures, monetary returns may have little meaning, and various attempts have been made to set up, for example, charitable foundations to distribute benefits in other ways (Mulholland & Wilman, 2003).

Various access and benefit-sharing agreements have been analysed (Castree, 2003; Mulholland & Wilman, 2003; Medaglia, 2004; Laird & Wynberg, 2008). Clear criteria for success, in either economic or conservation terms, are missing so that it is difficult to reach objective conclusions. It seems to be extremely unlikely that bioprospecting will be a sufficient economic driver to support conservation of biodiversity. Indeed, the case of paclitaxel has been cited to show that a bioprospecting success can lead to negative consequences (Frisvold & Day-Rubenstein, 2008). Bioprospecting should perhaps be examined as an activity with some opportunities for local benefits and one that has to be regulated to make sure that it does not endanger biodiversity or deny fair and equitable benefits to indigenous knowledge-holders.

4. Historical successes in pharmaceutical bioprospecting

Bioprospecting (in terms of seeking leads for new drugs from natural products) can follow two main approaches: use of leads from traditional medical uses (i.e., from "ethnopharmacology"), and use of natural products as a highly diverse set of chemicals for random screening.

4.1 Traditional medicines

Historically, nature was the origin of all medicines (Sneader, 1996), and ethnopharmacology has provided some very notable past successes, including morphine (isolated in 1804), quinine (isolated in 1820), digitoxin (isolated in 1841), ephedrine (isolated in 1897), and tubocurarine (isolated in 1935). These compounds, or their analogues and derivatives, are still in widespread use. A further 50 examples are given by Cox (1994). More recent developments with an association with traditional uses include artemisinin and derivatives for malaria and prostratin as an anti-viral (see, e.g., Kingston, 2011).

The development of a pharmaceutical product (as an appetite suppressant) from the traditionally used South African plants in the *Hoodia* genus was stopped, although the plant moved into development as a food supplement (Laird & Wynberg, 2008; van Heerden, 2008). Unfortunately, the early promise was not upheld and the commercial development rights have reverted to CSIR in South Africa. The approach to the commercial development of Hoodia was a case with inappropriate agreements between the South African research organisation CSIR and commercial development partners: the original agreements did not include the holders of the traditional knowledge, the San people. The absence of any benefits accruing to the San was subsequently successfully challenged by the South African San Council, and a mutually acceptable agreement was finally reached between the CSIR and the South African San Council after 18 months of negotiations (Wynberg et al., 2009). The more recent efforts to develop a medicinal or dietary supplement product from another plant (*Sceletium tortuosum*) originally used by the San may be instructive: the company involved, HG&H Pharmaceuticals (Pty) Ltd, successfully concluded a prior informed consent benefit-sharing agreement with the South African San Council and was also awarded the first bioprospecting and export permit to be issued by the South African Government (see http://www.zembrin.com/).

There have also been leads from traditional medicines used as the starting point for the development of analogues that become the active ingredients of the final medicinal product. An example is podophyllotoxin, a compound isolated from *Podophyllum peltatum*, a plant used traditionally in North America for treating warts: this stimulated the work that led to the anti-cancer agent etoposide. Other examples can be found in the review by Newman and Cragg (2007), and more details and examples of ethnopharmacological investigations can be found in the books by Chadwick and Marsh (1994) and Cotton (1997). As discussed in the preceding section, accessing natural products used as traditional medicines can lead to many challenges - relating to ownership of intellectual property and benefit-sharing. It is also not necessarily a successful strategy for developing pharmaceutical products, as evidenced by the failure to date of either Shaman Pharmaceuticals or Phytopharm to commercialise new drugs or botanical medicines from traditional medicines.

4.2 Lucky finds

Random screening of natural products does not presuppose the existence of particular biological activities in any set of natural products: it relies on the assay to detect the activity. The key to success is likely to be having the most chemically diverse collection of natural products, and this can be approached by using collections from diverse genetic sources.

Notable successes include the development of cyclosporine A from a fungus (*Tolypocladium inflatum*) collected in Norway, and the development of rapamycin from a microbe (*Streptomyces hygroscopicus*) from Easter Island. The anti-cancer agent paclitaxel was discovered as a result of the National Cancer Institute's large-scale screening of plant extracts (Frisvold & Day-Rubenstein, 2008).

Conservationists have highlighted the fact that 70% of the world's plant species and more than 60% of the world's vertebrate species are found on 1.4% of the land area of the world, and that some of the regions containing the greatest biodiversity are being threatened by development (Mittermeir *et al.*, 1999). The richest regions have been defined as 25 mega diverse "hotspots" on the basis of the number of species found there and the high proportion of endemic species, i.e. those that occur naturally only in that region. The hotspots are shown in Table 2 along with the number of plant species they contain. Collections from areas with high endemism would be expected to yield many unusual compounds. However, apart from INBio's efforts in Costa Rica, there have been no systematic and widespread collections of samples from the biodiversity hotspots for bioprospecting.

Seventy percent of the earth's surface is covered by sea, and the marine environment contains examples of most types of organisms. There may be more than 10 million species of marine macro fauna (Poore and Wilson, 1993) and many more species of marine micro-organisms. There are few marine-based collections for bioprospecting: possibly, those of

Magellan BioScience (www.magellanbioscience.com) in USA and MarBank in Norway. There has been relatively little work on bioprospecting such marine biodiversity, although salinosporamide from a marine bacteria is in early-stage clinical trials and ecteinascidin 743 (trabectedin) from a marine tunicate has been approved in Europe for the treatment of some cancers.

Hotspot	Plant species	Endemic species
Tropical Andes	45,000	20,000
Penisular Malaysia and Western Indonesia	25,000	15,000
Mediterranean basin	25,000	13,000
Mesoamerica	24,000	5,000
Atlantic Forest, Brazil	20,000	6,000
Indo-Burma	13,500	7,000
Madagascar and Indian Ocean Islands	12,000	9,700
Caribbean	12,000	7,000
Mountains of south-central China	12,000	3,500
Brazilian cerrado	10,000	4,400
Wallacea (Indonesia)	10,000	1,500
Choco-Darien-Western Ecuador	9,000	2,250
Guinean forest, West Africa	9,000	2,250
Cape floristic province, South Africa	8,200	5,700
Philippines	7,620	5,800
Polynesia/Micronesia	6,500	3,300
Caucasus	6,300	1,600
South-west Australia	5,500	4,300
Succulent Karoo, South Africa	4,800	1,860
Western Ghats and Sri Lanka	4,780	2,180
Californian floristic province	4,400	2,125
Eastern arc mountains, Tanzania and Kenya	4,000	1,400
Central Chile	3,400	1,600
New Caledonia	3,320	2,500
New Zealand	2,300	1,865

Table 2. The megadiverse hotspots and their vascular plants. Adapted from the information given by Mittermeier et al. (1999).

New and unusual biodiversity is still being discovered as unusual habitats are being sampled (Harvey 2007). Since this biodiversity has never been available for biological testing, it can be predicted that novel chemicals with potential as drug leads will be discovered if such biodiversity can be accessed for screening. However, this is largely driven by small-scale academic endeavours. Few companies are currently involved in providing access to biodiversity samples. Table 3 lists companies cited in 2001 as being active in this area (Harvey, 2002). Apart from Albany Molecular Research (AMRI), all have gone out of business or changed strategy. Analyticon still provides screening samples, but these are prepared as synthetic modifications of natural product scaffolds rather than purified natural products.

Company	Type of natural product	Description	Current status
Drug Discovery Ltd, UK	Plant extracts	Worldwide sources 87% of plant families 6,500 species	Not active
MicroBotanica, USA	Plant extracts	Peruvian Amazon 12,000 samples	Not active
BioProspect, Australia	Plant extracts	Western Australian 3,000 samples Untested species	Focusing on product development
Molecular Nature, UK	Plant compounds	Unusual compounds from relatively common plants	Not active
AnalytiCon Discovery, Germany	Compounds	Made to order libraries	Synthetic modifications of natural product scaffolds
bioLeads, Germany	Microbial extracts	45,000 actinomycetes and other microorganisms	Taken over by BioFrontera; now a dermatological company
InterLink Biotechnologies, USA	Microbial samples	33,000 samples	Assets sold; no longer active in bioprospecting
Albany Molecular Research, USA	Microbial extracts	Ex-PanLabs collection of 25,000 microbial species	Now AMRI; expanded to 300,000 samples including marine and plant species
Exalpha, USA	Microalgae compounds	Prefractionated	Now lab test company
Phytera	Marine and plant extracts	Neptune library and ExPAND tissue culture-derived samples	Taken over; no longer active in bioprospecting
Diversa, USA	Microbial gene products	Small molecules from unique gene expression pathways	Now Verenium Corporation; changed to product development based on enzymes
Cubist, USA	Microbial extracts	54,000 partially characterised extracts of fungi and actinomycetes	Focus on clinical development of antibiotics

Table 3. Companies previously active (in 2001) in providing access to biodiversity for drug discovery purposes and their current status.

5. Current practices in bioprospecting

Most bioprospecting is currently performed on a small scale by numerous academic groups throughout the world. There are some larger programmes based on multi-group collaborations. These include the various International Cooperative Biodiversity Groups (ICBG) funded by the NIH in the USA and efforts coordinated by individual universities such as Rutgers in New Jersey and Strathclyde in Scotland. ICBG programmes involve US institutions and commercial companies with overseas participants in Costa Rica, Fiji, Indonesia, Madagascar, Panama, Papua New Guinea, the Philippines, and Vietnam and Laos (Cao & Kingston, 2009; Kingston, 2011). Rutgers University hosts a relatively new initiative: the Global Institute for BioExploration, GIBEX. This is an international network that aims to promote successful drug discovery from biodiversity through developing pharmacological screening methods that can be readily transferred to groups in partner countries (see http://www.gibex.org/). The University of Strathclyde has a long history of research in phytochemistry in collaboration with research groups in institutes overseas. This formed the basis for the creation of a worldwide network for drug discovery based on natural products. A highly diverse collection of plant extracts was assembled (covering more than 90% of the world's plant families) and used in drug discovery screening assays, either with a commercial partner or through collaborations between members of the network (see http://www.sidr.org/).

Very few large pharmaceutical companies have maintained a strong presence in natural product-based drug discovery. Novartis has developed extensive collaborations with a few academic centres in the Far East, notably in China and Thailand. AstraZeneca had a long connection with a group based at Griffith University in Queensland. This continues within the Eskitis Institute (see http://www.griffith.edu.au/science-aviation/eskitis-institute-cell-molecular-therapies). There are smaller specialist companies involved in bioprospecting. These include MerLion Pharmaceuticals in Singapore (http://www.merlionpharma.com/) and Sequoia Sciences in St Louis, Missouri (http://www.sequoiasciences.com/).

A different approach has been developed to make use of the structural diversity of isolated natural products: *in silico* drug discovery or virtual screening. In this, the chemical structures and physico-chemical properties of compounds are gathered in a computerised database that can be searched to find matches either to complement the three-dimensional structure of a drug target or the chemical features of a compound with the desired activity. This has been used at the University of Strathclyde in its Drug Discovery Portal (www.ddp.strath.ac.uk). Chemists can submit structures to the Portal's database and biologists can propose therapeutic targets. Compounds are screened *in silico* by computational chemists within the Portal against the targets and the relevant chemists and biologists are notified of any predicted hits. The chemists then supply compounds to the biologists for testing on real assays against the target. Because chemists can only enter compounds into the Portal's database if they can guarantee that they can supply the compounds and because biologists can only suggest targets if they have relevant assays available, the reduction to practice can be very rapid after the initial *in silico* screening. The unique chemical database is rapidly expanding (currently over 14,500 compounds), with academic scientists from 21 institutions in five continents contributing, and the compounds have been shown to be highly diverse but still generally drug-like in their properties (Clark et al., 2010; Harvey et al., 2010; Schuster & Wolber, 2010). Another approach to using natural products and virtual screening has been developed at the University of

Innsbruck, Austria. Several databases of natural products have been created: a large general natural product database ("NPD") including over 110,000 compounds of molecular weights between 150 and 700, a small database of about 10,000 constituents from medicinal plants mentioned by Dioscorides ("DIOS" database) and a TCM database of about 10,000 compounds known from plants used in Traditional Chinese Medicine (Rollinger et al., 2004; 2009). The assumed advantage of using compounds from known medicinal plants is that they may be less likely to be toxic than randomly sampled constituents.

6. Future prospects

Bioprospecting has been proposed as a potential means to encourage the conservation and sustainable use of biodiversity. The legal framework under the auspices of the United Nations is slowly being implemented by biodiversity-rich countries, but much still needs to be done if there is to be a genuine facilitation of bioprospecting. Perhaps the implementation of the Nagoya Protocol will provide the necessary impetus.

However, the appetite for bioprospecting by pharmaceutical development companies has clearly diminished since the Rio Earth Summit in 1992, partly because of the complexities relating to access and benefit-sharing, often in the absence of adequate national regulatory clarity and institutional capacity. Despite the continuing appearance of successful drug development projects based on natural products, there is a sentiment that this approach may be too old-fashioned to be considered seriously: screening of natural products for new leads. Various technical problems undoubtedly exist with the screening and isolation of natural products, but the rewards for overcoming them would seem to justify the effort required, and technical solutions are being described in the literature. For example, purification and identification of natural products are believed to be difficult and slow: high throughput separation methods coupled with sensitive analytical techniques can resolve this (Bugni et al, 2008; Hu et al, 2008). Natural products are chemically complex: comparisons of the chemical properties of collections of natural products show that they more closely match the "chemical space" of successful drugs than collections of synthetic chemicals (Grabowski and Schneider 2007; Ganesan 2008). Natural products are reputed to give too many false positives on modern screening assays, but phenotypic assays are becoming more and more popular and it has been suggested that natural products, with their drug-like properties, are well-matched to such cell-based approaches, and extracts of natural products can be processed to remove reactive compounds or even convert them into novel drug-like structures (Rishton 2008). Natural products may only be available in small amounts: techniques for direct synthesis (Sunazuka et al, 2008) or production by molecular biology (Kennedy 2008) have been rapidly developing.

While there is certainly no single "best" way to conduct drug discovery, just as there is not a single panacea for all ailments, it is surely time for a fresh look at the relatively unexplored opportunities provided by modern approaches to applying natural products in drug discovery. Perhaps the lead will have to be provided by the numerous academic groups active in bioprospecting. However, these groups would stand more chance of success if they could pool resources and work towards finding validated lead compounds that are likely to be suitable for development into medicines for unmet therapeutic needs. The growth of translational research and the establishment of centres of translational research will enable academic groups to become essential partners in pharmaceutical innovation.

7. Appendix

7.1 Articles 15 and 16 from the United Nations Convention on biological diversity

Article 15. Access to Genetic Resources

1. Recognizing the sovereign rights of States over their natural resources, the authority to determine access to genetic resources rests with the national governments and is subject to national legislation.
2. Each Contracting Party shall endeavour to create conditions to facilitate access to genetic resources for environmentally sound uses by other Contracting Parties and not to impose restrictions that run counter to the objectives of this Convention.
3. For the purpose of this Convention, the genetic resources being provided by a Contracting Party, as referred to in this Article and Articles 16 and 19, are only those that are provided by Contracting Parties that are countries of origin of such resources or by the Parties that have acquired the genetic resources in accordance with this Convention.
4. Access, where granted, shall be on mutually agreed terms and subject to the provisions of this Article.
5. Access to genetic resources shall be subject to prior informed consent of the Contracting Party providing such resources, unless otherwise determined by that Party.
6. Each Contracting Party shall endeavour to develop and carry out scientific research based on genetic resources provided by other Contracting Parties with the full participation of, and where possible in, such Contracting Parties.
7. Each Contracting Party shall take legislative, administrative or policy measures, as appropriate, and in accordance with Articles 16 and 19 and, where necessary, through the financial mechanism established by Articles 20 and 21 with the aim of sharing in a fair and equitable way the results of research and development and the benefits arising from the commercial and other utilization of genetic resources with the Contracting Party providing such resources. Such sharing shall be upon mutually agreed terms.

Article 16. Access to and Transfer of Technology

1. Each Contracting Party, recognizing that technology includes biotechnology, and that both access to and transfer of technology among Contracting Parties are essential elements for the attainment of the objectives of this Convention, undertakes subject to the provisions of this Article to provide and/or facilitate access for and transfer to other Contracting Parties of technologies that are relevant to the conservation and sustainable use of biological diversity or make use of genetic resources and do not cause significant damage to the environment.
2. Access to and transfer of technology referred to in paragraph 1 above to developing countries shall be provided and/or facilitated under fair and most favourable terms, including on concessional and preferential terms where mutually agreed, and, where necessary, in accordance with the financial mechanism established by Articles 20 and 21. In the case of technology subject to patents and other intellectual property rights, such access and transfer shall be provided on terms which recognize and are consistent with the adequate and effective protection of intellectual property rights. The application of this paragraph shall be consistent with paragraphs 3, 4 and 5 below.
3. Each Contracting Party shall take legislative, administrative or policy measures, as appropriate, with the aim that Contracting Parties, in particular those that are developing countries, which provide genetic resources are provided access to and

transfer of technology which makes use of those resources, on mutually agreed terms, including technology protected by patents and other intellectual property rights, where necessary, through the provisions of Articles 20 and 21 and in accordance with international law and consistent with paragraphs 4 and 5 below.

4. Each Contracting Party shall take legislative, administrative or policy measures, as appropriate, with the aim that the private sector facilitates access to, joint development and transfer of technology referred to in paragraph 1 above for the benefit of both governmental institutions and the private sector of developing countries and in this regard shall abide by the obligations included in paragraphs 1, 2 and 3 above.
5. The Contracting Parties, recognizing that patents and other intellectual property rights may have an influence on the implementation of this Convention, shall cooperate in this regard subject to national legislation and international law in order to ensure that such rights are supportive of and do not run counter to its objectives.

8. References

Artuso, A. (1997). *Drugs of Natural Origin: Economic and Policy Aspects of Discovery, Development, and Marketing*. Haworth Press, Binghamton, New York.

BIO-1 (n.d.). Guidelines for BIO Members Engaging in Bioprospecting. Accessed 1 May 2011, Available from http://www.bio.org/ip/international/200507guide.asp

BIO-2 (n.d.). Suggested Model Material Transfer Agreement. Accessed 1 May 2011, Available from http://www.bio.org/ip/international/BIO_Model_MTA.pdf

Boyd, M.R. (1996). The position of intellectual property rights in drug discovery and development from natural products. J. Ethnopharmacol. 51, 17-27, ISSN: 0378-8741.

Brendler, T. & van Wyk, B.-E. (2008). A historical, scientific and commercial perspective on the medicinal use of *Pelargonium sidoides* (Geraniaceae). J. Ethnopharmacol. 119, 420-433, ISSN: 0378-8741.

Bugni, T.S. *et al.* (2008). Marine natural product libraries for high-throughput screening and rapid drug discovery. J. Nat. Prod. 71, 1095-1098, ISSN: 0163-3864.

Butler, M.S. (2008). Natural products to drugs: natural product-derived compounds in clinical trials. Nat. Prod. Rep. 25, 475-516, ISSN 0265-0568.

Cao, S. & Kingston, D.G.I. (2009). Biodiversity conservation and drug discovery: can they be combined? The Suriname and Madagascar experiences. Pharm. Biol. 47, 809-823, ISSN: 1388-0209.

Castree, N. (2003). Bioprospecting: from theory to practice (and back again). Trans. Inst. Br. Geogr. 28, 35-55, ISSN:0020-2754.

Chadwick, D. & Marsh, J. (Eds.). (1994). *Ethnobotany and the Search for New Drugs*. Wiley, Chichester, UK.

Clark, R.L., Johnston, B.F., Mackay, S.P., Breslin, C.J., Robertson, M.N., Harvey, A.L. (2010). The Drug Discovery Portal: a resource to enhance drug discovery from academia. Drug Discovery Today 15, 679-683, ISSN 1359-6446.

Conrad, A., Kolodziej, H., Schulz, V. (2007). *Pelargonium sidoides* extract (EPs® 7630): registration confirms efficacy and safety. Wiener Medizinische Wochenschrift 157, 331-336, ISSN: 0043-5325.

Costello, C. & Ward, M. (2006). Search, bioprospecting, and biodiversity conservation: comment. J. Envir. Econ. Mgmt. 52, 616-625, ISSN: 0095-0696.

Cotton, C.M. (1997). *Ethnobotany. Principles and Applications.* Wiley, Chichester, UK.

Cox, P.A. (1994). The ethnopharmacological approach to drug discovery: strengths and limitations. In: Chadwick, D. & Marsh, J. (Eds.). *Ethnobotany and the Search for New Drugs*. Pp. 25-36. Wiley, Chichester, UK.

Dias, C.C. & da Costa, M.C. (2007). International Cooperation and Bioprospecting in Brazil and Peru. RECIIS - Electronic Journal of Communication, Information & Innovation in Health 1, 121-125 DOI: 10.3395/reciis.v1i1.33en, ISSN: 19816278.

Frisvold, G. & Day-Rubenstein, K. (2008). Bioprospecting and biodiversity: what happens when discoveries are made? Arizona Law Rev. 50, 545-576, ISSN 0004–153x.

Ganesan, A. (2008). The impact of natural products upon modern drug discovery. Curr. Opin. Chem. Biol. 12, 306-317, ISSN: 1367-5931.

Grabowski, K. & Schneider, G. (2007). Properties and architecture of drugs and natural products revisited. Curr. Chemical Biol. 1, 115-127, ISSN: 1872-3136.

Harvey, A.L. (2002) *Natural Product Pharmaceuticals: a Diverse Approach to Drug Discovery*, PJB Publications, Richmond, Surrey.

Harvey, A.L. (2007). Natural products as a screening resource. Curr. Opin. Chem. Biol. **11,** 480-484, ISSN: 1367-5931.

Harvey, A.L. (2008). Natural products and drug discovery. Drug Discovery Today 13, 894-901, ISSN 1359-6446.

Harvey, A.L., Clark, R.L., Mackay, S.P., Johnston, B.F. (2010) Current strategies for drug discovery through natural products. Expert Opin. Drug Discovery 5, 559-568, ISSN 1746-0441.

Hu, J.F., Eldridge, G.R., Yu, Y.H., O'Neil-Johnson, M. (2008). High-throughput natural product chemistry methods and the application of the capillary NMR probe. Progr. Chem. 20, 429-440, ISSN: 1005-281X.

IFPMA (2007). Guidelines for IFPMA Members on Access to Genetic Resources and Equitable Sharing of Benefits Arising out of their Utilization. Accessed 1 May 2011, Available from http://www.ifpma.org/Issues/fileadmin/templates/ifpmaissues/pdfs/2008_05_22_Guidelines_Genetic_Resources_EN.pdf

IISD (2007). ABS-Management Tool. Best Practice Standard. Accessed 1 May 2011, Available from http://www.iisd.org/pdf/2007/abs_mt_standard.pdf

Kaitlin, K.I. (2010). Deconstructing the drug development process: the new face of innovation. Clin. Pharmacol. Ther. 87, 356–361, ISSN: 0009-9236.

Kennedy, J. (2008). Mutasynthesis, chemobiosynthesis, and back to semi-synthesis: combining synthetic chemistry and biosynthetic engineering for diversifying natural products. Nat. Prod. Rep. 25, 25-34, ISSN 0265-0568.

Kingston, D.G.I. (2011). Modern natural products drug discovery and its relevance to biodiversity conservation. J. Nat. Prod. 74, 496-511, ISSN: 0163-3864.

Laird, S. & Wynberg, R. (2008). *Access and Benefit-Sharing in Practice: trends in Partnerships across Sectors*. CBD Technical Series no. 38, Secretariat of the Convention on Biological Diversity, Montreal, Canada.

Lesser, W.H. & Krattiger, A. (2007). Valuation of bioprospecting samples: approaches, calculations, and implications for policy-makers. In *Intellectual Property Management in Health and Agricultural Innovation: A Handbook of Best Practices* (eds. A Krattiger, RT Mahoney, L Nelsen, et al.). MIHR: Oxford, U.K., and PIPRA: Davis, U.S.A. Accessed 1 May 2011, Available online at www.ipHandbook.org.

Li, J.W.-H. & Vederas, J.C. (2009). Drug discovery and natural products: end of an era or an endless frontier? Science 325, 161-165, ISSN: 0036-8075.

Mays, T.D. & Mazan, K.D. (1996). Legal issues in sharing the benefits of bioprospecting. J. Ethnopharmacol. 51, 93-102, ISSN: 0378-8741.

Medaglia, J.C. (2004). Biodiversity for (bio)technology under the Convention on Biological Diversity: bioprospecting partnership in practice. Accessed 1 May 2011, Available online at http://cmsdata.iucn.org/downloads/cel10_medaglia_1.pdf

Mittermeier, R.A., N. Myers, P. Robles-Gil, & C.G. Mittermeier (Eds.). (1999). *Hotspots. Earth's Biologically Richest and Most Endangered Terrestrial Ecoregions.* CEMEX/Agrupación Sierra Madre, Mexico City.

Mulholland, D.M. & Wilman, E.A. (2003). Bioprospecting and biodiversity contracts. Environ. Develop. Econ. 8, 417-435, ISSN: 1355-770X.

Newman, D.J. & Cragg, G.M. (2007). Natural products as sources of new drugs over the last 25 years. J. Nat. Prod. 70, 461-477, ISSN: 0163-3864.

Prathapan, K.D. & Rajan, P.D. (2011). Biodiversity access and benefit-sharing: weaving a rope of sand. Curr. Sci. 100, 290-293, ISSN: 0011-3891.

Poore, G.C.B. & Wilson, G.D.F. (1993). Marine species richness. Nature 361, 597-598, ISSN: 0028-0836.

Rausser, G.C. & Small, A.A. (2000). Valuing Research Leads: Bioprospecting and the Conservation of Genetic Resources. J. Polit. Econ. 108, 173-206, ISSN: 0022-3808.

Reid, W.V., Laird, S.A., Meyer, C.A., Gomez, R., Sittenfeld, A., Janzen, D.H., Gollin, M.A. & Juma, C. (Eds.) (1993). *Biodiversity Prospecting: Using Genetic Resources for Sustainable Development.* World Resources Institute, Washington, DC; Instituto Nacional de Biodiversidad, Santo Domingo de Heredia, Costa Rica; Rainforest Alliance, New York; African Centre for Technology Studies, Nairobi, Kenya.

Rishton, G.M. (2008). Natural products as a robust source of new drugs and drug leads: past successes and present day issues. Am. J. Cardiol. 101 (Suppl.), 43D-49D, ISSN: 0002-9149.

Rollinger, J.M., Haupt, S., Stuppner, H., Langer, T. (2004). Combining ethnopharmacology and virtual screening for lead structure discovery: COX-inhibitors as application example. J. Chem. Inf. Comput. Sci. 44, 480-488, ISSN: 0095-2338.

Rollinger, J.M., Schuster, D., Danzl, B., Schwaiger, S., Markt, P., Schmidtke, M., Gertsch, J., Raduner, S., Wolber, G., Langer, T. & Stuppner, H. (2009). *In silico* target fishing for rationalized ligand discovery exemplified on constituents of *Ruta graveolens*. Planta Med. 75, 195-204, ISSN: 00320943.

Schuster, D. & Wolber, G. (2010). Identification of bioactive natural products by pharmacophore-based virtual screening. Curr. Pharm. Des. 16, 1666-1681, ISSN: 1381-6128.

Secretariat of the Convention on Biological Diversity (2002). Bonn Guidelines on Access to Genetic Resources and Fair and Equitable Sharing of the Benefits Arising out of their Utilization. Montreal: Secretariat of the Convention on Biological Diversity.

Secretariat of the Convention on Biological Diversity (2011). Nagoya Protocol on Access to Genetic Resources and the Fair and Equitable Sharing of Benefits Arising from their Utilization to the Convention on Biological Diversity. Montreal: Secretariat of the Convention on Biological Diversity.

Simpson, D., Sedjo, R. & Reid, J. (1996). Valuing biodiversity for use in pharmaceutical research. J. Polit. Econ. 104, 163–185, ISSN: 0022-3808.

Sneader, W. (1996). *Drug Prototypes and their Exploitation*. Wiley, Chichester, UK.

Sunazuka, T., Hirose, T. & Omura, S. (2008). Efficient total synthesis of novel bioactive microbial metabolites. Acc. Chem. Res. 41, 302-314, ISSN: 0001-4842.

Van Heered, F.R. (2008). *Hoodia gordonii*: a natural appetite suppressant. J. Ethnopharmacol. 119, 434-437, ISSN: 0378-8741.

Wynberg, R., Schroeder, D. & Chennels, R. (Eds.) (2009). *Indigenous Peoples, Consent and Benefit Sharing. Lessons from the San-Hoodia Case.* Springer, Heidelberg, Germany.

Part 6

Morphological Disparity

16

Morphological Disparity: An Attempt to Widen and to Formalize the Concept

Igor Ya. Pavlinov
Zoological Museum, Moscow Lomonosov State University, Russia

1. Introduction

One of the key paradigms in the contemporary biology, now emerging, is based on the acknowledging diversity of organisms (aka *biodiversity*) as a fundamental property of the life. This property is immanent to the evolving biota and constitutes a special research domain with its own problematics, tasks and, in part, methods. *Diatropics*, a discipline dealing largely with specific properties and regularities of the diversity, was suggested to recognize not so far ago (Chaikovski, 1990). This conceptual framework serves as a kind of alternative to the classical physicalist paradigm having been absolutely predominating until recently which purports to reveal unified laws and so concentrates on the uniformity rather than on the diversity. From this physicalist standpoint, the diversity, if it falls out of certain overall trends, is treated as a kind of "noise" just preventing to reveal the uniformity expressed by those laws. The currently prevaailing concept of biodiversity was initially advanced to reflect its taxonomic aspect, more particular the species diversity (Norton, 1986). However, more recent development of this concept has led to understanding that the taxonomic aspect is far from adequate representation of the entire natural phenomenon called *the* biodiversity. Scientists working on diversity of organisms became to realize that taxa do not exist without their morphological (or any other) traits, that is, without those morphological (or any other) features that emerge in the evolution together with the taxa and constitute an essential part of the entire biodiversity. This yields renaissance of understanding of the latter as largely a diversity of morphological forms (Gould, 1989). And, consequently, the so called *morphological diversity* became recognized as one of the key aspects of the overall biodiversity deserving special attention and investigation. And this "dual" view of biodiversity was eventually fixed terminologically; its taxonomic aspect is now called *diversity* proper (multiplicity) while its morphological aspect merits the special term *disparity* (heterogeneity) (Erwin, 2007; Foote, 1996; Kaplan, 2004; Pavlinov, 2008; Wills, 2001). Investigations of this second aspect of biodiversity now draw much attention of both evolutionists and ecologists (Chakrabarty, 2005; Ciampaglio et al., 2001; Erwin, 2007; Faleev et al., 2003; Foote, 1993, 1996, 1997; Gerber et al., 2008; Hulsey & Wainwright, 2002; Mcclain et al., 2004; Kaplan, 2004; Roy & Foot, 1997; Wainwright 2007; Willis et al., 2005).

Notwithstanding that, morphological disparity (aka *morphodisparity*) still remains basically "tied" to the taxonomic aspect of the overall diversity; as a matter of fact, it is actually being studied as essentially a morphological dissimilarity among taxa (Foote, 1993, 1996, 1997; Kaplan, 2004; Roy & Foote, 1997). That is why it is infrequently still declared that it is the taxonomy that deals with biodiversity. As a result, morphological dissimilarities among many other kinds of biological groups, such as sex or age groups, casts of social insects, biomorphs, etc, appeared not covered by currently predominating understanding of the disparity. However, it is evident that these dissimilarities constitute a very significant portion of the latter and hence they are to be given no less attention than the differences among taxa (Pavlinov, 2008).

This means that the very notion of disparity delimited by taxic differences only is quite insufficient. Instead, disparity is to be understood as a *compound of the dissimilarities among all and any kinds of biological groups of organisms* and not only among taxa. Accordingly, analyses of these dissimilarities in their plenitude should constitute one of the principal tasks of investigations on disparity as a whole. This disparity is morphological one as far as morphological traits are involved, but there are also other trait-defined disparities such as ecological, physiological, ethological, etc.

In the present chapter, I shall analyze some fundamental issues concerning morphodisparity in the above outlined widened sense. I shall start with consideration of background models underlying that understanding (section 2). It will be followed by discourses on the objectives of morphodisparity investigations (section 3), a part of which is causal interaction between disparity forms (section 4). Thereafter, some key parameters and characteristics of morphodisparity will be overviewed and formalized (section 5). Some attention will be paid to operationalization of those formalisms by means of numerical techniques allowing analysis of respective parameters (section 6), and several examples will be given to illustrate their application to particular datasets (section 7). At last, I shall consider briefly several important problems concerning general properties of morphological disparity to be considered more closely in future studies (section 8).

2. The background models

Any conceptual construct exists and functions not by itself but within a certain wider scientific framework, or background knowledge. It is this framework that provides the general understanding of what that construct specifically is, why are there different ways of its treatment and how it is to be dealt with (described etc.). Therefore it seems to me quite reasonable to start our consideration of morphodisparity matters with a very short reference to rather metaphysical topic.

The ways biodiversity can be understood and defined are different. This leads to several general concepts of biodiversity which underlay different understanding of what is the morphodisparity (Pavlinov, 2008).

The simplest is the *organismal* concept according to which only organisms are observable entities and thus are to be laid down in any definition of biodiversity. Theoretically, it is based on acknowledging the organism as a focal point of the life; accordingly, the biota is an array of organisms and the diversity is an array of organismal dissimilarities. Such a position is deeply rooted in the classical biology of the 18–19th centuries and is now largely supported by the Evo–Devo concept (see Hall, 1998 on the latter). However, this standpoint seems to be insufficient in respect to the entire biodiversity problematics. Indeed, such a reductional treatment looks no more biologically sound than, say, understanding of the organism as just an array of its cells.

The neo-Darwinian evolutionary theory presumes that the biota is structured basically at the population and species level. This gives a *species* concept of biodiversity (Claridge et al.,

1997), which is popular among environmental protection experts for it is quite operational and makes it easy to measure the taxic diversity (e.g. Sarkar, 2002). Accordingly to this standpoint, differences among species and their populations constitute the core of the biodiversity (and hence morphodisparity) isuue.

More general is the concept of biodiversity which considers this phenomenon at the biotic level. In its rather simplified version developed basically within phylogenetic taxonomic framework, biodiversity is treated as an array of dissimilarities among monophyletic taxa of any rank, that is, dissimilarities within the phylogenetic pattern (Eldredge & Cracraft, 1980; Pavlinov, 2005). Its further development led to acknowledging two other types of supra-organismal living systems as equivalent components of biodiversity, namely ecosystems (Faith, 2003) and biomorphs (Krivolutsky, 1998; Pavlinov, 2007).

Such a *biotic* concept of biodiversity, as it was further developed, fits quite naturally the modern consideration of biota from the synergetic standpoint (Pavlinov, 2007). According to a general model elaborated by this approach, development of such type of systems implies their structuring, that is, the emergence of certain subsystems (groups) within them recognizable by their specific features (Barantsev, 2003). This process of structuring is brought to existence and controlled by an array of causes, each responsible for producing groups of certain kind. Historical causes produce phylogenetic pattern of monophyletic groups while ecological causes are responsible for many other biodiversity phenomena, beginning from the differentiation of coenoses and biomorphs and going down to various infraspecific groups such as sexes etc. At last, there are intrinsic causes of development of organisms that are responsible for differences between ontogenetic stages and, hence, for existence of morphologically (physiologically etc) different age groups in populations. It is to be stressed that none of these general causes can be classified as more or less "important" in structuring biota; instead, they are equivalent in a sense, which makes biodiversity forms equivalent, in the same sense, as components of the overall biota's structure generated by different causes (but see about "primary" and "secondary" forms in the section 4 below).

The most significant consequence of such causal treatment of biodiversity as a biotic phenomenon is that it presumes a causal relation between biota's development, its structure, and forms (manifestations) of its diversity (Pavlinov, 2007, 2008). According to this, the biota is non-accidentally structured into various groups, be they either taxa, or sex or age or other infraspecic groups, or biomorphs, etc., each taking certain place in the biotic total structure and having certain peculiar features. So, it is the entire pool of various kinds of the disparity forms that constitutes the biodiversity as a whole. As it was just stated, this idea presumes that all such forms are, in the above sense, equivalent to each other.

3. What is the aim of the morphological disparity analyses?

According to the above biotic concept of biodiversity, the latter could be considered as a kind of *macroparameter* of the biota allowing to characterize it as a whole, regardless of its particular elements described by respective microparameters (pairwise differences among particular organisms and groups thereof). It reflects the very fact that biota is differentiated into certain groups of organisms produced by certain causes. These groups are dissimilar in certain features, the entire array of these dissimilarities is registered and studied as *the* morphodisparity.

It is evident that the latter does not exist by *per se*; rather, it represents a kind of epiphenomenon of the biota structured into groups dissimilar by morphological (and any other) features. Thus, descriptors of morphodisparity also constitute an array of macroparameters of the biota allowing to characterize, explore and eventually explain certain aspects of its structure.

The latter makes quite obvious the answer to the question placed in the title of the present section. We study morphodisparity in order, before all, to reveal the structure of biodiversity manifested in the dissimilarities among various groups of organisms. As this structure is non-accidental relative to the structure of biota proper, the next level of generalization based on the analyses of morphodisparity will be about the biota's structure (pattern). And as the latter is non-accidental relative to the causes affecting the structuring process, the *ultimate goal of investigations of the morphodisparity is understanding of the array of causes structuring the biota* as the evolving and functioning whole.

Taking into consideration possible causes structuring the biota (historical, ecological, etc), several general approaches to the disparity analyses can be recognized (Pavlinov, 2008).

The first of them can be called the *structural* approach, which deals with the disparity structure as such, with its general properties, parameters and characteristics, with its basic elements and their interrelations, etc. Though just a "descriptive", it is the most fundamental approach underlying all others; it is evident that any causal consideration of the morphodisparity and then of the entire biodiversity is impossible without reconstruction of the disparity structure. The latter, as an object of investigation, is most usually being reduced to several particular forms subjected to a particular causal explanation, be it phylogenetic or ecological or other. Such a reductionism is presumed by limitations of paticular exploratory themes and so is unavoidable, but nevertheless the totality of the overall disparity structure is not to be forgotten. It is this structural approach that constitutes the main subject of the present chapter.

The next is the *adaptational* approach; it considers various forms of differentiation (and of disparity) as the results of particular adaptations to particular environments. Divergence between sexes within species with prominent sexual dimorphism, or between ontogenetic stages in insects with complete metamorphosis, or at last balanced polymorphism in populations - all of them could be treated as a reflection of the niche structure of respective natural communities. It is evident that this approach deals basically with ecological causes structuring the biota and making its diversity as it is.

Another approach can be baptized as the *evolutionary*; it eventually intends to reveal and understand evolutionary, or more precisely, historical causes of the biota's differentiation. It is this approach that serves as a framework for analysis of the above phylogenetic pattern in general and of morphological differences among monophyletic taxa (such as phylospecies) in particular.

At last, there is one more approach to the causal morphodisparity analysis worthwhile being recognized; it could be called the *ontogenetic* (or more strictly epigenetic). It considers mainly those dissimilarities which are produced by individual growth processes and explains interrelations between disparity forms in terms of the "growth factor" itself (Eble, 1999; Pavlinov, 2008). It occurs to me that, because the ontogeny itself cannot be interpreted directly in terms of historical or ecological causation, this approach cannot be reduced to the others just considered. However, the Evo-Devo concept is to be mentioned here as the one that might connect phylo- and ontogenetic approaches to the morphodisparity analyses.

Ontogenetic approach to understanding and analyzing morphodisparity is especially important in case of variation of the measurable traits. It is evident that, for the growing organisms, it is the ontogenesis that produces most of their differences at phenotypic level.

These differences may involve rate or/and duration of growth processes but the final result (other things being equal) will be the same: in some individuals the respective structure will be smaller in size than in others. Accordingly, morphological differences of such kind could be explained by growth pattern specific to respective groups of organisms, say, to the sex groups (Blackith, 1965; Mina & Klevezal, 1976).

4. A "succession" between disparity forms

Systemic understanding of the morphodisparity proper presumes implicitly some specific interactions between disparity forms, which are not isolated from each other. Such an interaction can be thought about as a kind of causal relation between these forms in a way that some of them might be, at least in part, causes of others. So the disparity forms could themselves be included in the system of causal relations defining the morphodisparity structure. The entire system of such causation between disparity forms can be designated as their *succession*. Accordingly, it is possible to classify disparity forms involved in such a causal interaction as "*primary*" and "*secondary*" ones, which means that certain portion of the latter is a consequence of the former (Pavlinov, 2008; Pavlinov et al., 1993, 2008).

In more explicit form, variation of measurable traits may serve as a particular case of such kind of interrelation between disparity forms. Accordingly to the above ontogenetic standpoint (see section 3), all differences between growing organisms by such traits could be explained in term of their linear growth. It makes age variation a kind of "primary" disparity form, just because variation of such traits is basically regulated by the growth process by definition. Correspondingly, other disparity forms can be thought about as "secondary" ones relative to the age variation; they are "superimposed" over the differences attributable to age variation while they emerge due to action of some other factors not reduced to the growth (say, sex or geographic ones).

Such an approach makes it possible to elaborate a kind of causal *null model* connecting various forms of disparity of measurable traits with the growth process (Pavlinov, 2008; Pavlinov et al., 2008). It differs principally from the one elaborated in respect to phylogenetic interpretation of the morphospace occupation (Pie & Weitz, 2005). The above so called "growth" null-model presumes that, other factors structuring the disparity not in effect, one has to anticipate some strong positive correlation between age and other disparity forms (sex, geographic, etc.) of the measurable traits involved in the growth process. Significant deviation from such a correlation means non-fulfillment of the condition implied by this null model and indicate a possible significant effect of other cause(s) irreducible to the "growth factor".

The same ranking principle could be applied to other disparity forms which could be interconnected by the "primary–secondary" relation. Positive correlation among individual and geographic variation constituting the so called "Kluge–Kerfoot phenomenon" (Kluge & Kerfoot, 1973; Mitton, 1997) seems to be another important case of such relation. Here, individual variation is the "primary" form and geographic variation is the "secondary" one; the latter could be considered as an "extrapolation" of the former over the territory (Sokal, 1976). (Note that I here do not concern possible mathematical aspects of this phenomenon considered by Rohlf et al., 1983.)

A concept of the "succession" relation between the disparity forms can also be applied to correspondence of within- and between-species differentiation of closely related species. The focal point here is that the original Darwinian concept of gradual speciation must mean, in the terms adopted here, high "succession" between subspecies and species differentiation

(Pavlinov, 2008; Pavlinov et al., 1993). This presumption may serve as another null model relevant to this kind of comparisons, according to which any significant deviation from it might indicate irrelevance of the Darwinian concept.

5. Basic notions

One of the most fundamental problems concerning morphodisparity is that the latter, if it is not reduced down to just a sum of dissimilarities among individuals but treated as a macroparameter of the biotic structure (see section 2 above), is not a directly observable matter. To the contrary, it is given to a researcher not as a such but only in the concepts, notions and estimates defined by the very researcher. This means that a sufficient thesaurus is needed in order to designate and to describe properly both the disparity itself, its forms and interrelations among them.

The most inclusive is notion of **disparity**, which designates *any and all manifestations of both dis- and similarities among organisms* by any kind of traits; as far as the latter are morphological, the disparity is also designated as the morphological one. Let any fixed aspect of the disparity be designated as a **disparity form** (or a *form of variation*); each disparity form could be considered as a **component** of morphodisparity. Any group of organisms homogenous in respect to all the disparity forms will be called *elementary*; respectively, the disparity observed within such elementary group corresponds by definition to the *individual variation*. Elementary groups arranged according to certain biologically meaningful variable defining certain disparity form (sex, age, species belonging etc) constitute a *composite group*; respectively, dissimilarities observed between the groups so arranged represent the *group variation* of the same name.

In modern researches of disparity, the latter is formalized by notion of **morphospace** (Eble, 2000; McGhee, 1991, 1999; Pavlinov, 2008; Stone, 1997). It can be defined as an *array of actual and potential states of a morphological system realized in an array of organisms.* With some reservation, it is equivalent operationally to the phenetic hyperspace (Eble, 2000; Sneath & Sokal, 1973).

Before going ahead with considering morphospace parameters and characteristics, it is worthwhile to stress especially that the morphospace is not identical to the morphodisparity proper. The latter, as it is here understood, is what does exist in the objective world, just because the organisms themselves and their features by which they are dissimilar are all objective (real). Unlike this, a morphospace does not exist without a researcher who defines it on the basis of respective notions, definitions and estimates by characters selected in some or other way, so it is largely subjective. One can think about morphospace as a kind of *model* of the disparity; it is used as a representation of the real disparity in certain operational form to which certain analytical methods and estimates could be applied properly. It is to be clearly understood, as well, that the morphospace is formed exclusively by morphological variables selected to characterize disparity within a particular group of organisms, so no "external" (physical) variables are taken into consideration when morphospace properties are analyzed. It is this "abstract" status of morphospace that allows to compare directly such different disparity forms as age, sex, and geographic variation regardless of their orderliness in the real (physical) world. It is evident that morphospace concept is analogous in some respects to the ecological niche concept, as the latter is based on analysis of certain ecological variables which do not actually exist out of certain formalized model of real ecological communities (Pianka, 2000; Vorobeichik, 1993).

Variables, by which morphological entities under comparison are described, define *axes* of the morphospace. This makes possible a geometric representation of the latter as a hypervolume. A unit observation corresponds to the morphospace *element* and is represented by a point in respective hypervolume. Position of each element in the latter is uniquely defined by a combination of variable states peculiar to it. Particular interpretation of both morphospace axes and elements depends on the morphospace consideration aspect, which can be twofold (Pavlinov, 2008).

Using terminology developed by numerical taxonomy (Sneath & Sokal, 1973), they could be denoted as *Q*- and *R-aspects*. In the case of morphodisparity, the *Q*-aspect corresponds to consideration of disparity forms in the hyperspace defined by the traits describing the morphological entities (individuals, morphotypes, etc.). Alternatively, the *R*-aspect corresponds to the consideration of the traits in the hyperspace defined by variables designating the disparity forms. As it is seen, the principal difference between them involves interpretations of morphospace axes and elements (points in the hypervolume). In the *Q*-considered morphospace, the axes correspond to original variables (traits), by which the objects under comparison (individuals, morphotypes, etc.) are described, these objects being morphospace elements. By this, the *Q*-considered morphospace is fully analogous to the standard phenetic hyperspace, positions of elements in which are defined by their respective traits states. Contrary to this, *R*-aspect provides a kind of *inverse* morphospace, which axes correspond to the variables designating disparity forms (sex, age, etc) and morphospace elements are not individuals (morphotypes, etc.) but their traits. Operationaly, the axes of *R*-considered morphospace are defined by some quantitative measure of disparity form, for instance by a portion of the total moprhodisparity attributed to this disparity form; respectively, positions of elements (traits) in moprhospace are defined by respective estimates of explained variance for these traits. Morphodisparity is rarely considered in such a manner, but it provides some interesting possibilities (see section 7 below).

In the geometric terms, components of morphodisparity can be identified as *subspaces* of respective overall morphospace. A strict correspondence between the above groups of organisms and their disparities, both elementary and composite, and respective subspaces in the given morphospace are to be postulated for the sake of operationality. Thus the entire morphospace is defined as consisting of elementary and composite subspaces corresponding to dissimilarities both within and among elementary and composite groups (Pavlinov, 2008; Pavlinov & Nanova, 2009). For two morphospaces, in one of which between-group dissimilarities are more prominent than in other, while within-group dissimilarities are the same, the total volume estimates in the Foot's approach will also be the same, which seems to be quite erroneous. This makes it clear that between-group interaction within a morphospace constitute quite important portion of the latter and cannot be ignored, so accentuation of this morphospace fraction in an explicit form is quite necessary. It is evident from this viewpoint that definition of morphospace as just a sum of (in the terms adopted here) its elementary subspaces (Foote, 1993, 1996, 1997; Zelditch et al., 2004) provides oversimplified morphospace concept.

The morphospace may be *empirical*, if it is defined by the observed data only, or *theoretical*, if at least some of its components are hypothetical or imaginary (Eble, 2000; McGhee, 1999, 2007); the latter can also be defined as a "space of logical possibilities" (Zavarzin, 1974). The theoretical morphospace may be *interpolated*, if the imaginary data fit strongly between the observed ones, or *extrapolated*, if the imaginary data exceed the boundaries defined by the observed data.

Analysis of correspondences between empirical and theoretical morphospaces is an important part of morphodisparity researches. It allows to reveal actually existing ("permitted") and non-existing ("forbidden") states of a morphological system under investigation, which is a significant task for some branches of theoretical morphology (Levinton, 1988; McGee 1999, 2007). It is to be noticed that such an analysis provides a kind of "bridge" between classical morphology dealing with organisms and exploration of diversity of these organisms.

If morphospace is defined by the original traits it can also be called *original*; if the traits are transformed some way, this gives *transformed* morphospace. An example of the latter is the hyperspace defined by principal components extracted from the original traits.

Morphospaces construing for the same dataset may differ not only due to consideration aspects or trait transformations but also because of use of different dissimilarity measures. For instance, Euclidean and correlation distances reveal different aspects of overall dissimilarity pattern (Sneath & Sokal, 1973) and thus structure the morphospace in different ways. Particular morphospaces thus obtained can hardly be classified as original or transformed; rather, they correspond to different *patterns* of the same morphodisparity. From such a standpoint, it looks incorrect to state (Foote, 1997) that a morphospace can be characterized by some general pattern regardless of metrics applied.

The most important and most general parameter of the morphospace is its **structure** defined as a relation between its elements and/or its subspaces (Pavlinov, 2008). Indeed, nearly any actual morphospace (even if it is delimited by individual variation) is structured; there are "clouds" of elements within morphospace and "gaps" among them which are caused by both evolutionary and natural history of organisms under investigation. It is clear that the structure refers, in most general sense, to differentiation of some group of organisms; the more it is differentiated, the more structured is its morphospace. In more restrictive sense, the structuredness reflects if there are any conspicuous subspaces within the morphospace studied and how distinct they are. The morphospace structure is quite multifold and includes several characteristics.

If the group variation is considered, the structure is characterized by morphospace *composition*, that is simply by a list of the disparity forms (composite subspaces) recognized. An important structural characteristic of the subspace is its *portion* in the entire morphospace; it is defined as a part of the latter occupied by respective subspace. Though being a characteristic of the structure parameter, it depends evidently on the volume estimates (see below).

Quite important characteristic of morphospace structure is the subspaces *overlap*. It could be defined most simply as a balance between within- and intergroup dissimilarities; a more sophisticated might be its definition by interrelation of partial subspaces within total morphospace (Pavlinov, 2008; Pavlinov & Nanova, 2009). Additional to overlap is the *hiatus* (gap) between (usually elementary) subspaces. These two characteristics, overlaps and gaps, yield a complete description of interrelations between subspaces within the given morphospace. This is one of the most "problematic" characteristics of the morphospace; it is similar in a sense to that known in ecology as the niche overlap (Pianka, 2000; Sohn, 2001; Vorobeichik, 1993) and so faces the same troubles.

Another characteristic of the morphospace structure is morphospace *occupation* which means unevenness of distribution of points within hyperspace (Ciampaglio et al., 2001; Erwin, 2007; McGhee, 1999). It is usually illustrated by some schemes showing either distribution of elements within a hypervolume (e.g. standard scatter-diagram) or certain trajectories connecting these elements to indicate probable transitions between them, which corresponds in most advanced versions to so called epigenetic landscape (McGhee, 1999, 2007; Shishkin, 1988). If subspaces are recognized, this characteristic can more formally be defined through combination of all overlaps and hiatuses.

Particular descriptors and characteristics of morphospace and its subspaces can reasonably be identified as *scalar* and *vector* ones. Informally speaking, scalar parameters provide "static" descriptions, while vector ones provide "dynamic" descriptions of morphodisparity.

One of the most general scalar descriptors of the *Q*-considered morphospace is its **dimensionality** understood in topological sense as number of axes defining the morphospace. This descriptor can be estimated in two ways. The *complete* dimensionality is defined by all variables (traits) that are used in the given study; it corresponds to the *complete* morphospace, where completeness is understood not in any "absolute" sense but in respect to the given trait set only. The *sufficient* dimensionality is defined by a number of variables which are selected from the initial set and are considered as sufficient, by some criteria, for adequate representation of morphospace as a disparity model. This smaller set of variables defines a subspace of complete morphospace, which could be called *reduced* morphospace. Certain correspondence exists between morphospace structure and dimensionality; the more structured is morphospace in some or other respect (other things being equal), the less is its sufficient dimensionality (Zelditch et al., 2004). So the morphospace dimensionality can be interpreted and used as an indirect measure of its structuredness.

The morphospace/subspace **volume** is one of the most general and informative scalar descriptors indicating a magnitude of disparity observed for the given set of individuals described by the given set of traits. The entire morphospace is characterized by *total* volume, its subspaces are characterized by respective *partial* volumes (Pavlinov & Nanova, 2009; Zelditch et al., 2004). The *absolute* volume is defined as a total sum of dissimilarities among the objects; geometric interpretation of morphospace allows to assess absolute volume as a sum of pairwise distances between points within respective hypervolume.

For the morphospace containing several subspaces, estimates of its total volume depend on interrelations (overlaps and hiatuses) between its subspaces corresponding to disparity forms. The morphospace total volume is equal to the sum of partial volumes only if the latter do not overlap and have no evident hiatuses; this corresponds to the above morphospace definition by Foot (1993, 1996). If there are overlaps of the subspaces, then the total morphospace volume is less than the sum of partial volumes of its subspaces. If there are hiatuses, the total morphospace volume exceeds the sum of partial volumes of its subspaces. An estimate of partial volume attributed to all possible combinations of overlaps and hiatuses may provide a kind of measure of overall morphospace occupation. It reflects to some degree, just as the morphospace dimensionality does, the morphospace structuredness: the greater this summary partial volume, the less uncertain (unexplained) variation, the more structured (by definition) the morphospace.

A particular value of absolute volume calculated for a particular dataset depends on amount of objects and traits and on dimensions of the latter. Due to this, estimates calculated for morphospaces defined for different datasets may not be compatible directly. To remove this effect, several corrected estimates of morphospace volume are introduced (Ciampaglio et al., 2001; Pavlinov & Nanova, 2009; Villier & Eble 2004; Wills 2001; Zelditch et al., 2004). One of them is a *relative* volume calculated to exclude effect of traits number and dimension, another is a *unite* (or specific) volume calculated to exclude effect of number of objects.

It is evident that any estimates obtained for these volumes inherently depend on particular definitions and numerical methods applied. Attention is to be drawn to an important difference in formal properties between various volume estimates as they are defined here. Absolute estimates for the entire morphospace and all of its subspaces are by definition (and calculation methods) strictly additive; one can say that the total absolute volume is "decomposed" into respective partial volumes without residue (Foot, 1993, 1996). The same is true for the relative volume estimates. Unlike this, the unit volume estimates are not additive: the arithmetic sum of partial unit volumes is greater than the total unit volume (Pavlinov & Nanova, 2009). This is because each partial unit volume is estimated as some "mean" value for an object (similarly to the deviate in dispersion analysis), and the total unit volume is roughly equal to an average of all partial unit volumes, some of which can be smaller and others can be larger than this average value.

Vector descriptors of morphospace allow characterizing both directions of predominating variation trends within each of the subspaces and their co-directionality. Vector representation of morphodisparity is not so popular as the scalar ones, although both its basic ideas and principles of analysis, at least in case of measurable traits, are quite simple and transparent for understanding (Blackith, 1965; Lissovsky & Pavlinov, 2008).

As a matter of fact, each disparity form involves certain differences between organisms or groups thereof by certain traits. With the morphospace defined by these traits as axes, these differences are represented as distribution of morphospace elements along those axes in accordance to the observed differences. It is obviously possible to fix a predominating trend for such a distribution (this is a routine procedure in many numerical ordination techniques) and to define it as a vector characterized by an angle relative to the fixed morphospace axes. If several disparity forms are analyzed at once, it is possible to compare them by this vector characteristic, that is, to explore co-directionality of trends of respective disparity forms in the given morphospace. The more similar are trend directions of disparity forms being compared, the less is the angle between respective vectors. It is clear that nothing like directions in the "physical" space is presumed by such a vector analysis; only abstract morphospace axes and vectors are considered. Similar approach is quite popular in researches in ecology and biogeography where various angular (correlation) measures of similarity are employed (Pesenko, 1982; Pianka, 2000; Sneath & Sokal, 1973).

Such a comparison of vectors can be conducted between both the same disparity form in various organismal groups and between various disparity forms within each of such groups (Pavlinov & Nanova, 2009). Say, one may wish to compare trends of sex differences in various species, or trends of sex and age differences in the same species. Operational interpretation of results of such comparisons in respect to causes of the similarity in question is quite simple: high co-directionality of trends means that the same traits are mostly involved in the disparity form(s) under comparison. However, more general (and more speculative) explanations for between- and within-species comparisons are different; high co-directionality might be interpreted as reflection of the above "succession" among disparity forms in some cases, while its interpretation as a kind of "parallelelism" would be more appropriate in others.

There is a restriction in these comparisons; they seem to be permissible only for the morphospaces defined by the same variables. By this, analysis of vector parameters differs markedly from that of the scalar ones. It is quite normal to compare, say, unit volumes of different morphospaces, one of which is defined by cranial and another by dental traits. But comparison of such morphospaces by vector parameter seems to be unsound both biologically and operationally, and there seems to be no methods for analyzing co-directionality of, say, sex differences in the morphospaces defined by different traits.

All the above considerations are relevant to the *Q*-aspect of morphospace consideration which is much developed in this respect. As to the consideration of morphospace in its *R*-aspect, it is not so intensively studied and thus its properties are not so evident. Some scalar characteristics of disparity forms are most easy to use in this case, and their interpretations are most clear (Pavlinov et al., 2008). First instance of such a consideration of *R*-aspect of morphodisparity that gave certain important results is the above Kluge–Kerfoot phenomenon, which concerns concordance between individual and geographic variation of a set of morphometric traits. Subsequently, similar approach was extended to comparison of several forms of group variation (Nanova & Pavlinov, 2009; Pavlinov et al., 1993, 2008).

In the *R*-considered morphospace, it is possible to compare various disparity forms in the same group of organisms or the same disparity form in various groups. It is based on scalar characteristics, but the tasks it deals with are quite similar to those accomplished by the above vector parameter. The basic idea (Nanova & Pavlinov, 2009; Pavlinov et al., 2008) is that the axes of *R*-considered morphospace may be interpreted as the vectors corresponding to certain trends of the disparity forms. Respectively, distribution of traits in such inverse *R*-considered morphospace indicates degree of concordance of these vectors: the closer the points of hypervolume are disposed to diagonal of the quartile between the axes, the more concordant are respective trends of the given set of traits.

Thus, we have two approaches to exploration of co-directionality of trends in the morphospace, one of which involves vector parameter of *Q*-considered morphospace and another works with scalar parameter of *R*-considered morphospace. They yield quite similar results, but yet there is an important difference; it is possible to analyze the traits individually in the *R*-considered morphospace only, this option being unavailable in the *Q*-considered morphospace. Such a possibility is of evident biological importance; to realize it, one has to abandon an idea of the traits set as a kind of statistical ensemble on which certain simple regularities (like the above Kluge–Kerfoot phenomenon) are to be fulfilled. The latter idea belongs to a physicalist standpoint according to which only such regularities are of scientific importance while deviations from them are non-significant. As it was stressed at the beginning of the present paper, a standpoint like this is not productive in the explorations of the morphodisparity. As far as the morphospace is defined by not randomly but reasonably selected traits, there is no biological reason in opposing statistically significant "regularities" and irrelevant "deviations" from them. Instead, equal attention is to be paid to all and any manifestations of overall disparity pattern "tied" to particular traits or subsets thereof, as they may equally be biologically sound and deserve causal analysis (Pavlinov, 2008; Pavlinov et al., 2008).

6. Operationalization

As it was stressed above, the morphospace "exists" in form of notions, definitions and estimates; it is the latter that make the entire theoretical construct operational and concrete.

There is a great diversity of approaches and methods of quantifying morphospace parameters and characteristics, beginning with quite simple indices of differentiation and finishing with sophisticated statistic measures of overall disparity and its particular properties. Here I consider certain methodological problems and then outline briefly some numerical techniques employed in my and my colleagues' investigations, which are based on the above morphospace parameters and descriptors.

First of all, it is to be stressed that numerical methods employed in exploration of the morphodisparity should and could not be independent of biologically sound premises underlying this concept. The latter is the biological one, so the methods of its investigation are to be biologically reasonable; this means they are to be at least compatible with those premises and at most deducible from them. A spectacular example of such interrelation between theoretical and methodical parts of a research program in biodiversity studies is provided by cladistics in which techniques of construing branching diagrams are directly deduced from a particular phylogenetic theory (Pavlinov, 2005).

Another quite evident requirement for the methods of analysis of morphospace parameters is that they have to be transparent in respect to their formal properties. Therefore, these methods are to have pretty good mathematical background and are easy to understand and interpret by biologists not experienced in mathematics. Without this it is hard to hope for a clear-cut understanding of the results obtained by these methods. However, it does not mean that such a criterion would overbalance the one of biological validity, as it is sometimes suggested (Abbott et al., 1985).

At last, it is important that the methods in question should provide commensurable numerical estimates of various disparity forms, be it individual variation, sex or age differences, etc. It is evident prerequisite for various disparity forms to be directly comparable by their quantitative characteristics.

Let us consider briefly some methods which allow numerical analyzes of morphospace parameters introduced in the previous section and satisfy just above properties.

The morphospace sufficient dimensionality is usually assessed by some ordination methods such as principal component analysis (PCA) or multidimensional scaling (MDS) (Faleev et al., 2003; Puzachenko, 2000; Zelditch et al., 2004). The main difference between these two is that PCA operates on covariation/correlation matrices while MDS operates on distance matrices and thus is free from some limitations inherent to correlation analysis. There are several possible ways to assess numerically this parameter. In the most simple case, it is conventionally defined as certain portion (for instance, 75 per cent) of the disparity studied which is "explained" by respective number of variables; this number is taken as a value of the sufficient dimensionality. More sophisticated is an approach based on analysis of distribution of stress values obtained by sequential iterations of MDS procedure. The serious limitation inherited in these methods is that they do not allow to discriminate various disparity forms and to compare them by sufficient dimensionalities of respective composite subspaces. Such a possibility could be provided by Variance Component Analysis (VCA) or Discriminant Function Analysis (DFA), both applied to a priory recognized subgroups corresponding to the disparity forms. It is to be noticed that numerical values of sufficient dimensionality obtained by any of these approaches can be both integer and fractional numbers. The latter might be of special interest, as it allows to reflect also supposed fractal properties of the entire morphospace and its subspaces.

The total morphospace volume and partial volumes of its subspaces are currently being analyzed by various dispersion-based or distance-based methods (Eble, 2000; Faleev et al., 2003; Foote, 1996, 1997; Van Valen, 1974; Villier & Eble, 2004; Zelditch et al., 2004). The main (but not principal) difference is that dispersion-based methods imply *a priory* decomposition of the entire sample dispersion into within- and between-group dispersions, while distance-based methods are free of such precondition, which provides some specific merits (see below). Both approaches allow analyzing (using the terminology adopted here) total and partial volumes, though they treat relation between these estimates in different ways. In the dispersion-based methods, the primary goal is the analysis of composite subspaces formed by between-group variation, while within-group variation is treated as just an unexplained variance which is not analyzed at all. In distance-based methods, either only elementary subspaces are initially analyzed or both elementary and composite subspaces are considered as equivalent components of the entire morphospace deserving equal attention. It is important that, within each of these approaches, all disparity forms are estimated in the same units and thus are comparable directly.

The nowadays most popular dispersion-based methods are the Multivariate Analysis of Variance (MANOVA) and the above VCA. The descriptors of disparity forms are considered by these methods as independent variables and morphological (and eventually any other) traits are considered as dependent variables. In MANOVA, the independent variables are uniformly taken as fixed, while VCA treats them as random. The both conditions seem to be too strong for the disparity forms actually involved in the biological investigations, so this provides certain problems (Leamy, 1983; Pavlinov, 2008). As a matter of fact, there are disparity forms, such as geographic or age variation, in which particular values (geographic localities or age groups) are actually non-fixed though hardly completely random, while other forms such as sex differences are certainly fixed. This inconsistency could conventionally be resolved by taking all the disparity forms as non-fixed independent variables in the dispersion analysis (Pavlinov et al., 2008).

Among algorithms of this analysis, Models I and III of both MANOVA and VCA, and maximum likelihood models of the latter are the most used (Ciampaglio et al., 2001; Leamy, 1983; Pavlinov et al., 1993, 2008; Zelditch et al., 2004). The Model I applied by Leamy (1983) seems to be least appropriate, as its dispersion estimates ascribed to particular independent variables are correlated significantly with the order they are entered in the analysis (StatSoft…, 2010). The maximum likelihood VCA was shown to be quite conservative in respect to variation of factor gradations and "sampling defects" producing more or less unbalanced design (Lissovsky & Pavlinov 2008). It could be applied as the prime method of *Q*-considered morphospace analysis while Model III of MANOVA is of use for the analysis of *R*-considered morphospace.

The subspaces overlap is impossible to explore directly by dispersion analysis, especially if overlap of the elementary subspaces is of interest. Meanwhile, it is principally possible to consider respective factor interaction as an indirect measure of composite subspaces overlap, but this question needs future special clarification (Pavlinov et al., 2008). Among dispersion-based approaches, standard DFA can also be employed here, in which disparity forms are again considered as classificatory (independent) variables and the traits are considered as dependent variables. Subspaces overlap could be estimated as per cent of posterior re-classification of objects allocated to the elementary groups recognized in respect to the given disparity form. This technique has been earlier employed in analysis of niche overlap (Krasnov & Shenbrot, 1998). Its results are quite easy to interpret, principal shortage of this method is that it does not allow considering the entire morphospace volume when overlap of its subspaces is evaluated.

Another approach to be considered here includes distance-based methods. Most usually they are based on calculations of Euclidean distance obeying the metric axioms, though correlation distances could also be used for some particular tasks (Sneath & Sokal, 1973). The first kind of distance makes the morphospace analysis faced at the problem of trait dimensions, which is easy to resolve by introducing respective correction factor (briefly discussed above). Unlike the dispersion-based methods, number of elements in the morphospace matters in this case; thus dimensionality of distance matrix is to be included in the formulas used to estimate respective morphospace scalar characteristics. This approach makes it possible to analyze both volumes of morphospace and its subspaces and their overlaps. Several techniques of calculations are in use (Ciampaglio et al., 2001; Foote, 1997; Pavlinov & Nanova, 2009; Villier & Eble, 2004; Zelditch et al., 2004).

In the most simple case (Foote, 1997; Zelditch et al., 2004), the distances among objects and a centroid are calculated for each elementary subspace separately to give an estimation of the latter's partial volume, and then the total morphospace volume is calculated as a sum of all the elementary ones (see notes on its insufficiency above). Another approach is based on calculation of distances among all elements and the centroid of the entire morphospace, after which volumes for both partial subspaces and the total morphospace can be calculated (Ciampaglio et al., 2001; Foote, 1997; Zelditch et al., 2004). This method is more advanced, as it includes both within- and between subspaces dissimilarities; its shortage includes impossibility to discriminate directly these two sets of dissimilarities and to analyze separately composite subspaces corresponding to the forms of group variation. In our approach (Pavlinov & Nanova, 2009), we calculate pairwise distances among all objects without detecting any centroids, and then the entire distance matrix is decomposed into several blocks each corresponding to the particular elementary and composite subspaces.

Three correction factors are to be taken into consideration in calculations of scalar characteristics of mophodisparity, which are (a) number of traits by which the objects are compared, (b) trait dimension related to the general problem of size/shape components in the analyses of measurable traits, and (c) the number of objects in the sample being analyzed. These factors are widely discussed in respective literature and so are to be just mentioned here.

Numerical techniques for analyses of vector parameters of morphodisparity first offered by Blackith (1965) was based on the above DFA. The latter however provides just an indirect evaluation of both directions and co-directionality of the vectors. These characteristics can be obtained straightforwardly from MANOVA based on the original Pearson's (1901) idea: the predominating trend ascribed to the given disparity form is defined and calculated as the first eigenvector of covariance matrix for the factor effect corresponding to that disparity form (Lissovsky & Pavlinov, 2008; Nanova & Pavlinov, 2009). Accordingly, similarity of two trends is defined and calculated as a cosine or arccosine of the angle between two respective eigenvectors (Lissovsky & Pavlinov, 2008).

It is evident that this operational definition of vector parameter is true for composite morphospaces and cannot be applied directly to the elementary ones. However, it is in principle possible to apply this concept to the latter using, for instance, PCA operating with respective covariance matrices and extracting eigenvectors from them (Eble, 2000).

This vector estimate deals with *Q*-considered morphospace and is hardly applied to the *R*-considered one. For the latter, Pearson's correlation analysis could be applied to estimate numerically concordance between values of explained variances ascribed to the particular traits in respect to the disparity forms being compared (Pavlinov et al., 1993, 2008).

To make my brief review more complete, information statistic measures of diversity/disparity are to be mentioned, which are popular among ecologists and are used sometimes in morphometrics (Faleev et al., 2003; Kupriyanova et al., 2003; Pavlinov, 1978; Pustovoit, 2006; Zelditch et al. 2004). As far as scalar characteristics of the morphospace are concerned, principal disadvantage of these measures is that they evaluate only evenness of a trait frequency distribution and do not consider magnitude of its variation; so they do not allow measuring morphospace/subspace volumes.

Although all the above morphospace estimates are treated as strictly quantitative, the task of evaluation of their statistical significance can be of interest. Several null hypotheses (not to be mixed with the above null model) are formulated and tested to reach this aim. One of them concerns significance of differentiation of groups fixed in respect to a particular disparity form under consideration. This kind of null hypotheses is formulated as follows: a characteristic (volume, vector, etc) of the given disparity form does not differ from the one caused by random variation. Another kind of null hypotheses concerns differences between these disparity forms by some characteristic used in their analysis (volume, overlap, co-directionality, etc). In this case, a relevant null hypotheses looks like that: difference between two disparity forms by respective characteristic is accidental. Null hypotheses of the first kind are easy to test by standard methods in case of dispersion-based approaches. Thus, both volume and vector estimates for any composite subspace can be tested by F-criterion. However, this statistical procedure cannot be applied for testing the second kind of null hypotheses above; nor can it be used in case of the distance-based estimates at all.

There are several distance-based methods of morphospace /subspace comparisons by their respective characteristics. The most simple is comparison of the morphospace /subspace volumes, for which standard parametric t-test seems to be appropriate. It allows to evaluate statistical significance of differences between any pair of morphospaces /subspaces by their volume estimates (Zelditch et al., 2004). However, more popular is a non-parametric approach that employs various resampling methods (Pavlimov & Nanova, 2009; Villier & Eble, 2004; Zelditch et al., 2004). They imply bringing some stochastic component in the original data matrix and subsequent analyses of thus generated matrices by standard algorithms. Comparison of the morphospace characteristic estimates for the original data matrix with stochastically generated distributions allows to calculate a probability of non-randomness of the original estimates.

To sum up the present section, I should like to stress that any formalized analytical approach is much simpler than the disparity being analyzed, so the latter is to be reduced some way to an operational state accessible to that approach. There are no universal quantitative methods which would provide reasonable decisions of all the tasks of biologically sound morphodisparity explorations. Each method is more or less narrow in its resolutive abilities and uncovers just particular properties of the overall disparity. For instance, of the two dispersion-based general methods, dispersion analysis makes it possible to measure volumes of morphospace and its subspaces but does not allow analyzing overlap of the latter, while DFA yields simple (though not complete) estimate of subspaces' overlap but provides no straightforwardly interpreted estimates of their volumes. As to the distance-based methods, they seem to be universal, but have their own limitations inherited in diversity of distance measures which choice is not a trivial task. Therefore, there should be a kind of "toolkit" with several methods included in it that would make it possible a more a less comprehensive analysis of various morphospace parameters and characteristics.

At last, it is wise to keep in mind that there are no "good" or "bad" methods by themselves; they become so due to their proper or improper application. This again turns us to the fundamental problem of interrelation between biologically sound backgrounds and technical tools of the morphospace exploration. One has to interpret a method properly within the background framework to comprehend if it is reasonably "good" or not.

7. Some examples from mammal skull variation

Below I shall present briefly some results of my recent investigations of morphodisparity fulfilled on skull variation in several mammal species. The principal "model" species in question are the pine marten (*Martes martes*), the red fox (*Vulpes vulpes*), and the polar fox (*Alopex lagopus*); some examples are borrowed from the data on the jirds (genus *Meriones*) and on the Przewalskii horse (*Equus przewalskii*). The skull is described by 14 standard measurable traits taken by caliper (the character list is not of particular relevance here). Most of these results were published elsewhere, together with description of the materials and particular methods (Lissovsky & Pavlinov, 2008; Nanova & Pavlinov, 2009; Pavlinov et al., 2008; Pavlinov & Nanova 2009). I shall consider here the following results: sufficient morphospace dimensionality and volumes estimates; some scalar and vector characteristics of *Q*-considered morphospace; scalar characteristics of *R*-considered morphospace.

Sufficient morphospace dimensionality was estimated by PCA for no less than 75 per cent of explained variance. It was found that the respective estimate is between 2 and 3 in the pine marten and the red fox, it is between 3 and 4 in the polar fox, and is reachs about 6 in the Przewalskii horse. So it can be concluded from the promises underlying this approach that the entire morphospace is much less ordered in the latter species as compared to others.

Estimations of subspace volumes were considered using data for pine marten and polar fox by means of dispersion analysis. The pine marten was represented by one geographic sample, sex and age differences were analyzed, four age groups were recognized. The polar fox was represented by four geographic samples, which allowed to consider spatial differentiation; sex and age differences were also considered, with two age groups recognized. These two species were shown to differ markedly from each other by the summary subspace occupied by all forms of group variation; the latter takes about 74 per cent of the morphospace volume in the pine marten and about 42 per cent in the polar fox.

These two species differ from each other also by ratios of partial volumes of the subspaces corresponding to the sex and age differences. The first disparity form takes much greater portion of the entire morphospace as compared to the second one (67 and 5 per cent, respectively) in the pine marten, while they are more similar in this respect in the polar fox (11 and 21 per cent, respectively).

Results of distance-based analysis of subspaces obtained for pine marten and polar fox indicate that the entire morphospace unit volume is higher in the former than in the latter (0.310 and 0.235, respectively). So the marten is more variable in general than the fox (of course, by the skull traits and in the units used here). As far as elementary subspaces are concerned, males appeared to be more variable than females in both species in all age groups, average values are 0.212 and 0.182 in the pine marten and 0.212 and 0.208 in the polar fox, respectively.

Analysis of composite subspaces indicate that overlap of age groups is evidently more expressed as compared to that of sex groups in the pine marten (0.843 and 0.437, respectively), while these values are almost equal in the polar fox (0.693 and 0.662, respectively). These results agree with well-known estimates of age and sex differences in these two species obtained previously by traditional approaches.

Estimations of co-directionality of disparity forms have never been considered any seriously before, so the results obtained on the skull variation in the above mammal species are quite new. Two principal approaches to analysis of this parameter were adopted in my studies, one dealing with integral (for all traits) vector characteristic of *Q*-considered morphospace and the other being based on scalar characteristic (partial volumes of respective subspaces per traits) of *R*-considered morphospace. Within-species disparity forms (geographic, age and sex variation) were considered in the carnivoran species, while subspecies structure of a polytypic species was compared with the differences between closely related species in the jird genus *Meriones*. The angles between respective vectors were measured by their arccosines (in rads), the lower arccosine value indicates less angle which thus means higher co-directionality.

Comparisons of vectors of age differences with those of sex and geographic differences within each of the carnivore species revealed the following. Co-directionality of age and sex variation was estimated as rather high in the canids (0.16–0.29 rad) and pretty low in the marten (0.75 rad). It is evident from these results that the two canid species are more similar to each other in respect to their age and sex variation than to the pine marten. Co-directionality of age and geographic variation was estimated as rather high in red fox (0.23 rad) and low in polar fox (0.72 rad).

Interpretation of this within-species vector comparisons presumes a kind of "succession" between disparity forms caused by the "growth factor". Thus, differences among males and females in canids by their skull dimensions may be interpreted as resulting basically from differences in rate/duration of their linear growth. To the contrary, sex differences in the pine marten cannot be explained in such a simple manner but involve noticeably different allometric trends. The same seems to be true for the comparison of age and geographic variation in the canids. In the red fox, cranial differences between animals from various geographic regions may be explained mainly by differences in their linear growth. Contrary to this, geographic differentiation in the polar fox cannot be explained by such simple regularity, differences in growth patterns are to be anticipated (Nanova, 2010).

Calculation of angles between vectors of particular disparity forms in the carnivores indicates that (*a*) predominating trends of sex differences appear to be quite similar in all species (0.11–0.28 rad), (*b*) those of age differences are more diverse (0.24–0.60 rad), and (*c*) the trends of differences between geographic samples in the morphospace are most different in two fox species compared (0.68 rad). It is of importance to note that canid species appeared again to be more similar to each other by the vector characteristics calculated for sex and age differences, than each of them to the mustelid.

The results of such between-species comparison were offered by me to interpret as reflections of a kind of "parallelism" of morphospace properties in various species (see Section 5 above). It might be reasonably supposed that such a "parallelism" is caused by similarities in mechanisms structuring these species' morphospaces. Thus, one may infer from the data just exposed that such a similarity in mechanisms regulating various disparity forms is most expressed in case of sex differences, less so in age differences, and the least so in geographic differences. These similarities and differences among species are possible to explain, at least at a very general speculative level, as a consequence of certain balance of intrinsic (developmental) and extrinsic (environmental) factors. From this standpoint, trajectories of sex differentiation by skull dimensions come out to be quite similar in all species compared because they are most dependent on just the intrinsic factors which seem to be quite stable by themselves. Contrary to this, drastic between-species differences in predominating geographic trends in the same morphospace could be explained by that they are mostly regulated by some extrinsic factors, which are quite variable and so their manifestations are quite different in the species under consideration.

Analysis of co-directionality of subspecies and interspecies differentiation by skull traits in several closely related jird species of the genus *Meriones* revealed quite unexpectedly opposite directionality of these two disparity forms, numerical estimate yielding negative correlation (about -0.59) between respective distributions of explained variances attributed to different traits. It is evident from this comparison that some traits vary between subspecies of *M. libycus* while others discriminate three close species (*M. libycus*, *M. shawi*, *M. grandis*). Based on the speculative suggestions underlying such kind of vector analysis (see section 4 above), it is possible to interpret this disparity pattern in *Meriones* as indirect indication of various traits being involved in subspeciation and speciation events, respectively. If this treatment is correct, then other than Darwinian speciation model could be in action in this particular case, for example the one supposed by punctuated equilibrium concept (Eldredge & Gould, 1972).

8. Some problems and tasks for the future

Investigations of differences among organisms have as a long history as the history of entire biology. However, the structural approach to the morphodisparity having been developed during the last decades seems to provide a new insight into this biological phenomenon. As a matter of fact, it means a new look at the disparity, namely a look at its own properties and eventually at the causes of these properties. In this respect, structural approach to the analysis of disparity of morphological objects may be of no less "revolutionary" significance to the morphometrics than geometric morphometric approach to description of these objects themselves (Rohlf & Marcus, 1993). In other words, the structural approach to the morphological disparity may constitute a basis of a new research program for the morphometrics.

This program is at the beginning of its elaboration and therefore contains a number of problematic positions concerning the very fundamentum of respective background theory. They are mostly biological rather than technical and so their decisions would appeal to a kind of "scientific metaphysics" of biodiversity (Pavlinov, 2007). Though metaphysical, discourses of this kind have actually a significant impact on operational definitions and, respectively, on the methods of analysis of morphological disparity. Some of these problems are briefly discussed here.

First of all, the very understanding of the central notion of morphospace is to be indicated. As I stressed above (see section 5), its comprehensive definition has to include subspaces corresponding to the dissimilarities both within- and between groups of organisms, which can either overlap or be separated by hiatuses. So the morphospace is to be defined operationally as a combination of all elementary subspaces minus subspaces of their overlaps plus subspaces of their hiatuses. It is evident that, generally speaking, any elementary subspace can overlap with some subspaces and be separated from others. This provides a rather complicated overall pattern of morphospace occupation which, nevertheless, is to be assessed by adequate numerical methods. Our efforts lead us to a pretty sophisticated index, which however appeared to allow to analyze only a narrow part of this occupation (Pavlinov & Nanova, 2009). It is evident that the entire problem deserves special investigations in the future.

Several other problems may be united under the same cover that might be called morphospace non-Euclidity (Pavlinov, 2008). The latter, in the particular case under consideration, means non-orthogonality of the morphospace axes, which means that angles between them may vary along their gradients. Translating these into biological terms faces us at the well known problems of between-trait correlation and size-dependent allometry, the latter providing a kind of specific developmental "allometric subspaces" within the entire morphospace (Gerber et al., 2008). There is a number of well known formal technical decisions of these problems, for instance replace of correlated original traits by respective uncorrelated principal components. However, content-wise consideration of this subject inclines me to suspect that such kinds of decisions seem, at least in part, to deprive them of biological sense, as they eventually lead to a loss of quite important and biologically sound fraction of morphodisparity.

Between-traits correlations deserve special consideration in this respect. As a matter of fact, these correlations are formed by the same growth processes as the traits themselves. So it seems quite reasonable to treat this "growth correlations" as a significant aspect of the morphospace defined by measurable traits. Therefore their elimination from explorations of the entire morphospace would reduce biological contents of their results. Instead of formal elimination of correlations from analyses of a morphospace by, say, defining the latter by principal components, it would be more reasonable to find the proper ways of special investigations of impact of correlations onto morphospace properties. For instance, it might be promising to divide traits into certain blocks ("pleiads") accordingly to their correlations (Rostova, 2002) and to analyze each block separately to reveal specific properties of the subspaces defined by respective traits. The evident background of such kind of investigations includes a supposition that the stronger is correlation between traits, the more similar are the factors structuring their variation. So comparative structural analysis of morphospaces defined by various sets ("pleiads") of traits may serve as another important source of information about causal mechanisms of the overall morphospace patterning.

9. Conclusion

1. Morphodisparity is a macroparameter of the structured biota, the latter being understood as a developing complex non-equilibrium system. The overall disparity thus understood includes various disparity forms corresponding to dissimilarities among all and any kinds of groups of organisms as components of the structured biota.
2. The ultimate objective of analysis of morphodisparity is uncovering mechanisms (causes) effecting both process and results of the biota's development and structuring. This objective requests prior investigations of properties of the disparity proper, as they are defined by interrelations among disparity forms. Such a structural approach to the investigations of morphodisparity may constitute the conceptual core of a specific research program in the morphometrics.
3. The key notion in analyses of morphodisparity is that of morphospace. The morphospace differs from the disparity itself in that the latter is objective (real) while morphospace is a theoretical construct largely depending on definitions and descriptors (variables and methods) employed in the given investigation. Basic components of a morphospace are elementary and composite subspaces corresponding to the particular disparity forms.
4. Two consideration aspects can be fixed for morphospace analysis, Q- and R-aspects. The former corresponds to consideration of disparity forms in the hyperspace defined by the traits, while the latter corresponds to consideration of the traits in the hyperspace defined by the disparity form descriptors.

5. Morphospace can be characterized by scalar and vector parameters and respective quantitative characteristics. Scalar parameters include morphospace dimensionality, volume, subspaces overlap and distinctness, the latter two defining morphospace occupation. The overall morphospace volume is to be defined as a sum of volumes of elementary subspaces minus subspace(s) of their overlap plus subspace(s) of the hiatuses separating them. Vector parameters include directions of predominating trends of particular disparity forms and their co-directionality in the same morphospace.
6. As far as measurable traits are involved, a kind of null model of causal relation between disparity forms can be elaborated, which allows to classify at least some of these forms as "primary" and "secondary" ones. This model presumes that certain properties of the secondary forms are explained by properties of the primary ones. In such a case, it is possible to interpret relation between primary and secondary disparity forms as a kind of "succession" which can be strong or week. Particular contents of such null model depends on the background theoretical construction. For instance, for the diversity of growing organisms, age variation can be considered as a primary form relative to sex and geographic variation, which provides a simple "growth model" for explaining interrelations among such disparity forms. Another example of such null model is a pretty strong succession between subspecies and close species divergence presumed by the Darwinian microevolutionary model.
7. Serious restriction of the methods considered herein follows from their being linear and additive, while morphospace, in at least some of its properties, is non-linear and "non-Euclidean". The latter quite significant property means the morphospace axes non-orthogonality caused by correlations and allometric relations among morphological traits. This problem is to be considered as before all a biological and not just a "technical" one, because at least some between-traits relations are of biological nature and so are also the part of the overall morphodisparity.

10. Acknowledgements

I discussed very productively various topics concerning morphodisparity and the methods of its analysis with Drs A. Lissovsky, O. Nanova, and W. Lebedev (all from Zoological Museum of Moscow State University, Russia), A. Puzachenko (Institute of Geography, Russian Academy of Sciences, Russia).
Preparation of this contribution was supported in part by Russian Foundation for Basic Research (grant 09-04-00283-a).

11. References

Abbott, L.A., Bisby, F.A. & Rogers, D.J. (1985). *Taxonomic analysis in biology.* Columbia Univ. Press, ISBN 0-231-04926-9, New York

Barantsev, R.G. (2003). *Synergetics in contemporary natural science.* URSS, ISBN 5-354-00201-X, Moscow (In Russian)

Blackith, R.E. (1965). Morphometrics. In: *Theoretical and mathematical biology.* Waterman, T.H. & Morowitz, H.J., (Eds), pp. 225–249, Blaisdell Publ, New York.

Chaikovski, Y.V. (1990). *Elements of the evolutionary diatropics.* Nauka, ISBN 5-02-008086-1, Moscow (In Russian)

Chakrabarty, P. (2005). Testing conjectures about morphological diversity in cichlids of lakes Malawi and Tanganyika. *Copeia,* Vol.2, No.2, pp.359–373, ISSN 0045-8511

Ciampaglio, C.N., Kemp, M. & McShea, D.W. (2001). Detecting changes in morphospace occupation patterns in the fossil record: characterization and analysis of measures of disparity. *Paleobiology,* Vol.27, No.4, pp.695–715, ISSN 0094-8373

Claridge, M.F., Dawah, H.A. & Wilson, M.R. (1997). *Species – the units of biodiversity.* Chapman & Hall, ISBN 0-412-63120-2, London

Collar, D.C., Near, T.J. & Wainwright, P.C. (2005). Comparative analysis of morphological diversity: does disparity accumulate at the same rate in two lineages of centrarchid fishes. *Evolution,* Vol.59, No.8, pp.1783–1794, ISSN 0014-3820

Eble, G.J. (1999). Developmental and non-developmental morphospaces in evolutionary biology. In: *Morphospace concepts and applications.* Chapman, R.E., Rasskin-Gutman, D., Wills, M., (Eds), Cambridge University Press, Cambridge (UK), Retrieved from http:// www.santafe.edu/media/workingpapers/99-04-027.pdf

Eble, G.J. (2000). Theoretical morphology: state of the art. *Paleobiology,* Vol.26, No.3, pp.520–528, ISSN 0094-8373

Eldredge, N. & Cracraft, J. (1980). *Phylogenetic patterns and the evolutionary process.* Columbia University Press, ISBN 0-231-08378-5, New York

Eldredge, N. & Gould, S.J. (1972). Punctuated equilibria: An alternative to phyletic gradualism. In: *Models in paleobiology.* Schopf, T.J.M., (Ed.), pp.82–115, W.H. Freeman & Co, ISBN 0-877-35325-5, San Francisco

Erwin, D.H. (2007). Disparity: morphological pattern and developmental context. *Palaeontology,* Vol.50, No.1, pp.57–73, ISSN 0022-3360

Faith, D.P. (2003). Biodiversity. In: *The Stanford Encyclopedia of Philosophy.* Zalta E.N., (Ed.), ISSN 1095-5054, Retrieved from http://plato.stanford.edu/archives/sum2003/ entries/biodiversity/

Faleev, V.I., Galaktionov, Yu.K. & Vasiljev, A.G. (2003). *Realization of morphological diversity in natural populations of mammals.* Publisher for Siberian Branch of Russian Academy of Sciences, ISBN 5-769-20612-8, Novosibirsk (In Russian)

Foote, M. (1993). Contribution of individual taxa to overall morphological diversity. *Paleobiology,* Vol.19, No.3, pp.403–419, ISSN 0094-8373

Foote, M. (1996). Models of morphological diversification. In: *Evolutionary paleobiology.* Jablonski, D., Erwin, D. & Lipps, J., (Eds), pp. 62–86, Chicago University Press, ISBN 0-226-38913-8, Chicago

Foote, M. (1997). The evolution of morphological diversity. *Annual Review of Ecology and Systematics,* Vol.28, pp.129–152, ISSN 1545-2069

Gerber S., Eble G.J. & & Neige P. (2008). Allometric space and allometric disparity: A developmental perspective in the macroevolutionary analysis of morphological disparity. *Evolution,* Vol.62, No.6, pp.1450–1457, ISSN 0014-3820

Gould, S.J. (1989). *Wonderful life: The Burgess Shale and the nature of history.* W.W. Norton, ISBN 0-393-30700-X, New York.

Hall, B.K. (1998). *Evolutionary developmental biology,* 2nd ed. Kluwer Academic Publishers, ISBN 0-412-78590-0, Dordrecht

Hulsey, C.D. & Wainwright, P.C. (2002). Projecting mechanics into morphospace: disparity in the feeding system of labrid fishes. *Proceedings of Royal Society B: Biological Sciences,* Vol.269, No.1488, pp.317–326, ISSN 0962-8452

Kaplan, P. (2004). Morphology and disparity through time. *Encyclopedia of life sciences*. Wiley Online Library, Article A0001637, Retrieved from http://onlinelibrary.wiley.com/doi/10.1038/npg.els.0001637/abstract

Kluge, A.G. & Kerfoot, W.C. (1973). The predictability and regularity of character divergence. *American Naturalist*, Vol.107, No.955, pp.426–442, ISSN 0003-0147

Krasnov, B.R. & Shenbrot, G.I. (1998). Structure of communities of ground dwelling animals at the junction of two phytogeographic zones. *Journal of Biogeography*, Vol.25, No.6, pp.1115–1131, ISSN 0305-0270

Krivolutsky, D.A. (1998). Life forms and biological diversity of animals. *Bulletin of Moscow Society of Naturalists, Biology Section*, Vol.104, No.1, pp.61-67, ISSN 0027-1403 (In Russian)

Kupriyanova, I.F., Puzachenko, A.Yu. & Agadzhanyan, A.K. (2003). Spatial and temporal variation of cranial parameters in the common shrew, *Sorex araneus* (Insectivora). *Russian Zoological Journal*, Vol.82, No.6, pp.839–851, ISSN 0044-5134 (In Russian, with English summary)

Leamy, L. (1983). Variance partitioning and effects of sex and age on morphometric traits in randombred house mice. *Journal of Mammalogy*, Vol.64, No.1, pp.55–61, ISSN 0022-2372

Levinton, J. (1988). *Genetics, paleontology, and macroevolution*. Cambridge University Press, ISSN 0-521-00550-7, Cambridge

Lissovsky, A.A. & Pavlinov I.Ya. (2008). Toward exploration of morphological disparity of measurable traits of mammalian skull. 2. Scalar and vector parameters of group variability form. *Journal of General Biology*, Vol.69, No.6, pp.508–514, ISSN 0044-4596 (In Russian, with English summary)

Mcclain, C.R., Johnson, N.A. & Rex, M.A. (2004). Morphological disparity as a biodiversity metric in lower bathyal and abyssal gastropod assemblages. *Evolution*, Vol.58, No.2, pp.338–348, ISSN 0014-3820

McGhee, G.R. (1991). Theoretical morphology: the concept and its applications. In: *Analytical paleobiology. Short courses in paleontology*, vol. 4. Gilinsky N.L. & Signor P.W., (Eds), pp. 87–102, Paleontol. Soc., ASIN B003ZN2JUS, Knoxville (TN)

McGhee, G.R. (1999). *Theoretical morphology: the concept and its application*. Columbia University Press, ISBN 0-231-10617-3, New York

McGhee, G.R. (2007). *The geometry of evolution. Adaptive landscapes and theoretical morphospaces*. Cambridge University Press, ISBN 0-521-84942-X, Cambridge.

Mina, M.V. & Klevezal, G.A. (1976). *Animal growth. Analysis at the organismal level*. Nauka, Moscow (In Russian)

Mitton, J. (1997). *Selection in natural populations*. Oxford University Press, ISBN 0-195-06352-X, Oxford

Nanova, O.G. (2010). Age variation of the skull morphometric traits in the mainland polar fox (*Alopex lagopus lagopus*) and the Commodor Isls polar fox (*A. l. beringensis, A. l. semenovi*). *Russian Zoological Journal*, Vol.89, No.7, pp. 871-881, ISSN 0044-5134 (In Russian, with English summary)

Nanova, O.G. & Pavlinov, I.Ya. (2009). The stucture of morphological disparity of skull traits in three carnivoran species (Mammalia: Carnivora). *Russian Zoological Journal*, Vol.88, No.7, pp.883–891, ISSN 0044-5134 (In Russian, with English summary)

Norton, B.G. (1986). *The preservation of species: the value of biological diversity*. Princeton University Press, ISBN 0-691-02415-4, Princeton (NJ)

Pavlinov, I.Ya. (1978). Use of elements of information statistic analysis in the studies of intraspecies variation. *Journal of General Biology*, Vol.39, No.6, pp.840–848, ISSN 0044-4596 (In Russian, with English summary)

Pavlinov, I.Ya. (2005). *Introduction to contemporary phylogenetics*. KMK Scientific Press, ISBN 5-8731-7225-0, Moscow (In Russian)

Pavlinov, I.Ya. (2007). On the structure of biodiversity: some metaphysical essays. In: *Focus on biodiversity research*. Schwarz J., (Ed.), pp. 101–114, Nova Science Publisher, ISBN 1-600-21372-3, London

Pavlinov, I.Ya. (2008). Morphological disparity: general conceptions and basic characteristics. *Archive of Zoological Museum of Moscow University*, Vol.48, pp.343–388, ISSN 0134-8647 (In Russian, with English summary)

Pavlinov, I.Ya. & Nanova, O.G. (2009). Toward exploration of morphological disparity of measurable traits of mammalian skull. 3. Distance-based analysis of the morphospace volume and occupation. *Journal of General Biology*, Vol.70, No.1, pp.35–45, ISSN 0044-4596 (In Russian, with English summary)

Pavlinov, I.Ya., Nanova, O.G. & Spasskaya, N.N. (2008). Toward exploration of morphological disparity of measurable traits of mammalian skull. 1. Interrelation between different forms of group variation. *Journal of General Biology*, Vol.69, No.5, pp.344–354, ISSN 0044-4596 (In Russian, with English summary)

Pavlinov, I.Ya., Puzachenko, A.Yu., Shenbrot, G.I. & Rossolimo, O.L. (1993). Comparing various forms of morphological diversity by means of dispersion analysis (examplified by studying measurable morphological traits variation in three rodent families, Mammalia). *Journal of General Biology*, Vol.54, No.2, pp.324–328, ISSN 0044-4596 (In Russian, with English summary)

Pearson, K. (1901). On lines and planes of closest fit to systems of points in space. *Philosophical Magazine*, Vol.2, No.6, pp.559–572, ISSN 1478-6435

Pesenko, Yu.A. (1982). Principles and methods of numerical analysis in the faunal researches. Nauka, Moscow (In Russian)

Pianka, E.R. (2000). *Evolutionary ecology*. 6th ed. Benjamin Cummings Publisher, ISBN 0-321-04288-3, San Francisco

Pie, M.R.. & Weitz, J.S. (2005). A null model of morphospace occupation. *American Naturalist*, Vol.166, No.1, pp.1–13, ISSN 0003-0147

Pustovoit, S.P. (2006). Morphological diversity of salmon (*Oncorhynchus nerka* Walbaum) in Kamchatka River. *Siberian Journal of Ecology*, Vol.13, No.6, pp.503–509, ISSN 0869-8619 (In Russian, with English summary)

Puzachenko, A.Yu. (2000). Use of multidimensional scaling in analysis of structure of morphological variation. In: *Taxonomy and phylogeny of rodents and lagomorphs*, pp. 137–140, All-Russian Theriological Society RAS, Moscow, (In Russian)

Rohlf, F.J., Gilmartin, A.J. & Hart, G. (1983). The Kluge–Kerfoot phenomenon – a statistical artifact. *Evolution*, Vol.37, No.1, pp.180–202, ISSN 0014-3820

Rohlf, F.J. & Marcus, L. (1993). A revolution in morphometrics. *Trends in Ecology and Evoluiton*, Vol.8, No.4, pp.129–132, ISSN 0169-5347

Rostova, N.S. (2002). Correlations: their structure and variability. *Proceedings of St.-Petersburg Society of Naturalists*, Ser. 1, Vol.94, pp.5–306, ISBN 5-283-01936-3

Roy, K. & Foot, M. (1997). Morphological approaches to measuring biodiversity. *Trends in Ecology and Evoluiton*, Vol.12, No.5, 277–281, ISSN 0169-5347

Sarkar, S. (2002). Defining "biodiversity"; assessing biodiversity. *The Monist*, Vol.85, No.1, pp.131–155, ISSN 0026-9662

Shishkin, M.A. (1988). Evolution as epigenetic process. In: *Contemporary paleontology*, Vol. 1, pp. 142–169, Nedra, ISBN 5-247-00152-4, Moscow (In Russian)

Sneath, P.H.A. & Sokal, R.R. (1973). *Numerical taxonomy: principles and practice of numerical classification*. W.H. Freeman & Co, ISBN 0-716-70697-0, San Francisco

Sohn, M.-W. (2001). Distance and cosine measures of niche overlap. *Social Networks*, Vol.23, No.2, pp.141–165, ISSN 0378-8733

Sokal, R.R. (1976). The Kluge–Kerfoot phenomenon reexamined. *American Naturalist*, Vol. 110, No.976, pp.1077–1091, ISSN 0003-0147

StatSoft, Inc. (2010). *Electronic statistics textbook*. StatSoft, Tulsa (OK), Retrieved from http://www. statsoft.com/textbook/

Stone, J.R. (1997). The spirit of D'Arcy Thompson dwells in empirical morphospace. *Mathematical Biosciences*, Vol.142, No.1, pp.13–30, ISSN 0025-5564

Villier, L. & Eble, G.J. (2004). Assessing the robustness of disparity estimates: the impact of morphometric scheme, temporal scale, and taxonomic level in spatangoid echinoids. *Paleobiology*, Vol.30, No.4, pp.652–665, ISSN 0094-8373

Vorobeichik, E.L. (1993). About some indexes of width and overlap ecological niche. *Journal of General Biology*, Vol.54, No.6, pp.706–712, ISSN 0044-4596 (In Russian, with English summary)

Wainwright, P.C. (2007). Functional versus morphological diversity in macroevolution. *Annual Review of Ecology and Systematics*, Vol.38, pp.381–401, ISSN 1545-2069

Wills, M.A. (2001). Disparity *vs.* diversity. In: *Palaeobiology II: A Synthesis*. Briggs, D.E.G. & Crowther, P.E.L., (Eds), pp. 495–500, Blackwell Publisher, ISBN 978-0-632-05149-6, London.

Willis, S.C., Winemiller, K.O. & Lopez-Fernandez, H. (2005), Habitat structural complexity and morphological diversity of fish assemblages in a Neotropical floodplain river. *Oecologia*, Vol.142, No.2, pp.284–295, ISSN 0029-8549

Zavarzin, G.A. (1974). Phenotypic taxonomy of bacteria: A space of logical possibilities. Nauka, Moscow (In Russian)

Zelditch, M., Swiderski, D., Sheets, D.H. & Fink, W. (2004). *Geometric morphometrics for biologists*. Academic Press, ISBN: 0-127-78460-8, Elsevier